AF518368

Nursery Production and Orchard Management

Volume 3: Arid and Tropical Fruits

The Editor

Dr Wasakha Singh Dhillon has significantly contributed in the field of education, research, extension and research management in horticultural sciences. He did his graduation in agriculture, masters and doctorate in horticulture from PAU, Ludhiana. He worked on various research management positions as Director, Fruit Research Station Gangian; Director, Punjab Horticultural Post-harvest Technology Centre, PAU campus; Assistant Director General Horticulture in ICAR, New Delhi, and after retirement working as Consultant NHB, GOI. He guided 12 M Sc and 7 Ph D students. He has written 9 books namely : Viticulture in north-west India; Impact of climate change on fruit crops; Fruit tree physiology; Fruit production in India; Doubling farmers income through horticulture; Horticulture for food and environment security; Plantation crops in India; Post-harvest management of Fruits and plantation crops (Two Volumes *i.e.* Basic concepts, and Applied concepts). His technological pursuits led to the development of 10 fruit varieties, 1 root-stock and 38 technologies. Dr Dhillon published 152 original research papers in national and international journals of repute, 25 book chapters, 4 review papers, 20 full proceeding papers, 60 technical articles, and 1 technical bulletin. He attended 60 national and international seminars. He has been awarded with prestigious Fellowships by Indian Academy of Horticultural Sciences (IAHS); Confederation of Horticulture Association of India (CHAI); and Society for Horticultural Research and Development. CHAI also awarded Dr Dhillon with Life Time Achievement Award-2024, and Life Time Achievement Award in Sports by PAU. He was also awarded with Dr JC Anand Gold Medal, and Shri GL Chadha Memorial Gold Medal by IAHS; Meritorious Teacher Award, S GS Nihalsingwala Award, Letter of Appreciation for Agriculture Development by PAU; Distinguished Scientist Award by High-tech Horticultural Society. Apart from research, academics and management, Dr Dhillon was an outstanding athlete of PAU during his student days. He was best athlete of the College, University and Ludhiana district, and won numerous medals like Gold, Silver and Bronze at College, University, District, State, and Open National level competitions. For recognition of achievements in sports he was honored with College and University colors, and merit certificates.

Nursery Production and Orchard Management

Volume 3: Arid and Tropical Fruits

– Editor –

W.S. Dhillon

Former Assistant Director General, Horticulture, ICAR, New Delhi

Former Director, Punjab Horticultural Post-harvest Technology Centre

Former Director PAU-Fruit Research Station, Gangian

Former Professor of Horticulture, PAU, Ludhiana

2026

Daya Publishing House®

A Division of

Astral International Pvt. Ltd.

New Delhi – 110 002

ISBN 978-93-5919-915-3

Publisher's Note:

Every possible effort has been made to ensure that the information contained in this book is accurate at the time of going to press, and the publisher and author cannot accept responsibility for any errors or omissions, however caused. No responsibility for loss or damage occasioned to any person acting, or refraining from action, as a result of the material in this publication can be accepted by the editor, the publisher or the author. The Publisher is not associated with any product or vendor mentioned in the book. The contents of this work are intended to further general scientific research, understanding and discussion only. Readers should consult with a specialist where appropriate.

Every effort has been made to trace the owners of copyright material used in this book, if any. The author and the publisher will be grateful for any omission brought to their notice for acknowledgement in the future editions of the book.

Published by : **Daya Publishing House®**
A Division of
Astral International Pvt. Ltd.
– ISO 9001:2015 Certified Company –
4736/23, Ground Floor, Ansari Road
Darya Ganj, New Delhi-110 002
Ph. 011-43549197, 23278134
e-mail: info@astralint.com
Website: www.astralint.com

Printed at : **Replika Press Pvt Ltd.**

This volume of the book is dedicated to

Dr Gurbachan Singh, FNAAS

Awardee of Rafi Ahmad Kidwai Award,
Hari Om Ashram Trust Award of ICAR

Former Chairman ASRB, GOI; Union Agriculture Commissioner, GOI; Director Soil Salinity Institute, Karnal; ADG, NRM, ICAR; and presently running a Foundation at Karnal.

He played very important mentor role to groom me professionally

भारतीय कृषि अनुसंधान परिषद
कृषि अनुसंधान भवन–I, पूसा, नई दिल्ली–110 012
INDIAN COUNCIL OF AGRICULTURAL RESEARCH
KRISHI ANUSANDHAN BHAVAN-I, PUSA, NEW DELHI 110 012

डॉ. राजबीर सिंह
उप महानिदेशक (कृषि विस्तार)

Tele : 91-11-25843277 (O)
E-mail : ddg-extn.icar@gov.in

Dr Rajbir Singh
Deputy Director General (Agricultural Extension)

Dated: 18th August, 2025

FOREWORD

The successful establishment and long-term productivity of a fruit orchard hinge on two interconnected factors: elite nursery and canopy management. Quality planting material forms the cornerstone of a successful orchard. Nursery production is focused on propagating high-quality, true to type and disease-free planting material. There are various means of producing elite nursery plants through field based vegetative propagation and micropropagation. The vegetative propagation involves developing plants through cuttings, layering, grafting/budding and micro propagation. The technique of grafting/budding involves uniting a desired cultivar (scion) onto a robust root system (rootstock) to form a composite tree. The choice of rootstock can influence overall performance of a tree by altering its size, fruit bearing duration, resistance to pests and diseases, and adaptation to specific soil conditions.

Over the period of time, India has become leader in horticultural produce at global level and producing 367.72 million tonnes and placed at second position. Horticultural crops contribute 33 per cent to agricultural GDP that is having 16 per cent share in the overall GDP of the country. Among various horticultural crops, fruits play an important role and are providing and ensuring nutritional security to a large population of the country. Fruits are vital component of daily diet because it contains all the essential nutrients, minerals, vitamins and fibre required to maintain a good human health. Presently, 114.51 million metric tonnes of fruits (2024) from 7.04 million hectares of area (IInd estimates, MOA&FW) are produced in India. In spite of producing large quantities of fruits, India is still lacking in per capita consumption that is required for nutritious diet recommended by the World Health Organization (WHO). To fulfil the minimum dietary requirement per head per day, the country requires more than 150 million tonnes of fruits every year. To achieve this mission concerted efforts by the scientists, Government agencies, policy support and fruit growers are being made to enhance the productivity from the same land area.

To achieve the above goals after quality plants once planted, the focus shifts to canopy management to maximize the tree production potential. Canopy management involves the strategic manipulation of the tree's structure. This begins with early training and continued with annual pruning, aimed to optimize light penetration, improve air circulation, and facilitate cultural practices. By balancing vegetative growth with flowering and fruit production. Effective canopy management directly impacts fruit size, quality, and overall yield, making it an indispensable practice for sustainable and profitable fruit farming.

Conclusively, the synergy between quality nursery production and scientific canopy management underpins modern fruit production. While nursery techniques supply robust planting material, proper canopy management ensures higher productivity, better fruit quality, and improved economic returns across diverse agro-climatic regions.

I am happy see that the 3 volume of the book on "Nursery Production and Orchard Management: Arid and Tropical Fruit Crops is published and edited by Prof. (Dr.) WS Dhillon, Former Assistant Director General Hort., ICAR, New Delhi. It consists of chapters on different crops will be useful to the students, teachers, extension workers, policy planners and orchardists. The various chapters contributed by experienced horticulturists are highly praiseworthy.

(Rajbir Singh)

Preface

Incorporation of fruit cultivation in horticulture farming systems is a cornerstone of global agriculture, providing nutrition, economic development, and ecological sustainability. At the centre of this endeavour lie two critical components: nursery production of fruit plants and orchard management. Nursery production mainly deals with propagation, production and supply of high-quality, disease free planting material. Nursery production is a science and art of propagating fruit plants under controlled conditions and best agro-horticultural practices until they are ready for transplantation. It serves as the foundation of orchard establishment by ensuring uniformity, vigour, and genetic fidelity in planting material. Poor planting materials usually lead to low quality yield. Whereas, orchard management encompasses the long-term care and to maintain proper health of fruit trees in the field. In unison, these high precision practices decides the success of fruit production systems.

India's diverse climate and topography contribute to its rich fruit diversity and have potential for feeding the growing global population. Despite the tremendous potential, challenges like the abiotic and biotic stresses due to climate change during the last two decades have resulted in productivity decline of fruit crops. Optimizing canopy management is essential for improving both the yield and quality of fruit.

The advances made in nursery production techniques, fast multiplication and production of disease free planting material and canopy architecture and management has totally changed fruit production systems. The third volume of the book solely deals with the Arid and Tropical Fruits.

I humbly acknowledge the contribution of contributors, who are experts on their own specialized fields that helped me getting this book (Volume: 3) in the present shape. No professional and academic journey is complete without constructive feedback and comments from the readers of this book. I urge upon the scholars and readers to share their critical views on wasakhasingh@yahoo.com regarding the coverage of chapters in this book so that it would value for the improvement in future editions.

W.S. Dhillon

Contents

Chapter 1

Basics and Applied Aspects of Propagation of Fruit Crops

M.I.S. Gill and Y.R. Chanana

Department of Fruit Science, PAU, Ludhiana–141004, Punjab

History and Principles of Plant Propagation

Domestication and consequently the need for food necessitated the development of innovative ways or means to multiply the plants having commercial importance to mankind. The movement of individuals or the group of individuals from one place to another within the country or across the globe in search of greener pastures has been a primary reason for movement of plants, linked to mankind either for food or ornamental purpose. Movement of plant/plant material around the world in this fashion has been a phenomenon even today, though with restrictions like patenting. The books on plant propagation written as early as 16th Century (Seminarium by Charles Estienne in 1530) indicate the awareness of the concept of plant propagation in much earlier times. Likewise, publication of a famous book on 'Grafting and Budding' in 1821 by Charles Baltet describing 180 methods of grafting (Hartmann *et al.*, 1997), also indicate the critical experimentation on plant propagation in start of 19th Century.

Since ages, the plants are primarily propagated by two basic means *viz.*, by sexual and asexual/vegetative methods. The sexual propagation is primarily a seed-seed method. While, in vegetative propagation, asexual plant parts such as stem cuttings, shoot tips, buds are generally used for plant propagation. In asexual method, seed production is primarily considered for propagation, when the resultant plant is expected to be true to type. It is true only in case of polyembryonic seedlings. Polyembryony, as the name indicates refers to the seeds having more than one embryo in the seed. One of the embryos arises from the union of male and female gametes and is called gametic and sexual embryo, whereas the others are produced by simple mitotic division of cells of nucellus without the help of male gamete in their formation. The phenomenon of nucellar embryo is of common occurrence in citrus and mango.

Several improvements have been witnessed over the period of time in the basic techniques of plant propagation, using either conventional or advanced biotechnological approaches. During the primitive period, seed-seed method was the only method considered viable for plant propagation. The idea was conceived from the natural germination of seeds into seedling plants in wild open places. Then the other methods like cutting, budding, grafting, layering and even in vitro propagation techniques have flourished over the period of couple of decades. Each method has its pros and cons. The techniques of plant propagation like meristem culture, shoot tip grafting, micro budding *etc.* are of recent origin and have been commercially used to produce large number of plants true to type and in a limited period of time. The propagation under protected environment creating artificial conditions such as controlled light, temperature, humidity under controlled irrigation through drip coupled with fertigation has further enhanced the potential of plant propagation, which too has made it possible to propagate the plants around the year.

The natural phenomenon of differentiation, dedifferentiation leading to plant propagation can be exploited to its full potential if the basic concepts involved in control of plant propagation are well understood. Some of the important components which contribute in propagation efficiency are genotype, temperature, light, humidity *etc.*

Genotype

All the plants do not respond in a similar way to plant propagation. In other words, some of the plants are easy to propagate while others are recalcitrant. Even though the external factors like light, temperature, humidity *etc.* are uniform, some plants respond better to propagation than others. In general, the herbaceous group responds better than woody plant species. The presence of lignified tissues, the chemicals like phenols and requirement of stratification are some of the factors which may limit the propagation potential in some plant genotypes. The inherent growth regulators or endogenous growth regulators also influence the response of a particular genotype to propagation.

Temperature

Success of propagation is primarily dependent on temperature surrounding the propagating material. Broadly the temperature requirement of evergreen and deciduous fruit plants is different. This is the reason that evergreen fruit plants are generally propagated during February-March or August-September. While, deciduous fruit plants are commercially propagated in the winter i.e during December-January. Likewise, the temperatures under controlled conditions play an important role in plant propagation. Not only do specific species require an optimum germination temperature, but some seeds may or may not require light. Temperature of air as well of soil is equally important for successful plant propagation. For instance, if air temperature is high, it results in more evapotranspiration, thereby are more chances for plant mortality if suitable soil moisture is not maintained in the rooting media. It is cheaper to maintain soil temperature as temperature of rooting media can be regulated by use of bottom heat cables, while to it is expensive to raise the temperature of glass house, especially during winter months. Bottom heat can be supplied in many ways. Bottom heat can be provided by use of propagation mats which have electrically heated wires having rubber coating, by use of warm water circulated in tubes beneath the rooting medium, and by use of steam pipes under work benches. Bottom heat treatment increases rooting to significant extent in several plants.

Light

The plant growth is highly dependent upon light intensity, duration and type. Thus the propagation potential is also varied in different light conditions. The most important activity the plant performs is photosynthesis in the presence of light. Light is vital, as the young plant continue to manufacture food during rooting. Light in excess is also not required, as it may burn the plants or raise the air temperature beyond the tolerance limit of plant. In such cases, the use of shade net houses is recommended. Shade net which can limit the light penetration upto 50 percent are successfully used for propagation of sensitive plants. Light duration has also to be as per the requirement of a specific plant part to be propagated. Type/quality of light correspond to specific range of wavelength. The differential response to red, far red and blue light has been observed in plants, especially those plants which are raised under protected conditions. The response to light quality varies from plant to plant.

Humidity

Though belonging to same genotype, the plant species respond differently to propagation if humidity is different. Even the age of plant material to be propagated determines the need of humid conditions. The requirement of humidity is more critical when plant parts are young such as shoot tips. Relatively high humidity is required for young/tender shoot tips or cuttings. The balance between transpirational losses and humidity around the plants play a critical role in the success rate of propagation. The mature cuttings to be used for plant propagation do not require high humidity, in general for initiation of roots or shoots. In general, until the young roots are not developed in the young/mature cuttings/shoot tips, relatively high humidity is required. Providing the cuttings with intermittent mist or fog maintains optimum humidity in air keeps the substrate moist and prevents the cuttings from wilting. Excess watering, may lead to shortage of oxygen in the soil, thereby can adversely affect root growth of fresh cuttings. Excess humidity may also lead to onset and spread of fungal and bacterial diseases in young plantlets. Hence, it is important to provide just optimum humidity, around the propagation material.

Propagation Media

Soil less media has recently replaced soil media for large scale plant propagation. Peat, pertlite and vermiculite are important components of soil less media, which are relatively sterile as compared to soil media. The presence of soil pathogens in soil media, which in turn affect the growth of plant in the nursery and field, is one of the primary causes for this drastic change of media for propagation of plants. However, soil less media (Plate 1) cannot be routinely used for all the plants. In other words, the economics of the plant to be produced determines the media for propagation of plants. The adequate media should provide required oxygen, retain moisture and should allow proper drainage. In addition, the availability of proper nutrition through media to the young plantlet is also an important factor. Good rooting media encourages fibrous roots, which are in turn useful for better uptake of nutrients by young plantlets, as well hold the plants in compact form, thereby also reduce risk of mortality during transplanting.

For rooting of cuttings and for germination of seeds a suitable medium is required. The desirable characteristics of the suitable rooting medium are as follows:

1. Rooting media should be loose and easily workable.
2. It should be firm and dense.
3. It should have good moisture holding capacity.
4. Well drained/sufficiently porous.
5. Sufficiently rich to provide nutrition.
6. It should be economical.
7. It should not have high level of salinity.
8. Free from pathogens, weed seeds and nematodes.
9. Capable of sterilization by heat/stream.

The commonly used rooting media are as follows:

- Soil/garden soil
- Sand
- Compost
- Peat
- Perlite

- ✰ Vermiculite
- ✰ Sphagnum moss

Soil/Garden Soil

1. Composed of solid, liquid and gasses.
2. Texture depends on proportion of sand, silt and clay.
3. Sandy loam/clay loam- preferred.

Sand

Generally river sand is used

1. Sand as a media is most widely used.
2. Well aerated.
3. Useful for extensive root growth.
4. Pathogen free- no organic matter.
5. It has low water holding capacity and thus need frequent watering.
6. It has low fertility level and should preferably be used in combination with organic matter.

Compost

1. Consists of decomposed cow dung, soil and some farm wastes.
2. Generally used in mixture with soil or sand as a propagation medium.
3. Useful as a high moisture holding material.
4. Should be mixed with soil for its beneficial effects.

Peat

It consists of remains of aquatic, marsh or swamp vegetation- preserved under water in partially decomposed stage. It has very high water holding capacity.

Perlite

1. Grey white, silacacious mineral.
2. High water holding capacity.
3. Neutral in pH.
4. No cation exchange capacity.
5. Generally used to increase aeration in potting mixtures.

Vermiculite

1. Micaceous material which expands on heating.
2. Light in weight.
3. pH neutral.

4. High cation exchange capacity- can hold more nutrients.
5. Chemically- Mg-Al-Fe-Silica compound.
6. Usually not mixed with soil- compressed.

Figure 1.1: Rooting Media (A) Perlite (B) Vermiculite.

Water Quality

Like other living beings, water is a basic component of life in plants. Water plays an important role in transport of nutrients from soil to the plants, under the influence of transpiration pull caused in the upper ground plant parts especially the leaves. Not only is the availability of water which is important for plants, but the water quality is equally important, especially for plant propagation. Important factors which determine water quality include pH, alkalinity and soluble salts. Water high in pH and high in salts/saline water hinder the growth of young plantlets. Chlorophyll content and soluble proteins are directly affected by pH of water, thus affecting the growth of plants. High pH of soils limits the plant growth and may result in iron and manganese deficiency, while acidic soils also restrict the plant growth. Thus pH must be optimum for normal growth of plant especially in the nursery. Likewise, excessive salinity leads to plant decline and nutrient deficiencies. Total dissolved salt (TDS) and sodium absorption ratio (SAR) are important parameters of salinity. TDS measures overall salinity of water, while SAR measures ratio of sodium to calcium and magnesium. High sodium uptake may lead to deficiency of calcium and magnesium in plants. Dissolved salt levels in water are measured through electrical conductivity (EC). Water high in EC can cause poor plant health due to nutritional imbalance. Poor quality water high in salts or even dirt/sand particles can clog the drip system used for irrigation and fertigation of nursery plants.

Sterilization Procedure

The primary component of sterility in nursery which needs attention to soil, which can be made sterile *in vivo* by soil solarization and/by use of chemicals.

Soil Folarization

The process involves covering soil during summer months with transparent polythene sheets that traps solar energy and helps to contain soil pathogens to great extent. Before soil solarization, irrigate the soil, as wet soil conducts heat better. FYM is also added to soil before solarization. Addition of organic matter increases the rate of decomposition, thereby generate more heat. The breakdown of organic matter also makes N, Ca, Mg and K more available to plants after solarization. Moreover, it is environment friendly method as no chemicals are used during this process. Soil solarization is done generally for 4-6 weeks.

Soil Fumigation

Generally formaldehyde and basamid are used for control of fungal diseases in soil. Soil fumigation is otherwise, undesirable as it causes residual toxicity in plants and soil. All fumigants operate by volatile action. To be effective, the released gas must penetrate through the soil and come into contact with the target pathogen. Thus as a pre-requisite to soil fumigation, we must make the soil loose and friable. The soil temperature and moisture must be optimum. Above all, there must be proper sealing of the soil to be fumigated. Although, it is an effective way to sterilize the soil, however, the high cost of chemicals and human health concerns are limiting factors in use of soil fumigation.

The other components of sterilization include sterile environment at each and every step of plant propagation, such as use of sterile tools like secateurs, budding/ grafting knives and use of gloves, while propagation.

In vitro Sterilization

The explants to be used for plant propagation *in vitro* can be surface sterilized by use of different surface sterilants like mercuric chloride (0.1-1.0%), sodium hypochlorite (0.5-5.0%), calcium hypochlorite (9-10%), hydrogen peroxide (3-12%), ethyl alcohol (70-95%), silver nitrate (1%) and benzalkonium chloride (0.01-0.1%). Mercuric chloride should be avoided for use as surface sterilant as it is toxic to both plants and humans and must be disposed off with care. To increase the effectiveness of surface sterilants, surfactants like Tween 20 can be used. Prior to using any of the surface sterilants, it is important to wash off external dirt/ dust particles by washing the explant clean with running tap water. The use of fungicides/bacteriocides on the field grown plants, from which explants are to be taken for *in vitro* culture helps to reduce contamination during *in vitro* culture of explants. The use of autoclaved/sterile forceps, scissors and blade handles can further help to maintain sterile cultures. Culturing relatively dry explants after sterilization also help reduce surface contamination. The explants can be dried with the help of sterile and dried filter papers.

Propagation Structures

The young seedlings having rudimentary leaves and tender cuttings without feeding roots need external support to show symptoms of life. The specialized structures like cold frame and hot beds are in use as a recent practice for survival of such plant material.

Cold Frames

To trap solar energy and create insulation in a need based fabricated structure/ box, and thereby creating a microclimate suitable for growth of nursery plants can be referred to as a cold frame.

Cold frames must be placed in such a location, where maximum light should come in, but avoid direct sunlight, as it will overheat the structure. Air circulation in also essential for healthy growth of young plantlets, under enclosed cold frames. Temperature also needs regular monitoring which must be gauged by installing thermometer inside the cold frame. The major advantage of cold frame over green houses in that it is convenient to move from one place to another as per the requirement of the young plants, due to their manageable size. However, the major disadvantage is that it is not a walk in type chamber like green houses, which makes rather difficult to handle young growing plants. Under our climatic conditions cold frames can be successfully used during winter months. An ideal size of cold frame is 6 feet by 6 feet, with tapering sides.

Hot Beds

A cold frame with a provision of bottom heat is termed as a hot bed. The bottom heat helps in better rooting, especially in case of tender cuttings. Heating cables can be placed below the propagating media to warm the soil, which help the roots to form quicker. It is advisable to have 2 inch layer of vermiculite below the soil warming cables to help radiate the heat upwards and insulate the propagation media from the soil underneath. A wire mesh is desired over the rooting media, to avoid damaging heating cables. Alternatively, stream carrying pipes can be used for providing bottom heat to the cuttings. Soil less media like peat, perlite, vermiculite or its mixture can be used for plant propagation in such kind of beds.

Use of Plant Regulators in Propagation

Plant growth regulators weather endogenous or exogenous play an important role in plant propagation. Plants high in endogenous growth regulators like auxins may not respond well if high exogenous application of auxins is provided. In other words, a balance between exogenous and endogenous growth regulators plays a key role in deciding the rate of propagation in plants.

Auxins

1. During 1934, IAA (Indole-3-acetic acid) was identified as the first naturally occurring compound having auxin activity, which takes part in inducing rooting in the cuttings.
2. Later on, IBA (Indole butyric acid) and NAA (Napthalene acetic acid) were found to be more effective in inducing rooting in cuttings.

3. For initiation of rooting in cuttings, a continuous supply of auxins is required for first 3-4 days (Auxin active stage).
4. Withholding auxin during root elongation process does not affect the rooting process (auxin in-active stage).
5. Cuttings do not respond to exogenous application of auxins once the rooting process has initiated.

Cytokinins

1. These are responsible for growth and differentiation of the cells.
2. Zeatin, Kinetin, 6-benzyl adenine have cytokinin activity.
3. Cuttings having high levels of cytokinin are more difficult to root than those having low cytokinin activity.
4. Root initiation is inhibited when cytokinin are applied artificially.

Gibberellins

1. These were first isolated in Japan in 1939 for their effect in promoting stem elongation.
2. At higher concentration, they inhibit root initiation process in cuttings but lower concentration promotes it.
3. GA_3 is most commonly used gibberellin for commercial uses.

Abscisic Acid

1. Its role in rooting of cuttings is not very clear.
2. Its affect depends upon concentration and nutritional status of the stock plant from where cuttings have been prepared.

Ethylene

1. At low concentration, it has a promoting effect on rooting of cuttings but at higher concentration, effect is inhibited.
2. Rooting promotion activity of ethylene is due to its synergistic effects on auxin synthesis, which leads to root initiation process.

Techniques of Propagation and Specialized Equipment, Mechanization

Propagation is primarily done through either sexual or asexual methods. In other words, propagation through seeds or using vegetative means are two main means of plant propagation, each having its advantages and disadvantages. Seed propagation is typically a product of sexual reproduction within species, as genetic recombination occurs. Thus the plants that originate from seeds, except nucellars, have different characteristics from its parents. In fruit plants like in some of citrus rootstocks, mango nucellar embryony prevails, which results in production of more than one plant from the seed, one being zygotic and other nucellars. While the plants propagated through asexual or vegetative means are all true to type of the mother plants. Moreover, it depends upon the fruit plant and the objective of plant propagation which decides the method to be chosen for plant propagation. For instance, in fruit plants like papaya, seed is the only used commercial means of plant propagation. If rootstock raising is the objective, then propagation through seeds is a method of choice, as it leads to development of plants having healthy and vigorous root system. While, in case of propagation for scion purpose, vegetative means of propagation can only ensure trueness to the type of its mother plant.

Specialized equipment used for plant propagation includes different types of pots/containers, seed plug trays, propagation mat used for bottom heat *etc.* Pots are available in different forms, shapes and are made up of different material such as earthern or of plastic. These days biodegradable pots are also available in the market. Type of pot/container depends upon the type of plant to be propagated i.e routine or recalcitrant, propagation material i.e seed or cutting, duration for which propagation material is to be kept, with or without assured buy back guarantee. Above all economics of plant propagation is an important component. Some of the desired requirements for choosing a right container includes: light weight, preferably degradable, relatively durable and free from inherent microbial contamination.

The prime objective of mechanization is efficient, but economical plant propagation. Use of bioreactors, seed germinators, bottom heat treatment and *in vitro* propagation are different components of mechanization with varied results depending upon plant species.

Seed Dormancy and Germination

Dormancy refers to lack of growth of viable seed either due to internal or external factors. Seeds of tropical and sub-tropical fruits (citrus, mango) germinate immediately after extraction from the fruits under favourable conditions of moisture, temperature and aeration. However, the seeds extracted form temperate fruit plants like pear, peach and plum do not germinate even under favourable environmental conditions, even if they are viable-this condition is termed as dormancy. Nature has provided protection to such fruit plants, as otherwise they may not survive in harsh winters. Therefore it is essential to understand the dormancy and the factors which are responsible for enhancing or overcoming the dormancy of such seeds.

The dormancy in seeds might be due to exogenous or endogenous factors. Most commonly the exogenous seed dormancy is either due to hard seed coat, which does not allow the entry to water and gases to the developing viable embryo and thus inhibits seed germination. While in case of endogenous seed dormancy, physiological immaturity of embryo, deficiency of some endogenous growth promoters or excess of endogenous growth inhibitors are responsible for low seed germination.

Types of Dormancy

Exogenous

If the dormancy is imparted by external factors, which may either be due to physical, mechanical or chemical reasons, we term it as exogenous factors.

Physical

Dormancy may be due to hard seed coat, which does not allow the solutes/ the water uptake and the aeration in the seed. It is worth to mention that the seed germination do not occur in such cases even if the embryo is viable in such seeds. Thus the seeds having seed coat dormancy do not germinate until hard seed coat is either cracked or removed. For example peach or almond seeds have hard seed coat, which needs either rubbing with sand paper or acid scarification to thin the outer seed layer. In nature such hard seed coats are either weakened by continuous weathering or by attack of microorganisms or by passage of seeds through digestive tracts of animals.

Mechanical

It refers to seed coverings that are too hard to allow the embryo to expand during germination even though water can penetrate it. Hardening of seed covering is never a sole cause of dormancy in any species but the hardening in combination with other factors, may delay seed germination. For example, the seed coat are apparently permeable to water and gases, but have very high mechanical resistance to allow expansion of embryo, as a result the germination does not occur unless seed coat are softened. So the removal of seed covering is must to overcome such dormancy.

Chemical

Sometimes the seeds do not germinate due to some chemicals or commonly known as inhibitors present in the seed coverings. It is caused by germination inhibitors which accumulate in the fruit and seed coverings during development (*e.g.* phenols, ABA). Such inhibitors are accumulated in either fruits or seed coats during fruit development. Leaching of inhibitors is essential to ensure normal seed germination in such cases.

Endogenous

Unlike external factors, in this case dormancy is due to internal factors such as morphological, physiological factors.

Morphological

This type of dormancy is imparted if the embryo remains under developed at the time of fruit ripening or maturity. Such seeds do not germinate if planted immediately after harvesting of fruits. Warm stratification can help to overcome this type of dormancy.

Physiological

It is thought to be controlled by endogenous plant growth substances and environmental factors such as temperature and light. Factors within the embryo inhibit germination. This type of dormancy is common in temperate fruits like apple, pear and peach which last for one to six months and it disappears on storage. The seeds of these species have dormant embryos and they do not germinate even under favourable conditions. This type of dormancy may also be due to inappropriate temperature or light conditions at the time of seed germination. This type of dormancy can be overcome by storage of seeds in between layers of moist sand at chilling temperature for few days to months, as per the requirement of a fruit crop. The low chill requiring fruit crop seeds may require few days as compared to those varieties which have high chilling requirement to break dormancy.

Germination

Germination refers to the process of formation of a new plant from the mature seed. In other words, germination means emergence of the radicle (root) and plumule (shoot), leading to production of seedling.

For the germination to be initiated the seed must be viable, that is, the embryo must be alive and capable of germination and the environmental conditions must be appropriate. Seed germination involves different phases such as i) rapid increase in water uptake ii) followed by a period where there is little water uptake called the lag phase and iii) there is increase in fresh weight of seed, which leads to the emergence of roots.

Seed Viability

Seed germinates only if it is viable. In other words, only those seeds germinate which have live embryo. Seed viability may range from few hours to several years depending of plant species, the surroundings and several other factors.

Factors Affecting Seed Viability

There are several factors which affect seed viability:

Environmental Conditions

The weather conditions prevailing at the time of maturation, ripening and harvesting of seeds affect seed viability. A warm and dry weather at the time of seed maturity usually favours the production of quality seeds, whereas, adverse weather conditions like occurrence of frost, excessive rainfall or even drought at the time of seed harvest may affect their viability adversely.

Mechanical Injuries

If the seed suffers mechanical injury, at the time of harvest or during transit, it adversely affects seed viability. In particular when developing embryos are affected by injury, it affects seed viability vis a vis seed germination.

Genetic Factors

Seeds belonging to different species or of different varieties within species may have different viability patterns.

Storage Conditions

The temperature, light and humidity at the time of seed storage play an important role in maintaining seed viability. In general high relative humidity shortens life of the seeds. Seeds can be kept viable for longer durations of time, if stored at ultra low temperature (cryo preservation). However, all types of seeds cannot be stored at ultra low temperature. Aeration is another important factor affecting seed viability.

Method to Detect Seed Viability

Seed viability can be tested by several methods; the commonly used methods are as follows as:

1. Germination test
2. Excised embryo test
3. Tetrazolium test

Germination Test

This includes germination of seed on artificial media and counting the number of seeds germinated to the seeds cultured. This is routinely used for ensuring viability of seed in several plant species including horticultural crops. Seeds are germinated either in filter paper layers, Petri dishes or in sand, soil, perlite, vermiculite contained in plastic trays or boxes.

Excise Embryo Test

The embryo excised from mature seed is cultured on synthetic media or in petri dishes lined with moist filter papers. Excised embryo test is generally used to test the seed viability in tree species whose embryo require long periods of after ripening before true germination takes place.

Tetrazolium Test

The conventional methods for testing the germination capacity of seeds in Petri dishes or in sand are time consuming and laborious. Most of the non dormant fruit seeds require 7-21 days and the dormant seeds require few months to carry out germination tests. Citrus seeds loose their viability within a short period and are stored under special storage conditions. Therefore, a quick test which will give rough estimate of seed germinability is desirable. The tetrazolium test makes it possible to determine the seed viability within 4-24 hours. Its special advantage over other biochemical tests is that it can be performed on individual seed organs and tissues vital for growth. The embryo of seeds capable of normal germination and growth contain reducing enzymes which convert the almost colourless triphenyltetrazolium chloride solution into a clearly visible red stain (Figure 1.2) upon the cut surfaces of the embryo tissue. During this test seeds are soaked in 2, 3, 5-triphenyltetrazolium chloride (TTC) solution (0.1-1.0%) for few hours. Living tissue changes the TTC to an insoluble red compound. Thus the viable seeds turn red partly of fully depending upon the time of incubation, while non viable tissues/ seeds do not change colour.

Figure 1.2: Viable Citrus Seeds
(a) Without TTC; (b) After TTC Treatment.

Propagation by different Specialized Plant Parts

Cutting

It is a method of vegetative propagation in which a part of stem, root or other plant part is used for the multiplication of plants. Plants produced by cuttings are true to type to mother plants. This method is commonly used in plants, which root easily and readily, thus, multiplication of plants is very quick and cheap. The fruit plants like phalsa, baramasi lemon and grapes are commercially propagated by cuttings.

The hardwood cuttings are the common method of propagation, which are prepared from fully mature tissues. Round cuttings are preferred over angular immature cuttings. The shoots of about one year old or more can easily be used for preparing hardwood cuttings. In case of decidous fruit plants such as grape, pomegranate, phalsa and fig the cuttings are made after pruning. While in evergreen fruit plants like baramasi lemon, the cuttings can be prepared during the spring

(Feb –Mar) and rainy season (Aug-Sep). Generally the cuttings of 15-20 cm length and having 3-5 buds are made. The lower cut is given in a slanting manner just below the bud to increase the absorption of nutrients. The upper cut is given at right angles to reduce the size of the wound and as far as possible away from then upper bud to avoid its drying. After the cuttings are prepared they should be allowed to dry. These cuttings are usually tied in small bundles (20-25 cuttings) and buried in moist soil/sand for a certain period for healing of wounds, which is known as callusing.

Budding

Budding is a method in which only one bud is inserted in the rootstock. This method is very easy and fast. This method saves budwood as compared to grafting. As soon as the bark starts slipping both on the stock and scion, this is considered to be the optimum time for budding. This shows that the cambium, which is the tissue responsible for union, is active. This method is generally employed during spring and rainy season. The common methods of budding are T-budding, patch budding, and chip budding.

Advantages

1. It is quick and efficient method of propagation than grafting.
2. It requires less scion material.
3. It is useful in plants which releases excessive wound gum (stone fruits) from injury carried to xylem at the time of grafting.
4. It results into stronger union than grafting thus injury to the budded plants due to storms and strong winds is less.
5. It is a comparatively simple method of propagation than grafting.

Methods of Budding

T-Budding

This is also known as shield budding. A horizontal cut about $1/3^{rd}$ the distance around the stock (Figure 1.3) is given on the stock 15-20cm above then ground level. Another vertical cut 2-3 cm in length is made down from the middle of the horizontal cut and flaps of the bark are loosened with ivory end of the budding knife to receive the bud. After the 'T' has been made in the stock the bud is removed from the bud stick. To remove the shield of bark containing the bud, a slicing cut is started at a point on the bud stick about 1.25 cm below the bud, continuing underneath about 2.5 cm above the bud. A second horizontal cut is then made 1.25 to 2 cm above the bud, thus permitting the removal of the shield piece. The shield is removed along with a very thin slice of wood. The shield is then pushed under the two raised flaps of bark until its upper horizontal cut matches the same cut on the stock. The shield should fix properly in place, well covered by the two flaps of bark, but the bud itself exposed. The bud union should be wrapped with polythene strip to hold the two components firmly together until the union is completed. T budding can be performed at any time of the year provided cell sap flows freely.

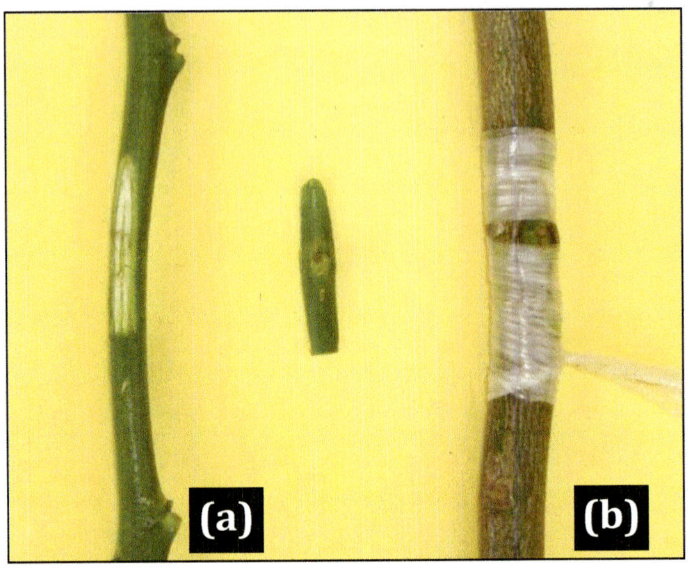

Figure 1.3: Steps Showing T-budding
(a) Removal of bud and (b) Fixing in rootstock and tying with polythene strip.

In most fruit trees it is performed either in the spring (March-April) or in rainy season (July-September) period. This is the most common method of propagation of citrus plants.

Patch Budding

This type of budding is quite successful in guava and it gives 60 to70 percent success during May and June. Freshly cut angular bud-wood from current seasons growth should be used as scion. A rectangular or square patch or piece of bark about 1.5 cm broad and 2.5 cm long is removed from the rootstock (Figure 1.4) at about 15 to 20 cm from ground level. A similar patch with a bud on it is removed from the bud stick taking care not to split the bark beneath the bud. This patch is then transferred to rootstock and fixed smoothly at its new position and tied immediately with polythene strip. To have better success, a patch having two buds is used as scion instead of a single bud. This method is termed as improved patch budding method.

Chip Budding

This method is usually employed when the stock and scion are still dormant, that is just before the start of new growth. In this method, one about 2.5 cm long slanting cut is given into the stock followed by another cut at lower end of this first cut, in such a way that a chip of bark is removed from the stock.

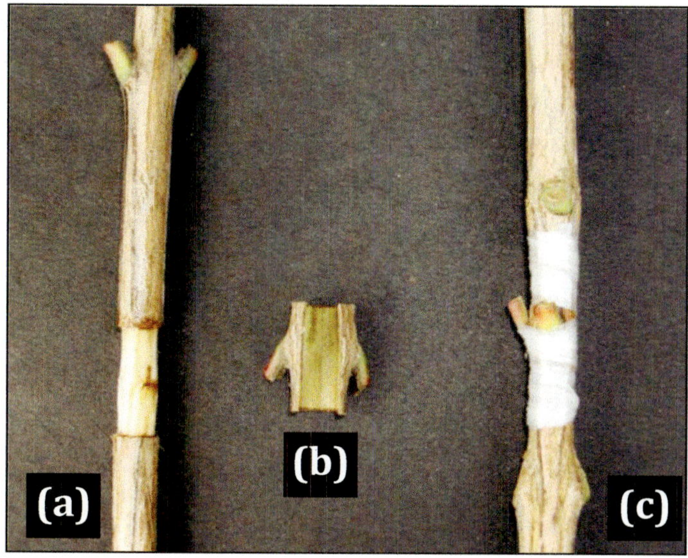

Figure 1.4: Steps Showing Patch Budding
(a) Removal of patch from rootstock; (b) Taking similar size patch from scion and (c) Fixing.

The bud from the scion wood is removed in the same way so that it matches the cuts given in the rootstock. This chip with a bud on it is fitted smoothly into the cut made in the rootstock taking care that the cambium layers of the stock and scion unite at least on one side. The bud is then tied and wrapped with polythene strip, to prevent drying up of the bud.

Layering

Layering is a method of vegetative propagation, in which roots are induced on the shoots while they are still attached to the mother plants. This is a alternate method of propagation in fruit plants which do not root easily when detached from the mother plants. Most commonly used methods of layering are air layering, ground and mound layering.

Advantages

1. Easy to perform and does not require much facility.
2. Reliable method for those plant species which do not root easily.
3. Does not require precise environmental control conditions.
4. Possible to produce large sized plants within a short time.

Disadvantages

1. Requires relatively more scion wood than other methods.
2. Not possible to produce large number of plants within a short time.

3. Plants produced have small and brittle roots.
4. Beneficial effects of rootstocks cannot be utilized.
5. Mortality rate of layers is usually high.

Methods of Layering

Air Layering

As name refers, in this method the layering is done in air. To be more precise the rooting is done on the shoot itself when it is still attached to the mother plant. In this method one year old, healthy and straight shoot is selected and ring of bark (Figure 1.5) measuring about 2.5 cm just below a bud is removed. Moist sphagnum moss is placed around this portion and is wrapped with a polythene strip. It is light in weight, and has a very high water holding capacity. If sphagnum moss is not available, any other material, which can retain moisture for long period of time, can be used for this purpose. The polythene covering does not allow the moisture to come out but permit gas exchange. Moreover the layers need not be watered afterwards, which saves a considerable labour. This method of layering is also known as goottee method.

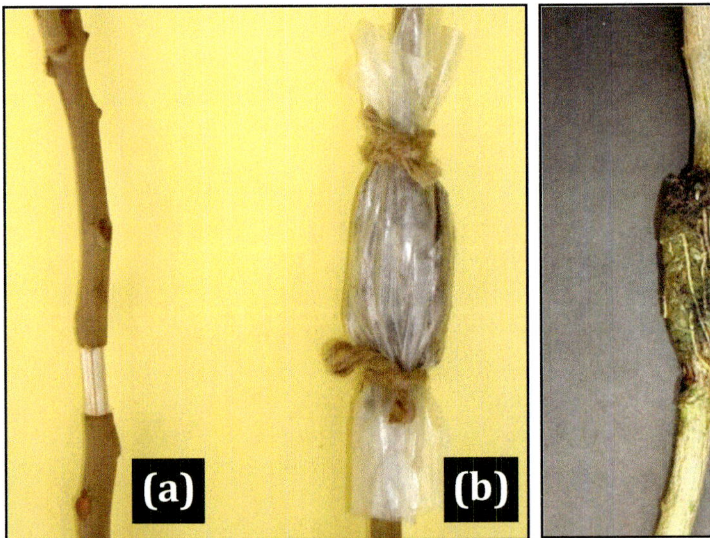

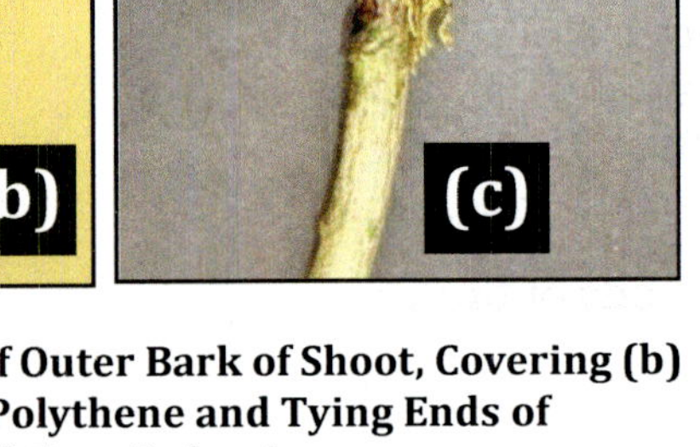

Figure 1.5: Steps Showing Removal (a) of Outer Bark of Shoot, Covering (b) with Moss Grass, Wrapping with Polythene and Tying Ends of Polythene with Rope (c) Root Induction.

Air layering can be practiced during February–March and July-August in guava, litchi, sapota, loquat *etc.* After a few weeks the roots are developed which are visible through the polythene covering. Then a half way cut should be given to the rooted layers on the parent branch at least 15 days prior to their permanent removal from the mother plant. At the time of separation, a few leaves or small shoot is retained. It is also advisable to plant these rooted layers in nursery for close attention than to plant them directly in field. These layers can be planted in the fields during the following year in February or September–October.

Ground Layering

In this method, a branch of plant, which is near the ground, is chosen and a ring of bark about 2.5 cm is diameter is removed just below the bud. This branch is then bended and buried in soil when still attached to the mother plant. The soil is regularly watered to keep it moist. Within a few weeks, the roots are formed and new plant is separated from the mother plant. Separation should be done in such a way that the roots formed also go with the detached plant. These new plants should preferably be planted in pots or nursery rows for development of better root system and shoot system before planting in the fields. This method is commonly followed for propagation of baramasi lemon.

Mound Layering

In this method, plant is headed back either in February or in July. The new shoots come out during April and September, from ground level. A ring of bark is removed from these shoots and they are covered with moist soil. The rooted stools of April stooling are separated during rainy season and those of August are removed in the following spring. These stools, after separating from the parent plant are planted in the nursery fields. This method is also known as stool layering and is used for propagation of guava.

Grafting

Grafting is another method of vegetative propagation, where two plant parts are joined together in such a manner that they unite and continue their growth as one plant. In this method, the scion twig has more than two buds on it. Grafting is commonly done in pear, peach, plum, almond, mango *etc.* In temperate fruits like peach, plum and almond grafting is done when the plants are dormant while, in mango it is done when the trees are in active growth. The different methods of grafting are tongue grafting, cleft grafting, approach grafting, side grafting and veneer grafting.

Methods of Grafting

Tongue Grating

This method is commonly used when the stock and scion are of equal diameter. First, a long, smooth, slanting cut of about 4 to 5 cm long is made (Figure 1.6) on the rootstock. Another downward cut is given starting approximately 1/3rd from the top and about centimeter in length. Similar cuts are made in the scion wood exactly matching the cut given in the rootstock. The scion having 2 to 3 buds is then tightly fitted with the rootstock taking care that the cambium layer of at least one side of the stock and scion unites together. This is then wrapped with polythene strip.

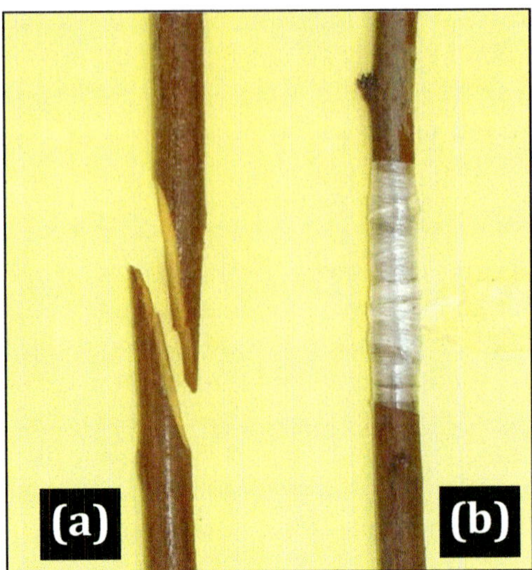

Figure 1.6: Steps Showing Tongue Grafting
(a) Making tongue in stock and scion; (b) Grafting and tying with polythene strip.

Cleft Grating

This is also known as wedge grafting. This method is useful in the nursery where the rootstock is quite thicker (Figure 1.7) than scion and tongue grafting cannot be employed successfully. The stock up to 8 cm in thickness can be grafted with this method.

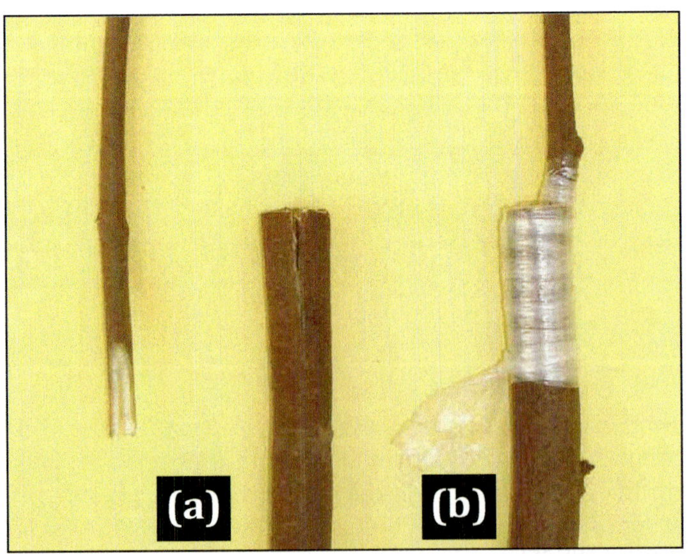

Figure 1.7: Steps Showing Cleft Grafting
(a) Preparation of scion and rootstock; (b) Grafting and tying with polythene strip.

The rootstock to be grafted is cut smoothly with a secateur or saw. It is then split in the middle down to about 4 cm. The bud stick having 3 to 4 buds is trimmed like a wedge at the lower end with outer side slightly broader than the inner side. The lower bud on the scion should be located just well in to the stock making sure that the cambium layers of both the stock and scion are perfectly matched. Cleft grafting is done during dormant period.

Approach Grating

This method of grafting is termed approach grafting, as the rootstock is approached (Figure 1.8) to the scion, while it is still attached to the mother plant. Alternatively the mother plants are trained to be low headed and the stock is sown under their canopy. Last week of July or the first week of August is the best period for approach grafting. In this method the diameter of rootstock and scion should be approximately the same. A slice of bark along with a thin piece of wood about 4 cm long is removed from matching portions of both the stock and the scion. They are then brought together making sure that their cambium layers make contact at least on one side. These grafts are then tied firmly with polythene strip or any other tying material. The stock and scion plants are watered regularly to hasten the union. The union is complete in about 2 to 3 months. A cut is then given to the scion shoot about half way through its thickness. If the shoot does not show any sign of wilting for a week or so it is completely detached from the mother plant. In case the scion starts wilting it shows that the union is not complete. In such cases the scions are detached from the mother plants after some days when the union is complete. The method is commonly followed in mango. This method is also called inarching.

Figure 1.8: Approach Grafting in Mango.

Side Grafting

A three sided rectangular cut about 4 x 1.25 cm is made on the rootstock at a height of about 15-20 cm from the ground level and the bark of the demarcated portion is lifted away (Figure 1.9) from the rootstock. A matching cut is also made on the base of the scion to expose cambium. The scion should be prepared well before the actual grafting is done. The healthy scion shoots from the last mature flush are selected for this purpose. The selected scion shoots should have plump terminal buds. After the selection of the scion shoots, remove the leaf blades, leaving petioles intact. In about 7 to 10 days the petioles shall drop and terminal buds become swollen. At this stage the scion stick should be detached from the mother tree and grafted on the stock. The prepared scion is inserted under the bark flap of the rootstock so that the exposed cambia of the two components are in close contact with each other. The bark flap of the rootstock is resorted in its position. The graft union is then tied firmly with polythene strip.

Figure 1.9: Steps Involved in Side Grafting
(a) Preparation of rootstock: (b) Preparation of scion;
(C) Grafting and securing the graft with polythene.

After the completion of the grafting operation, a part of the top of rootstock is removed to encourage growth of the scion. When the scion has sprouted and its leaves turned green, the root stock portion above the graft union should be cut away. Side grafting can be carried out successfully from March to October; but success during the May and October is rather low. This method of propagation is commonly used in mango.

Veneer Grafting

In this method, a shallow downward cut of about 4 cm long is given on the rootstock at a height of about 15-20 cm from the ground level. At the base of this cut, a second short downward (Figure 1.10) and inward cut is made to join the first cut, so as to remove a piece of wood and bark. The scion is prepared exactly as in side grafting. The cuts on the rootstock and scion shoot should be of the same length and width so that the cambial layers of both components match each other. Then, the prepared scion is inserted into the rootstock and tied security with polythene strip. After the union is complete the stock is cut back, leaving time for doing veneer grafting.

Figure 1.10: Steps Involved in Veneer Grafting (a) Preparation of rootstock; (b) Preparation of scion; (c) Grafting and securing the graft with polythene.

Propagation of Annuals

Most of the fruit plants are characterized as perennial plants, the life cycle of which is not completed in one year; rather the plants stay in field for years. Even in some instances like pear, the plant takes years to come even into 1st bearing. There are some exceptions; some fruit plants like strawberry, banana take less than a year to come into commercial bearing. Such plants are propagated by special structures like runners/suckers.

Runners

Runners are the young shoots, which the mature plant produces during its life cycle. The runners may contain 8-10 nodes/buds per young shoot. Each node if in touch with the soil will root easily and can develop into a complete plant, which is essentially true to type of the parent plant. Runners are produced in strawberry.

Suckers

Some plants like banana produce off springs commonly called suckers, which are used as one of the methods of plant propagation. The plants produced are true to type. On the other hand frequent suckering is not desirable as it affects the development and maturity of main crop. Delayed maturity has special significance in North Indian conditions as it may adversely affect the yield of banana crop due to frost damage.

Rootstocks through Seeds

Propagation through seeds in other words sexual propagation is otherwise not desirable, as it leads to variations in propagated plants as compared to propagation through asexual means. However, seed propagation in some instances in the only means of propagation in fruit plants such as papaya. Moreover, in rootstocks this is most desired method of propagation because of several advantages over asexual methods of propagation.

- ☆ The resultant plants are most vigorous.
- ☆ Have strong root system.
- ☆ Plants raised through seeds have better tolerance to diseases.
- ☆ More tolerant to biotic and abiotic stresses (For salt tolerance *e.g.* in grapes-Dogridge is used as rootstock).

Plug Plants Production

Production of plants on a commercial scale in an individual cell of a tray, under containerized conditions is known as plug plant production. Usually plants produced through seeds are raised under such conditions. Large number of plants can be accommodated in a small tray. Soil less media comprising of peat, perlite, and vermiculite is used for raising such kind of plants. Apart from seeds, tender cuttings can also be raised under controlled conditions in such trays. Once established, the plants raised in plug trays, when uprooted have least disturbance to their roots. Such plants are raised under controlled environmental conditions such as in green houses or in cold/hot frames. Such plants can also be produced around the year, as they do not encounter any seasonal constraints. Such plants are also easy to transport from one place to another. Such trays can also be used for *ex vitro* rooting of tissue culture raised material. As soil less media is used, plug plant production is relatively pathogen free. However, to raise such plants is a expensive proposition, if buy back guarantee is not assured for such plants.

Physiological, Anatomical, Biochemical and Genetical Basis of Rooting

Rooting like other plant growth components is not simply the formation of radicle and its elongation into visible roots. However, several invisible factors like physiological, anatomical, biochemical and genetical factors control rooting. Thick lignified walls of sclerenchyma tissues are one of the physiological barriers of rooting in cuttings. Thus in some instances wounding is done to remove sclerenchyma and suberized cortical tissues. Age of the parent plant/cutting affects rooing response in different plant species.

Root hairs, epidermis, endodermis and vascular tissues like xylem and phloem are different anatomical components of a root. Vascular tissues transport water absorbed by root to other plant parts. Dedifferentiation leads to formation of root initials from cells near vascular bundles or vascular tissues, thereby leading to formation of root primorida followed by visible roots. In woody perennials, adventitious roots usually originate from living parenchyma cells, primarily phloem. The response to rooting from juvenile and mature shoots is different. Main difference is that phloem and xylem are more developed in mature than in juvenile shoots. Root primordial develop more in juvenile shoots, followed by adventitious root formation.

Like physiological and anatomical differences, the biochemical components, also play an important role in rooting. Plant growth substances known as hormones like auxins, gibbrellins and cytokinins play an important role in formation of adventitious roots. Auxins are natural auxins, and regulate cell elongation and root initiation. Gibbrellins also favour cell elongation and inhibit rooting. Unlike, auxins and gibbrellins, cytokinins promote cell division. On the contrary, inhibitors like ABA may inhibit rooting in most of the instances; hence play a negative role in rooting. Changing levels of CHO and the rate of photosynthesis also play an important role in rooting. In leafless, hardwood cuttings, CHO levels are initially high, but falls rapidly as roots, callus is formed. Like auxins and cytokinins, polyamines are also associated with rooting process. Phenols also play an important role in rooting. Total phenolic content was high in IBA treated cuttings in initiation and expression phases. Phenolics play key role in induction of adventitious rooting.

Role of transcription and translation in competent cells to dedifferentiation and form root primordia need to be understood during study of biochemical aspects of rooting.

Stionic Influence

As most of the saleable fruit plants have two components stock and scion. Scion is generally considered important because we get visible fruits from this upper portion of the plant. Likewise, stock to which we commonly name rootstock has an important role in rooting, in other words is the real strength of the healthy plant. The perfect scion stock combination is visibly seen as a smooth union between stock and scion after grafting. While, abnormal scion influence is generally visualized as a abnormal growth of either stock or scion resulting in either bottleneck or inverted bottleneck type symptoms. These kind of abnormalities generally lead to improper movement of metabolites across the plant.

Graft Incompatibilities

Grafting is being done in fruit crops since time immoral and it is well understood as well that all the grafts are not likely to be successful. There may be several reasons for the same; however, incompatibility between stock and scion primarily due to genetic differences is quite apparent and has been reported in several fruit plants. The primary visible symptoms of graft incompatibility are bottle neck or inverted bottle neck at the site of union between stock and scion. In other words successful union is not formed between stock and scion. In some instances, even after the formation of visible symptoms like bottle neck or inverted bottle neck, resultant plant survives and gives commercial bearing like in case of grapes, when Dogridge is used as a rootstock, inverted bottle neck is a common and visible symptom, but grapes are in commercial bearing in Maharashtra on Dogridge. In other typical instances, the combination does not work at all, unless some solution is made to the problem.

In Non-translocatable/localized graft incompatibility, use of mutually compatible interstock can solve the problem, as there is no direct contact between stock and scion. The classical example is of Pear x Quince which are otherwise incompatible, but use of Old Home as interstock solves the problem of incompatibility between Bartlett pear and Quince. While, in case of translocatable incompatibility, the use of interstock do not solve the problem. The mobile chemical, causes phloem degradation hence the combination is unsuccessful. The chemical is a toxin, which is either produced by stock or scion. Punasin, a cyanogenicglucoside is one of the classic examples of a toxin responsible for causing incompatibility in stock and scion in Pear x Quince.

There are several factors which affect the grafting success such as genetically diverse plant species, endogenous chemicals in either stock or scion-the movement of which leads to success/failure, art of grafting, type of graft, endogenous growth regulators *etc.*

Incompatibility can be immediate or delayed. Immediate incompatibility leads to sudden death of the scion characterized anatomically by total lack of even initial events of union formation, such as formation of callus *etc.* In delayed incompatibility, plant survives for months or even years, as already explained above the case of Dogridge as a rootstock of grapes. One of the reasons for delayed incompatibility may be the differential susceptibility of the stock and scion genotypes to a particular virus. Tristeza virus of citrus has been noted to cause delayed incompatibility in citrus.

Rootstocks for Fruit Crops

Most of the commercial fruit crops are grown commercially as grafted plants. In other words, most of the commercial fruit crops are grafted onto rootstocks (Table 1.1). Plants to be used as rootstocks may belong to same or different species. There are several advantages for the use of rootstocks in commercial fruit crop cultivation, as has been discussed in the previous section.

Selection of Root Stock

1. Root stock should have a proper vigour and growth habits.
2. Root stock should be resistant to soil born diseases and other pests.
3. Root stock should be tolerant/resistant to toxic salts like Na, Mg, and Ca *etc.*

4. It should have wide range of edaphic adaptability (all soil and climatic factors).
5. Should have wide range of graft compatibility.
6. Should be easy to propagate.
7. It should not go under any mutation.
8. Its age should be one to one and half years and not more than 2years.
9. Its diameter should be greater than 1- 1.5 cm (pencil thickness).

Table 1.1: Most Commonly Used Rootstocks for Commercial Fruit Crops

Sl.No.	*Fruit Crop*	*Rootstock most Commonly Used*
1	Almond	Almond, peach seedlings
2	Aonla	Desi aonla seedlings
3	Apple	Crab apple seedlings
4	Apricot	Wild apricot, wild peach seedlings
5	Bael	Bael seedlings
6	Ber	Desi ber/Malah ber
7	Cherry	Mahaleb seedlings
8	Citrus	Rough lemon
9	Grapes	Dogridge
10	Guava	L- 49
11	Jamun	Jamun seedlings
12	Mango	Seedling of desi mango
13	Olive	Seedlings of wild olive
14	Peach	Flordaguard
15	Pear	Kainth
16	Plum	Kabul Green Gauge
17	Sapota	Rayan
18	Walnut	Walnut seedlings

Note: Clonal rootstocks of apple M9 (dwarf), M4, M7, MM 106 (semi dwarf), semi vigorous (MM 111), Vigorous (Merton 793) and for cherry *viz.*, colt, Mazzard F12/1 are also in commercial use.

Techniques of Micropropagation

To understand different techniques, it is important to know what is micropropagation and what are the basic steps involved in micropropagation of plants.

Micropropagation

It refers to propagation from small plant tissues using tissue culture method. In other words, plants are propagated from small plant tissues under aseptic conditions on synthetic nutrient media.

Steps Involved in Micropropagation

Micropropagation (Figures 1.11 and 11.12) can be accomplished through four stages *i.e.* explant establishment, explant proliferation, *in vitro* rooting and acclimatization.

Figure 1.11: Micropropagation in Strawberry
(a) Shoot proliferation; (b) *In vitro* rooting; (c) Acclimatization.

Explant Establishment

Any part of a plant such as shoot tip, root tip, nodal segment *etc.* used for culturing on synthetic nutrient media is termed as explant. Establishment of these tissues is done on basal media containing macro nutrients, micro nutrients, vitamins, amino acids and a source of carbohydrates. Most commonly Murashige and Skoog (MS) media is used as basal media for raising most of tissue cultures.

Shoot Proliferation

The in vitro established cultures are subsequently cultured on MS media supplemented with appropriate growth regulators. For instance cytokinins like benzyl amino purine (BAP) or benzyl adenine (BA) is most commonly used for shoot proliferation.

In vitro Rooting

The individual shoots are separated out of proliferated shoot cultures and then are cultured on MS media supplemented with suitable auxins for root induction. Most commonly indole butyric acid (IBA) or naphthalene acetic acid (NAA) is used for *in vitro* root induction in the proliferated shoots.

Acclimatization

The rooted plantlets are then acclimatized (hardened) before transfer to the field. Acclimatization is done by planting *in vitro* rooted plantlets in appropriate rooting media and by maintaining desired relative humidity during hardening.

Figure 1.12: Micropropagation of Banana
(a) Shoot proliferation; (b) *In vitro* rooting.

Advantages of Micropropagation

1. Rapid- the rate of multiplication is much faster and quicker than the conventional methods of propagation.
2. True to type plants can be produced through this method.
3. Disease free plants can be produced, especially if meristem (0.1-0.5 mm tip of shoot) is the source tissue for multiplication.
4. Year around production- As the uniform temperature and humidity is maintained in the incubation room throughout the year, tissue culture plants can be produced around the year.
5. Easy exchange of germplasm- Using the technique of tissue culture propagation, germplasm can be easily carried across the counties.

6. Saves time and space- Large number of plants can be produced in a very little space as small as a culture vessel.
7. Exibit better quality and yield- Micropropagated plants come into bearing early and exhibit better yield and quality.

Disadvantages of Micropropagation

1. Needs expertise and expensive set up for *in vitro* plant production.
2. All plant species/varieties cannot be propagated at will-some are difficult to propagate (recalcitrant).

Thus micropropagation should only be used as supplemental technique for plant propagation, especially in case if it is not possible to conventionally propagate the plants. The different techniques of micropropagation are detailed below:

Shoot Tip Grafting

Shoot tip grafting (STG) is also commonly known as micrografting. The technique of shoot tip grafting has its basis in spain, where Navarro for first time in 1970's gave the concept of micrografting in citrus. Since then, STG has been used in several fruit crops for production of disease free plants.

It is difficult to regenerate complete plants from meristem in woody species like most of fruit and forest plants, thus as an alternative micrografting is done to produce virus free plants. The various steps in micrografting include scion preparation, rootstock preparation, *in vitro* grafting and acclimatization/hardening of the plants. The *in vitro* raised nucellar seedlings are used as rootstocks. The scion (meristem 0.1-0.4 mm) is obtained either from young growth of field grown trees, defoliated glass house grown plants, or *in vitro* proliferated nodal segments obtained from mature trees. The grafting is done with the help of stereomicroscope, under aseptic conditions. Several viruses have been eliminated via micrografting in fruit plants such as citrus tristeza virus, peach latent mosaic viroid and pear vein yellow virus.

The technique of STG involves several steps:

1. Preparation of stock
2. Preparation of scion
3. *In vitro* grafting
4. Hardening

Preparation of Stock

In vitro rootstock preparation involves several steps. As most of the research work on the aspect of shoot tip grafting has been done in citrus, the necessary steps will be discussed as a case study of citrus. First of all the seeds, preferably with testa removed are cultured *in vitro* on basal MS media. It is important to keep the cultures under dark conditions; this will help to raise etiolated seedlings with elongated epicotyls. It is easy to graft a stereoscopic scion having dark green colour on an etiolated pale yellow coloured epicotyl. About 4-6 weeks old etiolated seedlings are used for *in vitro* grafting.

Preparation of Scion

Meristem to be used as scion for *in vitro* grafting can be obtained from three sources. The first being mature field grown plants, second from *in vitro* grown cultures and thirdly the scion source can be glass house grown plants. Out of three sources of meristem, the scion from glass house grown plants is most practical solution for achieving high success rate. The scion obtained from field grown mature plants is difficult to make pathogen free, while, it is difficult to isolate meristem from *in vitro* grown proliferated cultures. In glass house, plants can be made pathogen free, as well as can be made to grow as and when required by practicing need based pruning.

In vitro Grafting

The first step is to decapitate the etiolated rootstock raised *in vitro* under dark conditions. The second step involves dissection of meristem from proven scion source using stereomicroscope, and under sterile aseptic conditions. The third vital step involves *in vitro* grafting of meristem on decapitated etiolated rootstock. Meristem can either be placed on top of decapitated rootstock or at an inverted cut position.

In vitro Growth and Hardening

The *in vitro* grafted cultures are cultured preferably under liquid culture media, with the support of filter paper bridge. The establishment and growth of cultures is better under liquid media, as there is better uptake of nutrients from liquid media as compared to semi solid media. Hardening process starts during *in vitro* stage. In this step gradually remove plugs from culture vessels and thus thereby acclimatizing them step by step.

Meristem Culture

The technique of culturing meristem (0.1-0.5 mm) - comprising of an extreme shoot tip and one to two leaf primorida on a synthetic nutrient media under asceptic conditions is known as meristem culture. The technique has been used on the basis that plants regenerated through meristem are primarily virus free. The possible reason for production of virus free plants through meristem culture is that cell division is fast and there are no plasmodesmata in the extreme tip of plant called meristem. Moreover, the resultant plants are expected to be virus free due to trauma of excision of meristem from the original plant or a scion source.

This technique is very useful for the elimination of viruses from infected plant material. Rapid multiplication of the plants, which are otherwise not easily propagated by vegetative means, is also possible through meristem culture. This technique is widely used in horticultural plants like potato, dahlia, carnation and orchids. As meristem culture is not much successful in woody plant species, as an alternative shoot tip grafting (STG) was introduced to produce disease free plants.

Microbudding

This technique involves budding young rootstock about 3-4 month age old with small scion buds *in vivo*. A unique way to keep micro bud on place involves the use of micro tips/plastic tips at initial stages after budding and later on removing them on seeing the scions of visible bud growth. Apart from holding the micro bud at its original place, the use of micro tip also maintains humidity at the site of union. This method has several advantages over *in vitro* grafting, namely its ease to perform budding operation, better success rate, less labour intensive, practically more number of plants can be produced. However, if the scion is not taken from pathogen free plant, the resultant plant can also be infected, while in case of meristem culture, having the size of meristem 0.1-0.5 mm is in most of instances free of viruses. It is need to understand that even plants raised through meristem culture are free of viruses but are not virus resistant.

Synthetic Seed

The somatic embryo produced through tissue culture and encapsulated with an artificial coating is known as synthetic seed or artificial seed. Somatic embryos are bipolar structures with both apical and basal meristematic tissues which are capable to form both root and shoot. The *in vitro* induction of embryo (Figure 1.13) requires a high level of auxin in culture medium, followed by low concentrations of auxin and cytokinin in the medium. Somatic embryos may arise in culture directly on explants or via callus formation or liquid suspension cultures. The induction of somatic embryos is the best technique for rapid and true to type multiplication of plants. The somatic embryos originate from somatic or vegetative cells, and are bipolar structures, which possess both shoot and root meristem. The somatic embryos thus produced can be encapsulated, thereby forming artificial or synthetic seeds, which is an attractive alternative for propagation of plants.

Among two methods *viz*., hydrated or dessicated, for artificial seed production, the production of hydrated seeds is more popular. In this method, the individual somatic embryo is encapsulated in a water based gel (hydragel; such as calcium alginate). Embryos developed through tissue culture technique are mixed with sodium alginate and dropped with pippette into a calcium salt (calcium chloride) solution to form calcium alginate capsules. The capsules are washed in water and then placed on culture medium for germination. The other gelling agents like agar, guar gum, sodium pectate *etc.* can also be used for encapsulation of somatic embryos.

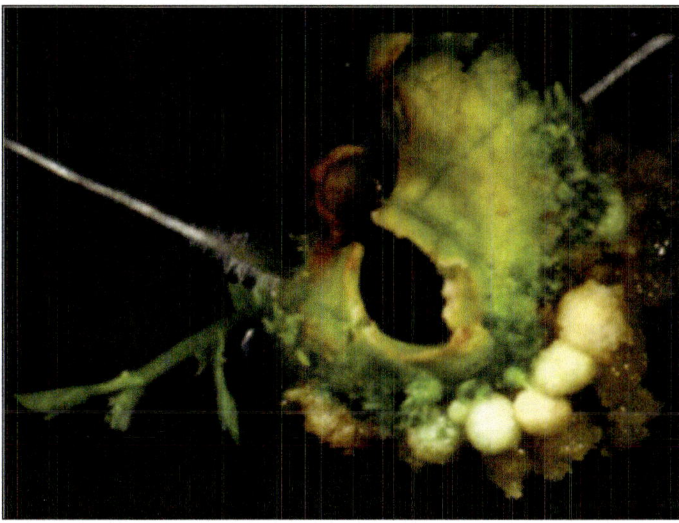

Figure 1.13: *In vitro* Induction of Somatic Embryos.

The basic steps involved in seed production includes: Identification of suitable explant, induction of somatic embryos *in vitro*, maturation of somatic embryos, protocol for their mass production *in vitro/in vivo*, Encapsulation of mature somatic embryos, desiccation and transfer to field/pots. Since somatic embryos lack seed coat and endosperm, the encapsulation medium must have nutrients and growth regulators, which will serve as an artificial endosperm.

Synthetic/Artificial seeds have been produced in banana, citrus, mango, apple, olive and kiwi fruit. The technique is useful in fruit plant, as due to cross pollination the seed is heterozygous and thus cannot be used as the methods of propagation, particularly where true to type plants is the objective and there is monoembryonic seed development. Also this technique can be adopted in those fruits where seeds in general are not formed and it is difficult to propagate the plants by vegetative means. Other advantages of this technique include large scale propagation and genetic uniformity of the resultant plants. For production of genetically modified plants/transgenics, this technique has also great potential. However, large scale production, asynchronus development of somatic embryos and improper maturation are some of the limitations of this method, which have unique protocols and need standardization on crop to crop basis. This technique can be of immediate help to a plant breeder for *in vitro* conservation of valuable germplasm.

Commercial Methods of Propagation in different Fruit Crops

Table 1.2: Commercial Methods of Propagation in different Fruit Crops

Sl.No.	*Fruit Crop*	*Commercial Methods of Propagation*	*Time of Propagation*
1	Almond	Shield budding	July-Aug
2	Aonla	Patch budding	June-Sep
3	Apple	Tongue grafting	Feb-March
4	Apricot	Tongue grafting	February
5	Bael	Patch budding	June-July
6	Banana	Suckers (sword)	March
7	Ber	T-budding	June-Sep
8	Cherry	Tongue grafting	Feb-March
9	Citrus	T-budding	Feb-March, Aug-Sep
10	Datepalm	Offsets (sucker)	Feb-March
11	Fig	Cutting	Jan-Feb
12	Grapes	Cuttings	Jan-Feb
13	Guava	Patch budding	May-June
14	Litchi	Air layering	July-Sep
15	Loquat	Inarchig	July-Aug
16	Mango	Side grafting	July-Sep
17	Olive	Cuttings	Jan-Feb
18	Papaya	Seeds	July-Sep
19	Passionfruit	Cuttings	July-Aug
20	Peach	Tongue grafting	Dec-Jan
21	Pear	Tongue grafting	Dec-Jan
22	Plum	Tongue grafting	Dec-Jan
23	Pomegranate	Cuttings	Jan-Feb
24	Sapota	Inarching	Dec-Jan
25	Strawberry	Runners	June-Sep
26	Walnut	Patch budding	July-Aug

Commercial Tissue Culture Laboratories and Nursery Sector in the Country

Till the recent date 88 commercial* Tissue Culture Production Units have been "Recognized" by the Department of Biotechnology (DBT), Govt. of India under the "National Certification System for Tissue Culture Raised Plants (NCS-TCP)". The list is as follows:

1. A.G. Bioteck Laboratories (India) Ltd., Hyderabad;
2. ABC Agrobiotechnology Pvt. Ltd., Mahemdabad
3. ACE Agro Technologies, Secunderabad
4. Aditya Biotech Lab and Research Pvt. Ltd., Raipur
5. Agri Vitro Tech Laboratories, Secunderabad
6. Ajeet Seeds Ltd., Aurangabad
7. Anantha Biotechnologies, Ananthapur
8. Arcadia Agro, Mogar Anand
9. ATGC Biotech, Raipur
10. Bejo Sheetal Bio-Science Foundation, Jalna
11. Bhoomiputra Biotech, Chopda, Maharashtra
12. Biotech Park, Lucknow
13. Biotechnology Centre, Department of Horticulture, Govt. of Karnataka
14. Blossom Tissue Culture Nurseries, Hosur
15. Boinchi Bioplants Pvt. Ltd., Hooghly
16. BrookFileds Biotech Pvt. Ltd., Kadapa
17. Cadila Pharmaceuticals Ltd.(Agro Division), Ahmedabad
18. Callus Biotech Pvt. Ltd., Kolhapur
19. Devleela Biotechs, Raipur
20. Elegant Flower Company Pvt. Ltd., Kolkata
21. Excel Plant Link Pvt. Ltd., Bhubaneswar
22. Futura Bioplants Pvt. Ltd., Pune
23. Ganapati Biotech, Kalahandi, Orissa
24. Geeta Agro Biotech, Pune
25. Genewin Biotech, Hosur
26. Godrej Agrovet Ltd.(Plant Biotech Division), Secunderabad
27. Greeno Agrotech India Pvt. Ltd., Anathapur
28. Growmore Biotech Ltd., Hosur
29. H.U Gugle Agro Biotech Co., Ahmadnagar
30. Harshita Biotech Pvt. Ltd., Hyderabad
31. Hindustan Bio energy Ltd., Lucknow
32. Hindustan Paper Corporation Ltd., Kagajnagar, Assam
33. Hithkari Nursery, Bangalore
34. Hosur Hortitech, Hosur
35. HU Gugle Biotech Pvt. Ltd., Bangalore
36. *In Vitro* International Pvt. Ltd., Bangalore

37. ITI Biotech Tissue Culture Lab, Betul
38. Jain Irrigation Systems Ltd. Jalgaon
39. K.F. Bioplants Pvt. Ltd., Pune
40. K.F. Biotech Pvt. Ltd., Bangalore
41. Kala Biotech Pvt. Ltd., Pune
42. Kisan Agri Biotech, Hyderabad
43. Kodeesvaran Biotech, Hosur
44. Kore Biotech, Solapur
45. Krishiraj Nursery, Jalna
46. Kshitij Biotech Corporation, Karad
47. Kutch Crop Services Ltd., Kutch
48. La Chandra Tissue Culture Laboratory, Banaskantha
49. Lakshmi Biotech, Bangalore
50. Mauli Hitech Nursery, Pune
51. Micropropagation Technology Park (TERI), New Delhi
52. Microsun Bioplants (India) Pvt. Ltd., Secunderabad
53. MSR Biotech Pvt. Ltd., Bangalore
54. Mysore Organic Farms Pvt. Ltd., Mysore
55. Navsarjan Biotech, Bhuj, Kutch
56. Neesa Agritech and Foods Limited, Ahmedabad
57. Nishant Biotech, Bilaspur, Himachal Pradesh
58. Novel Biotech, Bangalore
59. PepsiCo India Holdings Pvt. Ltd., Hoshiarpur
60. Ram Biotech, Jalgaon
61. Ramrich Biotech Pvt. Ltd., Pune
62. Rise N Shine Biotech Pvt. Ltd., Pune
63. Sagar Agrisciences Pvt. Ltd., Barabanki
64. Sai Lara Biotechnologies, Hyderabad
65. Sarjan Biotech Pvt. Ltd., Bhuj-Kutch
66. Secon Agriventures Pvt. LTd., Bangalore
67. Seema Biotech, Kolhapur
68. Shaili Biotech (P) Ltd., Ahmedabad
69. Sheel Biotech Ltd., Gurgaon
70. Shivashakti Biotechnologies Ltd., Medak Dist., Hyderabad
71. Shree Biotech, Pune
72. Shree Ganesh Biotech (India) Ltd., Kolkata

73. Shree Samarth Biotech, Pune
74. Shriram Prathishtan Mandal's Lokmangal Tissue Culture Lab, Solapur
75. Siddhi Plantek, Anand
76. SPIC Agro Biotech Centre, Coimbatore
77. Sri Soma Biotech, Guntur
78. Sri Venkateshwara Agro Technologies, Hyderabad
79. Sristi Agro Biotech Pvt. Ltd., Howrah
80. Sunglow Biotech, Coimbatore
81. Synergy Agri Products Pvt. Ltd., Durgapur
82. Technico Agri Sciences Ltd., Chandigarh
83. Thulasi Biotech, Jolarpet, Vellore
84. Turlapati Bio Tech Pvt. Ltd., Secunderabad
85. Vasantdada Sugar Institute, Pune
86. Vitrogold Tissue Culture Pvt. Ltd., Mehsana
87. Vitroplant, Hyderabad
88. Palm Grove Nurseries (K.Lakshmansa and Co.) Nursery Division, Bangalore

**Source*: http://dbtncstcp.nic.in/html/content/Recognized_TCPU's.pdf (June 23, 2014).

Scion Bank and Hi-tech Nursery

Scion is an important component of a fruit plant. The upper portion of all the grafted plants is termed as scion or the bud inserted while budding onto rootstock develops into scion. In other words, a grafted fruit plant has two physical components *viz.*, Scion and rootstock/stock. Scion proceeds to form a fruit bearing portion of a plant. While, rootstock is a root bearing portion of plant. Scion/bud wood can be obtained either from field grown plants, plants grown under protected conditions/screen houses, or from *in vitro* sources. An institute or division or any unit area private or government controlled, having an authentic collection of bud wood obtained from known sources, having a full pedigree record is known as scion bank. National Bureau of Plant Genetic Resources (NBPGR) is one of the most authentic official scion banks of India having more than 500 accessions of fruit plants. To have a authentic source of fruit plant and with known pedigree is one of the preliminary parameters for production of quality plants and vis-à-vis to have high yield potential per unit area. Most of the production problems such as related to production, infestation of insect pest and diseases are related to dubious germplasm having unknown pedigree. Getting scion from known sources will solve day to problems and will certainly enhance fruit production in country.

To multiply/propagate the plants on a commercial scale and more importantly around the year, high tech nursery production is the only solution. Though, the initial cost of establishment of high tech nursery is high, but large scale production of difficult to propagate plants aim at getting higher net returns per unit area of nursery. The high tech nursery includes production of plants around the year via mist propagation, net house/green house and tissue culture propagation. Drip, sprinkler, foggers are important other components of this high tech nursery production, which must be under control of an automated system. Several advantages of high tech nursery production includes off season production, having no seasonal constraints thus plant production is expected around the year. There is optimum use of resources such as irrigation water and fertilizers *etc.*

Nursery and Seed Acts

Act is important to monitor and regulate the quality nursery plant production. The nursery act has been on record since more than 5 decades, formulated as The Punjab Fruit Nurseries Act 1961. The main aim of the act is to provide for the regulation and licensing of fruit nurseries in the state of Punjab. Likewise, The Maharashtra Fruit Nurseries and sale of Fruit Plants (Regulation) act 1969 was formulated in late seventies. As per the liability of the act, state government has to ensure production and supply of quality planting fruit plants in the state. It is worthwhile, to note here that not all the states have nursery acts.

Likewise, to regulate the quality of seed for sale to farmers of the state, The 'Seeds Act, 1966' was passed and is commonly known as 'Act no 54 of 1966'. This act unlike state dominant nursery acts is applicable for whole of India. Now this act has become redundant and has been replaced by 'Seed Bill 2004', which subsequently has been amended as 'Seed Bill 2010, 2011'.

Nursery Registration

Considering the existence of Nursery Act to be precise as 'The Punjab Fruit Nurseries Act 1961', all the nurseries in the state weather private or public must be registered. The application for recognition of horticultural nursery needs to be submitted to NHB, Ministry of Agriculture, Government of India, Department of Agriculture and Cooperation, Gurgaon. Though the act is there, but it needs strict implementation, which is in the interest of the state. The nursery act is more important to implement in Toto in fruit plants, as most of the fruit plants are perennials and are likely to stay in field for even decades. Once the non-pedigreed plant is in the field, may produce havoc to the overall fruit production vis-à-vis economic returns to the fruit grower. When all the fruit plant nurseries in a particular state are registered, the advantages accrued upon will be: stoppage of illegal sale of fruit plants by street vendors/small nurseries, production of quality-indexed planting material will be ensured, quality fruit production with reduced juvenile period-as illegal plants are mostly on their own roots and take long time to come into commercial bearing and thus making endless wait for the farmer waiting for its commercial fruit production, ensured remunerative return to the farmers and above all the faith in government deeds and its commitment for diversification from ongoing wheat-rice cropping system to high output yielding fruit crops will be realized. Hence, it is in the interest of public in general and farming community in particular that all the nurseries of India, must be registered. To ensure this, nursery act of India must be formed, with uniform guidelines applicable to all the states.

IPR Issues Related with Vegetatively Propagated Horticultural Crops

Intellectual property rights (IPR) issues are of recent concern in plants and off late in horticulture. Manual interventions in growing plants including fruit crops, being a natural phenomenon is not considered on the whole as a new patent and also a part of IPR. Ever since the first-generation patent acts, patent protection for plants has always been questioned. The new products developed in plants including horticultural crops are not considered, as a patent, as the plants are considered not to be obtained as a result of creative process, which is the main criteria of patentability and hence not considered as inventions. To understand IPR issues, it is first important to understand what a patent and what are its implications for a owner. A patent is a legal monopoly granted to the owner of any new invention, which is capable of being used for limited period of time. The owner of a patent on a product/variety has full right to prevent others from making, using or importing this product without his consent.

Now the point of concern in this part of a chapter is that, if a plant is propagated by vegetative means such as by budding, grafting, layering or even through tissue culture, is it a patentable act or a routine manipulation of nature. In such instance it depends upon the case to case basis. In simple words, if the plant- to be used as stock or scion for further propagation, is protected/patented, any method used for its vegetative propagation is illegal and on the whole, needs permission from the owner of the patent of such plants.

To conclude the chapter on plant propagation, it can be clearly stated that propagation is an art, the techniques of which must be practiced selectively on case to case basis. For instance, if normal seed can be used a method of propagation, synthetic seed production must not be avoided. Likewise, the tissue culture approaches like meristem culture, must only be used, when virus elimination is the prime objective. IPR issues must also be considered, while propagating horticultural crops.

REFERENCES

Aldriance GW and Brison FR, 2000. Propagation of horticultural plants. Mc Grow Hill Book Company. Inc, New York.

Bose TK, Mitra SK, Sadhu MK and Das P, 1997. Propagation of Tropical and Sub tropical Horticultural Crops. IInd Edition, Naya Prakash, Calcutta.

Sharma RR, 2002. Propagation of Horticultural crops: Principles and Practices. Kalyani Publishers, New Delhi.

Hartmann HT, Kester, DE, Davies, FT Jr., and Geneve, RL 1997. Plant propagation principles and practices. 6th ed. Prentice Hall, Englewood Cliffs, N.J. pp. 770.

Whitehead GE, 1970. Grow fruit in your green house. Faber and Faber, London.

Chapter 2

Propagation Techniques in Arid Fruits

P.L. Saroj[1] and A.K. Singh[2]

[1]ICAR–Central Institute for Subtropical Horticulture, Rehmankhera, PO Kakori, Lucknow–226101, Uttar Pradesh
[2]ICAR–CIAH (Central Horticultural Experiment Station), Vejalpur–389340, Panchmahals, Godhra, Gujarat

For improving production, productivity and profitability in horticulture sector, availability of clean planting materials to the doorsteps of growers at reasonable price can play pivotal role in area expansion under perennial horticulture. In spite of all efforts, there is a wide gap between demand and supply of quality planting materials of various fruit crops. It is also pertinent to mention that the development of arid horticulture in India is comparatively late than tropical, subtropical, temperate and coastal horticulture; hence availability of quality planting materials of arid fruits is in high scarcity. Though, a vast land resource in India (12.02 per cent of total geographical area) is categorized as arid region. Out of 39.54 million hectares of arid land; 31.71 million hectares is under hot arid region and 7.84 million hectares is under cold arid region. The major portion of hot arid region is spread over north-western part of the country covering the state of Rajasthan (19.60 million hectares) followed by Gujarat (6.20 million hectares) and Punjab (1.50 million hectares) and only 3.10 million hectare is in the peninsular region of the country (Dhandar and Saroj, 2004). The major constraints of arid region are poor and erratic rainfall, extremes of temperature, sandy soil with poor fertility, high wind velocity, very deep-water table and brackish underground water resource, problem of wild animals *etc.* The major challenges in arid region are complex diverse and risk-prone farming situations owing to poor productivity and profitability, vulnerable to climate change, genetic erosion of important biological diversity, poor input use efficiency, poor infrastructural facilities, limited and traditional post-harvest management *etc.* (Saroj *et al.*, 2022)

Despite these constraints, there are various strengths in arid region *viz.*; vast land resource, intense solar radiation, increasing canal command area, low incidence of pest and diseases, surplus family labourers, good animal population, suitability to establish agro-based and allied industries *etc.*; the arid region is gaining importance to promote as horticulture bowl of the country. The stress conditions coupled with wide variations in day and night temperature greatly favours production of high-quality fruits, vegetables, spices and medicinal and aromatic plants. For sustainable production in arid environment, the crops either having xerophytic adaptation or its varieties are resistant/tolerant to biotic and abiotic stresses. There are various fruit crops like Ber (*Ziziphus mauritiana*), Aonla (*Emblica officinalis*), Pomegranate (*Punica granatum*), Fig (*Ficus carica*), Date palm (*Phoenix dactylifera*), Bael (*Aegle marmelos*), Tamarind (*Tamarindus indica*), Jamun (*Syzigium cumini*), Khirni (*Manilkara hexandra*), Lasoda (*Cordia myxa*), Ker (*Capparis decidua*), Phalsa (*Grewia subinaequalis*), Pilu (*Salvadora oleoides* and *S. persica*), Karonda (*Carissa carandus*), Wood apple (*Feronia limonia*), Chironji (*Buchnania lanzan*), Cactus pear (*Opuntia ficus indica*), Mulberry (*Morus* spp.), Custard apple (*Annona* spp.) *etc.* which are climate resilient and can be successfully grown in arid region with very low inputs (Saroj *et al.*, 2017). Therefore, there is a need for mass multiplication of clean planting materials of arid fruits by adopting appropriate propagation techniques.

Principles of Propagation

Propagation is an important process by which every kind of living organism multiplies to form new individuals of its own kind. This process maintains the continuity of life from one generation to next. It can be defined as the science and art of multiplication of plants either by sexual or asexual means. The major objective of fruit tree propagation is to multiply population that will be similar to the parent plant. Thus, the best plant propagation method is one that will help in transmission of all characteristics of the mother plant to its offspring without transmitting any kind of disease. This is necessary to preserve the new varieties/hybrids being developed by modern techniques either by sexual (seed) or by asexual (vegetative) methods. Since, most woody plants are highly heterozygous; therefore, a relatively high number of genes on one chromosome of a Mendelian pair differ from those on the other chromosome. Thereby, sexual reproduction involves meiotic cell division, which ultimately produces progeny (seedlings) with new or differing genotypes relative to their male and female parents. Consequently, the progeny of woody plants grown from seed tend to exhibit a relatively high amount of genetic variation. Whereas, in asexual propagation, reproduction from the vegetative parts of the donor plant involves mitotic cell division in which the chromosomes duplicate and divide to produce two nuclei, which are genetically identical to the original nucleus. This clonal process, in which the genotype of the parent plant is exactly duplicated, is made possible because of two unique plant characteristics *viz.*, totipotency and dedifferentiation. Totipotency is the property of vegetative plant cells to carry all of the genetic information necessary to regenerate the original plant, while dedifferentiation is the ability of mature (differentiated) cells to return to a meristematic condition and produce a new growing point.

Production of clean/quality planting materials refers i) plant must be true to-type, ii) multiplied by appropriate technique, iii) free from pest and diseases, iv) as per standard and v) subject to traccability. To bring additional area under arid fruits, huge quantity of quality planting materials of arid fruits will be required. To cater the demands, these fruit crops need to be multiplied by an efficient means of propagation. There are two primary forms of propagation of fruit crops: sexual (seeds) and asexual (vegetative plant parts). The most commonly known vegetative propagation of plant includes propagation by cuttings, layering, budding, grafting, by modified specialized organ or by micropropagation methods. To perform this, a sound knowledge and skills of identification of plant organs, which has to be used for plant propagation, is of paramount importance for fruit crops in order to multiply them as per requirements. The protocol for mass multiplication through tissue culture techniques has been standardized in some arid fruits like date palm and pomegranate.

The inordinate bottleneck in area expansion under arid fruits is the non-availability of genuine and quality planting material from reliable sources. Often, growers are procuring planting materials from non-accredited nurseries and this practice is causing great loss to the Indian fruit industry. For production of quality planting material, fruit nursery needs a great deal of expertise for planning, execution and monitoring. Provisions of certain basic pre-requisites are required for raising a fruit nursery on scientific ground, so as to perpetuate true-to-type planting materials. Nursery hygiene is another area in order to prevent harmful pathogens, pests and weeds from entering a nursery and minimizing their spread will reduce crop wastage and economize nursery input cost towards labour and plant protection measures. Also, waste management in perennial fruit crop nursery is very important. All waste biomass should be properly decomposed and recycled for further use.

Requirement of Fruit Nursery Establishment

Nursery Site Development

Nursery is an area where plants are raised either for own plantation or for sale purpose. Therefore, appropriate nursery site should be identified and developed for best quality planting materials production. Roadside nursery is always better for display and disposal of produce. Nursery area must be protected by proper fencing. In arid region suitable wind breaks like Khejri, *Pongamia pinnata*, *Albizia lebbeck*, Mulberry, Bael *etc.* should also be planted in order to save inside plants from hot winds. The area should be properly leveled and after removal of perennial tree roots if any, different blocks should be developed for establishment of mother block, raising rootstocks, propagation bed, shifting beds, protected structures, display site, sale office, store, water storage tank, irrigation channels, path *etc.* Proper irrigation facility is essentially required. Different types of growing structures like shade net house, semi controlled polyhouses with mist system and fully auto-controlled polyhouses with fogging system should also be established as per need. For better branding, nursery should be named with a visible sign board and all data must be maintained for its accreditation.

Establishment of Mother Block

The planting stock that maintained as a source for commercial propagation is referred to as a mother block. Therefore, for preservation of pathogen free planting stock, proper isolation, sanitation, periodical monitoring and proper management are essentially required. The mother plant should be healthy, precocious, prolific bearer, disease free, genetically true-to-type and free from insect-pests attack. As far as possible, mother block should be in the close proximity of the nursery site. The scion shoots should be taken only from bearing plants. There should be proper record and certification of planting stock. The plants should have known pedigree records regarding bearing potential, fruit quality and problems associated, if any. Commercial planting stock can be referred as certified stock if it is grown under supervision of a legally designated agency with prescribed regulations designed to maintain minimum standards of cleanliness and clonal identity.

Raising of Rootstocks

Rootstocks are used where plants are propagated on the roots of other stock. Rootstocks used in plant propagation are generally raised by seeds but clonal multiplied rootstocks are true-to-type and homogenous. Further, there are certain rootstocks which are moderately tolerant to salinity, suitable for moisture stress situations, impart dwarfism in scion vigour, tolerant to diseases *etc.* upon which scion of arid fruits be grafted. For raising of rootstocks, polyethylene bags with different sizes *viz.*, 25cmx10cm, 25cmx15cm and 30cmx15 cm as per the requirement are being used. Small holes are made in the bottom and sides of polyethylene bags for drainage and aeration and filled with porous rooting medium or pot mixture for raising rootstocks. The locally available seeds (preferably one tree seedling) should be sown in the polythene bags, filled with soil and well rotten FYM (3: 1) for raising the rootstocks. Seeds can also be sown in the raised nursery beds or polyhouses and transplanting of very young seedlings in the polythene bags at the 2-4 leaf stage gives about 80 per cent success. About 9-12 months old seedling of uniform size having stem of pencil thickness should be used as rootstock. Plants raised in the polythene bags can easily be transported to distant places and resulted into higher transplanting success. Though, costly, but biodegradable bags should be used to minimize environmental pollution.

Propagation Techniques

Sexual Propagation

Since beginning, seeds are used to raise seedlings and subsequent plantation. Even today, fruits like chirounji, karounda, khirni, phalsa, lasoda, wood apple, bael, lime, tamarind, manila tamarind *etc.* are propagated by seeds. For raising of seedlings, nursery beds are prepared properly to ensure better germination and survival of seedlings. The seeds of most of the arid fruits have no dormancy and they germinate within 8-25 days after sowing. Seeds are sown in the polythene tubes during rainy season for better germination and survival under open conditions. These seedlings can be directly transplanted in pit in the field at 3-5-leaf stage during rainy season for *in-situ* patch budding under rainfed semi-arid conditions. Under semiarid conditions, sowing is done during February-March so that the seedling should attain buddable size during May–June. Whereas in arid conditions, the sowing is done in July-August and plants are ready for budding/grafting in June-July. Foliar spray of GA3 (1000 ppm) has been found effective to accelerate the growth of seedling with the maximum fruit: shoot ratio in jamun (Singh *et al.*, 2010c). Seeds are also sown in polythene bags which enable easy handling of seedling rootstocks and grafted plants during transportation. Sometimes coiling of roots become problem, hence use of root trainer and timely disposal of polybag raised plant is suggested. Some seed propagated arid fruit crops and related information is given in Table 2.1.

Table 2.1: Seed Propagated Arid Fruits and Related Information

Fruit Crop	*Related Information*
Jamun (*Syzygium cuminii*)	Jamun seeds have no dormancy; hence fresh seeds can be sown within 10-15 days and seeds germinate 10-15 days after sowing (Singh *et al.*, 2011c). Studies revealed that seed extraction from jamun after heaping the fruits for a single day was better for getting good quality seeds in comparison to the extraction of seeds immediately after collection. Sasthri *et al.* (2001) recorded that large size seeds had higher germination (99-98%) than small sized seeds (79- 89%).
Tamarind (*Tamarindus indica* L.)	Tamarind is normally grown from seeds. Seed germination can be accelerated by soaking seeds in 10 per cent cow urine or cow dung solution (500 g in 10 l water) for 24 hours and germination was increased from 37 per cent in untreated to 72.60 and 82.80 per cent, respectively in treated seeds (Singh and Singh, 2007).
Chironji (*Buchanania lanzan* Spreng)	Chirounji is cherished for its almond-flavoured seeds, are generally propagated by seed giving a long gestation period (15-20 years) and wide range of variability (Singh *et al.*, 2019g). Seed treatment with sulphuric acid (5-7%) was found very effective for promotion of seed germination in chironji (Singh and Singh, 2014b).

Fruit Crop	*Related Information*
Khirni (*Manilkara hexandra* Roxb.)	Khirni is generally propagated by seed giving a long gestation period (10-12 years) and large variability. Seeds can be sown in polythene bags during June-July and it germinates within 25-35 days. The seedlings become ready for planting after one year from the date of sowing (Singh *et al.*, 2021). When one-year-old seedlings of khirni raised for rootstock commence putting on new growth and the leaves are of light green colour, these are ready for soft wood grafting (Singh *et al.*, 2019). Khirni seeds have tendency to lose viability quickly (Bajaniya *et al.*, 2018).
Lasoda (*Cordia myxa* Roxb.)	Lasoda is generally propagated by seeds in spite of low and irregular germination of seeds (30-35%), however the germination percentage in *Cordia gharaf* (a related species, used primarily as rootstocks) is observed about 70-80 per cent. The seed has hard seed coat which does not absorb water easily. Therefore, pre-treatment of seeds with cow dung for 24 hours and soaking in water is advisable to enhance the germination and seedling growth. The higher germination percentage can be obtained when the seed is treated with H_2SO_4 for 30 minutes (Singh *et al.*, 2019).
Mahua (*Bassia latifolia* Roxb.)	Indian Butter Tree or mahua is a plant of economic importance, particularly, in tribal belts due to its uses in brewing beverages. Mahua is generally propagated by seeds. It takes 12-15 days to germinate completely (Singh and Singh, 2015). Mahua seeds have no dormancy. Fresh seeds after storage of 5-6 days at room temperature show 80 per cent germination (Singh and Singh, 2014). Seeds can be sown on raised beds about 3 cm deep during June-July; thereafter, shifted in polythene bags with 3-4 leaves. Polyembryony is found in mahua seeds. Since mahua is cross-pollinated crop, propagation through seed lead to lot of genetic variability in term of growth, biomass, seed yield and oil content (Singh *et al.*, 2008). The recalcitrant nature of fruit limits the sexual propagation.
Phalsa (*Grewia subinaequalis* DC)	Owing to its sweetness exquisitely counterbalanced by astringent, acidic and sour tones, is commercially propagated by seed. Being a self-pollinated crop, it does not show variability when propagated through seeds. Seeds require 15-20 days for germination and seedlings get ready in 3-4 months for transplanting in the field (Mishra *et al.*, 2019). Singh *et al.* (2019) reported that seeds sown immediately after extraction gave the highest germination percentage (75%) with the shortest germination period (16 days).
Cactus pear (*Opuntia ficus indica*)	Cactus pear is commonly called prickly pear, when raised through seeds show variability in growth. Seed viability in cactus pear is longer but the rate of germination is poor. Its germinability can be improved by mechanical breaking and/or after immersion in GA, hot water (90-95 0C) treatment for 5 to 20 minutes and chemical scarification with concentrated H_2SO_4 followed by soaking in 100 ppm gibberellic acid (GA).

Fruit Crop	*Related Information*
Wood apple (*Feronia limonia* Swingle)	Wood apple is much hailed for its blended tamarind and raisin jam like taste, is usually propagated by seed. Freshly extracted seeds should be used for raising seedlings. Seeds sown fresh usually show a high percentage of germination (66%). Pre-treatment of the seed is not necessary (Yadav, 2017b). Polyembrony to the extent of 63 per cent is observed in wood apple.
Manila tamrind (*Pithecellobium dulce* Roxb.)	The seeds of manila tamarind are dispersed by birds, owing to natural regeneration. The pods are collected during June-July, cleaned and dried seeds are used for sowing. Seed viability of dried seeds is longer and after sowing germinates within few days. Owing to hard seed coat, there is some problem in germination which can be improved by treating the seeds with 50 ppm BA and GA_3 (Singh *et al.*, 2011).

Asexual Propagation

Aonla

In case of aonla, Amin (1978) recorded 70 per cent success following softwood grafting at the site of the terminal shoot of the rootstock raised *in-situ* for a year or more. Softwood grafting under controlled conditions using six bud scion shoot has given 85 per cent success in aonla (Ram and Kumar, 2005). The rootstock matrix diameter (graft/budding height) height has direct relation on ultimate success in aonla.

Figure 2.1: Patch Budding-Aonla.

The minimum size of rootstock should be at least 0.5 cm at the time of budding/grafting (Saroj *et al.*, 2001). It would be advisable to raise the seedling rootstocks *in-situ* and bud them with superior types, particularly in dry land conditions. Similarly, Tewari and Bajpai (2002) reported that graftable stem diameter is available at about 3.0 cm height in 5-6 months old seedling grown in polybags. At this height average diameter is around 5 mm. The scion shoots should be semi-hard and equal or slightly thicker than rootstock at graft height. Aonla propagation through stem cuttings, Rajan and Ram (1882) recorded 87.30 per cent rooting in cuttings using IBA at 15000 ppm with bottom heat (33 °C).

Budding has been found most successful method of aonla propagation. Patch budding showed better response followed by modified ring budding. Buddeing operation should be performed from mid May to mid August for better success, which may be continued upto September with appreciable success (Pathak *et al.*, 1991). Seedling rootstocks are generally used. For raising seedlings, seeds should be soaked in water over night before sowing. For better germination and healthy seedling, seeds may be soaked in 500 ppm GA_3 solution. Seeds should be sown in poly bags during March-April. Early sowing in February can also be done in polyhouse. Seedling of late sown seeds in May-June in arid region do not reach to budable size in same season; thereby, they are used in the next season. Seeds sown in March-April in poly bags (40 x 15 cm) under shade net germinate within 7-10 days and seedlings of about 10 cm height (after 35- 40 days after sowing) should be transferred in field beds followed by light irrigation for better establishment under hot arid ecosystem. Seedling of about pencil thickness (90 days old) should be used for patch budding. More than 90 per cent success was obtained under irrigated hot arid ecosystem when patch budding was performed from mid-June to mid-July. By this method 6 month time can be saved than conventional method *i.e.* budding on one year old rootstocks (Saroj *et al.*, 2001).

Ber

The commercial method of ber propagation is by budding. Different types of budding *viz.*, 'T', 'I', ring, patch and forkert budding have been tried but the best method widely adopted is shield or T budding (Pareek, 1978). Ring budding is cumbersome, as it requires equal freshness and thickness of rootstock and scion. The *Ziziphus rotundifolia* (bordi) is most widely used rootstock for ber propagation. *Ziziphus nummularia* can also be used as dwarfing rootstock but slow growing and form inverted bottle neck incompatibility. The seed germination takes place about 4-6 weeks due to hard endocarp, therefore seeds should be separated by breaking of endocarp and sowing of such seeds take only 7-10 days for germination. Dipping the seeds in 500 ppm GA_3 for 24 hours recorded 90 per cent germination (Krishnan and Kulasekaran, 1984). Murthy and Reddy (1989) found optimum temperature is 30 °C for seed germination. Pareek (1978) suggested a method of raising rootstocks in polytubes, which cut down the period of four months. The seeds are sown in polytubes (25 x 10 cm, 300 gauge) filled with mixture of sand, clay and FYM in first week of April. The budding was done on 90 days old seedlings and ready for transplanting a month later. Monsoon period is best time for budding in western Rajasthan. Patch budding technology for *in-situ* establishment of ber orchard in hot arid ecosystem was also given by Nath *et al.* (2000).

Figure 2.2: Patch Budding-Ber.

Custard Apple

Custard apple can be grafted on their own seedling rootstocks. However, *Annona muricata* can be grafted on *A. reticulata* and *A. glabra*. Whereas *A. reticulata*, which promotes vigour with good graft compatibility, commonly used as a rootstock for most of the *Annona* spp. Shield or T-budding carried out in spring are equally effective. Patch and chip budding are other methods (Singh, 2018). Propagation studies were undertaken to find out the best method at CHES (ICAR–CIAH), Godhra and soft wood grafting was found the most successful method for commercial propagation of *Annona*. Softwood grafted plants showed the maximum increase in the growth parameters compared to budded plants six months after grafting (Hiwale *et al.*, 2009).

Bael

In recent years *bael* is gaining popularity because its medicinal attributes, hence people have started its block plantation. Softwood grafting and patch budding are the commercial methods of bael propagation. Time of budding influences the survival of plants under field condition. One year old, 0.8-1.20 cm thick rootstocks budded in May-June showed better success. About 15-20 cm long mature shoots (4-6 months old) obtained from the plants which have completely shed their leaves or new shoots (2-3 months old) defoliated 10-12 days prior to grafting operation, are used for softwood grafting and budding, respectively (Singh *et al.,* 2019d).

Figure 2.3: Softwood Grafting-Custard Apple.

Figure 2.4: Softwood Grafting-Bael.

Patch budding was found successful when performed in the month of May-June before the onset of rain under Gujarat conditions (Singh, 2014b). Patch budding in the month March can give 70-75 per cent success (Singh *et al.,* 2021). This method has been found better for *in-situ* grafting under rain fed conditions. Among different grafting methods tried, whip grafting had given good response (70 per cent success). Recently softwood grafting during June-July has also been found successful with more than 85per cent success in *bael* under semi-arid ecosystem of western India (Singh, 2018a). Bael can be propagated by inarching as well stem and root cuttings but it is not adopted commercially. Singh *et al.* (2019f) also studied stooling in *bael* and reported very low success rate under field condition in western India.

Datepalm

Date palm is a dioecious plant, where male and female palms are separate. It is mainly propagated by off shoots/ground suckers taken from the female mother palm. The offshoots about 8-10 kg with well-developed root system is detached carefully with mother palm for further plantation. Direct planting of these suckers are not done instead they should be planted first in nursery and surviving suckers should be planted in the field during July-August. Slow multiplication of date palm through offshoot is the major bottleneck in area expansion of date palm. Therefore, for mass multiplication, tissue culture technique is the only alternative in date palm. In India, some work on tissue culture protocol was done but it is variety specific, hence more intensive work on tissue culture is required. Even today, India is importing hardened tissue culture planting materials and extending subsidy to farmers for area expansion under date palm.

Jamun

Chovatia and Singh (2000) recorded 41.67 per cent success in jamun through soft wood grafting during the month of June under Gujarat conditions. Singh and Singh (2006) and Singh and Singh (2014b) reported that softwood grafting in months of March- April with 35-45 days old scion shoot of jamun gave more than 80 per cent success and recommended as an ideal method for commercial propagation. Bose *et al.* (2001) recorded 31 per cent success in veneer grafting when one-year-old seedlings were used as rootstock. They further reported 45 per cent rooting with 100 ppm IBA and 40 per cent with 100-ppm IAA treatments.

Figure 2.5: Patch Budding-Jamun.

Tamarind

Patch or shield budding is one of the common methods of vegetative propagation in tamarind since it is easier, more economical and is ideal for establishing *in-situ* orchards (Singh and Singh, 2007). For soft wood grafting, rootstocks are defoliated and their tops are cut off at 15 cm height immediately before grafting (Singh and Singh, 2007). Pathak *et al.* (1991) reported patch and modified ring budding with more than 94 per cent success in tamarind and recommended these methods for mass multiplication. Soft wood grafting in April resulted in highest graft success (82.50%) followed by March (76.60%) under sub humid conditions of Bastar, Chattishgarh (Awasthi and Shukla, 2003). However, in arid region, availability of active buds and higher temperatures during July–August helped in early sprouting, while lower temperatures during the preceding winter months must have inhibited sprouting in tamarind as reported by Awasthi *et al.* (2005).

Figure 2.6: Patch Budding-Tamarind.

Chironji

Softwood grafting was proved to be better for establishing *in-situ* chironji orchard in gravelly soils and drier tracts where mortality of nursery raised grafts is very high. Defoliation of pencil-thick shoots is done 10-12 days before detaching them from the parent tree for grafting operation (Singh and Singh, 2014d). Better graft success 65-70 per cent is recorded in the month of July- August. Singh *et al.* (2014) reported the highest percent of bud sprout (68.00%) and graft success (66.66%) in July in chironji (Singh *et al.,* 2010d). Two-year old root cuttings having thickness of 3.6-5.5 cm, treated with IAA 1600 ppm recorded 67.66 per cent rooting (Singh *et al.,* 2014b). However, in this method, taproot is absent in the plants that affect the longevity of the plants, particularly, in degraded lands.

Wood Apple

Wood apple can be propagated asexually through budding, softwood grafting, and also by cuttings and inarching. It would be advantageous to go for *in- situ* budding in this crop under semi-arid conditions. Under dryland conditions, patch budding in the month of April-June was found the most successful for commercial multiplication of wood apple. Ghosh *et al.* (2012) reported that grafting was better as compared to budding and 85 to 95 per cent success in grafting was recorded when operation was performed in the month September. Singh (2018) reported in situ softwood grafting in the month of March-April with 6-8 months old scion shoots with 85.13 per cent success on 10-12 months old rootstock under semi-arid conditions of Western India.

Pilu

Conservation of pilu (*Salvadora* sp.) germplasm both by *in-situ* and *ex-situ* is pre-requisite to prevent loss of genetic diversity. *Salvadora* can be raised through air layering or cuttings inserted in sand in a glass house for rooting. The rooting in stem cutting is low (Singh, 2018).

Manila Tamarind

Different asexual propagation methods have been recommended by several workers from different parts of country with variable degree of success *viz.*, air chip budding, softwood grafting, hardwood and semi-hardwood cutting, wedge grafting (Singh, 2018; Singh *et al.*, 2018b). Propagation studies in manila tamarind conducted at CHES, Godhra revealed that it can successfully be multiplied through patch budding in the month of March (Singh, 2018).

Karonda

Regarding karonda, it is known to be difficult-to-root; therefore, rooting does not occur easily. Stem cuttings treated with 5000 ppm IBA as quick dip was effective for rooting. Karonda should be budded during the month of June-July under semi-arid conditions of Western India (Ghosh *et al.*, 2014).

Khirni

One-year-old seedlings of khirni, raised for rootstock commencing new growth with light green leaves, are suitable for soft wood grafting (Singh *et al.*, 2019a). The vigorous seedlings, attaining pencil thickness after one year are soft wood grafted with scion sticks of the suitable genotypes (Singh and Singh, 2015). Better graft success was recorded in the month of March-April.

Lasoda

Singh *et al.* (2003) under arid conditions of Bikaner reported that lasoda (*Cordia myxa*) can be successfully propagated through budding in middle of August (95.59%), though it can be propagated starting from middle of July to end of August with about 80 per cent success. Cleft grafting in lasoda has been found successful and best time for grafting is August, followed by July and September (Singh *et al.*, 2019g). For quality planting material production, a vivid account of seed, rootstock and plant standard of lasoda have been standardized by Kumar *et al.* (2022).

Mahua

In mahua, softwood grafting is practiced. The highest percentage of graft success in mahua was recorded in the month of March (70.00) closely followed by April, July and August, while the least percentage of success was noted in the month of September (8.33%) (Singh *et al.*, 2015). Shukla (2005) reported that one year old shoots on mahua trees were ringed and upper end of the cuts were treated with IBA 1000 ppm. Rooted shoots were removed from the mother plant 60 days after the treatment and planted.

Figure 2.7: Softwood Grafting-Khirni. **Figure 2.8: Softwood Grafting-Mahua.**

Phalsa

Phalsa is conventionally propagated by hard wood cuttings, which is hard to root. Therefore, it poses problem in faster multiplication of elite planting materials.

Figure 2.8: Hardwood Cuttings-Phalsa.

It was found that types of cuttings and time of wood cutting collection both have significant influence on rooting in phalsa (Singh *et al.,* 1961 and Singh, 2017). Treatment with auxins like indole acetic acid (IAA), IBA and naphthalene acetic acid (NAA) improved rooting of difficult-to-root hardwood cuttings of phalsa (Yadava and Rajput, 1969). Cuttings from 6 months old shoots during July-August gave profuse rooting when treated with 1000 ppm IBA through quick dip method (Singh and Kumar, 1967).

Mulberry

Mulberry is propagated mainly by semi-hard stem cuttings. Hardwood, softwood and apical shoot cuttings are also used for propagation of mulberry. Hard wood cuttings of 20-30 cm length with 4-5 buds taken from mature stems treated with IBA (50ppm) are planted in nursery in the month of December and January (Fatma and Eylem, 2003). Softwood cuttings of white mulberries root easily when taken in midsummer and treated with rooting hormone. Red mulberries are less easily rooted. Black mulberries are difficult to propagate. Budding is also reported to be an efficient means of multiplication (Singh *et al.,* 2008a).

Cactus Pear

For commercial multiplication, prickly pear is only vegetatively propagated using at least one-year old cladodes. The cladodes are removed from the mother plant and are allowed to dry in shade for 4 6 weeks before planting in the field (Singh *et al.,* 2003). For better rooting, the basal portion of cladodes should be dipped for 15 minutes in 500 ppm NAA and 1000 ppm IAA (Singh and Singh, 2001). More number of cladodes can be produced by cutting them into 4-5 pieces and planting in nursery bags/tubes of 25x10 cm size filled with sand, clay and FYM in equal proportion (Singh and Singh, 2003).

Ker

Attempts to propagate ker through vegetative means yielded low rate of success so far. Under mist system of nursery shade net house, Bhargava *et al.* (2006) reported that semi-hardwood cuttings taken in the month of March, treated with 7500 ppm IBA + 1000 ppm Thymine had given maximum sprouting (50%). The time and technique for propagation of various arid fruits are given in Table 2.2.

Use of Rootstocks

The rootstock is basal apportion of a component plant upon which scion shoot/ bud is grafted. In modern fruit orchards, features like; controlled and uniform tree vigour, precocity, efficient canopy with high yield efficiency and adoptability to various biotic and abiotic stresses are essentially required, where rootstock can play vital role. The use of rootstock is utmost essential in arid ecosystem which is subjected to various biotic and a biotic stresses. The basic criteria for selection of rootstocks are as under;

Table 2.2: Arid Fruit Crops and their Methods of Propagation

Fruit Crops	*Period of Multiplication*	*Propagation Methods*
Aonla	June-August	Patch budding
Bael	April-June	Soft wood grafting and patch budding
Chironji	July-August	Soft wood grafting
Jamun	April -May	Soft wood grafting, patch budding
Lasora	June-July	Patch budding
Tamarind	July-August	Soft wood grafting and patch budding
Mahua	March-April	Soft wood grafting
Custard apple	April- May	Soft wood grafting
Wood apple	May- June	Soft wood grafting, patch budding
Karonda	June -July	Cutting and layering
Fig	October-November	Cutting
Gangana	July-August	Seed and cutting
Manila tamarind	March- May	Patch budding
Khirni	March-April	Softwood grafting
Phalsa	July -August	Seeds and Cutting
Mulberry	December- January	Stem cutting
Ker	July -August	Suckers and cuttings
Pilu	July-August	Cuttings, Root sucker and layering
Cactus pear	July-August	Cladodes

Source: Saroj *et al.,* 2017.

- ✰ Rootstock must be compatible with scion genotypes.
- ✰ Easy to propagate clonally.
- ✰ Least suckering habit.
- ✰ Well adapted to wide range of soil types, pH, and tolerance to salinity/ alkalinity as well as moisture stress.
- ✰ Ability to improve production potential and fruit quality attributes of scion cultivars.
- ✰ Tolerant to pest, pathogen and nematode.

Some rootstocks of arid fruits are given in Table 2.3.

Precautions in Vegetative Propagation

The planting material is an important input for orchard establishment. Therefore, following precaution should be taken while multiplying the plants.

Table 2.3: Rootstocks Recommended for Arid Fruits

Crops	*Rootstocks*	*Rootstock Attributes*
Ber	*Ziziphus mauritiana* var. *rotundifolia*	Well compatible, vigourus tree frame and suitable for dry region.
	Ziziphus nummularia	Drought tolerance and dwarf tree stature. Extensive root system and precocious. Inverted bottleneck incompatibility.
	Ziziphus xylopyrus, Ziziphus spinachristi	Drought tolerance and dwarf tree stature.
	Ziziphus jujube var. *spinosa*	Resistant to low temperature.
Fig	*Ficus glumerata*	Nematode resistant.
Jamun	*Syzygium fruticosum*	Resistant to termite.
Sapota	*Manilkara hexandra*	Highly compatible and efficient.
Guava	*Psidium molle, Psidium cujavillis and Psidium cattleianum*	Tolerant to guava wilt and moderately tolerant to moisture stress.
	CISH inter-specific hybrid (*Psidium molle* x *Psidium guajava*)	Dwarfing effect and resistant to wilt and nematode.
	Pusa Srijan	Dwarfing effect
Orange	Rangpur lime	Large tree size, high yield with medium fruit quality. Moderately tolerant to moisture stress but susceptible to foot rot.
Custard apple	*Annona reticulata*	Compatible and vigourus

- ☆ Scion shoots should be selected from healthy plants having better fruiting record so as to evade the infestation of pests and diseases during the establishment of orchard.
- ☆ Under dry land conditions, mother plant should be irrigated 2-3 days before detaching scion shoots for better success and survival.
- ☆ The contact between stock and scion/scion shoot should be very fine and there should not be any gap between them as it adversely affects the success.
- ☆ The bud union should be properly tied with polythene band, so that the rain water could not enter into it.
- ☆ Avoid splitting of bark during separation of bud from scion shoots to ensure better success.
- ☆ Size of bark of patch containing bud should properly be matched with the cut portion of the bark on the root stocks for perfect and quick union.
- ☆ Polythene strip should not be removed unless it is ensured that scion shoots has started growing.

- ✰ Sprouts emerging from the rootstock should be removed except the budded one to promote the scion shoots growth.
- ✰ New emerging shoots are often damaged by leaf eating caterpillars and hence, the pest management should be done timely.
- ✰ Standards for production of quality planting material should be followed.
- ✰ Proper handling, packaging and levelling is also essential while transporting planting materials.
- ✰ Proper nursery record should be maintained.

Micropropagation of Arid Fruit Crops

Bael

Pathirana *et al.* (2018) reported that the micropropagation of *bael* using leaf and twig explants could be successful if the explants are harvested during April to May, immediately after the fruiting season of the plant. The full MS and ½MS media provided significantly similar performance in shooting. The most successful rooting (60 per cent success) was spotted with full MS, 1mg/ml NAA, and 3 per cent sugar (Pathirana *et al.,* 2020). Both leaf and twig explants exhibited similar success rates in undergoing direct organogenesis and axillary bud proliferation. However, the explants axillary bud and shoot tip from *in-vitro* germinated seedlings were also found to be suitable for successful micropropagation of *bael* (Gandhi *et al.,* 2018). Raj *et al.* (2020) suggested the use of woody plant medium (WPM) using nodal segment explant for shoot proliferation and MS medium using leaf explant for callus induction in order to achieve large scale micropropagation of bael.

Jamun

An efficient micropropagation system was developed for jamun using nodal explants excised from 15-d-old aseptic seedlings (Naaz *et al.,* 2019). The explants were employed on a Murashige and Skoog (MS) medium supplemented with different concentrations (1.0 - 10.0 μM) of cytokinins, such as benzyladenine (BA), kinetin (Kin), meta-topolin (mT), or 2-isopentyl adenine (2ip), either alone or in combination with different concentrations (1.0-3.0 μM) of auxins, such as indole-3-acetic acid (IAA), indole-3-butyric acid (IBA), or α-naphthalene acetic acid (NAA). Addition of mT into MS medium is a much better option than addition of other cytokinins for *in-vitro* growth, morphogenesis, rooting and survival in jamun. Later, Naaz *et al.* (2021) described the effectiveness of Thidiazuron (TDZ) as potent cytokinin for bud breaking, shoot multiplication, healthy culture, and acclimatization in *S. cumini* (L.). Sirsat *et al.* (2019) observed supplementation of culture medium with 6-benzylaminopurine (BAP) + kinetin (Kn) @ 8 μM/l resulted in lower culture period and higher survival rate of micropropagated plants of jamun. Carmelita and Prabhuling (2015) noted that cocopeat alone proved superior for primary hardening. Likewise for secondary hardening, perlite+red soil (1: 1 v/v) and vermiculite+red soil (1: 1 v/v) was found superior.

Tamarind

A protocol for *in-vitro* regeneration of tamarind plants via adventitious bud formation from mature embryo axis of tamarind was standardized by Mehta *et al.* (2000). Explants consisting of longitudinal section of embryo axis with attached cotyledon were cultured in MS medium with various combinations and concentrations of NAA, BAP and sucrose. The shoots developed roots in a half strength MS medium with 2 per cent sucrose following a 72-h treatment with auxin mixture in the dark. However, Pattepur *et al.* (2016) noted that among the various combinations of growth regulators, BAP 0.5 mg/l + NAA 0.1 mg/l was the best combination for shoot proliferation of cotyledonary nodes derived from axenic seedling of tamarind. Recently, Ferreira *et al.* (2019) observed that *in-vitro* establishment of tamarind nodal segments is viable when using the MS medium with 75 per cent of salts plus 2 g L-1 of activated charcoal.

Chirounji

Sharma *et al.* (2005) first reported a protocol for somatic embryogenesis and plantlet regeneration of chironji from calluses initiated from immature zygotic embryos. The medium supplemented with 15 mM abscisic acid (ABA) was most effective for maturation and germination of somatic embryos. Recently, Singh *et al.* (2022) developed a protocol for organogenesis and *in-vitro* multiplication of chironji using young leaf and nodal segments. The *in-vitro* raised plants were acclimatized with approximately 70 per cent survival rate.

Custard Apple

Panchal and Patel (2016) standardized the sterilization protocol of explants for *in-vitro* culture establishment of custard apple cv. Balanagar. Sterilant HgCl2 @ 0.1 per cent for 8 minutes and HgCl2 @ 0.1 per cent for 5 minutes were found to be better surface sterilizing treatment for axillary bud and shoot tip explants, respectively, for better culture establishment. Later, Minipara *et al.* (2019) eliminatedendophytic bacterial contamination from explants of cv. Anand selection-1 using gentamicin (96mol/l) in the culture medium. Panchal and Patel (2019) devised the remedy of leaf abscission of custard apple cv. Balanagar under *in-vitro* propagation by supplementing silver nitrate @5 mg/l in the culture medium. Kudikala *et al.* (2020) developed an efficient and reproducible *in-vitro* plant regeneration protocol for *Annona reticulata* L. using nodal explants cultured on MS medium supplemented with 2.5 mg/l 6-BAP and 0.3 mg/l IBA. Rhizogenesis occurred on half-strength MS medium supplemented with 4 mg/l IBA. Earlier, Padilla and Encina (2004) described a micropropagation procedure for the adult cherimoya tree (*Annona cherimola* Mill.) cv. 'Fino de Jete'. Roots were induced after pre-incubation of shoots for 3 d in light followed by culturing for 10 d (7 d dark and 3 d light).

Wood Apple

In-vitro culture of wood apple from hypocotyl and internodal explants was described by the Hiregoudar *et al.* (2005). They observed that the optimum response was recorded on the MS medium containing 2 µM BA, while rooting was noted with 1 µM 1 NAA. The developed plantlets were successfully hardened and later transplanted in the field. Later, Vyas *et al.* (2008) reported bio-acclimation protocol for *in-vitro*-raised plantlets of wood apple using an endosymbiotic root fungus *Piriformospora indica*. As a result of inoculation, improved growth has been observed. More than 90 per cent of such plants survived. In the recent past, Saini and Tyagi (2019) optimized the protocol for establishment of callus and *in-vitro* regeneration of wood apple. Recently, Anis *et al.* (2022) observed the stimulatory effect of meta-Topolin (mT) on *in-vitro* shoot induction and proliferation of cotyledonary explants of wood apple. mT @ 5.0 µM gave the best response for inducing maximum shoots in 90 percent cultures after 12 weeks. The regenerated plantlets were successfully acclimatized and hardened off.

Manila Tamarind

Goyal *et al.* (2012) observed shoot bud induction from nodal explants of manila tamarind on MS basal medium supplemented with 4.4 µM 6-BA. Multiplication was achieved on MS medium supplemented with BA + phenylacetic acid (PAA). Addition of adenine sulphate (AdS) to this medium further enhanced the number of shoot buds. *In-vitro* rooting was achieved with 2.46 µMIBA + 41.63 µM activated charcoal. The micro-propagated shoots with well-developed roots were acclimatized in green house in pots containing sand, soil and manure (1: 1: 1).

Karonda

Rai and Misra (2005) cultured shoot tips from field grown, mature plants of karonda cv. Pant Sudarshan on medium with BA and IBA and noted the maximum sprouting rate from explants collected in spring season (February–March). The rooted plantlets were successfully acclimatized in vermiculite: sand: soil (1: 1: 1) potting mixture. In the recent past, Bhadane *et al.* (2018) developed an *in-vitro* organogenesis protocol for karonda using an auxin transport inhibitor (quercetin) and silver nitrate ($AgNO_3$), an inhibitor of ethylene action, in association with cytokinins in the culture medium

Lasoda

A protocol for *in-vitro* propagation of lasoda was standardized using nodal segments by Krishna and Singh (2013). Single node segments, prepared from the new vegetative shoots, were cultured on MS medium with 4.0 mgl-1 kinetin. The regenerated shoots from nodal segments rooted on medium supplemented with NAA and IBA along with activated charcoal. Inoculation of arbuscular mycorrhizal fungi during *ex-vitro* hardening stage resulted in higher survival (97%) and improved growth of micro-propagated lasoda plantlets (Krishna *et al.*, 2015).

Mahua

Micropropagation of mahua (*Madhuca latifolia* Macb.) was achieved by culturing excised nodes on WPM supplemented with different plant growth regulators (Bansal and Chibbar, 2000). A combination of IBA with BA or kinetin was found suitable for axillary shoot initiation within a week of culture from explants collected in May. Excised micro-shoots treated with IBA produced roots. Likewise, Shirin *et al.* (2020) successfully attempted to establish *in-vitro* cultures of mahua from nodal segments. Axillary bud break (64.44%) was successfully achieved by culturing nodal segments on MS medium supplemented with 3 mg l-1BA in nodal explants collected during the months of July-September (rainy season). The plantlets were acclimatized successfully firstly in a mist chamber and then in shade house.

Phalsa

In-vitro propagation of phalsa through both organogenesis and somatic embryogenesis was achieved by Biswas *et al.* (2005). Later, Wani *et al.* (2016) noted multiple shoot induction from nodal explants with axillary buds, on culturing in WPM fortified with Kinetin and IBA. Healthy regenerated shoots rooted on WPM with IBA. Rooted plantlets, were transferred to soil-sand (1: 1; v/v) with about 92 per cent success.

Mulberry

A single-step protocol using PGR-free culture medium for the *in-vitro* shoot induction, multiplication, and rooting was optimized for *Morus nigra* using apical and axillary bud's explants (Durate *et al.,* 2019). However, Rana *et al.* (2022) suggested the use of nodal explants of *M. alba* based on axillary shoot proliferation, culture multiplication and root formation of the *in-vitro* shoots. Taha *et al.* (2020) described a protocol using auxiliary buds on the MS medium fortified with 1.5 mg/l BA. Culturing of derived mulberry shoot on the MS medium with 1 mg/l NAA enhanced root system formation. *In-vitro* derived mulberry plantlets successfully acclimatized using peat moss + sand + perlite at the ratio of 1: 1: 1: 1 v/v/v/v.

Ker

Gianguzzi *et al.* (2019) proposed a new propagation method of *in-vitro* regeneration of *Capparis spinosa* L. by using a temporary immersion system Plant Form bioreactor, in which the shoots undergo temporary immersion in a liquid medium in order to avoid the accumulation of gas through forced ventilation. Ahlawat *et al.* (2020) optimized the micropropagation of *C. decidua*. Culture medium supplemented with 1 mg/L BAP,1 mg/L NAA; 15 mg/L adenine sulphate, and 0.2 per cent activated charcoal was found the most suitable for bud outbreak and shoot induction. Further, supplementing the media with adenine sulphate favoured shoot differentiation, while silver nitrate improves bud outbreak percentage.

Cactus Pear

Luciana *et al.* (2019) also suggested the use of temporary immersion bioreactors for *in-vitro* culture of cactus pear. Recently, Bouzroud *et al.* (2022) reviewed the micropropagation of cactus pear.

Pilu

Shrivastava and Agnihotri (2019) described an *in-vitro* regeneration protocol for pilu (*Salvadora oleoides*). Nodal segments were used as explant and cultured on MS medium. Combination of NAA (2 mg/L)+ BAP(1 mg/L) were observed to be the best suited medium for direct regeneration. IBA (4 mg/L) induced maximum roots in regenerated shoots. Likewise, Fuke *et al.* (2022) attempted *in-vitro* propagation of *Salvadora persica* L. from the nodal segment. The regeneration of multiple shooting was achieved on MS medium along with 0.5 mg/L of IBA in combination of various concentrations of BAP and kinetin. Earlier, Kumar *et al.* (2019) standardized micropropagation protocol for *Salvadora oleoides*. *In-vitro* raised plantlets successfully acclimatized in pots containing soil in natural conditions.

Commercial Varieties of Arid Fruits

It is suggested that only recommended scion varieties of arid fruits should be multiplied for production of planting materials as given in Table 2.4.

Table 2.4: Commercial Scion Varieties of Arid Zone Fruit Crops

Crops	*Scion Varieties*
Aonla	NA-6, NA-7, Banarasi, Chakaiya, Kanchan, Krishna, Balwant, NA-10, Anand 1, Anand 2.
Ber	Gola, Umaran, Banarasi Karaka, Kaithali, Mundia, Seb, Goma Kirti, Katha, Aliganj, Mundia, Thar Savika, Thar Bhuvraj,
Bael	GomaYashi, Thar Divya, Thar Neelkanth, Thar Srishti, Thar Prakriti, Thar Shivangi, Thar Gauri, Thar Bhavya, NB-5, NB-7, NB-8, NB-9, NB-10, NB-11, NB-16., NB-17, CISHB-1, CISHB-2, CISHB-3, Pant Aparna, Pant Shivani, Pant Sujata and Pant Urvashi
Date palm	Halawy, Barhee, Medjhool, Khuneizi, Khadrawy, Zahidi, Zagloui, Khalas, Sewi, Shamran
Jamun	Goma Priyanka, Thar Kranti, Konkan Bahadoli, Jamwant, Paras, Rajendra Jamun-1
Custard apple	Thar Amrit, Washington PI 98797, Washington PI 107005, British Guinea, Island Gem, Bullocks Heart, Pink Mammoth, Balanagar, Mammoth, Red Sitaphal, Yellow Sitaphal, Phule Janki and Sindhan
Mulberry	Thar Lohit, Thar Harit, Victoria-1, China White, Saharanpur Local-1, Saharanpur Local-2, S-13, S-34, S-146, S-7999, S-1635, Chak Majra
Karonda	Pant Manohar, Pant Sudarshn, Pant Suverna, Konkan Bold, Thar Kamal, Maru Gaurav
Tamarind	Goma Prateek, Thar Rashmi, Prathisthan, PKM-1, T 263, Urigam, Ajanta, Yogeshwari, DTS 1 and DTS 2, Anant Rudhira, Thettu Amalika
Lasoda	Thar Bold, Paras Gonda, Maru Samridhi, Karan Lasoda.
Khirni	Thar Rituraj
Phalsa	Thar Pragati

Crops	*Scion Varieties*
Chironj	Thar Priya
Mahua	Thar Madhu, NM-2, NM- 4, NM-7, NM- 9
Manila Tamarind	PKM (MT) 1
Fig	Poona Fig, Dianna, Dinkar, Conadria, Excel, Brown Turkey
Wood apple	Thar Gaurav, Thar Prabha

Disposal of Planting Materials

Timely disposal of planting materials are very essential, otherwise, if left for longer period in the nursery, there is every possibility of root coiling. To ensure the quality of planting materials used in Mission for Integrated Development of Horticulture (MIDH) programmes; the Govt. of India has made it mandatory that the planting material used for schemes like area expansion/replanting/rejuvenation should be procured from accredited nurseries only. The accreditation is spearheaded by National Horticulture Board (NHB), Govt. of India. The detailed guideline for nursery accreditation and rating has been formulated by NHB and it is done online also.

Before disposal of nursery plants, proper hardening is required to allow the plants for transition from a protected indoor or greenhouse environment to the harsh outdoor conditions of fluctuating temperatures, wind, and full sun exposure; so as to acclimatize them for better establishment in field. Hardening off is a term applied to any treatment that makes the tissues firm to ensure better establishment in the unfavorable environmental conditions. With holding of watering, removal of undesired biomass followed by shifting of plants in partial shade is the best practical method for hardening of nursery plants. Now a day, bio-hardening of plants like pomegranate is also done for better establishment of plants in field.

Before sale, planting stock should be labeled. Labeling provides critical information about the name of nursery with full address, fruit-variety, propagation technique, date of packing and other necessary information/instructions/precautions related to planting materials. This information helps growers to establish right kind of orchard.

Proper packaging is important for safe transportation of planting materials to long distance. While packing the plants in container, it should neither over-packed nor loose enough allowing the plants to move about. All space should be filled up by some packing materials like straw, dried grass, *etc.* The polycontainer raised plants are often safe during transportation. For planting materials also, there should be electronic bar-coding system for the purpose of traceability to track labeled product throughout supply chain.

There is 'Online Digital Platform for Nurseries' in India, enabling easy access to information on availability of quality planting material in the nearby locations for farmers/growers and other stake holders. Through this portal, the nursery owners can showcase their profiles and post sale offers. Similarly, buyers of planting material can post enquiries and view matching sale offers. There should be minimum time gap between lifting of the planting materials from the nursery and planting them at desired site.

REFERENCES

Ahlawat J, Sehrawat AR, Choudhary R, Samarina L, Bandaralage JH and Chaudhary R, 2020. Quantifying synergy of plant growth hormones, anti-oxidants, polyamines and silver nitrate for optimizing the micro propagation of *Capparis decidua*: an underutilized medicinal shrub. *The Nucleus* **63(3)**, 313-325.

Amin RS, 1978. Softwood grafting-a new technique for hardwood plants. *Current Science*, **47**: 468-469.

Anis M, Alam N, Khanam MN and Ahmad N, 2022. Potential of meta-Topolin (mT) in improved *in-vitro* regeneration of *Feronia limonia* (L.), Swingle. *Vegetos* 1-7.

Awasthi OP, Saroj PL and Dhandar DG, 2005. Standardization of time on success of patch budding in tamarind (*Tamarindus indica*) under arid conditions. *Progressive Hort.* **37(2)**: 294–297.

Awasthi OP and Shukla N, 2003. Effect of time on success of softwood grafting in tamarind (*Tamarindus indica* L). *Range Management and Agroforestry*, **24(1)**: 31-34.

Bajaniya VG, Karetha KM, Parmar LS, Purohit VL and Chotaliya BM, 2018. Influence of pre-soaking treatment on seedling growth of khirni (*Manilkara hexandra* Roxb) seedling cv. Local. *International Journal of Pure and Applied Biosciences*, **6(1)**: 1668-1672.

Bansal YK and Chibbar T, 2000. Micropropagation of *Madhuca latifolia* Macb. through nodal culture. *Plant Biotechnology*, **17(1)**: 17-20.

Bhadane BS, Maheshwari VL and Patil RH, 2018. Quercetin and silver nitrate modulate organogenesis in *Carissa carandas* (L.). *In-vitro Cellular and Developmental Biology-Plant* **54:** 600-605.

Bhargava R, Verma P, Saroj PL and Chauhan N, 2006. Propagation of *Capparis decidua* through cuttings. *Indian Forest*er, **132(6)**: 737-745.

Biswas B, Joshee N, Yadav A and Yadav AK, 2000. *In-vitro* propagation and somatic embryogenesis in *Phalsa*. *Hort Science*, **40(4)**: 1104A-1104.

Bose TK, Mitra SK and Sanyal D, 2001. Fruits: Tropical and Subtropical, Pub. Naya Udyog, 206- Bidhan Sarani, Culcutta, India. pp. 643-56.

Bouzroud S, El Maaiden E, Sobeh M, Devkota KP, Boukcim H, Kouisni L and El Kharrassi Y, 2022. Micropropagation of *Opuntia* and other Cacti species through auxiliary shoot proliferation: A comprehensive review. *Frontiers in Plant Science*, **13**: 926653.

Carmelita LP and Prabhuling G, 2015. Studies on hardening of tissue culture propagated plants of Jamun cv. AJG-85. *International Journal of Tropical Agriculture* **33**(2 (Part I)): 343-349.

Chovatia RS and Singh SP, 2000. Effect of time on budding and grafting success in jamun (*Syzygium cuminii* Skeels). *Indian Journal of Horticulture,* **57(3)**: 255-258.

Dhandar DG and Saroj PL, 2004. Changing scenario in arid horticulture. In: Advances in Arid Horticulture, Vol.I (Eds. P.L. Saroj *et al.*), Pub. International Book Distributing Company, Lucknow, India. pp. 1-27.

Duarte WN, Zanello CA and Cardoso JC, 2019. Efficient and easy micropropagation of *Morus nigra* and the influence of natural light on its acclimatization. *Advances in Horticultural Science,* **33(3)**: 433-439.

Fatma K and Eylem S, 2003. Rooting of black mulberry (*Morus nigra* L.) hard wood cuttings. *Journal of Fruit and Ornamental Plant Research,* **11**: 53-57

Ferreira AFA, da Silva MSAC, Monteiro LNH, Faria GA, Boliani AC, Rodrigues MGF and Pereira FD, 2019. Nodal segments and zygotic embryos in culture media for the *in vitro* propagation of tamarind. *Comunicata Scientiae,* **10(4)**: 438-447.

Fuke PS, Chavan SG and Chavan AM, 2022. Tissue culture protocol of *Salvadora persica* L. *International Journal of Current Science,* **12(1)**: 406-413.

Gandhi K, Rajesh E, Saravanan S and Elamvaluthi M, 2018. Micropropagation of *Aegle marmelos* L.-A medicinal tree through culture of auxiliary bud and shoot tip explants from *In vitro* germinated seedlings. *International Journal of Scientific Research in Biological Sciences,* **5(3)**: 65-71.

Ghosh SN, Bera B and Roy S, 2011. Propagation studies in karonda (*Carissa carandas*) under Paschim Midnapore condition of West Bengal. *Journal of Crop Weed,* **7(2)**: 247-248.

Ghosh SN, Banik AK, Banik B., Bera B and Roy S, 2012. Conservation, multiplication and utilization of wood apple (*Feronia limonia*)-a semi-wild fruit crop in West Bengal (India). I International Symposium on Wild Relatives of Subtropical and Temperate Fruit and Nut Crops **948**. 2011.

Gianguzzi V, Inglese P, Barone E and Sottile F, 2019. *In-vitro* regeneration of *Capparis spinosa* L. by using a temporary immersion system. *Plants* **8**, 177.

Goyal P, Kachhwaha S and Kothari SL, 2012. Micropropagation of *Pithecellobium dulce* (Roxb.) Benth—a multipurpose leguminous tree and assessment of genetic fidelity of micropropagated plants using molecular markers. *Physiology and Molecular Biology of Plants,* **18:** 169–176.

Hiregoudar LV, Ashok Kumar HG and Murthy HN, 2005. *In-vitro* culture of *Feronia limonia* (L.) Swingle from hypocotyl and internodal explants. *Biologia Plantarum,* **49(1)**: 41-45.

Hiwale SS, More TA and Bagle BG, 2009.Vegetative propagation in *Annona*. (*Annona squamosa* L) in National Conference on Production of Quality Seeds and Planting Material-Health Management in Horticultural Crops at New Delhi from 11 to 13 March, 2009.

Krishna H, Singh RS, Singh UV, Sharma BD and Sharma SK, 2015. *Lasoda*: The cherry of desert. CIAH/Tech/Pub 59, pp. 26, ICAR–CIAH, Bikaner.

Krishna H and Singh D, 2013. Micropropagation of *lasora* (*Cordia myxa* Roxb.). *Indian Journal of Horticulture,* **70(3)**: 323-327.

Krisnan BN and Kulasekaran M, 1984. Studies on seed germination in wild ber (*Ziziphus rotudifolia*). *South Indian Horticulture,* **32**: 153-154

Kudikala H, Jogam P, Sirikonda A, Mood K and Allini VR, 2020. *In vitro* micropropagation and genetic fidelity studies using SCoT and ISSR primers in *Annona reticulata* L.: an important medicinal plant. *Vegetos,* **33**: 446–457.

Kumar R, Najar RA, Gupta KB and Saini RG, 2019. Micropropagation protocol for *Salvadora oleoides. Journal of Forestry Research,* **30:** 87–93.

Kumar K, Sarolia DK, Saroj PL, Samadia DK and Singh D, 2022. Standardization of seed, rootstock and plant standards of lasoda (*Cordia myxa* L.) for commercial orcharding in India. *J. Environ. Biol.,* **43**, 267-275.

Luciana CNL, Wander SV, Selma SR, Pedro GMDA, Jssica GC and Jlio CGP, 2019. Use of temporary immersion bioreactors on *in-vitro* culture of cactus pear. *African Journal of Agricultural Research,* **14(32)**: 1487-1492.

Mehta UJ, Krishnamurthy KV and Hazra S, 2000. Regeneration of plants via adventitious bud formation from mature zygotic embryo axis of tamarind (*Tamarindus indica* L.). *Current Science*, 1231-1234.

Minipara D, Dhaduk H, Patil G, Narayanan S and Kumar S, 2019. Identification of best surface sterilization treatment and control of endophytic bacterial contamination in *Annona squamosa* L. *International Journal of Plant and Soil Science,* **29***:* 1-10.

Mishra DS, Singh S and Singh AK, 2019. Plant growth regulators in phalsa. In: *Plant Growth Regulatos in Tropical and Sub Tropical Fruits* Part I (Eds. Ghosh, S. N., Tarai, A. K. and Ahlawat, T. R.), Pub. Jaya Publishing House Delhi, India, pp. 520-526.

Murthy BNS and Reddy YN, 1989. Effect of different methods of seed germination in ber. *Journal of Maharashtra Agricultural Universities,* **14**: 296-298.

Naaz A, Hussain SA, Anis M and Alatar AA, 2019. Meta-topolin improved micropropagation in *Syzygium cumini* and acclimatization to *ex-vitro* conditions. *Biologia Plant*arum, **63(1)**: 174-182.

Naaz A, Siddique I and Ahmad A, 2021. TDZ-Induced efficient micropropagation from juvenile nodal segment of *Syzygium cumini* (Skill): A recalcitrant tree. In:

Propagation and Genetic Manipulation of Plants (Ed. Siddique, I.). Springer, Singapore. pp. 163-175.

Nath V, Saroj PL, Singh RS, Bhargava R and Pareek OP, 2000. *In-situ* establishment of ber orchard under hot arid ecosystem of Rajasthan. *Indian J. Hort.*, **57(1)** : 21-26

Padilla IMG and Encina CL, 2004. Micropropagation of adult cherimoya (*Annona cherimola* Mill.) cv. Fino de Jete. *In Vitro Cellular and Developmental Biology-Plant*, **40(2)**: 210-214.

Panchal R and Patel R, 2016. Effect of different sterilization treatments in combating *in-vitro* contamination problem and establishment of various explants of custard apple (*Annona squamosa* L.) cv. Balanagar. *The Biosca*, **11(4)**: 2503-2506.

Panchal RK and Patel RM, 2019. Effect of silver nitrate on leaf abscission in culture during establishment of explants of custard apple (*Annona squamosa* L.) cv. Balanagar. *International Journal of Agriculture, Environment and Biotechnology*, **10(2)**: 263-266.

Pareek OP 1978. Quiker way to raise ber (*Ziziphus mauritiana*) orchard. *Indian Horticulture*, **28(2)**: 13-15.

Pathak RK, Ojha CM and Dwivedi R, 1991. Adopt patch budding for quicker multiplication in tamarind. *Indian Horticulture*, **36930**: 17.

Pathak RK, Ojha CM, Dwivedi R and Hari Om, 1991. Studies on the effect of method and duration of budding in aonla. *Indain J. Hort.*, **47**: 207-212.

Pathirana C, Attanayake U, Dissanayake U, Gamlath L, Ketipearachchi K, Madhujith T and Eeswara J, 2020. Establishment of a micropropagation protocol for elite accessions of *bael* (*Aegle marmelos* (L.) Corr.), a tropical hardwood species. *Advances in Agriculture*.

Pathirana CK, Attanayake AUMRK, Dissanayake DMUSK, Gamlath JGKL, Ketipearachchi KW, Madhujith T, Bandaranayake PCG and Eeswara JP, 2018. Effect of phenological growth stage on establishment of *in-vitro* cultures of *bael* (*Aeglemarmelos* (L.) Corr.). *Tropical Agricultural Research*, **29 (3)**: 268 – 275.

Pattepur SV, Mokashi AN and Hegde RV, 2016. Effect of cytokinins and auxin on shoot proliferation of cotyledonary nodes derived from axenic seedling of tamarind (*Tamarindus indica* L.). *In*: Malhotra SK, Kandiannan K, Mini Raj K, Neema VP, Prasath D, Srinivasan V, Homey Cheriyan and Femina (Eds.) Proceedings - National Seminar on Planting Material Production in Spices, Directorate of Areca nut and Spices Development, Kozhikode, Kerala. pp. 162-165

Rai R and Misra KK, 2005. Micropropagation of *karonda* (*Carissa carandas*) through shoot multiplication. *Scientia Horticulturae*, **103**: 227-32.

Raj P, Jakhar ML, Ahmad S, Mtilimbanya KY and Jat SCHR, 2020. Effect of culture media on shoot proliferation and callus induction of *bael* (*Aegle marmelos*

L.). *International Journal of Current Microbiology and Applied Sciences,* **9(3)**: 243-248.

Rajan S and Ram S, 1982. A new approach towards vegetative propagation of aonla through cuttings. *Progressive Horticulture,* **14**: 190-191.

Ram RB and Kumar A, 2005. Performance of softwood grafting in aonla (*Emblica officinalis* Gaertn.) during off Season. *Indian J. Agroforestry,* **7(1)**: 68-69.

Rana MS, Amin MN and Azad MAK, 2022. *In-vitro* regeneration protocol for mulberry (*Morus alba* L.) through tissue culture techniques. *International Journal of Research and Innovation in Social Science,* **7(1)**: 124-128.

Saini C and Tyagi S, 2019. *In-vitro* regeneration of under exploited ethno-medicinal plant *Limonia acidissima* L. *International Journal of Creative and Innovative Research,* **2(6)**: 1-8.

Saroj PL, Nath V and Vashishtha BB, 2001. Effect of polycontainers on germination, seedling vigour, root characters and budding success in aonla. *Indian J. Hort.,* **57(4)**: 300-304.

Saroj PL, Bhargava R, Singh RS and Hare Krishna, 2017. ICAR–CIAH-an overview. CIAH/Tech/Pub./No. 60, pp: 1-29.

Saroj PL, Sarolia DK and Sharma BD, 2022. Seventy five years of research and development in arid fruit crops. *International Journal of Innovative Horticulture,* **11(1)**: 24-35.

Sasthri G, Srimathi P and Malarkadi K, 2001. Effect of seed size on seed quality in jamun (*Syzygium cuminii* Skeels). *Madras Agric. J.,* **88(7-9)**: 524-526.

Sharma P, Koche V, Quraishi A and Mishra SK, 2005. Somatic embryogenesis in *Buchanania lanzan* Spreng. *In-vitro Cellular and Developmental Biology-Plant,* **41(5)**: 645-647.

Shirin F, Bhadrawale D, Singh S, Panika S and Gupta T, 2020. Effect of environmental factors on successful *in-vitro* morphogenesis of *Madhuca longifolia. Advances in Research,* **21(11)**: 9-19.

Shrivastava R and Agnihotri R, 2019. *In-vitro* approach for plant regeneration and conservation of *Salvadora oleoides* Decne.: a potential medicinal plant. *Indian Research Journal of Genetics and Biotechnology,* **11(02)**: 21-29.

Shukla, S. K. and Solanki, K. R. 2000. Studies on seed germination, plant survival and growth of Chironji (*Buchanania lanzan* Spreng). *J. Trop. Forestry,* **6**: 44-49.

Singh S, Singh AK, Joshi HK, Bagle BG and Dhandar DG, 2007a. Jamun-A fruit for future. *Technical Bulletin,* CHES (ICAR–CIAH), Godhra, pp.1-18.

Singh S, Singh AK, Mishra DS, Appa Rao VV and Saroj PL, 2019a.The Khirni (*Manilkara hexandra* Roxb.Dubard).*Technical Bullettin*: 79, ICAR–CIAH, Bikaner, pp.1-24.

Singh JP, Godara PS and Singh RP, 1961. Effect of type of wood and planting dates on the rooting of phalsa. *Indian J. Hort.*, **18(1)**: 46-50.

Singh KK, 2017. Multiplication of Phalsa (*Grewia asetica* L.) Cv. Dwarf type through hardwood stem cutting under Srinagar Garhwal Himalayas. *Int. J. Curr. Microbiol. App. Sci.* **6(2)**: 1173-1178.

Singh SM and Kumar H, 1967. Influence of ringing source shoot and concentrations of indole-3-butyric acid on the performance of phalsa (*Grewia asiatica*L.) stem cutting. *Indian Journal Agricultural Science,* **37**: 151-154.

Singh AK, Pandey D, Singh S, Singh RS, Misra AK and Singh RK, 2019e. The Bael in India. Published by DKMA, ICAR, New Delhi, pp. 1-129

Singh AK, Singh RS, Kumari L, 2010. Propagation techniques in underutilized fruits. In. *Underutilized Fruits of India* (eds. R. S. Singh, A. K. Singh, S. K. Maheshwari, R. Bhargava and P. L. Saroj), Pub. Brillion Publishing, New Delhi, pp. 149-164.

Singh AK, Singh S and Saroj PL, 2018a. Bael (Production Technology). *Technical Bulletin No.67,* Pub. ICAR–CIAH, pp. 1-53.

Singh AK, Singh S, Saroj PL, Appa Rao VV and Singh RS, 2019f. Studies on stooling in bael (*Aegle marmelos* Correa). *Indian Journal of Arid Horticulture,* **1(1)**: 60-62.

Singh AK, Singh S, Saroj PL, Krishna H, Singh RS and Singh RK, 2019d. Research status of bael in India: A review. *Indian Journal of Agricultural Sciences,* **84(10)**: 1563-71.

Singh AK, Singh S, Singh RS, Sharma BD and More TA, 2011a. The bael- fruit for dryland. *Technical Bulletin No.33,* CHES (ICAR–CIAH), pp.1-46

Singh AK, Singh S, Singh RS, Contractor K and Makwana P, 2014b. *In-situ* patch budding for better establishment of bael in rainfed areas, *Indian Horticulture,* **59(5)**: 24-25.

Singh AK, 2018. Propagating arid fruit commercially. *Indian Horticulture,* **63(5)**: 82-88.

Singh HP, Sharma KD, Subba Reddy G and Sharma KL, 2004. Dryland Agriculture in India. In Challenges and Strategies of Dryland Agriculture (eds Rao SC and Ryan J). https: //doi.org/10.2135/cssaspecpub32.c6

Singh RS and Singh V, 2003. Growth and development influenced by size, age and planting methods of cladodes in cactus pear (*Opuntia ficus-indica* (L.) Mill.). *J. Prof. Assoc. Cactus Dev.,* **5**: 47–54.

Singh S and Singh AK, 2014d. Chironji. In: *Propagation of Horticultural Plants of Arid and Semi-Arid region* (Eds. Singh, R. S. and Bhargava, R.). New India Publishing Agency New Delhi, pp. 363-370.

Singh S and Singh AK, 2014. Mahua. In: *Propagation of Horticultural Plants of Arid and Semi-Arid Region* (Eds. Singh, R. S. and Bhargava, R.), New India Publishing Agency New Delhi pp.395-402.

Singh S and Singh AK, 2006. Standardization of method and time of propagation in jamun (*Syzygium cuminii*) under semi-arid environment of western India. *Indian Journal of Agricultural Sciences*, **76(4)**: 242-5.

Singh S and Singh AK, 2007. Standardization of method and time of vegetative propagation in Tamarind (*Tamarindus indica* L) under semi- arid environment of western India. *Indian Journal of Horticulture*, 64(1): 45-49.

Sing S and Singh AK, 2014d. Jamun. In: *Propagation of Horticultural Plants of Arid and Semi-Arid Region* (Eds. Singh, R. S. and Bhargava, R.). New India Publishing Agency New Delhi, pp.315-324.

Singh S and Singh AK, 2014e. Standardization of softwood grafting in chironji under semi-arid environment of western India. *Indian Journal Horticulture*, **71(1)**: 120-22.

Singh S and Singh AK, 2015. Standardization of time of softwood grafting in mahua and khirni under semi-arid environment of western India. *Indian Journal of Agricultural Sciences*, **85(2)**: 166-70.

Singh S, Saroj PL, Mishra DS and Singh AK, 2019. Underutilized Fruit Crops: crop improvement and agro-techniques. KAAV Publication, New Delhi, pp: 1-306.

Singh S, Singh AK and Bagle BG, 2007b. Propagating chironji by *in-situ* soft wood grafting. *Indian Horticulture* **52(3)**: 35-36.

Singh S, Singh AK, Bagle BG and More TA, 2008. Mahua-A multipurpose tree for tribals. *Technical Bulletin*, Pub. CHES (ICAR–CIAH), Godhra, pp.1-22.

Singh S, Singh AK, Bagle BG and More TA, 2010c. The Jamun. ICAR–DKMA, New Delhi. pp. 1-48.

Singh S, Singh AK, Bagle BG and More TA, 2010d. Chironji-A potential dry fruit for dry lands. *Technical Bulletin No.33,* CHES (ICAR–CIAH), Godhra, pp.1-22.

Singh RS, Bhargava R and Pal G, 2011. Effect of seed treatment on germination and growth behavior of manila tamarind (*Pithecellobium dulce*) under hot arid conditions. *Journal of Tropcal Forestry*, **27(4)**: 25-37.

Singh S, Singh AK, Mishra DS and AppaRao VV, 2017. Effect of size of polythene bag on seedling growth and budding success in jamun. *Indian Journal of Arid Horticulture*, **12 (1 and 2)**: 58-61.

Singh S, Singh AK, Saroj PL, Mishra DS, 2019b. Research status for technological development of Jamun (*Syzygium cuminii*). *Indian Journal of Agricultural Sciences*, **84(11)**.

Singh S, Singh AK, Mishra DS and Appa Rao VV, 2018b. Effect of stem cutting and IBA concentration on rooting and percentage success in karonda (*Carissa carandus* Linn.). *Indian Journal of Arid Horticulture*, **13(1 and 2)**: 116-117.

Singh S, Singh HP, Singh AK and Sisodia PS, 2011c. The Jamun (fruit for future).Pub. Agro-tech Publishing Agency, Udaipur, Rajasthan, pp.1-100.

Singh SV, Sharma R, Ramakrishnan RS, Sharma S, Dwivedi N, Kumar A and Gaur VS, 2022. *In-vitro* mass multiplication of economically endangered forest plant *Chironji* (*Buchanania lanzan* Spreng.). *Indian Journal of Genetics and Plant Breeding* **82(01)**: 116-120.

Sirsat VM, Rathod DR, Gahukar SJ, Akhare AA, Gawande VL, Maske NC and Rathod AV, 2019. *In-vitro* shoot proliferation from nodal segments of Indian Blackberry (*Syzygium cumini* L.). *International Journal of Current Microbiology and Applied Science,* **8(3)**: 2464-2469.

Taha H, Ghazy UM, Gabr AMM, 2020. Optimization of *in-vitro* culture conditions affecting propagation of mulberry plant. *Bulletin of the National Research Centre,* **44**: 60.

Tewari RK and Bajpai CK, 2002. Studies on quality parameters of scion and rootstock on success of grafting in aonla. *Indian J. Agroforestry,* **4(2)**: 138-140.

Vyas S, Nagori R and Purohit SD, 2008. Root colonization and growth enhancement of micropropagated *Feronia limonia* (L.) Swingle by *Piriformospora indica*-a cultivable root endophyte. *International Journal of Plant Developmental Biology,* **2(2)**: 128-132.

Wani TA, Rana S, Bhat WW, Pandith SA, Dhar N, Razdan S, Chandra S, Sharma N, Lattoo SK, 2016. Efficient *In-vitro* regeneration, analysis of molecular fidelity and *Agrobacterium tumifaciens* - Mediated Genetic Transformation of *Grewia asiatica* L. *Journal of Plant Biochemistry and Physiology,* **4**: 167.

Yadav V, Singh AK, Appa Rao VV, 2017a. Effect of GA_3 and cow urine on seed germination and seedling growth of custard apple. *Indian Journal of Arid Horticulture,* **12 (1 and 2)**: 71-74.

Yadav V, Singh A.K, Appa Rao VV, Mishra DS and Singh Sanjay, 2017b. Efficacy of GA_3 on seed germination and seedling vigour of wood apple. *Indian Journal of Arid Horticulture,* **12 (1 and 2)**: 53-55.

Yadava UL and Rajput CBS, 1969. Anatomical studies of rooting in stem cuttings of phalsa (*Grewia asiatica* L.). *Hort. Science,* **1**: 19–22.

Chapter 3

Orchard Establishment and Canopy Management in Arid Fruits

Lal Chand, D.K. Sarolia and Jagadish Rane

ICAR–Central Institute of Arid Horticulture, Sri Ganganagar Rd, Bichhwal, Bikaner–334006, Rajasthan

In India's arid regions, primarily in Rajasthan and parts of Gujarat, Haryana, and Punjab, the harsh climatic conditions are well-suited for the cultivation of fruit crops such as ber, pomegranate, custard apple, aonla, and date palm. Establishing an orchard is a long-term investment that requires meticulous planning, including site selection, spacing, species and choice of variety, planting material quality, and design. Effective orchard management is crucial for maintaining high productivity and extending the orchard's lifespan. Canopy management and the application of plant growth regulators (PGRs) are essential for optimizing fruit production. PGRs, including auxins, gibberellins, cytokinins, ethylene, growth retardants, and growth inhibitors, are chemical compounds that regulate vital metabolic activities in plants. They enhance seed germination, root development in cuttings and air layers. They also play crucial role in canopy structure, flowering, fruit set, and hence ultimately improve yield and fruit quality.

The arid regions of India, primarily spanning Rajasthan and parts of Gujarat, Haryana, and Punjab, are known for their harsh climatic conditions. Fruit trees in these regions exhibit specialized adaptations (xerophytic traits) such as deep taproot systems, thick cuticles, waxy coatings or pubescence on leaves, leathery and small leaves, reduced leaf area, sunken and covered stomata, summer dormancy or leaf shedding during water stress, and synchronized flowering and fruiting with optimal conditions. These species also have high water content in tissues and demonstrate resilience to drought, salinity, and alkalinity (Chandra and Pareek, 1992; Choudhary *et al.,* 2020). Moreover, many fruit species in these arid regions thrive in shallow, rocky, gravelly, and undulating wastelands.

Potential Fruit Crops for Arid Region

This region have potential to grow fruits includes ber (*Zizyphus mauritiana* Lam.), date palm (*Phoenix dactylifera* L.), pomegranates (*Punica granatum* L.), Mulberry (*Morus* sps.), figs (*Ficus carica* L.), bael (*Aegle marmelos* Correa), aonla (*Phyllanthus emblica* L.), various citrus species (sweet orange, mandarin, and lime), lasoda (*Cordia myxa* L.), karonda (*Carissa carandas* L.), and phalsa (*Grewia asiatica* L.). Additionally, minor fruit species like ker [*Capparis decidua* Edgew (Forssk.)], Mithi Jal (*Salvadora oleoides* Decne.), gundi (*Cordia sinensis* Lamarck), *etc.* are found in arid region and possess nutraceutical potential with medicinal, and economic value and adapted well to the extreme conditions of arid environments.

Climatic Conditions

The arid regions of India experience distinct summer and winter seasons. Summers are characterized by extreme heat, often surpassing 48°C, and in some areas, temperatures can reach up to 50°C. Winters feature cold nights, occasionally dropping below freezing, with daytime temperatures ranging from 20°C to 25°C. Annual rainfall is very low, typically less than 250 mm per year, and in some places, as little as 100 mm. Rainfall is erratic and unevenly distributed, mostly occurring during the short monsoon season from July to September, with monsoon onset around July 1st and withdrawal by September 15th (Bhansali, 2009).

During summer, humidity remains extremely low, often below 20 per cent, while in winter, mornings may see higher humidity due to cooler temperatures, but overall humidity remains low. Evaporation rates are exceptionally high due to intense heat and low humidity, often exceeding precipitation levels. Pre-monsoon periods (April to June) are marked by frequent dust storms, strong winds that contribute significantly to soil erosion. Although monsoon winds provide some relief from the heat, they are less consistent compared to humid regions.

Certain fruit crops in these arid regions, like date palms, have specific climatic requirements. Date palms require extremely low humidity with rain free period and high heat summation units (HSU) during fruit maturation period with assured irrigation for tree growth and productivity. Ideal locations for quality harvest of date palm include Jaisalmer and the western parts of Barmer, Bikaner, and Jodhpur districts, as well as arid coastal areas of Kachchh and parts of Saurashtra (Bhansali, 2009).

Edaphic Conditions

Arid soils across India typically exhibit sandy texture, poor soil physical properties, low organic matter, and limited water-holding capacity, resulting in generally low plant productivity. Furthermore, the reliance on groundwater and high evapotranspiration rates contribute to salt accumulation, which further degrades soil quality. Despite these challenges, certain arid fruits such as ber, pomegranate, date palm, figs, bael, aonla, lasoda, mulberry (Morus spp.), and ker exhibit tolerance to saline and alkaline soils and water conditions.

For successful cultivation of arid fruits, sandy loam soils with a pH range of 7.0 to 8.5 are considered ideal. In soils with higher pH and electrical conductivity (EC), proper application of organic amendments becomes crucial for sustaining fruit production.

Orchard Establishment

The fruit production in the arid region requires careful planning for irrigated and rainfed conditions and accordingly implementation of practices that maximize water efficiency, soil health, and overall sustainability. In hot arid regions, where environmental conditions are particularly harsh, careful selection of crop species and their varieties is crucial for successful growth and production. In commercial orchards, it is advantageous to plan a suitable combination of varieties (early, mid, and late season) to extend the harvest window. For crops like aonla, ber, and date palm, selecting compatible pollinizer varieties is essential to enhance fruit set, fruit quality, and overall yield. For instance, Narendra Aonla-7 (NA-7) is known for its abundant fruiting and high-quality fruits, but it requires cross-pollination due to self-incompatibility. Therefore, varieties like NA-6 and Krishna can serve as suitable pollinizer with NA-7 to improve fruit set, retention, and quality when planted alongside NA-7 (Singh *et al.,* 2001). Similarly, date palm is a dioecious tree in nature where male and female flower bear on separate tree, require hand pollination for commercial production. Suitable pollinizer variety needs to be planted for pollen collection. Likewise, in ber, a combination of varieties need to planted for optimum productivity.

Different arid fruit genotypes have been evaluated, and high-yielding varieties have been identified and recommended by various institutes, including NDUAT, Ayodhya; ICAR–CISH, Lucknow; ICAR–CIAH, Bikaner; AICRP on Arid Zone Fruits, Bikaner; CIAH-CHES, Godhra, Gujarat; and ICAR–CAZRI, Jodhpur (Table 3.1).

Optimization of Planting Density

Optimizing planting density in orchards, especially in arid regions, is crucial for maximizing yield, resource use efficiency, and sustainability. Proper planting density balances the number of trees per unit area with the availability of water, nutrients, sunlight, and space, ensuring each tree has optimal growing conditions. Density optimization depends on growth habit, soil fertility, use varieties and rootstocks dwarfing rootstocks, adoption of training systems, canopy management and use of plant growth regulators. Some of recent advances on density optimization in arid fruits are given in Table 3.2.

Table 3.1: Important Fruit Crops and their Varieties for Arid and Semi-Arid Region of the Country

Crops	*Suitable Varieties*
Ber	Thar Sevika, Gola, Thar Bhuvraj, Goma Kirti, Umran, Illaichi, Apple Colour
Aonla	Goma Aishwariya, NA-6, NA-7, NA-10
Jamun	Goma Priyanka, Thar Kranti,
Pomegranate	Bhagwa, Jalore Seedless
Bael	Goma Yashi, NB-7, NB-9, CISH-B-1, CISH-B-2
Date palm	Barhee, Hillawi, Medjool, Khunezi, Khalas, Dhamas
Citrus	Sweet orange: Blood Red, Sathgudi, Mandarin: Kinnow, Fremont, Daisy
Tamarind	Goma Prateek and local land races
Lasoda	Thar Bold, Karan Lasoda, Maru Samridhi
Phalsa	Thar Pragati
Mulberry	Thar Harit and Thar Lohit
Karonda	Thar Kamal
Fig	Dianna, Poona

Table 3.2: Advances in Density Optimization in Arid Fruits

Sl.No.	*Crop*	*Plant Spacing*		*Special Techniques for Density Optimization*	*Reference*
		Normal Density	*High Density*		
1.	Aonla	10m × 10m	5m × 5m	Double hedge row planting system	Singh *et al.* (2011)
2.	Beal	8m × 8m	5m × 5m	Dwarf variety-Goma Yashi	Singh *et al.* (2020)
3.	Ber	7-8m × 7-8m to 11-12m×11-12m	6m × 6m	Rootstock-scion combinations: Umran budded on *Zizyphus mauritiana* (Coimbatore)	Bal and Gill (2016)
4.	Date palm	10m × 10m	9m × 9m10m × 8m	Use of tissue cultured plants	Al-Yahyai *et al.* (2023)
5.	Lasoda	8m × 8m	5-7 m	Vegetative propagation, canopy management	Meghwal *et al.* (2021)
6.	Citrus	6 m × 6m	Suitable for high density	Pectinifera rootstock induced dwarfism in Fremont mandarin	Gora *et al.* (2022)

Sl.No.	*Crop*	*Plant Spacing*		*Special Techniques for Density Optimization*	*Reference*
		Normal Density	*High Density*		
7.	Pomegranate	5m x 5m	4m × 4m	Bush training system for cv. 'Kandhari'	Gill *et al.* (2011)
		5m x 5m	4 m x 3 m	For cv. 'Bhagwa'	Kumar *et al.* (2018)

Canopy Management in Arid Fruits

Canopy architecture management involves altering tree canopies by means of training and pruning to achieve the desired shape, develop a strong framework, maintain an appropriate height and to maximize fruit production and quality (Chand *et al.,* 2021). This practice facilitates inter-cultivation operations and enhances productivity of orchards. In arid regions, where water availability is limited, efficient canopy management is essential for sustainable fruit production.

Principles of Canopy Management

The principal of canopy management (Srivastava, 2023) include

- ☆ **Light Interception**: Manipulating canopy architecture to optimize light penetration.
- ☆ **Pruning physiology**: Understanding how pruning affects growth and yield.
- ☆ **Source-sink relationships**: Balancing vegetative growth and fruit production.
- ☆ **Tree architecture development**: Considering tree shape and form.
- ☆ **High-density planting systems**: With the use of dwarf rootstocks, dwarf scion and plant growth inhibitors to enhance per unit productivity.

Methods and Practices in Canopy Management

- ☆ Training
- ☆ Pruning
- ☆ Use of dwarf scion varieties
- ☆ Use of rootstock to manage canopy size

Canopy Management in Date Palm

Canopy management in date palms is crucial for optimizing fruit production, ensuring tree health, facilitating easier pollination, harvesting and maintenance. The effective canopy management in date palms involves a combination of training, pruning, thinning, and de-suckering. These practices not only enhance fruit quality and yield but also contribute to the overall health and longevity of the date palm trees. By implementing these management techniques, growers can optimize the productivity and sustainability of their date palm orchards.

- **Training:** Training the canopy involves directing the growth of the fronds and fruit clusters to ensure the structural stability of the canopy structure to bear crop load and shaping the canopy to facilitate easier access for maintenance and harvesting.
- **Pruning:** Pruning is essential for maintaining a healthy canopy and involves the removal of old, dead, or diseased fronds (leaves). The key objectives of pruning in date palm are:
- **Ease of pollination:** By removing lower and excess fronds, access to the inflorescence for facilitating pollination.
- **Enhance air circulation:** Removing excess fronds improves air circulation, reducing the risk of fungal diseases and pest infestations.
- **Light penetration:** In general, fractional ground cover of date palm orchard should be higher than 0.7 and light interception should be greater than 0.5 (Al-Muaini *et al.,* 2019). Pruning allows more light to reach the inner parts of the canopy and the trunk, promoting better fruit development.
- **Ease of harvesting:** By removing lower and excess fronds, access to the fruit bunches is easier, facilitating harvesting and other maintenance activities.
- **Pruning time and intensity:** Leaf pruning can be carried out thrice in the year, including before pollination, adjusting the ratio of leaf to bunch within 6–8 weeks after pollination, and after harvest as the priority time (Alikhani-Koupaei and Aghdam, 2021). The leaves provides photosynthetic materials, their transfer inside the plant, and help in flower induction. Leaf starch reserves and leaf-to-bunch ratio during flower induction *i.e.* July to October affect its leaf pruning intensity. Date palm leaves are pruned from a distance of 30–40 cm above the leaf sheath to maintain the ratio of leaf to bunch to produce with appropriate quantity and quality. In date palm, leaf-to-bunch ratio of 9–10 considered optimum for fruit quality and source-sink balance (Alikhani-Koupaei and Khosravinejad, 2024).
- **Pollination management:** Date palms are dioecious, meaning male and female flowers are on separate trees. Ensuring effective pollination involves:
- **Hand Pollination:** This is commonly practiced to ensure good fruit set. Male flowers are collected and their pollen is manually dusted onto female flowers.
- **Timing:** Pollination should be done at the right time when the female flowers are receptive to maximize fruit set.

- **Fruit thinning:** The pruning of female spathes is important to harvest marketable size dates, generally eight fruiting spathe per palm considered optimum for balancing the leaf-area to fruit-mass ratio (Al-Muaini *et al.,* 2019). Fruit thinning involves the removal of some of the developing fruit clusters. The benefits are improved fruit quality and preventing overbearing. Reducing the number of fruit clusters ensures that the remaining fruits receive more nutrients, leading to larger and higher-quality dates. Thinning prevents the tree from being overburdened with too many fruits, which can stress the tree and reduce its lifespan.
- **De-suckering:** De-suckering is a process of removal of offshoots/suckers in date palms arising from base or on trunk of the tree. De-suckering is essential to promote vigorous growth of main trunk and timely removal of suckers prevents overcrowding and competition for nutrients and water between the mother tree and the suckers.

Canopy Management in Ber

Training ber trees during the initial 2 to 3 years is crucial for establishing a robust framework. A budded ber plant produces vigorous growth during its first year, so vertical growth is encourage until the plant reaches a height of about 1 to 1.5 meters. Once the desired height is achieved, lower buds are allowed to sprout on the main stem by pruning the tip. Allow only 3-4 well-spaced main branches to develop on the main trunk at the desired height. It is necessary to keep the main trunk (up to 75 cm from ground level) free from side sprouts and water sprouts. During the second year, retain 3-4 well-spaced secondary branches on each main branch by clipping the rest. Continue this process to develop tertiary branches. This will create a modified central leader-based framework, maintained by annual pruning and removing water sprouts (fast-growing long shoots) as they appear (Shukla *et al.,* 2011).

In ber trees, fruit is borne on the axils of leaves on current season shoots. Therefore, annual pruning is required to encourage new shoots and maintain fruit bearing load and quality. Pruning opens the canopy to maximize sunlight distribution and penetration, improve air circulation, and ease spraying operations for harvesting regular, high-quality fruits. The severity of pruning varies with plant vigour and growth gain. In general, annual pruning consists of heading back the previous season's growth by 25 to 50 percent (Bal and Gill, 2016, Sarolia *et al.,* 2021) or to the level of the sixth secondary branches (Krishna *et al.,* 2014), especially in Umran retaining six buds (Bal and Gill, 2016), and thinning out weak, diseased, and inter-crossing branches. The ideal time for pruning ber trees is during the summer season, after harvesting the crop (15th April to 30th May), with mid-May being optimal under Bikaner conditions (Sarolia *et al.,* 2021).

Canopy Management in Pomegranate

- **Training:** Training operations for developing a structural framework in fruit trees begin 6-8 months after planting and continue for 2-3 years. The lowest branch should be developed at 30-40 cm above the ground, using either single stem or multi-stem training methods. In arid regions, a multiple stem training system (3-5 stems) (Kumar *et al.,* 2018) or bush training system (Gill *et al.,* 2011) is preferred to mitigate the risk of stem or plant loss due to termite attacks, which are a significant problem during the initial stages of orchard development (Kumar *et al.,* 2018).

- ☆ **Pruning:** Pruning aims in pomegranate to: (1) balance between vegetative and reproductive wood to optimize fruit yield, (2) reduce non-productive period, (3) prolong longevity of the trees, (4) delay senescence (Hakimi *et al.,* 2018). Pruning practiced twice a year to remove dried twigs and branches and to maintain a balance between vegetative and reproductive growth. Major pruning is performed just after the winter harvest, while light pruning is done in May-June to regulate flowering. In orchards severely affected by bacterial blight, heavy pruning is carried out after harvest to remove as much of the diseased portion as possible. After training and pruning, apply a 10 per cent Bordeaux paste to the cut ends of the plants (Kumar *et al.,* 2018).

Flower Regulation in Pomegranate

Pomegranates continuously flower under arid and semi-arid conditions, resulting in irregular and small crops throughout the year, which is not commercially desirable. To manage this, flower regulation or bahar treatment is applied. Irrigation is withheld for 1-2 months before the bahar, and ethrel (1-2 ml/liter) is applied to facilitate leaf shedding. Trees are lightly pruned 40-45 days after stopping irrigation. Fertilizers are applied immediately after pruning, and irrigation is resumed, leading to profuse flowering and fruiting (Kumar *et al.,* 2018).

Pomegranates bear bisexual, intermediate, and male flowers on both new and old growth in three flushes: February-March (ambe bahar), July-August (mrig bahar), and September-October (hasta bahar). The choice of bahar depends on irrigation availability, market demand, and pest/disease incidence.

- ☆ **Ambe Bahar:** Requires assured irrigation; fruits are ready for harvest from June to August but have poor size and color, and higher pest incidence. It is not recommended for hot arid regions.
- ☆ **Mrig Bahar:** Harvested from December to February. Favored in arid and semi-arid regions due to limited water needs and favorable climate during fruit development.
- ☆ **Hasta Bahar:** Harvested from March to April, producing attractive fruits with dark arils, fetching high market value. However, poor flowering and fruit set due to coinciding rainy season affect yield.

In hot arid regions, inducing flowering in August-September is beneficial as fruits mature in the milder climate from November to February. For this, irrigation is withheld in May-June, followed by light pruning (20 cm of new growth) and applying ethrel (1-2 ml/liter) mixed with 5 g DAP. This improves fruit size and aril colour, especially in light-coloured varieties like Jalore Seedless. Deblossoming flowers before and after the main blooming period helps improve yield and fruit quality by reducing competition for nutrients and water. Approximately 80-120 fruits are retained by removing weak, deformed, and clustered fruits/flowers. Proper canopy management and flower regulation are essential for high yield and sustainable production (Kumar *et al.*, 2018).

Canopy Management in Bael

Beal is propagated through vegetative for orchard establishment, start bearing forth year onwards. It is a medium sized tree and under narrow spacing (5-6 m) requires canopy management with aims to optimize tree structure for maximum productivity and fruit quality.

- ✰ **Training:** Training of young Bael plants for tree structure in which 4-6 well-spaced branches encouraged for growth in all directions. These branches will form the main scaffold structure of the tree.
- ✰ **Pruning :** Pruning intensity depend on bearing habit, annual extension growth typically a varietal character, planting density, soil and climatic conditions, and orchard management practices. The bearing habit of bael is typically ramiflorous, fruit set occur on both current-season shoots and older branches (up to 9 years old) (Singh *et al*, 2019). In general in bael, pruning of approximately 25 per cent of the annual extension growth during the leafless stage (typically in month of April-May) is recommended (Singh *et al.*, 2018). This practice stimulates the emergence of new shoots and contributes to the development of a dense canopy. A dense canopy helps protect against sun scald, especially in drier conditions. Removal of water shoots arising from base of tree is an essential operation in Bael.

Canopy Management in Fig

Fig is fairly tolerant to drought through avoidance mechanism by shedding leaves under stress and quick recovery with water availability (Ammar *et al.*, 2020). Fig is also considered moderately tolerant to saline soil and irrigation water (Golombek and Lüdders, 1990; Qrunfleh *et al.*, 2013). Therefore, fig could be grown successfully in drought-prone areas with certain level of salinity with supplementary irrigation to ensure optimal yields. Fig trees flourish in the regions with long, hot summers. Frost is a common feature in North and Central India during winter. Fig trees undergo into dormancy during winter and ability to shed leaves makes it a suitable crop for frost-prone areas.

Fig trees are usually trained as open center or modified leader system, with either a single or multi-stem structure. Since fig trees bear fruit on current season growth twice a year, effective canopy management is essential to induce the growth of flower-bearing wood.

- ☆ **Training:** Training is crucial in the initial year of planting to establish the desired tree framework. After planting, training begins in the first dormant season. During this time, the tree is headed back to a height of 60-70 cm, and side branches are also headed back or thinned out to shape the canopy, preferably in an open center. In the second dormant season, the laterals in the upper portion of the tree are headed back, leaving 2-3 basal buds to encourage more shoots.
- ☆ **Pruning:** Fig trees (*Ficus carica*) require strategic pruning to optimize fruit production. Fig trees bear fruit in the axils of leaves on current season shoots and produce a small crop on old wood. Pruning is essential to induce new flower-bearing shoots. The timing and extent of pruning depend on growth habits, location, variety, and annual crop yield. Annual pruning helps to harvest a single marketable crop.

In fig trees, pruning mainly involves heading back laterals during the dormant season *i.e.* in late winter from January to February in Northern and Central India. For the upper portion of the tree, laterals are headed back, leaving 2-3 basal buds to stimulate more shoots (Chand *et al.,* 2021). In the lower portion, less severe heading cuts are made to maintain tree shape, and some thinning cuts may be used to open up the canopy. All dried, diseased, broken, interfering, and interfacing laterals are thinned out.

The summer crop of dormant-pruned Poona fig trees remain hard and fail to mature, require advance pruning by 2-3 months to ensure good quality fruit. Light pruning is practiced in Poona Fig in October in parts of Maharashtra. Dormant-pruned Deana fig trees produces a high-quality summer crop from the end of June to mid-July (Chand *et al.,* 2021). Further, Zare (2021) suggested that a combination of thinning and heading cuts during winter pruning found effective in mitigating drought damage in fig orchards under rainfed conditions. This strategy improves vegetative characteristics, reduces negative leaf water potential, lowers leaf temperature, minimizes syconia drop, increases syconia production, enhances soluble solids content, improves fruit colour, and increases fruit weight quality.

Notching may be used to stimulate the production of laterals on vigorous upright branches. In post-pruning care, apply blitox paint to cut surfaces to prevent pathogens and promote wound healing. Farmers can adopt these techniques based on local conditions and fig varieties.

Canopy Management in Aonla

- ☆ **Training and pruning:** In the initial year of planting, trees are trained to promote the growth of four to six well-spaced branches with wide angles, forming the tree's framework. Two main training systems are employed: the open centre and modified central leader systems. The open canopy structure is preferred to ensure ample light penetration and air circulation. Specifically, the modified central leader system is commonly used for training aonla trees.

Initially, young plants are allowed to grow straight upwards to a height of 0.75-1 meter above ground level. During March-April, any unwanted branches are pinched off. In subsequent years, the focus is on allowing 4-6 branches to develop, establishing a strong framework. Regular pruning is crucial, involving the removal of dead, infested, broken, weak, or overlapping branches after fruit harvest.

It's noted that in the aonla cultivar NA-7, limb breakage has been observed in several instances due to overbearing and narrow branch angles with the main stem. Therefore, ensuring a balanced canopy is essential for this variety (Pathak *et al.,* 2006; Singh *et al.,* 2014).

Canopy Management in Lasoda

- ☆ **Training and pruning:** Canopy management in Lasoda plays a crucial role in regulating crop growth and improving fruit yield and quality. For budded plants, it is recommended to manage trees with a low-headed approach to facilitate easier fruit harvesting. This involves retaining upright growing shoots and pruning the others as necessary.

Over time, the focus shifts to developing 3-4 well-spaced upright limbs as the main scaffold. Additionally, leaf defoliation is essential to promote early and uniform fruiting. According to Roy and Meghwal (2012), optimal defoliation should occur from the end of December to the beginning of January. This period coincides with natural leaf yellowing and shedding after irrigation is withheld during November-December.

Defoliation can be achieved manually or through a chemical spray. Roy and Meghwal (2012) suggest a foliar spray of ethrel at 1000 ppm during the first week of January to facilitate easier leaf fall. Their research indicates that defoliating lasoda leaves during this period promotes early and uniform fruiting, which not only avoids the adverse effects of late-season high temperatures but also enhances the market value of early harvested fruits.

Optimization of Water Use

Effective water management is crucial for sustainable agriculture, especially in regions facing water scarcity. Proper scheduling, monitoring, and adjusting irrigation based on the needs of arid fruit crops helps optimize water use. The water requirements of different fruit crops can vary based on factors such as climate, crop type, and growth stage. The critical stage of water requirement in different fruit crops is at any stage during growing season and water stress indicated by dulling of leaf colour and drooping of growing points but major demand of watering is in flowering, fruit set and fruit development stages. For instance, Kinnow mandarin is particularly sensitive to water stress during the flowering and fruit set stages, which occur in March in northern India (Panigrahi *et al.,* 2014).

Various techniques are available to enhance water use efficiency, including use of adaptive rootstocks, drip irrigation (surface and subsurface), deficit irrigation, specialised techniques like cocoon technology, and mulching. Employing drought-tolerant scion-rootstock combinations—where drought-tolerant rootstocks can convey their drought tolerance to the scion—could enhance water use efficiency (Chand *et al.,* 2024). Encouraging deep root systems through the use of rootstocks and sub-surface irrigation (Phene *et al.* (1990) further enhances water efficiency. Drip irrigation delivers water directly to plant roots, minimizing wastage and maximizing efficiency. Subsurface drip irrigation systems offer substantial advantages over surface drip irrigation, including increased crop yield, improved crop quality, efficient water and nutrient utilization, and lowered agronomic costs associated with weed control, fertilization, tillage and salinity management (Phene *et al.,* 1987; Ayars *et al.,* 2015, Sun *et al.,* 2016).

Deficit irrigation (DI) involves strategic water application during critical growth stages, minimizing water stress while maintaining fruit quality. This induces minimal stress, which does not significantly affect yield, and can be more profitable for farmers by maximizing crop water productivity rather than harvest per unit land (English, 1990). In the hot sub-humid climate of central India, DI has proven effective for citrus, enhancing water productivity and fruit yield while producing superior quality fruits (Panigrahi and Srivastava, 2016). In date palms, applying deficit irrigation (DI) at 70 per cent of the crop evapotranspiration (ETc) based on 100 mm cumulative evaporation—compared to 50 mm and 150 mm—over a four-month period, starting two months before flowering (early January) and continuing until two months after flowering, which coincides with the middle of the Kimri stage (late April), resulted in the highest bunch weight, yield, and water use efficiency (Alikhani-Koupaei *et al.,* 2018). Similar studies are needed in arid zones to develop water-saving strategies for fruit crops.

Innovative water-saving methods like Cocoon technology provide additional solutions in establishment phase of the orchard in area lacking irrigation facilities. Cocoon™ reservoirs, for instance, supply water during the dry season, eliminating the need for continuous irrigation (Land Life Company, 2021, Carabassa *et al.,* 2022).

In-situ water harvesting, achieved by capturing rainwater in tree basins, is especially valuable in rainfed areas of arid regions. Designing tree basins in shapes like half-moons and full moons helps collect and retain runoff water. These structures enhance soil infiltration and restore soil fertility.

Additionally, applying organic mulch around tree bases conserves soil moisture, reduces evaporation, and moderates soil temperature. Mulching is an effective practice that contributes to water conservation and improved soil health in arid regions.

Soil Amendment and Nutrient Management

Organic amendments, such as compost and humic acid, play a crucial role in enhancing plant growth by improving the physical, chemical, and microbial properties of soil (Muhammad *et al.,* 2023). The addition of organic matter to soil not only enhances moisture retention capacity but also serves as a substrate for microbial proliferation, facilitating nutrient mobilization and improving nutrient availability to plants.

In India's arid regions, where longer and more severe droughts are anticipated in the near future, these amendments are increasingly vital for optimizing cropping strategies. They contribute significantly to sustainable agriculture by mitigating the impact of water scarcity and improving soil fertility, thereby supporting resilient and productive agricultural practices.

In arid regions, the primary salt-affected soils are saline and sodic soils. Soils with high calcium carbonate content are categorized as calcareous. These soils, common in arid climates, often have >15 per cent calcium carbonate, which raises soil pH and reduces zinc, potassium and magnesium availability. Thus, crops planted on such soil require above normal levels of potassium, zinc and magnesium for optimum nutrition. Fertilizer guidelines for fully mature arid fruit trees are provided in Table 3.3.

Table 3.3: Fertilizer Recommendations for Full Bearing Trees of different Arid Fruit Crops

Crop	*FYM and NPK (kg per tree per year)*				*Other Nutrients*	*Reference*
	FYM	*Nitrogen*	*Phosphorus*	*Potassium*		
Aonla	30-40	1.0	0.5	0.5	100-500 gm B and ZnSo4 in sodic soil	Pathak *et al.,* 2006; Singh *et al.,* 2020
Date palm	20–40	0.5–1.0	0.5–1.0	0.25–0.5	Ammonium sulphate (1–2 kg per palm)	Chandra *et al.,* 1992; Pareek and Nath 1996; Bhansali 2009
Pome-granate	40	0.6	0.25	0.25	-	Kumar *et al.,* 2018
Lasoda	40	-	-	-	-	Meghwal and Singh, 2019
Bael	50-80	0.5	0.25	0.5	0.6 per cent each $ZnSO_4$, borax and $FeSO_4$ during July, Oct. and Nov.	Singh *et al.,* 2019; Singh *et al.,* 2011

Fruit Drop and its Control

Flower and fruit drop is a significant issue in fruit cultivation. The shedding of flowers and fruits, known as fruit drop, can indeed be a major problem for crops like Bael, Citrus, Ber, and Pomegranate. This phenomenon occurs in distinct waves, such as post-bloom drop, mid-size drop, and pre-harvest drop (Rattanpal *et al.*, 2019). Several factors contribute to this problem, including:

- **Imbalance of growth regulators:** Hormonal imbalances can lead to premature fruit drop.
- **Extreme temperatures:** Both high and low temperatures can stress the plant, causing it to drop fruits.
- **Water stress:** Both drought and excessive water can disrupt the plant's water balance, leading to fruit drop through disruption of the plant's physiological processes.
- **Nutrient deficiencies:** Lack of essential nutrients like nitrogen, potassium, and calcium can weaken the plant and triggers fruit drop.
- **Frost for long periods:** Prolonged exposure to frost can damage the plant tissues, leading to fruit drop.
- **Poor health of the tree:** General poor health due to disease, pests, or other factors can make the tree more susceptible to fruit drop.
- **Disease and insect pest attack:** Pathogens and pests can damage the plant, leading to fruit shedding.

When a plant is stressed due to any of the aforementioned reasons, it produces ethylene. This ethylene production leads to formation of the abscission layer between the branch and the fruit, ultimately resulting in fruit drop. Addressing these factors through proper agricultural practices, such as balanced fertilization, pest management, and irrigation control, can help mitigate the issue.

Saleem *et al.* (2008) found that application of GA @ 45 mg/l in Blood Red Orange reduced flower and fruit drop both. Eman *et al.* (2007) investigated the effects of GA_3 and zinc sprays on improving the yield and fruit quality of Washington Navel Orange trees grown in sandy soil conditions. They concluded that applying GA_3 at 10 ppm along with zinc reduces fruit drop percentage and increases plant yield.

Use of Plant Growth Promoting Substances

Various plant growth regulators are employed in managing fruit production in arid regions to modify physiological processes. Recommendations from research for arid fruit crops are summarized in the Table 3.4.

Table 3.4: Plant Growth Regulators, their Concentrations, Response in Arid Fruit Crops

Crop	*PGR and Concentration*	*Response*	*Reference*
Aonla	GA_3@500 ppm+ thiourea	Breaking seed dormancy	Lilabati and Sahoo, 2015
	GA_3@500 ppm	Early and higher seed germination, decrease in mortality, increase in seedling height and seedling girth, higher buddable seedling.	Rashmi *et al.,* 2007; Supe *et al.,* 2012
	NAA@10 ppm	Fruit retention, fruit quality (TSS, total sugar, higher vitamin C) and improved yield.	Ghosh *et al.,* 2009
	Calcium nitrate @1.5per cent or GA3 @ 50ppm	To reduce physiological weight loss of fruits.	Singh *et al.,* 2020a
Ber	Potassium sulphate (1.5%) and NAA @ 20 ppm at fruit set stage.	To reduce fruit drop and improve fruit set.	Singh and Bal, 2006
Custard apple	GA_3 @ 50 ppm through seed soaking for 48 hrs	Improve seed germination and seedling growth.	Jadhav *et al.,* 2015
	Foliar spray of GA3 @ 50 ppm coupled with organic mulching	To widen flowering duration and fruit retention .	Patel *et al.,* 2010
Pome-granate	IBA@2500ppm	Inducing fibrous root in hard wood cutting.	Saroj *et al.,* 2008
	Ethrel @1-2 ml/litre + 5 g DAP	Used in bahar regulation and applied to facilitates leaf shedding.	Kumar *et al.,* 2018
	NAA@25 ppm	Fruit set and fruit retention.	Ghosh *et al.,* 2016
	Ethrel @ 200ppm	Improved fruit quality (TSS, reducing sugar,non-reducing sugar, total sugar and ascorbic acid content).	Goswami *et al.,* 2013
	Salicylic acid (SA) and its derivatives@ 10 mM	Pre-harvest treatment to improve aril colouration, quality parameters and fruit yield.	García-Pastor *et al.,* 2020
	Salicylic acid @ 1 mM	To reduce bacterial blight.	Lalithya *et al.,* 2017)
Lasoda	ethrel @ 1000 ppm during the first week of January	Induction of defoliation to achieve uniform flowering and fruit maturity.	Roy and Meghwal (2012)
Bael	NAA @20-30ppm	Minimize fruit drop.	Uniyal and Mishra, 2015

Way Forward

Developing successful orchards in arid regions necessitates innovative strategies to address water scarcity and challenging soil conditions. Effective orchard management in such climates should focus on the following approaches:

1. **Identification of drought-hardy, dwarf cultivars**: Select drought-hardy, dwarf fruit tree varieties with dense canopies to maximize production per unit area, which can help optimize space and water usage.
2. **Rootstock development**: Identify or develop rootstocks that enhance traits such as crop water use efficiency, water capture, and transpiration efficiency while maintaining fruit quality and productivity under drought conditions.
3. **Genetic approaches for drought tolerance**: Explore genetic modifications, such as gene overexpression or gene silencing, to enhance the drought tolerance of fruit trees.
4. **Optimized irrigation scheduling**: Develop and implement irrigation scheduling strategies to improve water use efficiency and productivity, ensuring that water resources are used most effectively.
5. **Climate-adaptive farming practices**: Adopt farming practices that are specifically tailored to adapt to the changing climate, including techniques for managing microclimates and soil conditions.
6. **Use of Unmanned aerial vehicle (UAV), remote sensing and AI tools:** for continuous observation on plant phenology, disease and pest surveillance and in resource optimization.

By focusing on these approaches, orchard management in arid regions can be made more sustainable and resilient to environmental challenges.

REFERENCES

Alikhani-Koupaei M and Aghdam MS, 2021. Defining date palm leaf pruning line in bearing status by tracking physiological markers and expression of senescence-related genes. *Plant Physiology and Biochemistry*, **167**: 550-560.

Alikhani-Koupaei M and Khosravinejad A, 2024. Optimizing the leaf-to-bunch ratio in 'Mazafati' date trees by considering source-sink limitation and physiological-molecular responses. *Journal of Agriculture and Food Research*, p. 101-225.

Alikhani-Koupaei M, Fatahi R, Zamani Z and Salimi S, 2018. Effects of deficit irrigation on some physiological traits, production and fruit quality of 'Mazafati' date palm and the fruit wilting and dropping disorder. *Agricultural Water Management*, **209**: 219-227.

Al-Muaini A, Green S, Abou Dahr WA, Kennedy L, Kemp P, Dawoud M and Clothier B, 2019. Water use and irrigation requirements for date palms on commercial farms in the hyper-arid United Arab Emirates. *Agricultural Water Management*, **223**: 105702.

Al-Yahyai R, Khan MM, Al-Kharusi L, Naqvi SA and Akram MT, 2023. Date Palm Plantation Establishment and Maintenance. Date Palm, p. 179.

Ammar A, Aissa IB, Mars M and Gouiaa M, 2020. Comparative physiological behavior of fig (Ficus carica L.) cultivars in response to water stress and recovery. *Scientia Horticulturae*, **260**: 108881.

Ayars JE, Fulton ALAN and Taylor B, 2015. Subsurface drip irrigation in California—Here to stay?. *Agricultural water management*, **157**: 39-47.

Bal JS and Gill KS, 2016. Canopy management, nodal pruning and planting density in Indian jujube (ber). Acta Hortic. 1116, 99-104. https: //doi.org/10.17660/ActaHortic.2016.1116.17

Bhansali RR, 2009. Date palm cultivation in the changing scenario of Indian arid zones: challenges and prospects. Desert plants: biology and biotechnology, pp.423-459.

Carabassa V, Alba–Patiño D, García S, Campo J, Lovenstein H, Van Leijen G, Castro AJ, González F, Viera G, Kourkoumpas DS and Zioga AA, 2022. Water–saving techniques for restoring desertified lands: Some lessons from the field. *Land Degradation and Development*, **33(1)**: 133-144.

Chand L, Patidar J, Anuragi H, Taria S, Handa AK, Dev I. and Tiwari RK, 2021. Fig (Ficus carica): A potential fruit crop for agroforestry systems in arid and semi arid regions. Kerala Karshakan e-journal. **8(8)**: 4-11.

Chand L, Sharma N, Sharma RM, Pandey R, Sathee L and Dubey AK, 2024. Physio-Biochemical and Growth Response of Contrasting Reciprocal Grafting in Citrus Under Water Deficit and Rehydration. *Journal of Plant Growth Regulation*, **43(4)**: 1230-1246.

Chandra A and Pareek CS, 1992. Lasoda (*Cordia myxa* Roxb.)a potential fruit crop in Jaisalmer district of western Rajasthan. *Agricultural Science Digest*, **12**: 11-12.

Chandra A, Chandra A, Gupta IC, 1992. *Date palm research in Thar Desert*. Scientific, Jodhpur

Choudhary SM, Chavan DL and Singh R, 2020. Use of plant growth regulators in dry land fruit crops: A Review. *Indian Research Journal of Genetics and Biotechnology*, **12(02)**: 128-135.

Eman AA, Abd El-moneim Abd El Migeed, Omayma MMM and Ismail MM, 2007. Effect of GA3 and Zinc sprays for improving yield and fruit quality of Washington Navel Orange trees grown under sandy soil conditions. *Research Journal of Agriculture and Biological Sciences*, **3(5)**: 498-503.

English,M, 1990. Deficit irrigation. I: analytical framework. *J. Irrig. Drain. Eng. (ASCE)*, **116(1)**: 399–412.

García-Pastor ME, Zapata PJ, Castillo S, Martínez-Romero D, Guillén F, Valero D and Serrano M, 2020. The effects of salicylic acid and its derivatives on increasing pomegranate fruit quality and bioactive compounds at harvest and during storage. *Frontiers in Plant Science*, **11**: 668.

Ghosh SN, Bera B, Kundu A and Roy S, 2009. Effect of plant growth regulators on fruit retention, yield and physico-chemical characteristics of fruits in ber 'Banarasi Karka' grown in close spacing. *Acta Horticulturae*, **840**.

Ghosh SN, Bera B, Roy S and Kundu A, 2016. Effect of plant growth regulators in yield and fruit quality in pomegranate cv. Ruby. *Journal of Horticultural Science*, **4(2)**: 158-160.

Gill PPS, Dhillon WS and Singh NP, 2011. Influence of training systems on growth, yield and fruit quality of pomegranate 'Kandhari'. *Acta Horticulturae*, **890**: 305-310.

Golombek S and L€udders P, 1990. Gas exchange of *Ficus carica* L. in response to salinity, p. 487–493. In: M.L. Beusichem (ed.). Plant nutrition physiology and applications. Kluwer Academic Publishers, Dordrecht, the Netherlands.

Gora JS, Kumar R, Sharma BD, Ram C, Berwal MK, Singh D, Bana RS and Kumar P, 2022. Performance evaluation of Fremont mandarin on different rootstocks under the hot arid environment of India. *South African Journal of Botany*, **144**: 124-133.

Goswami JD, Patel NM, Bhadauria HS and Wankhade VR, 2013. Effect of plant growth regulators on quality traits of pomegranate cv. Sindur. *The Asian Journal of Horticulture*, **8(1)**: 361-363.

Jadhav AC, Bhagure YL and Rajeswari MR, 2015. Effect of PGR, chemicals and plant extract on seed germination and seedling growth of custard apple (Annona squamosa). *The Asian Journal of Horticulture*, **10(1)**: 184-186.

Krishna H, Parashar A, Awasthi OP and Singh K, 2014. Ber. In: Tropical and Sub Tropical Fruit Crops (Ed. Ghosh, S.N.): Crop Improvement and Varietal Wealth (Part-I). Jaya Publishing House, Delhi, pp. 137-156.

Kumar R, Meena R, Sharma BD and Saroj PL, 2018. Production technology of pomegranate in arid region. ICAR–Central Institute for Arid Horticulture, pp.1-34.

Lalithya KA, Manjunatha G, Raju B, Kulkarni MS and Lokesh V, 2017. Plant growth regulators and signal molecules enhance resistance against bacterial blight disease of pomegranate. *J. Phytopathol.* **165**: 727–736. doi: 10.1111/jph.12612

Land Life Company. (2021, May 28). Projects. https: //landlifecompany.com/ projects

Lilabati L and Sahoo UK, 2015. Effect of pre-treatments on seed germination and seedling vigor of *Emblica officinalis* Gaertn. *Global Journal of Advanced Research*, **2(10)**: 1520-1526.

Meghwal PR and Singh A, 2019. Maru Samridhi: New lasora variety. *Indian Horticulture*, **64(2)**: 32-33.

Meghwal PR, Singh A and Singh D, 2021 Research status of lasora (*Cordia myxa* L.) in India — a review. *Current Horticulture*, **9(1)**: 15–19.

Muhammad S, Shaukat M, Yasin M *et al.*, 2023. Compost and humic acid amendments are a practicable solution to rehabilitate weak arid soil for higher winter field pea production. *Sci Rep.*, **13**: 17519. https: //doi.org/10.1038/s41598-023-44921-x

Panigrahi P and Srivastava AK, 2016. Effective management of irrigation water in citrus orchards under a water scarce hot sub-humid region. *Scientia Horticulturae*, **210**: 6-13.

Panigrahi P, Sharma RK, Hasan M and Parihar SS, 2014. Deficit irrigation scheduling and yield prediction of 'Kinnow' mandarin (*Citrus reticulate* Blanco) in a semiarid region. *Agricultural Water Management*, **140**: 48-60.

Pareek OP and Nath V, 1996. CFR – Indian arid zone. Coordinated fruit research in Indian arid zone – a two decades profile (1976–1995). NRC for Arid Horticulture, Bikaner, pp 55–66.

Patel NM, Patel DK, Verma LR and Patel MM, 2010. Effect of cultural and chemical treatments on fruit set and fruit yield of custard apple (*Annona sqamosa* Lin.) cv. Sindhan, *Asian J. Hort.*, **5(2)**: 498-502.

Pathak RK, Singh Sanjay and Saroj PL, 2006. Aonla (*Emblica officinalis*), pp 1–20. Advances in Arid Horticulture Vol. 2. (Eds) P L Saroj and O P Awasthi. IBD Co. Lucknow.

Phene CJ, Davis KR, Hutmacher RB and McCormic RL, 1987. Advantages of subsurface irrigation for processing tomatoes. In: Proceedings, International Symposium on Integrated Management Practices for Tomato and Pepper Production in the Tropics, Shanhua, Taiwan, ROC, 325-338.

Phene CJ, Davis KR, Hutmacher RB, Bar-Yosef B and Meek DW, 1990. Effect of high frequency surface and subsurface drip irrigation on root distribution of sweet corn. *Irrig. Sci. J.*, **12**: 135-140.

Qrunfleh IM, Shatnawi MM and Al-Ajlouni ZI, 2013. Effect of different concentrations of carbon source, salinity and gelling agent on in vitro growth of fig (Ficus carica L.). *Afr. J. Biotechnol.* **12**: 936–940.

Rashmi K, Sindhu SS, Sehrawat SK and Dudi OP, 2007. Germination studies in Aonla (*Emblica officinalis* G.). *Haryana Journal of Horticultural Sciences*, **36(1-2)**: 9-11.

Rattanpal HS, Singh G, Arora A and Singh S, 2019. Management of fruit drop in citrus. Published by: Additional Director of Communication for Punjab Agricultural University, Ludhiana, pp. 1-22.

Roy MM and Meghwal PR, 2012. Central Arid Zone Research Institute, Jodhpur. In: Chadha K L, Singh S K and Singh AK (Eds.) Farmer Friendly Technologies in Horticulture, The Horticultural Society of India, New Delhi-110012, pp. 83-86.

Saleem AB, Malik AU, Pervez MA and Khan AS, 2008. Growth regulators application affects vegetative and reproductive behaviour of 'Blood Red' sweet orange. *Pak. J Bot.* **40(5)**: 2115-2125.

Saroj PL, Awasthi OP, Bhargava R and Singh UV, 2008. Standardization of pomegranate propagation by cutting under mist system in hot arid region. *Indian Journal of Horticulture*, **65(1)**: 25-30.

Sarolia DK, Haldhar SM, Kumar L and Pareek PK, 2021. Production Technology and Insect Pest Management of Ber Crop. In Pest Management in Dryland Horticultural Crops (Eds. Haldhar, S. M. and Maheshwari, S. K.). Biotech Books, New Delhi, India, pp. 133-142.

Shukla AK, Bhargava, Singh RS and Sharma SK, 2011. Germplasm resources and production technology of ber. Technical Bulletin No. 41. ICAR–Central Institute for Arid Horticulture, Bikaner. pp. 21-22.

Singh AK, Singh Sanjay and Saroj PL, 2018. Exploring morphovariations in bael (*Aegle marmelos*). *Current Horticulture*, **6(1)**: 52–7.

Singh AK, Singh Sanjay and Saroj PL, 2020. Kinetics of productivity and economics of Goma Yashi bael under high density. Indian Horticulture Summit-Mitigating Climatic Changes and Doubling Farmers Income through Diversification, 14-16 February, MGCGV, Chitrakoot, Madhya Pradesh, p. 73.

Singh AK, Singh Sanjay, Appa Rao VV, Bagle BG and More TA, 2011. Effect of high density planting systems on the productivity of NA-7 aonla (*Emblica officinalis* Gaertn) under rain fed conditions of semi-arid ecosystem. *Indian Journal of Horticulture*, **68(4)**: 461–65.

Singh AK, Singh Sanjay, Hiwale SS, Appa Rao VV and Joshi HK, 2014. Production technology of aonla under rainfed conditions of western India. Technical Bulletin, Pub. CHES (ICAR–CIAH), pp 1–36.

Singh A., Singh S and Saroj PL, 2020a. Scientific cultivation of aonla. KERALA KARSHAKAN e-journal. September, 2020. pp. 40-47

Singh AK, Singh S, Saroj PL, Krishna H, Singh RS and Singh RK, 2019. Research status of bael (*Aegle marmelos*) in India: A review. *Indian Journal of Agricultural Sciences*, **89(10)**: 1563-1571.

Singh C and Bal JS, 2006. Effect of nutrients and growth regulators on fruit drop, size and yield of ber (*Zizyphus mauritiana* Lamk.). *International. J. Agric. Sci.*, **2(2)**: 358-360.

Singh HK, Srivastava AK and Dwivedi R, 2001. Effect of pollinizer on fruit set and fruit quality of 'Narendra Aonla 7' aonla (*Emblica officinalis*). *Indian Journal of Agricultural Science*, **71(1)**: 65-66.

Srivastava KK, 2023. Canopy Management of Fruits, 2nd Fully Revised and Enlarged Edition. NIPA. ISBN: 9789390591053.

Sun SM, An OX, Yang PL, Lu XB and GU KK, 2016. Effect of irrigation depth on root distribution and water use efficiency of jujube under indirect subsurface drip irrigation. *J. Trans. Chinese Soc. Agric. Machinery*, **47(8)**: 81-90.

Supe VS, Dhanshree Patil, Anju Bhagat A and Rashmi S and Bhoge. 2012. Seed germination and seedling growth in aonla (*Emblica officinalis* Gaertn.) *Bioinfolet*, **9(2)**: 206-208.

Uniyal S and Mishra KK, 2015. Effect of plant growth regulators on fruit drop and quality of bael under Tarai conditions of Uttarakhand. *Indian J. of Horticulture*, **72(1)**: 126-129.

Zare H, 2021. Effects of Different Methods of Pruning Intensity on Old Fig (Sabz Cultivar) Trees under Rainfed Conditions. *International Journal of Fruit Science*, **21(1)**: 379–391. https: //doi.org/10.1080/15538362.2021.1892011

Chapter 4

Nursery Production, Canopy and Orchard Management in Arid Unexploited Minor Fruit Crops

Jagan Singh Gora, Lal Chand, S.M. Haldhar and Akheram

Scientist, ICAR–Central Institute of Arid Horticulture, Bikaner, Rajasthan

The arid and semi-arid regions of India, which include the *Thar* Desert, Rann of Kutch, Deccan Plateau, and Leh, are significant drought-prone areas impacting approximately 265 million people across 99 districts in 14 states. These regions are characterized by extreme temperature fluctuations, hot dry winds, low and erratic rainfall, high evaporation rates, and intense solar radiation, coupled with saline and nutrient-poor soils (Ahamad *et al.,* 2023; Sathyakumar and Sivakumar, 2007). Such conditions render them particularly vulnerable to commercial fruit crop cultivation. To enhance crop productivity in these harsh climates, integrating underutilized arid-zone fruit crops is a promising strategy. These crops possess unique morphological, physiological, anatomical, and biochemical xerophytic traits that enable them to thrive under challenging conditions. For instance, their deep root systems and high root-to-shoot ratios allow them to access moisture and nutrients from deeper soil layers; leaf shedding/dormancy during adverse periods, spines instead of leaves, and small-sized, thick leaves, fur/hairiness and waxy coating on the leaf surface and sunken and deep stomata help minimize water loss and protect against extreme temperatures, and synchronized flowering and fast fruit development during the season characterized by larger moisture availability are increased the overall resilience and resistance to suboptimal environmental conditions of arid region (Chundawat, 1990; Pareek *et al.,* 2007).

Numerous arid unexploited minor fruit crops thrive in the natural habitats of arid and semi-arid regions, including jamun, fig, bael, karonda, tamarind, phalsa, mulberry, custard apple, wood apple, pilu, tendu, mahua, khirni, khejri, and prickly pear. Earlier, these unexploited minor fruits served as environmental service plants due to their use in firewood, timber, agricultural implements, housebuilding, shade, animal and human food, *etc.* Over time, these crops warrant exploration for enhancing domestic nutritional food security and promoting commercial orchard cultivation. These fruits are rich in vitamins, proteins, minerals, crude fiber, phenolic compounds, and antioxidants, all of which contribute significantly to human health (Satya *et al.,* 2011; Krishna, *et al.,* 2019; Meena *et al.,* 2022). They offer various health benefits, including improved digestion, anti-tumor properties, anti-helminthic effects, diuretic and demulcent qualities, and expectorant properties (Gopalan, 2002; Gautam *et al.,* 2012; Meena *et al.,* 2022). By cultivating these underutilized crops, we can not only secure nutritional food sources but also enhance the livelihoods of farmers in arid regions.

In this context, arid unexploited fruit crops are ideally suited for organic cultivation. These crops are primarily consumed as processed products, require minimal inputs, and are eco-friendly, making them highly adaptable to local conditions. While their yields may be lower compared to traditional crops, their high market demand can significantly enhance farm income and diversify nutritional offerings, particularly in an era increasingly focused on sustainable practices over chemical farming. To harness this potential, ICAR institutes/SAUs have been actively engaged for the past two decades in identifying superior genotypes and cultivars. Their efforts encompass optimizing factors such as spacing, canopy management, water and nutrient application, weed management, and plant protection technologies to maximize fruit yield and quality. This chapter provides detailed insights into plant propagation, canopy management, and effective orchard management practices. These strategies aim to equip farmers with the scientific knowledge necessary to increase fruit production sustainably, ultimately supporting their livelihoods in arid regions.

Varieties and Plant Propagation

The cultivation of unexploited arid fruit crops is increasingly recognized as a vital strategy for addressing nutritional insecurity and poverty, particularly in rural and tribal areas where access to diverse fruits is limited. Recent advancements by public sector research organizations like ICAR–CIAH, Bikaner and AICRP on arid zone fruit centers have led to the development of improved varieties through selection and hybridization, tailored for better adaptability, yield, and quality in arid and semi-arid environments.

These locally adapted species not only contribute to dietary diversity but also act as reservoirs of stress-related genes, which can be harnessed to breed future varieties that are more resilient to environmental stressors. The details of developed varieties, along with their specific features and standardized plant propagation techniques for unexploited arid fruits in the arid and semi-arid regions of India, are summarized in Tables 4.1 and 4.2.

Table 4.1: List of Unexploited Arid Fruit Crop Varieties with their Trait-Specific Characteristics

Crops	*Varieties*	*Remark*
Bael	NB-5	Late maturing and large size fruit.
	NB-9	Late maturing and large size fruit.
	Goma Yashi	Semi dwarf tree, thornless and suitable for semi-arid dry land areas, and high yielding (90kg per plant).
	Thar Divya	Comparatively earliest in maturity and tolerant to drought, and fruit yield 85.20kg/plant.
	Thar Nilkanth	Highly drought tolerant under aberrant agro-climatic conditions.
	Thar Srishthi	Highly centric locule arrangement of fruit, attractive pulp colour with no off flavor, rich in fine fibre, less seeds, and suitable for table and processing purpose.
Lasoda	Thar Bold	Prolific and early bearing, cluster bearing with bold fruitshigh yielding (1.5-2.0 q tender fruits/plant).
	Maru Samridhi	Plant canopy in drooping in nature, cluster bearing (14 fruits per bunch with 10 g fruit weight), and high yield (90 kg per plant.
	Karan Lasora	Bold size and cluster bearing fruits.
Tamarind	Goma Prateek	Quality pods with 26.7g pod weight, high TSS (71°Brix), acidity (14.06%) and pod yield up to 50-55 kg/tree.
Wood apple	Thar Gaurav	Drought hardy, early maturing and high yielder (115 kg/tree).
Karonda	Thar Kamal	Dwarf, early, precocious bearer and suitable for HDP.
Custard apple	APK (Ca) 1, Balanaga, Arka Sahan, Phule Janki	Balanagar and Arka Sahan found superior in quality.
Jamun	Goma Priyanka	Semi dwarf, suitable for high density planting and thin seed.
Khirni	Thar Rituraj	Semi-spreading, regular bearer, semi-dwarf, early fruit ripens in 120-125 days after fruit set and fruit yield (10-16 kg/tree).

Crops	*Varieties*	*Remark*
Mahua	Thar Madhu	Dwarf, ideal for HDP and rainfed hot semi-arid ecosystems, precocious bearing, early fruits ripen within 90-105 days after fruit set, and high fruit yield (20.1 kg per tree) and dry flowers (6.30 kg per tree).
Khejri	Thar Sobha and Thar Amruta	Dwarf and dropping canopy nature, thornless, ideal for high density planting, quality fodder and pod (sangari) .
Mulberry	Thar Lohit (red), Thar Harit (light green)	Frost (-2°C) and high temperature (49°C) tolerant.
Chironji	Thar Priya	Dwarf, precocious bearer and suitable for high density planting.
Fig	Poona Fig, Dinkar, Dianna, Conadria, Excel	Dianna known for large size fruit and performing well under arid and semi-arid conditions.
Phalsa	Thar Pragati	Dwarf, precocious, drought tolerant and suitable for HDP and both table and processing purpose.

Natural multiplication in arid fruit crops primarily occurs through seeds, and in some species, through root suckers, as seen in Pilu, Kair, and Karonda. Seed propagation is generally used in fruit crops to create genetic variability, for regeneration of hybrid progenies and in fruit crops in which vegetative propagation protocols are not available. Seed propagation does not produce true-to-type plants. To maintain true-to-type characteristics, vegetative propagation methods are essential. Fruit trees can be propagated vegetatively by cuttings, air layering, and budding or grafting. In most of arid fruits, a well-established vegetative propagation protocols are available with high success rate at nursery and field establishment for semi-arid and arid conditions. Though species like Kair [*Capparis decidua* Edgew (Forssk.)] and Mithi Jal or Pilu (*Salvadora oleoides* Decne.), still largely rely on seed propagation as rapid multiplication through vegetative means has met with limited success. Kair demonstrates a strong seed germination rate of 72 per cent when seeds are sown in the last week of August in poly bags filled with a mixture of soil and farmyard manure (FYM) at a 3: 1 ratio (Mahla and Singh, 2013). Likewise, *Salvadora oleoides* seeds have a very short viability window, remain viable for just 4–5 days after extraction, and delayed sowing significantly reduces germination rates (Kumar *et al.,* 2016). For seed propagation, *Salvadora oleoides* fruits are typically harvested between May and June, de-pulped, washed, and dried. Fresh seeds sown in May, without pre-treatment, can achieve up to 87 per cent germination within 15 days (Jindal *et al.,* 2006) and 20 per cent germination after three months of storage (Kumar *et al.,* 2014).

Among vegetative propagation, propagation through cutting is the simplest and most effective, especially for species like fig (*Ficus carica* L.), phalsa, pomegranate (*Punica granatum* L.), mulberry (*Morus spp.*), karonda (*Carissa carandas* L.), gundi

(*Cordia sinensis* Lamarck) and prickly pear (*Opuntia spp.*), while Kair and pilu showed very limited success in propagation through cutting. The success in rooting percentage in Kair multiplication through cutting is about 20-25 per cent (Meghwal and Vashishtha, 1998). Kair is considered very hard to root species, take at least 7-8 months in rooting. In Kair rooting occur mainly from nodes without forming callus under *ex-vitro* conditions. Efforts are going on to improve the success rate and reduce time of rooting. Likewise, Pilu has the potential to regenerate vegetatively through *in-situ* layering by root suckers (Shekhawat *et al.,* 2012), but *in-situ* layering is a cumbersome process and limited success achieved yet. In pilu, stem cuttings and air layering respond poorly to rooting (Kumar *et al.,* 2014).

The budding and grafting protocols have been standardized in various unexploited arid minor fruit species including jamun (*Syzygium cumini* Skeels), bael (*Aegle marmelos* Correa), khejri (*Prosopis cineraria*), lasoda (*Cordia myxa*), tamarind (*Tarminalia indica* L.), custard apple (*Annona squamosa* L.), wood apple (*Limonia acidissima* L.) *etc.*, for arid and semi-arid conditions with variable success rates (Table 4.1).

Budding is the second most preferred method for propagating of these fruit crops. The specific techniques include patch budding for Aonla, ring, T, and I budding for Ber, patch budding for Bael, Jamun, and Tamarind, and I budding for Khejri (*Prosopis cineraria*). In grafting, the wedge or cleft grafting is most common and easiest method used in arid fruit trees using either softwood or hardwood bud sticks, depending on the season and climatic conditions. The fruit trees shows quiescence/dormancy, the dormant hard wood used for cleft grafting and give rise good success. Wedge grafting using soft-wood scion during July-August in jamun (Kumar and Chander, 2024), while, using hard-wood in bael during April-May and during January-February in aonla found most suitable and easiest method to propagate. Bench grafting (cleft method) has also been standardized for ber propagation under semi-arid conditions using bare rooted seedling rootstock and dormant scion with 85 per cent in summer season (May) on *Z. rotundifolia* rootstock using pruned material for scion purpose to save time and resources (Tewari *et al.,* 2011).

The demand for planting material of these unexploited arid fruits is rising due to their nutraceutical benefits, low input demands, and climate resilience. A significant challenge to expanding the area planted with arid fruit crops is the lack of reliable planting material. Registered nurseries meet only 20-30 per cent of the demand; the rest relies on importing planting material for crops like date palm (*Phoenix dactylifera* L.) and on unorganized sectors for other crops. The availability of quality planting material at lower costs provides ample opportunity for large-scale planting. The biggest challenges in propagation of these fruit crops include their hardy nature, unsupportive climate conditions, and lack of techniques for multiplying true-to-type rootstocks. Additionally, the absence of identified rootstock genotypes and hi-tech nursery infrastructure further complicates the process.

Consequently, the seedling plants sold for these fruit crops, as well as the budding and grafting practices often rely on non-descriptive seedlings.

Figure 4.1: Pictures of Unexploited Arid Fruit Crops of Arid and Semi-Arid Region of India.

Canopy Management

Training and pruning is an essential practice to standardize to obtain desired canopy in the fruit crops for easy intercultural operations, high yield, quality fruit, less incidence of disease and pest, and longevity of orchards. In phalsa, it is multi-branches thornless plants, bear flowering and fruiting on current season growth. Phalsa plant allows growing naturally as bush in first year of planting and later, selective branched were headed back at height of 40-50 cm, and mounding of soil for the promotion of maximum shoots. All the branches are pruned annually from a height of 60-80 cm from ground during last week of December to 10th January

Table 4.2: Standardized Propagation Techniques and their Remarks of Unexploited Arid Fruit Crops

Fruit Species	*Propagation techniques*	*Ideal Time and Success Rate*	*Remark*	*Reference*
Bael	Cleft/wedge grafting	March-April with use dormant leaf less hard wood scion on one year old seedling rootstock and achieved more than 80 per cent success rate. No need of defoliation of scion.	100 micron caps used to prevent scion desiccation.	Lal Chand (2019) unpublished data Singh (2018)
	Patch budding	Patch budding on 9-10 months old rootstock performed in May-June (before onset of rain) under Gujarat conditions gives highest success (80-90%).	Equally suitable under nursery and *in-situ* conditions during May-June.	Singh (2018)
Lasoda	Patch budding	Middle portion from current season's growth used as scion and budding on 9-12 month old seedling rootstocks during May-June give rise 80-85 per cent successful budded plants under arid condition.	Freshly extracted seeds used for rootstock raising..	Kumar *et al.* (2022) and Meghwal (2007)
Tamarind	Soft-wood grafting (cleft grafting)	Higher graft success in April-May.	Ideal condition is semi-arid region of Gujarat and Maharashtra.	Singh and Singh (2007)
	Patch budding	July-August.	Ideal condition is semi-arid region of Gujarat and Maharashtra.	Singh and Singh (2007)
Wood apple	Soft wood grafting	Maximum success obtained when grafting done in the month of February on one-year-old seedling rootstocks.	Higher survival in in-situ grafts than polybag grafted plants at nursery.	Hiwale *et al.* (2008)
Kair	Cutting (Hardwood)	Planting of Hard-wood cutting in August give better result.	Low success in rooting (<25%).	Meghwal and Vashishtha (1998)
Karonda	Seeds	Seeds have short viability and freshly extracted seeds from fully mature fruits sown in August-September.	The highest rooting (60%) with IBA (5000 ppm) treated cuttings during November planting.	Singh (2018)

Fruit Species	*Propagation techniques*	*Ideal Time and Success Rate*	*Remark*	*Reference*
Custard apple	Cleft/wedge grafting	Use of dormant shoot in Feb-March with Capping and softwood grafting in March-April using 6-8 months old scion shoots gives 91.45 per cent success.	10-12 months old rootstocks ideal for grafting and budding.	Singh (2018)
	Patch budding	Use of 3-4 months old scion shoots during April-May with gives higher success (about 60%).	Ideal condition is semi-arid region of Gujarat and Maharashtra.	Singh (2018)
Jamun	Cleft/wedge grafting	With graft survival rates (73-76%) in grafted 1st week of March under semi-arid climate. Use previous season shoot with 1 week prior defoliation.	Covering grafts with 100 micron polytubes (RRS, Abohar, Punjab).	Kumar and Chander (2024)
	Patch budding	40 per cent bud success in March under semi-arid climate.	Ideal condition is semi-arid region of Gujarat.	Singh and Singh (2006)
	Soft wood grafting	Grafting through cleft method give 36 per cent graft success in July-August under semi-arid climate.	Ideal condition is semi-arid region of Gujarat.	Singh and Singh (2006)
Tendu/ Timroo	Soft wood grafting (cleft grafting)	May-June.	Ideal condition is semi-arid region of Gujarat.	Singh (2018)
Khirni	Soft wood grafting (cleft grafting)	with 76 per cent graft success in March, 70-71 per cent in July and August.	Ideal condition is semi-arid region of Gujarat.	Singh and Singh (2015)
Mahua	Soft wood grafting (cleft grafting)	High graft success achieved in March (70%), April (65%) and July (64%).	Ideal condition is semi-arid region of Gujarat.	Singh and Singh (2015)

Fruit Species	*Propagation techniques*	*Ideal Time and Success Rate*	*Remark*	*Reference*
Khejri	Patch budding	Mother trees pruned during December to induce new growth-bearing juvenile buds. Bud-grafting by patch method gives rise 76 per cent budding success. This technique can also be used for top-working the wild seedling trees and *in-situ* budding. Four to five month old seedlings with 2-3 mm diameter at bud union used for budding.	Use of 50 ppm GA_3 improve height, collar diameter and stem diameter of rootstock seedling.	Pareek and Purohit (2002); Purohit *et al.* (2003).
Mulberry	Cutting	Semi-hard and hard-wood cutting with 15-25 cm length preferred give best rooting during January-February and July-September.	Treated rooting hormone enhanced root induction.	Sarolia *et al.* (2018)
Chironji	Soft wood grafting (cleft grafting)	>80 per cent graft success *In-situ*) and 60 per cent at nursery during July (best time).	Ideal condition is semi-arid region of Gujarat.	Singh (2018)
Fig	Hard wood cutting	Age (9-15 month old wood cutting) and size of cutting (20 to 30 cm length and 1.5 cm diameter) during January-February under North Indian conditions to achieve 90-100 per cent rooting by using Cocopeat-Vermiculite-Perlite (1: 1: 1).	Use 100-200 ppm IBAGrowing in polyhouse or in warm place enhance rooting.	Chand *et al.* (2021)Antunes *et al.* (2003)
Phalsa	Hard wood cutting and seed	Cuttings treated with IBA (100 ppm) give 65 per cent success under hot semi-arid region and freshly extracted seeds give also more than 90 per cent germination.	Seeds loose viability within 90-100 days under ordinary storage.	Singh (2018)
Prickly pear	Cladodes	Direct field planting of full size cladodes (12 months old, 8-15 cm width) during spring (February) planting give better plant establishment.	Plant cladode in upright position.	Singh and Singh (2003)

when plant shed their leaves under north Indian conditions. Additionally it is recommended that do not attempt either light or severe pruning which is reduced fruit yield. Phalsa plant can be trained on single or double stem at a height of 3.5-4.0 feet and light pruning is done under kitchen gardening. Phalsa is bears yellow colour flowers in axil of leaves in the cluster, and fruit mature in the month of May first week (Gora *et al.,* 2022).

Karonda plant is allowed training and pruning only in initial stage for the development of single or double stem up to 50 cm from the ground, and don't allow again any suckers below this height which facilitate easy to weeding, cleaning and manuring under the canopy of this thorny bushy plant. Further, bearing plant don't required any pruning annually except suckers, pest and disease infested, dry or any other unwanted shoots. Similarly, Jamun plants required training for the first 2-3 years after planting for architecturing the desired canopy frame. Plants should maintain a single stem up to three feet from the ground and topped, then allow 3-4 shoots to grow in all directions from this height. Further, trims these shoots to 50 per cent of their current season growth to encourage more lateral shoots and bushier growth for better fruiting. The jamun plant should be topped annually from 2.5 to 4.0 m height to maintain a dwarfing canopy in the month of September (Chadha, 2009).

In figs, training and pruning are critical techniques for tree care, frost protection, and increased fruitfulness. Its plants can be trained standard (tall and clear stem with a branched crown on top) or multi-stem (multiple branches from the ground). Fig produces two crops each year; the first, known as breba, develops on the previous year's growth while the second crop arises on the current season's growth. Therefore, the pruning is done twice in year one as severe pruning when plants shed their leaves due to dormancy during December to January, leaving 2-3 buds on shoots of previous year's growth to obtain production from July to October with quality fruits and dwarfness of tree, and second as light pruning or notching (only the terminal bud with one leaf) on more than six-month-old shoots during July for opening up the crown of the tree to let plenty of sun and air through, as well as encouraging the tree to branch evenly and devote its energy, and this pruning gives higher production but poor quality fruits. This combination of training and notching promotes lateral branches and produces more fruits on new growth than pruning or notching alone. In addition to pruning and notching, growers can adopt stalking technique for the development of better canopy frame and higher production with quality fruits. In Purandar taluka (Maharashtra), pruning is done in the month of May for Kahtta Bahar (mrug bahar) and in the month of August-September for mitha bahar (haste bahar) (Singh *et al.,* 2015). Likewise, mulberry is utilized for both fruit production and sericulture, necessitating different training and pruning methods for each purpose. For fruit production, train the plant on a single stem up to 1 meter high and thereafter allow developing 4-5 well-oriented branches in all directions to create an ideal canopy. Regular pruning is generally not needed, but light pruning can be performed in January to maintain a dwarfing habit, and these terminal cuttings can be used in propagation as stem cutting. Further, in the case of leaf production for sericulture, mulberry should be planted at closer spacing and pruned by three methods: bottom pruning (cut plants at ground level, leaving a 10-15 cm stump, and this is done once a year in June), middle pruning (trim branches to 40-60 cm above ground level), and kolar or strip system (In ultra-high density planting, cut branches to ground level five times in a year, and necessitating heavy fertilization and irrigation to support growth) (Krishnaswami, 1986 and Bharathi and Basavaiah, 2022).

Custard apple plants are trained with four to five scaffolds that branch outward from a single stem, starting at a height of 50-60 cm. This structure encourages the development of secondary and tertiary shoots through careful training and pruning during the initial years. Annual light pruning is essential for enhancing light penetration, air circulation, and overall plant health. This is best done when the plants shed their leaves in winter, about 30-35 days before flowering, or 40-60 days after fruit harvest in humid regions. Custard apple trees primarily bear fruit on current season growth, with minimal fruit from previous seasons. During pruning, it's important to remove 20-30 cm or 50 per cent of previous season growth from the top tips of shoots, as well as any water sprouts, root suckers, and dead or crisscrossing branches. This maintenance fosters healthy growth and maximizes quality fruit production (Yadav, 2018).

The Khejri tree has long been valued as a sacred and multipurpose resource, known for its nutritious pods, quality fodder, and valuable fuel, as well as its positive impact on ecology and soil fertility. Historically, it played a crucial role as a lifeline in the *Thar* desert ecosystem. With the initiative from ICAR–CIAH, a thornless and dwarf cultivar named *Thar* Sobha has been developed, transforming how Khejri is cultivated. Farmers are now cultivating this cultivar in orchard-like plantations, utilizing proper training and pruning techniques to achieve an optimal canopy and enhance the quality of both pods and fodder. In the first three to four years, pruning is performed twice a year once in November and again in June to shape the desired canopy. For optimal pod and fodder quality and synchronized flowering, Khejri growers are suggested to prune 70-80 per cent of the current year's growth in June. This timing allows the new growth to mature over 6-7 months, enabling the differentiation of flower buds, which leads to improved flowering and crop yields. Traditionally, looping was done primarily for fodder and fuel in November, but this new approach enhances both productivity and sustainability (Samadia *et al.,* 2021).

Lasora trees, like other fruit crops, require training and pruning in their initial year to achieve the desired shape and size. Since grafted plants tend to grow laterally, it's important to prune shoots from the base to encourage 3-4 upright growing shoots that spread in all directions. In early February, growers should also remove any damaged branches, including those affected by termites or shoot borers, as well as dried and overcrowded branches (Yadav *et al.,* 2020).

Prickly pear cacti are popular for their aesthetic appeal and use as boundary hedges, and its poor and incorrect pruning can lead to more serious problems, including premature death of the plants. Prickle pears should be pruned in the late spring to early summer while they are actively growing, rather than during dormant, full bloom and extreme heat. When pruning, use tongs to grasp the pads and a sharp knife to cut them off at the joint where they connect to the next pad. It's crucial to promptly clean up any fallen cholla pieces, as they can root and grow into new plants under ideal conditions, which can impede mobility in the garden (Liguori and Inglese, 2015).

Bael, tamarind, and wood apple trees can be effectively trained under a modified central leader system to achieve an optimal canopy and maximize fruit production. These trees are initially allowed to grow up to 1 meter after planting. At this height, they are pruned to select 3-5 well-spaced lateral branches from all directions. These branches are then pruned back to 50 per cent of their growth each year up to the 4th year, during April to May when the trees are leafless. The modified central leader system for canopy architecture will be fully established within the first 3-4 years thereafter dead, diseased, crisscross and infested twigs should be removed in March-April month as well as whenever necessary, to maintain tree health and encourage optimal growth (Singh *et al.,* 2018 and Hari Prasath *et al.,* 2019). Pilu and khirni trees are also required training for ideal canopy framing with single stems up to 1 m and allow 4-5 scaffold branches from this height during the initial stage. Thereafter, pruning is not required in both trees except removing dead, hanging, and wanted shoots.

In the case of mahua and tendu, the training is essential practice to develop the desired canopy framework, so follow the modified centre leader system as suggested in the bael tree. Further, the tendu tree requires light pruning for high yield and quality fruits because it bears fruits on shoot terminals. In additions to fruit production, the tendu plant leaves are valuable for bidi industries. Trees are planted at 2x2 m spacing and undergo severe pruning to promote vegetative growth and produce larger and thinner leaves. Therefore, coppicing and pollarding practices are applied when trees reach a girth of 15 cm and a height of 90 cm. This encourages multiple coppicing shoots, resulting in high-quality leaves about 40-45 days post-coppicing (Chadha, 2009).

Water and Nutrient Management

Water is a precious resource, particularly in arid ecosystems where underground sources are often brackish. To maximize its use, it's crucial to cultivate drought and salinity-tolerant crops, ensuring every drop is utilized wisely. Traditionally, unexploited minor fruit crops have thrived as environmental service trees under rainfed conditions. To establish these fruit crops orchards effectively, they require regular irrigation weekly in hot summers and monthly in winters, adjusting as needed during the rainy season. Bearing trees require 4-5 frequent irrigation during fruit growth and ripening stages. Recent research, particularly from the AICRP on Arid Zone Fruits (AZF), shows promising results with modern irrigation methods like drip irrigation and fertigation. For example, in custard apple cv. Arka Sahan, optimal growth was achieved with a drip irrigation system at an IW/CPE ratio of 0.6 and full RDN, resulting in a maximum plant height of 298.1 cm. Additionally, using a combination of bio-fertilizers like *Azotobacter* and *phosphobacteria*, along with a reduced rate of chemical fertilizers, enhanced fruit yield and quality. The combined application of bio-fertilizers such as *Azotobacter, Azospirillum, VAM*, and *phosphobacteria*, along with 50 per cent standard NPK, significantly enhances the yield and quality of custard apple. Under Aruppukottai (AP) conditions, this approach resulted in an impressive fruit yield of 14.9 kg per tree, with individual fruit weights averaging 133.6 g and a Total Soluble Solids (TSS) level of 19.0°B. Additionally, replacing 25 per cent of the RDF with a bio-fertilizer consortium containing *Azotobacter tropicalis, Bacillus aryabhattai*, and *Pseudomonas taiwanensis*, along with AM fungi demonstrated potential without negatively impacting fruit yield or quality. In terms of optimal growth, a combination of 75 per cent RDF, 20 kg of FYM, and specific NPK applications (100: 50: 50 g) yielded the best results in Jabalpur (MP) and Jhalawar (Rajasthan). For fig cv. Poona, applying 1125 g of nitrogen, 325 g of P_2O_5, and 415 g of K_2O, along with 50 kg of FYM per plant under Jadhavwadi conditions, maximized yield and net income.

In bael cultivation, the recommended integrated nutrient management (INM) approach from NDUAT, Ayodhya, involved using 50 kg of FYM, 100 per cent NPK, and 200 g each of *Azotobacter* and *PSB*, resulting in a remarkable fruit yield of 127.93 kg per plant. For lasora, applying 40 kg of FYM and NPK (1000: 500: 150 g) yielded the highest fruit production of 39.20 kg per plant under Jobner conditions. For tamarind and manila tamarind, a balanced nutrient regimen is crucial, requiring 50 per cent RDF of NPK per tree, along with additional organic inputs such as 2.5 kg of FYM, 100 g of phosphorus, and 2.5 kg of vermicompost for tamarind, and 25 kg of FYM with NPK (200: 150: 250 g) and 2 kg of neem cake for manila tamarind.

Weed Management

In arid and semi-arid regions, the prevalence of weeds significantly hampers plant growth and reduces potential yields. Perennial weeds such as dob grass dob grass (*Cynodon dactylon*), motha (*Cyperus rotundus*), dachhab (*Cyperus difformis*), dharba (*Desmostachya bipinnata*), aak (*Calotropis procera*), sonamukhi (*Cassia angustifolia*), usnakantaka (*Echinops echinatus*), bhatal (*Launaea nudicaulis*) pose ongoing challenges, while seasonal weeds like bhurut (*Cenchrus biflorus*), tumba (*Citrullus colocynthis*), water primrose (*Ladwigia hyssopifolia*), wild jute (*Corchorus tridens*), wild sunflower (*Verbesina encelioides*), utran (*Pergularia deemia*), skunkvine (*Paederia ovate*), morning glory/tiger's foot morning glory (*Ipomoea pestigridis*), dove milk (*Euphorbia hirta*), mexican poppy/kantili (*Argemone mexicana*), red chickweed (*Anagallis arvensis*), fleabane (*Conyza bonariensis*), congress grass (*Parthenium hysterophorus*), bitter apple (*Solanum sisymbrifolium*), and gokharu (*Tribulus terrestris*) can quickly proliferate.

Many rainy season weed plants are thorny and allergenic, making frequent visits and orchard operations quite cumbersome. To effectively manage these issues, integrated weed management (IWM) approaches like mulching, hoeing, ploughing, and herbicide application are essential. Ploughing is a crucial practice for controlling weeds in orchards, especially in arid and semi-arid regions. It is recommended to plough three times at 20-day intervals after the second rain, starting around August. This practice not only disrupts the growth of weeds but also helps conserve runoff rainwater and retain soil moisture for longer periods. Additionally, remaining weeds near or under the canopy should be managed through hoeing or hand weeding. Farmers are also advised to avoid flood irrigation, as it can promote weed dispersal and germination. Using local flora as organic mulch, such as bui (*Avera tomentosa*) and kheep (*Leptodenia pyrotechnica*), has shown significant benefits, with studies indicating increases in lasora fruit yields of up to 66 and 58 per cent, respectively (Awasthi *et al.,* 2006). Furthermore, incorporating materials like paddy straw, maize straw and grasses can help reduce weed populations, conserve moisture, and improve soil properties, such as pH and electrical conductivity by enhancing microbial activity and promoting earthworm populations in crops like bael (Singh *et al.,* 2012). But, these plant based mulches can also create certain risks, especially in areas where reptiles may inhabit therefore special attention is needed during work activities in the orchards. This holistic approach not only addresses weed management but also contributes to overall soil health and crop productivity.

Flowering and Fruiting

Most of these crops are vegetatively propagated, beginning to bear fruit within 2-4 years after transplanting in the field. They also respond well to dormancy, leaf shedding, and pruning, which contribute to abundant flowering and high-quality fruit production. The bearing behavior of these unexploited minor arid fruits includes both terminal and axillary types. For example, bael exhibits terminal bearing on new growth, while fig, phalsa, aonla, and karonda show axillary bearing on new growth, and axillary bearing on old growth in kair and tamarind. Most of these fruit crops begin flowering in spring, with fruit ripening occurring from June to August. These fruits tend to have lower yields and non-synchronous ripening. Additionally, they are highly perishable, with a shelf life of just one to four days, necessitating either quick marketing or processing to increased farm income. The summary details of flowering and fruiting of these crops are given in Table 4.3.

Plant Protection

Recently, many insect- pests and diseases were observing serious damage at the foliage, flowering, and fruiting stages, like lemon butter fly (*Papilio demoleus*) and fruit fly (*Bactrocera zonata*) in bael, aphid (*Aphis craccivora*), pink mealybug (*Maconellicoccus hirsutus*) hairy caterpillar (*Eupractis fraternal*) and falsa caterpillar (*Giaura scptica*) in phalsa, butterfly (*Anaphaeis aurota*) in kair, tinged bug (*Dictyla cheriani*) in lasora and khejri, small salmon arab (*Colotis amata*) in pilu, fruit borer, (*Dudua aprobola*) in jamun, karonda moth (*Digama hearseyana*) in karonda, shoot borer (*Derolus discicollis*) and galls (*Eriophyes prosopidis*) in khejri, termite (polyphagous nature) *etc.* in arid unexploited minor fruit crops. These

Table 4.3: Flowering and Fruiting Information's of Arid Unexploited Minor Arid and Semi-Arid Fruit Crops.

Crops	*Flowering Period*	*Fruit Harvesting and Yield per Plant*	*Remarks*	*References*
Bael	☆ Bloom in May-June and bisexual flowers	☆ Fruit mature from March to April. ☆ Fruit yield - 65 kg.	☆ Use organic mulch and regular water for prevention of fruit cracking in arid and semi-arid region.	Singh *et al.,* 2018
Fig	☆ Flowering occurs in mrig bahar and haste bahar. ☆ Common fig is parthenocarpic (don't required pollination). ☆ Smyrna Fig is gynodioecious (required pollination by wasp).	☆ Fruits are ready to harvest in 120-140 (April to June) days fruit setting in N. India and in W. and Central India (March May). ☆ In Purander (MH) two crops are taken in year from July–September and Feb-March. ☆ General yield 30-40kg, and Purandar fig -60-70kg.	☆ Hast bahar fruit is superior quality (extremely sweet and attractive skin violet colour). ☆ Parthenocarpic fruit set can be enhanced by spraying 25 ppm of NAA. ☆ Fully ripe fresh figs can be kept only for about a week at 0oC with a 90 per cent relative humidity.	Chadha (2009) and Singh *et al.,* 2015
Jamun	☆ Flowering occurs in March–April. ☆ Hermaphrodite (bisexual) flowers.	☆ June-July with the onset of monsoon. ☆ Fruit yield -80-100kg.	☆ Its seed is an excellent and natural medicine for diabetes patient.	Chadha (2009)
Karonda	☆ Bisexual flower and flowering in March–April.	☆ Fruit ripen in 55 days after fruit setting (July to September). ☆ Fruit yield -8-10kg.	☆ Fruit drop reduced by two sprays of GA3 @ 60 ppm during full bloom and 15 days after fruit setting. ☆ Fruit can store for 3-4 and 7 days at RT and 70C, respectively. Fruits used for making jam, candy and pickles.	Chadha (2009)

Crops	*Flowering Period*	*Fruit Harvesting and Yield per Plant*	*Remarks*	*References*
Lasora	☆ Flowering in March-April	☆ Fruit harvesting in May month. ☆ Fruit yield – 60-100kg.	☆ Very short period for fruit harvesting.	Yadav *et al.*, 2020
Ker	☆ Flowering during spring, rainy and autumn season. ☆ Bisexual and self-pollinated and cross (insect) but open pollination improve fruit set considerably.	☆ Green fruit harvested during May-June. ☆ Fruit yield - 5-10kg.	☆ Quality fruit production in spring flowering and summer fruit harvesting. ☆ Fruit astringency can removed by immerses in 5 per cent solution of common salt/butter milk for 4–5 days in earthen pot, and dehydrated. ☆ Green dried fruits are important ingredient of Panchkuta.	Singh and Singh (2011) and Sharma (2020)
Tamarind	☆ Peak flowering occurs in May-June. ☆ Bisexual and cross-pollinated crop.	☆ Fruit Matured in 240-260 days fruit setting. ☆ Fruit yield - Local cultivars (165kg), PKM-1 (260kg), DT-28 (450kg) and DT-1 (500kg).	☆ It has strong tendency of alternate bearing.. ☆ Tamarind lime pickle (TLP) and kernel oil are recently introduced product by Sri Lankan markets. ☆ TSS: acid ratio of brown (acidic pulp), Red pulp and sweet pulp type are 5.02-9.50, 4.91-8.28 and 11.05-11.84, respectively.	Chadha (2009)

Crops	*Flowering Period*	*Fruit Harvesting and Yield per Plant*	*Remarks*	*References*
Phalsa	☆ Flowering time is March-April.	☆ Fruit ripe in 55-60 days after fruit set (end of May). ☆ Fruit yield -2-4kg.	☆ Rich source of anthocyanins and iron.	Chadha (2009) and Gora *et al.*, 2022
Mulberry	☆ Flowering in March-April. ☆ Unisexual flower and pollinated by wind.	☆ Fruit ripe in April -May. ☆ Fruit yield- 20-30kg fruits.	☆ Rich source of sugar and highly perishable fruit.	Kumar and Haldhar (2021)
Custard apple	☆ Flowering occurs in July. ☆ Hermaphroditic flower and limited self-pollination due to protogynous, so required hand pollination for enhancing fruit yield.	. ☆ Fruit ripe in October to November. ☆ Fruit yield - 20-30kg.	☆ Fruits are harvested in two to three pickings and high TSS: 28-32°Brix.	Chadha (2009)
Wood apple	☆ Full blooming in March-April Flower with mainly staminate and hermaphrodite flowers.	☆ Fruit harvesting is done in November (more than 200 days after fruit setting). ☆ Fruit yield -100-110kg.	☆ Fruit rich in pectin, phosphorous, potassium, calcium and iron. ☆ Highly suitable for making pickles, RTS, jelly and powder *etc.*	Yadav *et al.* (2020)
Pilu	☆ Peak blooming in February-March. ☆ Bisexual flower and self-pollinated crop.	☆ Harvesting from mid May. ☆ Fruit yield -30-40kg. ☆ Fruit consumed as both fresh and dried.	☆ 100 fruit weight- 19.1 to 36.1g. ☆ 100-seed weight- 2.27 to 3.49g. ☆ Seed oil content- 31.2 to 43.8 per cent.	Jindal and Singh (2005)

Crops	*Flowering Period*	*Fruit Harvesting and Yield per Plant*	*Remarks*	*References*
Tendu	✰ Flowering starts from April to June. ✰ Both perfect and dioceous flowers.	✰ Fruit ripening start from May to June. ✰ Fruit yield – 30-70kg.	✰ Rich source of phenols and fibres and classical product is wine. ✰ Medicinal importance in diarrhea, cholera, dysentery, intermittent fevers, bleeding gums. ✰ Its leaves used for making bidi and paatal.	Kumar *et al.*, 2017
Mahua	✰ Flowering occurs in March to April and cross pollinated crop.	✰ Dry flower (corolla) collected from 15-20 days after flowering.. ✰ Fruit ripening in June-July. ✰ Yield of dry flower (100-150kg) and kernel yield (60-80kg).	✰ It is a multipurpose tree and its flower, fruit and seed oil are used in various ways.	Singh *et al.* (2008)
Khirni	✰ Flowering occurs in November–December. ✰ Bisexual flowers.	✰ Fruit harvesting in May month. ✰ Fruit yield -15-40 kg.	✰ Fruits are rich in minerals, sugars, protein, carbohydrate and vitamin A (carotene).. ✰ Dried fruits used by the tribal's. ✰ Seeds contain approximately 25 per cent oil and used for cooking purpose.	Singh *et al.* (2015) and Malik *et al.* (2012a)

Crops	*Flowering Period*	*Fruit Harvesting and Yield per Plant*	*Remarks*	*References*
Chironji	☆ Flowers occurs in January–February. ☆ Bisexual flowers and cross pollinated crop.	☆ Fruit harvesting in April–May. ☆ Yield - fruit (20-25 kg) and its kernels weight is 0.03-0.25g.	☆ Seed kernel (taste like pistachio)- a high value product fetches price of Rs. 400–500 kg^{-1}. ☆ Value added products are Chironji ki barfi and seed oil.	Malik *et al.* (2012b) and Singh *et al.* (2006)
Khejri	☆ Flowering occurs in March–April. ☆ Bisexual flower but poor pollination due to self-incapability.	☆ Pod mature in April-May. ☆ Yield - Sangria (tender pod) 4-5kg, and dry fodder (15-20kg).	☆ Tender pods (fresh or dried) used for vegetable purposes and important ingredient of Panchkuta-a popular delicacy of Rajasthan. ☆ Very short period for harvesting of tender pod.	Samadia *et al.* (2021)
Prickly pear cv. clone 1308	☆ Flowering occur in July. ☆ Polygamous sex form (male, female and hermaphrodite).	☆ Harvesting in August. ☆ Yield -tender nopale (1.5kg) and fodder (fresh biomass) is 20-25 ton/h.	☆ Future plant for CBG and animal fodder.	Liguori and Inglese (2015)

pests can be controlled by using of IPM approaches such as resistant cultivars, scientific orchard operations, biopesticides (*Thar jaivik*), and systemic insecticide applications like Cypermethrin at 1.5 ml, Spinosad at 0.5 ml or Dimethoate at 2.0 ml per liter at the beginning of insect infestation (Haldhar and Maheshwari, 2018 and Haldhar *et al.,* 2022).

Fungal and bacterial diseases were also causing economic loss to arid unexploited minor fruit growers, as in lasora, *Phytophthora spp.* (gummosis), and drying of branches occurs invariably on plants during March-April. The gum flow in the phloem region of branches impedes the sap flow, which causes the terminal branches to dry. Scrape the affected bark portions of the tree trunk, branches, and limbs along with some healthy green bark and paste with copper oxychloride twice in February-March and July-August. However, powdery mildew (*Phyllactinia corylea*), cercospora leaf spot (*Cercospora moricola*) were observed in mulberry and these diseases were causing economic loss to crops by affecting plant foliage as leaves gradually turned yellow and brown, followed by the entire leaf area covered by fungus, and the leaf dried and fell down. These diseases can be controlled by Dinocap (0.1%) and Indofil Z-78 (0.2%) on the young leaves at 15 day intervals. Further, bacterial leaf spot (*Xanthomonas campestris*) were also significant in affecting the plant's health, as its leaves turned yellow, necrotic, and died of severe infestation. It can be effectively managed by two sprays of Phytomycin at 200 ppm/lit of water at 10-day interval. Bael plantation was mostly affected by fungal and bacterial diseases like bacterial leaf spots (*Xanthomonas bilvae*), leaf spots (*Alternaria spp.),* and fruit rot (*Aspergillus nidulans* and *A. nigerum*). Two spraying with Streptomycin Sulphate 250 ppm or Bordeaux mixture 1 per cent at 12-15 days interval are required for effectively controlling these diseases during the peak rainy season (Maheshwari and Haldhar, 2018). Moreover, leaf spots and fruit rot fungal disease caused by anthracnose (Glomerella cingulata) in jamun and its affected leaves were shown as small scattered spots, light brown or reddish brown in colour, and fruits shrivelled and rotten, and this fungal infestation can be managed by two-three sprayings of Dithane Z-78 @ 0.2 per cent or Bordeaux mixture at 4: 4: 50 concentration at a weekly interval for effective control (Banerjee and Rana, 2019). Foot rot (*Phytophthora nicotianae*) disease was reported as a as a major hindrance in the establishment of cactus pear during August and November months, and soil drenching with 0.1 per cent Ridomil and dipping of cut ends of cladodes in fungicide at the time of planting found effective control measured (Nallathambi *et al.,* 2005).

In addition to insect- pests and diseases, physiological disorders like fruit drop, sunscald, fruit cracking, *etc.* were also causing severe losses to these fruit crops due to extremely adverse climatic conditions such as high heat with blowing of desiccating wind (*loo*) and lower relative humidity. Therefore, growers should be keep regular watering, canopy management and orchard operations timely for obtaining high yields with quality fruits.

REFERENCES

Ahamad F, Tyagi SK, Singh M, Sharma AK, 2023. Groundwater in Arid and Semi-arid Regions of India: A Review on the Quality, Management and Challenges. In: Ali, S., Armanuos, A.M. (eds) Groundwater in Arid and Semi-Arid Areas. Earth and Environmental Sciences Library. *Springer*, Cham.

Antunes MDC, Chalfun NNJ, Pasqual M, Dutra LF, Cavalcante-Alves JM, 2003. Factors affecting rooting of figs (Ficus carica L.) cuttings. *Acta Hort*, **605**: 141–146.

Awasthi OP, Singh IS and Sharma BD, 2006. Effect of mulch on soil hydro thermal regimes, growth and fruit yield of brinjal under arid conditions. *Indian J. Hort. Sci*, **63(2)**: 192-194.

Banerjee A and Rana D, 2019. Leaf Spot Disease of Syzygium cumini Caused by *Neopestalotiopsis clavispora* Newly Reported in India. *Plant Disease*, 104. 10.1094/PDIS-09-19-2018-PDN.

Bharathi VP and Basavaiah, 2022. Growth and Yield Performance of Mulberry (*Morus Spp.*) Cultivar V-1 as Small Tree in Karnataka. *Mysore Journal Agriculture Science*, **56(2)**: 83-91

Chadha KL, 2009. Handbook of Horticulture. Indian Council of Agricultural Research (ICAR), New Delhi.

Chand L, Patidar J, Anuragi H, Taria S, Handa AK, Dev I and Tiwari RK, 2021. Fig (*Ficus carica*): A potential fruit crop for agroforestry systems in arid and semi-arid regions. *Kerala Karshakan e-journal,* **8(8)**: 4-11.

Chundawat BS, 190. Arid Fruit Culture; *Oxford and IBH Publishing* Co. Pvt. Ltd.: New Delhi, India.

Gautam GK, Vidyasagar G and Dwivedi SC, 2012. Study on Medicinal Plants from Indian Origin-A Text Book of Indian Medicinal Plants; *Lambert Academic Publication*: Chisinau, Republic of Moldova, pp. 5–7.

Gopalan C, 2002. Nutritive Value of Indian Foods; National Institute of Nutrition, *Indian Council of medical research*: Hyderabad, India, p. 53.

Gora JS, Kumar R, Kumar A andChahal VP, 2022. Phlasa cultivation. In Book: Phal Phool, published by Datta, T. K., ICAR–CBRI, Hisar, Haryana. pp. 19-22.

Haldhar SM, Maheshwari SK, 2018. Insect-pests management in arid and semi-arid horticultural crops. Technical Bulletin No: 64, ICAR–CIAH, Bikaner.

Haldhar SM, Gupta A, Gora JS and Mani M, 2022. Pests and Their Management in Other Arid Zone Fruit Crops. In: Mani, M. (eds) Trends in Horticultural Entomology. *Springer*, Singapore. doi.org/10.1007/978-981-19-0343-4_34.

Prasath H, Balasubramanian CN and Radhakrishnan S, 2019. Effect of Canopy Management (Pruning) on Fruit Yieldin Tamarind Plantation at Harur Taluk, Dharmapuri, Tamil Nadu, India. *Indian Journal of Plant Soil.* **6(1)**: 21-24.

Hiwale SS, Dhandar DG and Bagle BG, 2008. Vegetative propagation of wood apple (*Feronia limonia* Correa). *Indian Journal of Agroforestry*, **10(2)**: 58-61.

Jindal SK and Singh M, 2005. *Salvadora* in the degraded saline habitat of arid region. In Shrubs of Indian Arid Zone (Eds. Pratap Narain, Manjit Singh, M.S. Khan and Suresh Kumar), pp. 83-86, CAZRI, Jodhpur.

Jindal SK, Singh M and Sivadasan R, 2006. Variability for Seed Size, Oil Content, Seed Germination and Juvenile Traits in Germplasm Collections of *Salvadora oleoides* Decne from Arid Parts of Gujarat. *Annals of Arid Zone*, **45(1)**: 53-58.

Krishna H, Saroj PL, Maheshwari SK, Singh RS, Meena RK, Chandra R and Parashar A, 2019. Underutilized fruits of arid and semi-arid regions for nutritional and livelihood security. *Int. J. Minor Fruits Med. Aromat. Plants,* **5**: 1–14.

Krishnaswami S, 1986. Mulberry Cultivation in South India. Central Silk Board, Govt. of India, Ministry of Textiles, Bangalore.

Kumar A and Chander S, 2024. Effect of grafting on success and survivability of jamun (*Syzygium cumini* Skeels.) varieties. *Journal of Horticultural Sciences, 19*(1).

Kumar K and Haldhar SM, 2021. Suitable genotypes of underutilized fruit crops for pest management. In: Pest management in dry land horticultural crops (Eds. Haldhar, S.M. and Maheshwari, S.K.). Biotech Books, Darya Ganj, New Delhi. pp. 155-171.

Kumar K, Sarolia DK, Saroj PL, Samadia DK and Singh D, 2022. Standardization of seed, rootstock and plant standards of lasoda (*Cordia myxa* L.) for commercial orcharding in India. *J. Environ. Biol.,* **43**: 267-275.

Kumar S, Deswal M and Laura J, 2014. Study on seed germination and vegetative propagation of *Salvadora oleoides* Decne. *International Journal of Current Research*, **6**: 8058-8062.

Kumar S, Lal N and Nath V, 2017. Tendu (*Diospyros melanoxylon* (roxb). In book: Underutilized Fruit Crops: Importance and Cultivation written by Ghosh et., published by Jaya Publishing House Publisher and Distributor 27-B, Pocket-B, Dilshad Garden, New Delhi. Part-III, chapter 62, pp. 1327-1338.

Kumar S, Laura JS and Singh N, 2016. A comparative in vitro propagation studies on different explants of *Salvadora oleoides* Decne. An endangered plant. *Int J Curr Microbiol Appl Sci,* **5**: 699–706.

Liguori G and Inglese P, 2015. Cactus pear fruit production: orchard planting and management of *Opuntia ficus-indica*. Special Issue 14. Department of Agricultural and Forest Sciences, Università Degli Studi di Palermo, Viale delle scienze, ed. 4, ingr. H, 90128 Palermo, Italy.

Maheshwari SK and Haldhar SM, 2018. Disease management in arid horticultural crops. CIAH/Tech./Pub. No. 68, ICAR–Central Institute for Arid Horticulture, Bikaner, Rajasthan, India, P. 42.

Mahla HR and Singh JP, 2013. Assessment of in-situ Variability in Kair (*Capparis decidua*) Germplasm for Utilization in Genetic Improvement through ex-situ Conservation. *Ann. Arid Zone*, **52**: 109-112.

Malik SK, Chaudhury R, Panwar NS *et al.* (2012b). Genetic resources of Chironji (*Buchanania lanzan* Spreng): a socio-economically important tree species of central Indian tribal population. *Genet Resour Crop Evol,* **59**: 615–623.

Malik SK, Choudhary R, Kumar S, Dhariwal OP, Deswal RPS and Chaudhury R, 2012a. Socio-economic and horticultural potential of Khirni [Manilkara hexandra (Roxb.) Dubard]: a promising underutilized fruit species of India. *Genetic Resources and crop evolution*, **59**: 1255-1265.

Meena VS, Gora JS, Singh A, Ram C, Meena NK, Pratibha Rouphael Y, Basile B and Kumar P, 2022. Underutilized Fruit Crops of Indian Arid and Semi-Arid Regions: Importance, Conservation and Utilization and Strategies. *Horticulturae*, **8(2)**: 171.

Meghwal PR, 2007. Propagation studies in lehsua (*Cordia myxa*). *Indian J. Agri. Sci.*,**77(11)**: 765–767.

Meghwal PR and Vashishtha BB, 1998. Effect of Time of Planting and Auxins on Rooting of Cutting of Ker (*Capparis decidua* Forsk.). *Annals of Arid zone,* **37(4)**: 401-404.

Nallathambi P, Umamaheswari C and Singh RS, 2005. Foot rot in cactus pear (*Opuntia spp.*) genotypes under arid conditions. *Annals of arid zone*, **44(1)**: 59-63.

Pareek OP and Purohit AK, 2002. Patch budding in Khejri (*Prosopis cineraria*). *Indian Journal of Horticulture*, **59**: 89-94.

Pareek S, Mukherjee S and Paliwal R, 2007. Floral biology of ber—A Review. *Agric. Rev.*, **28**: 277–282.

Purohit AK, Samadia DK and Pareek OP, 2003. Studies on growth enhancement of Khejri (*Prosopis cineraria* (L.) Druce) seedling and in-situ budded plants. *Annals of Arid Zone*, **42(1)**: 69-73.

Samadia DK, Haldhar SM, Verma AK, Gurjar PS, Berwal MK, Gora JS, Kumar R and Ram H, 2021. Khejri (*Prosopis cineraria*) research for horticultural harnessing and environmental services: an appraisal. *Journal of Agriculture and Ecology*, **12**: 1-26.

Sarolia DK, Samadia DK, Choudhary BR and Singh D, 2018. Production of quality seed and planting materials. CIAH/Tech./Pub. No. 75. ICAR–Central Institute for Arid Horticulture, Bikaner, Rajasthan (Indian).

Sathyakumar S and Sivakumar K (Eds.), 2007. Galliformes of India. ENVIS Bulletin: Wildlife and Protected Areas, Vol. 10 (1). Wild life Institute of India, Dehradun, India. 252.pp.

Satya SS and Narina Catanzaro CJ, 2018. Tamarind (*Tamarindus indica* L.), an Underutilized Fruit Crop with Potential Nutritional Value for Cultivation in the United States of America: A Review. *Asian Food Sci. J.,* **5**: 1-15.

Sharma S, 2020. Reproductive Biology of *Capparis decidua* (Forsk.) Pax. *Indian J. Applied and Pure Bio.,* **35(1)**: 71-78.

Shekhawat N, Mohnot S, Phulwaria M and Harish Shekhawat S, 2012. Micropropagation of *Salvadora oleoides*-an oil yielding tree of arid forests. *J. Sustain For.,* **31**: 620–632.

Singh AK, 2018. Propagating arid fruits commercially. *Indian Horticulture,* **5**: 82-88.

Singh AK, Singh S and Saroj PL, 2018. The Bael (Production Technology). Technical Bulletin No. CIAH/Tech./Pub. No. 67 published by Director, ICAR–CIAH, Bikaner. Pp. 1-55.

Singh AK, Singh S, Singh RS, Joshi HK Sharma SK and Sisodia PS, 2012. Production technology of bael under rainfed conditions of western India. In: Compendium of Winter School on Exploitation of Underutilized Horticultural Crops for Sustainable Production, held at CIAH, Bikaner, pp. 124-135.

Singh A, Prakash J, Meghwal P and Ranpise SA, 2015. Fig (*Ficus carica* L.). In book: Breeding of Underutilized Fruit Crops. Part I, Edition: I, Chapter: 12. Publisher: Jaya Publishing House New Delhi. Pp. 149-179.

Singh D and Singh RK, 2011. Kair (*Capparis decidua*): A potential ethnobotanical weather predictor and livelihood security shrub of the arid zone of Rajasthan and Gujarat. *Indian Journal of Traditional Knowledge,* **10(1)**: 146-155.

Singh RS and Singh V, 2003. Growth and development influenced by size, age and planting methods of cladodes in Cactus pear (*Opuntia ficus-indica* (L.) Mill.). *Journal Proceeding of Association on Cactus Development,* **5**: 47-54.

Singh S, Singh AK, Apparao VV and Bhargava R, 2015. Thar Returaj: a new khirni variety. *Indian Horticulture,* **60(6)**: 14-15.

Singh S and Singh AK, 2006. Standardization of method and time of propagation in jamun (*Syzygium cuminii*) under semi-arid environment of western India. *Indian Journal of Agricultural Sciences,* **76(4)**: 242-5.

Singh S and Singh AK, 2015. Standardization of time of softwood grafting in mahua (*Bassia latifolia*) and khirni (*Manilkara hexandra*) under semi-arid environment of western India. *Indian Journal of Agricultural Sciences,* **85(2)**, pp.166-70.

Singh S, Singh AK and Appara VV, 2006. Genetic diversity in chironji (*Buchanania lanzan*) under semi-arid ecosystem of Gujarat. *Indian Journal of Agricultural Sciences,* **76(11)**: 695-698.

Singh S, Singh AK, Bagle BG and More TA, 2008. Mahua - A multipurpose tree for tribals. Technical Bulletin, CIAH, Bikaner, pp. 4.

Tewari RK, Bajpai CK and Kumar S, 2011. Bench grafting in ber (Zizyplzus spp). *Indian Journal of Agricultural Sciences*, **77(4)**: 104-112.

Yadav PK, Gora JS and Kumari L, 2020. Cordia. In book: Production Technology of Tropical and Subtropical Fruits. Publisher: New India Publishing Agency 101, Vikas Surya Plaza, CU Block, LSC Market Pitam Pura, New Delhi - 110 034, India. Pp. 360-367.

Yadav V, 2018. Cultivation of custard apple fetches more. *Indian Horticulture*. V. **63(5)**: 42-46.

Yadav V, Hiwale SS, Singh AK and Rao VA, 2020. Thar Gaurav: High-yielding wood apple variety for dry land. *Indian Horticulture*, **65(5)**: 36-39.

Chapter 5

Propagation of Major Tropical Fruit Crops

Maheswarappa H.P., Prakasha D.P., Pavan, P.R. and Akshatha, H.

Directorate of Research, University of Horticultural Sciences, Bagalkot, Karnataka

Fruit plants are a crucial component of the development of the fruit industry. The lengthy gestation time of fruit crops, however, necessitates extreme caution in the choice of planting material and the adoption of appropriate technologies for its mass replication, as any errors made during the early construction of an orchard could cause significant financial loss down the road. The artificial or natural dispersal of plants is sometimes referred to as plant propagation. When propagated from seeds, the majority of fruit plants do not yield true to type progeny. The plants are housed in a designated location known as the nursery before being planted in the field. Fruit plants are therefore fostered in fruit plant nurseries during their formative stages until they are robust enough to be planted in orchards. High-quality, consistent, and true-to-type planting material is essential for commercial fruit production. The act of producing new plants from seeds, cuttings, bulbs, and other plant parts is known as propagation. In essence, all plants in this cosmos reproduce either sexually (via seeds) or asexually (by vegetative growth). However, more methods for the quicker and more effective multiplication of these plants have now been devised by man. This chapter deals with propagation of major fruit plants grown in the tropics of the country.

Plant propagation involves producing new plants from a selected parent plant. It is crucial for the cultivation of tropical and subtropical fruit crops, as it preserves desirable traits and enhances disease resistance. Clonal methods like grafting and tissue culture ensure genetic uniformity and allow the selection of disease-resistant rootstocks, leading to healthier crops. These techniques also help conserve biodiversity by preserving genetic material for breeding new varieties. (Kaviani *et al.,* 2022)

Additionally, propagation boosts the economic viability and sustainability of fruit production. Techniques such as grafting shorten the time to fruiting, providing quicker returns for farmers. Tissue culture enables mass production of disease-free plants, improving yields and reducing the need for more agricultural land, thus supporting sustainable farming practices (El-Sherif *et al.,* 2019). There are two main methods of plant propagation: asexual and sexual.

Sexual propagation is a natural process where a parent plant produces seeds, resulting in offspring that are usually not genetically identical to the parent. This method is seldom used for fruit trees unless there are no known superior cultivars, you are growing rootstocks for grafting, or the seeds yield plants very similar to the parent. Growing tropical fruit trees from seeds typically requires many years before the tree matures enough to bear fruit, although there are exceptions, such as the papaya (McKey *et al.,* 2010).

Asexual propagation is used to preserve specific selections of known identity and quality. This method includes techniques such as division, cuttings, air-layering, and grafting, all of which produce plants genetically identical to the parent plant. Because fruit trees often possess desirable traits, they are commonly propagated asexually to maintain these characteristics. These trees with particular traits are known as distinct cultivars. It is important to label the plants you are propagating to keep track of their cultivars as they grow (Singh *et al.,* 2021). Asexually propagated trees can start producing fruit as soon as they are physically capable of supporting the fruit weight.

The commercial methods of propagating various fruit trees depend not only on the availability of planting materials but also on the potential of these materials to produce quality fruits within a short timeframe. Fruit trees propagated from seeds generally take longer to bear fruit compared to those derived from asexual propagation (Teskey *et al.,* 2012).

Mango

Mangoes are among the most widely cultivated tropical fruits globally due to their economic and nutritional value, appealing appearance, and exotic flavour. Successful mango cultivation relies on vigorous and productive cultivars, which necessitate effective propagation techniques and advanced grafting plant management practices (Sivakumar *et al.,* 2011).

Mango trees can be propagated through both sexual and asexual methods. Sexual reproduction occurs when male and female sex cells unite during flowering. In mango trees, this involves pollen from the male plant transferring to the stigma of the female plant, leading to fertilization. The pollen tube carries male gametes to the embryo sac in the ovule, where one gamete fertilizes the female gamete to form a diploid zygote, while the other fuses with polar nuclei to create the triploid endosperm (Galán *et al.,* 2021). This process results in seeds that may have varying genotypes due to meiosis, which involves chromosome reduction.

A seedling grown from a single sexual embryo of a mono-embryonic mango cultivar does not guarantee a true-to-type mango tree, a fact well understood before vegetative propagation methods were developed. In contrast, asexual or vegetative propagation, which uses mitosis for nuclear cell division, produces an identical clone of the parent plant. The maintenance of purity is easy in asexually propagated fruit crops as compared to seed propagated ones (Rajmohan *et al.,* 2005). Still, it requires a close monitoring at different stages in the nursery to avoid mixing with other varieties.

Mango nurseries have historically varied in their approach to seedling establishment and propagation. There are three main types:

1. **Field-based nurseries:** Seedlings are grown directly in the field, with minimal protection and often using outdated irrigation systems.
2. **Partially protected nurseries:** Seedlings are grown in beds covered with wood laths or coconut leaves, offering some shade and using older irrigation methods.
3. **Modern nurseries:** Seedlings are established in soil substrates within polyethylene bags, in controlled environments with advanced irrigation systems.

The transition from the first to the third type of nursery reflects the growing demand for high-quality mangoes in international markets, particularly in arid tropical countries like India. Wealthier producers have adopted more advanced nursery methods. In the early 80s, most nurseries used the first type, resulting in lower-quality grafted mangoes. Many Asian countries still use field-based nurseries to cut costs, though this approach often leads to significant plant losses (>20%) during transplanting due to poor root systems. In the two nursery types mentioned, all stages from sowing to grafting occur in seedling beds. For new mango nurserymen, it is advisable to use sandy soil, as it facilitates easy removal of grafted plants without damaging the roots (Nath, 2014). Sometimes, nurserymen select the best seedlings from these beds and transplant them into polyethylene bags, where they continue to grow until they reach the appropriate height and stem diameter for grafting.

Mother Plant Selection

When selecting bud sticks or graft wood, choose from healthy, true-to-type progeny trees of commercial or new varieties, ensuring they are free from viruses, diseases, and pests. A nurseryman should maintain a progeny block with promising cultivars suitable for the local area, and this orchard should receive meticulous care and supervision (Singh *et al.*, 2021).

Criteria for selecting mother plants:

- The plants must be genetically true to type.
- They should be healthy, free from diseases and pests.
- Known pedigree history should confirm their bearing potential, fruit quality, and any issues.
- They should be precocious and prolific bearers.
- Commercial fruiting should be avoided to ensure a continuous supply of scions.

Criteria for selecting rootstocks:

- Prefer dwarfing or semi-dwarfing rootstocks that are vigorous.
- Ensure compatibility with known commercial varieties.
- They should be resistant or tolerant to diseases, pests, and abiotic stresses.
- The rootstock should have a well-developed root system.
- It should be easy to propagate vegetatively.

Raising Rootstocks

Use mango seedlings from stones of local, dwarf, disease-free, high-yielding trees collected during July-August. Due to their recalcitrant nature, mango seeds lose viability quickly if not sown soon after removal. Store seeds in a moist, shaded environment if needed. Before sowing, immerse the stones in water and discard any that float. Sow stones from May to August in beds with well-decomposed farmyard manure. After about two months, transplant seedlings into polybags, ensuring proper irrigation. Watch for shoot borer infestations, which can damage young leaves and affect seedling growth. Plant is allowed to grow as a single stem for six to eight months and such type of plant is mostly preferred as rootstock. When the plant attains pencil thickness or a little more, it should ready for grafting (Kumar, 2012).

Collection of Scions

Proper selection and preparation of scions are crucial for successful grafting. Ideally, scions should be terminal shoots that are 3 to 4 months old and not yet flowered. Defoliate the scion on the mother plant 7 to 10 days before detaching, leaving part of the petiole on the shoot (Garner, 2013). Ensure the thickness of the scion matches that of the stock.

Methods of Propagation

Initially, mango propagation relied on seeds, but since they don't produce true-to-type plants, vegetative methods have become preferred. Polyembryony, though capable of producing true-to-type seedlings, is rarely used. Traditional inarching has been largely replaced by veneer grafting, which is highly successful in North and Central India. Stone (epicotyl) grafting works well in humid, moderate climates like Eastern India. Softwood grafting is now common in South and North India for commercial purposes. Both softwood and veneer grafting are effective for large-scale propagation, and the use of polyhouses and net houses has extended propagation periods across various regions (Naik *et al.,* 2014).

Softwood grafting is becoming a popular method for mango propagation in many states, involving grafting on 3–8-month-old rootstocks. Scion shoots, 10-15 cm long and matching the rootstock thickness, are defoliated 7-10 days before grafting. A longitudinal cut with wedge-shaped cuts on the scion is made, and the grafts are secured with polyethylene strips. After 45-50 days, when the unions are complete, the strips are removed. This technique thrives in high humidity and moderate temperatures (May-September) and is favoured for its ease, high success rate, efficiency, and cost-effectiveness. However, it requires skilled labour and may have limited success with certain varieties like Himsagar.

Inarching and veneer grafting in North India are time-consuming, but stone grafting offers a quicker, simpler, and more cost-effective alternative. Fresh mango stones are sown in nursery beds and, after germination, 10–15-day-old seedlings with tender stems are used. The seedlings are treated with a Carbendazim solution, and their stems are trimmed and cut. A wedge-shaped cut is made on a 4–5-month-old scion stick, which is then inserted into the seedling cleft and secured with a polyethylene strip. The grafts are kept in polyethylene bags under shade until the scions sprout and leaves turn green, after which they are planted in nursery beds. July is the optimal time for stone grafting (Kumar, 2012).

Characters	*Standards*
Type of root stock	Straight and Vigorous growth
Raising of root stock	Polyethylene bag
Age of root stock	10 to15 days
Diameter of root stock	0.2-0.3cm
Age of scion shoots	3-4 months old
Diameter of scion	0.3-0.4 cm
Length of Scion stick	10-15 cm
Plant Height	20 cm
Stem girth	1.5-2 cm
Union Height	5-7 cm
Time of grafting	June-August (vary with the availability of fresh seed)
Union maturity for strip removal	40-45 days

Source: Prakash and Singh (2015).

Care of Nursery and Grafted Plants

Transplant young mango seedlings into 6"x5" polybags, ensuring they are pest, disease, and weed-free. High light intensity, especially from March to April, can harm young plants, so provide shade using thatches, polythene, or shade nets. Irrigate every 4-6 days from February until the monsoon starts. Keep grafted plants in a clean area, regularly inspecting for new bud and shoot growth. After a successful graft union, remove the polythene strip for proper growth. Remove any leaves or buds that emerge below he graft union to avoid hindering the grafted plant development.

Figure 5.1: Grafted Mango Saplings.

Banana

Banana cultivation is prevalent in tropical and subtropical regions, serving as a vital staple food and income source for small-scale growers. Known as one of the oldest cultivated fruit crops, bananas offer the advantage of being harvestable year-round. They are often called "Kalpataru – the plant of virtues" due to their socio-economic significance. However, banana productivity and lifespan have declined due to improper cultivation, poor-quality planting materials, and pest and disease stress (Gowen, 2012).

Farmers typically rely on natural regeneration for planting materials, which fails to meet high demands. Quality planting material is crucial for successful cultivation, with suckers being the primary propagation method. A persistent issue for growers is the insufficient supply of quality planting materials. Although tissue culture techniques can propagate bananas efficiently and eliminate pests and diseases, the high cost and complexity of establishing tissue culture labs make it inaccessible for many farmers, particularly in India and the northeastern region. To increase production, there is a need for affordable and simple techniques like macro-propagation, which can rapidly multiply plant materials and boost income for small-scale farmers, ensuring sustainability through a banana-based farming system (Chazdon *et al.,* 2016).

Macropropagation Technique

Macro-propagation is a relatively easy technique that can be performed in a shed or even in an open shed. It involves generating suckers from clean planting material by removing the apical dominance. This process includes the following methods (Angami *et al.,* 2019):

1. Decapitation (False and complete decapitation)
2. Corm technique (whole, split and excised corms)

Decapitation

False Decapitation

- ✰ Cut a small hole in the pseudostem of six-month-old plants to destroy the meristem.
- ✰ The hole, made about 20 cm above the ground, should slope downwards to collect water and sap, further killing the meristem.

Complete Decapitation

- ✰ Completely cut down the pseudostem of six-month-old plants at ground level, leaving the emerging suckers.
- ✰ Destroy the meristem with a clean knife, removing a 5 cm diameter growing part in the middle of the pseudostem.
- ✰ Ensure the meristems do not sprout again. (Bhende *et al.,* 2016)

Sprouting

- ✰ Both false and complete decapitated plants are left to stand for at least one month to allow sprouting.
- ✰ About 4-8 suckers, depending on banana cultivars, will sprout in three weeks after removing apical dominance.
- ✰ Sprouted suckers are detached once they attain 3-4 leaves and are transferred directly to the field.

Figure 5.2: Procedure for Macropropagation in Banana.

2. Corm Techniques

Whole Corm

- Remove roots and cut leaf sheets one by one to expose the buds.
- Apply fungicides if necessary.
- Scarify the top of the corm by cutting an X.
- Plant the entire corm at 30 cm spacing in a propagator (constructed of bamboo and polythene sheet of 1.5 m x 5.0 m x 0.5 m size) well-covered with sawdust and water immediately after planting (Singh *et al.,* 2021).

Split Corm

- Fragment the corm into two or more pieces and plant in the propagator for buds to sprout.
- Plant prepared corm pieces at 10 cm intervals and cover with 2 cm sawdust. Water immediately after planting.

Excised Corm

- Directly cut buds from the corm into pieces of 50-100 g as minisetts and plant in the propagator or polyethylene bags to sprout.

Benefits

- Macro-propagation offers an affordable, simple, and relatively rapid technique for vegetative multiplication of bananas.
- With proper training, it can help farmers overcome challenges such as the unavailability of planting materials, ensuring affordable, quality, true-to-type, and preferred cultivar planting materials (Opata, 2020).

Micropropagation Technique

Shoot tip culture has been routinely employed for the rapid clonal propagation of banana genetic resources since 1985. Micro propagated plants tend to establish more quickly, grow more vigorously, have a shorter and more uniform production cycle, and yield more compared to conventional propagules. Although there are numerous reports on banana micropropagation via shoot tip culture, plants can exhibit significant variation under in vitro conditions. This variation in shoot establishment, shoot proliferation, and regeneration of shoots and roots is influenced by several factors, including genotype, explant type, culture media composition, plant growth regulators (PGRs), and the culture environment (Vuylsteke, 1989).

Culture Media Preparation and Sterilization (Amente *et al.*, 2022)

- Modified Murashige and Skoog (MS) basal medium used.

- ☆ Macronutrients, micronutrients, vitamins, 30 g/L sucrose, and 7 g/L agar-agar type-I dissolved in double distilled water.
- ☆ pH adjusted to 5.8 with 1N NaOH and/or 1N HCl.
- ☆ Medium poured into 60 ml culture jars, autoclaved at 1.2 KPa and 121°C for 20 min, cooled, and checked for microbial contamination over 72 hours.

Explant Preparation and Surface Sterilization (Singh *et al.*, 2011)

- ☆ Young suckers of three banana varieties used as explants.
- ☆ Trimming and washing in tap water with detergent for 15 min.
- ☆ Explants dipped in mancozeb fungicide solution (3 g/L) for 10 min, rinsed in autoclaved distilled water.
- ☆ Sterilized with 2 per cent sodium hypochlorite for 15 min, followed by 70 per cent ethanol for 1 min, rinsed in autoclaved distilled water.
- ☆ Explants excised to final size (5 mm) under a laminar air flow hood.

Shoot Tip Culture Establishment (Dagnew *et al.*, 2012)

- ☆ Surface sterilized explants placed on MS medium with BAP (1.5 and 2.5 mg/L) and IAA (0.0 and 0.5 mg/L).
- ☆ 30 culture jars per treatment, incubated for 8 weeks at 25±2°C under 16/8 hr light cycle.

***In vitro* Shoot Proliferation (Al-Amin *et al.*, 2009)**

- ☆ Initiation medium with BAP and IBA resulted in hard meristematic ball-like structures.
- ☆ Shoot tips turned brown shortly after inoculation, but external leaf primordia turned green after four weeks.
- ☆ Globular hard coat mass developed into adventitious plantlets.
- ☆ Maximum shoot tip initiation (100 per cent sprout) observed in MS medium with 3.0 mg/L BAP in Grand Naine variety.
- ☆ After 8 weeks, shoots with at least one leaf transferred to multiplication medium.
- ☆ Maximum shoots per explant obtained in MS medium with BAP/IBA concentrations of 3.5/0.0, 3.0/0.2, and 3.5/0.3 mg/L for Grand Naine.

***In vitro* Rooting of Shoots (Alagumanian *et al.*, 2015)**

- ☆ Fine roots appeared from the basal portion of shoots after 9-11 days.
- ☆ Highest number of roots per shoot:
- ☆ 8.0 roots with 1 mg/L IBA

- ☆ 7.67 roots with 1.5 mg/L IBA
- ☆ 8.0 roots with 2 mg/L IBA

Acclimatization and Hardening of Plantlets (Parvin, 2015)

- ☆ Rooted plantlets washed to remove rooting medium, disinfected with Ridomil (2.0 g/L) for 3 min.
- ☆ Transferred to polybags with sterile soil mixture (forest soil, farmyard manure, sand in 2: 1: 1/2 ratio).
- ☆ Soil mixture sterilized at 180°C for an hour.
- ☆ Transferred to larger polybags with forest soil and manure (2: 1 ratio).
- ☆ Hardened plants transferred to the field for establishment and evaluation.

Citrus

India is the second-largest fruit producer globally, following China, with citrus being the third most important commercial fruit after mango and banana. The citrus genus includes mandarin, lime, lemon, grapefruit, pummelo, and citron, along with intergeneric and interspecific hybrids such as Fortunella and Poncirus. Mandarin leads with a 43 per cent share in India's citrus production, followed by sweet orange (25%) and acid lime (7%). Citrus is cultivated in over 49 countries, predominantly in tropical and subtropical regions between 40°N and 40°S. Commercial citrus production faces several challenges, including low seed germination, minimal success in budding and grafting, issues with flowering, fruit set, size, quality, premature fruit drop, pest and disease occurrence, and physiological disorders (Kamatyanatti and Singh, 2019). These issues can be addressed using advanced techniques such as quality planting materials, virus-free planting materials and high-density planting (HDP).

Advanced Nursery Techniques for Citrus

Budwood Certification System

Citrus crops are highly susceptible to viral diseases and are primarily propagated through budding (T budding) and seeds. Producing virus-free plants requires the use of registered and certified budwood from disease-free mother plants. Common diseases in citrus include those that are graft and vector transmissible. Essential tools for quality citrus production include early virus detection (clean stock programs), and effective disease management. Virus cleaning can be achieved through shoot tip grafting, budwood certification programs, and tissue culture to provide disease-free planting materials. Cross-protection for Citrus Tristeza Virus (CTV) is managed by isolating and inoculating milder virus strains. Citrus plants also require micronutrients for long-term sustainability. Maintaining nursery hygiene and identifying disease-free mother trees for scion wood are critical. The budwood certification program, as suggested by NRC for Citrus in Nagpur, plays a vital role in this process (Vidalakis *et al.,* 2010).

Method of Budwood Certification (Norman, 1957)

- Selection of mother plant: Based on pedigree, yield, quality, and pest and disease incidence.
- Detection of virus/viroid: Using ELISA methods.
- Protection of mother plant: Virus-free trees should be covered with insect nets.
- Safe propagation: Ensure cleanliness of implements and gardeners; tools should be sterilized before use.
- Protection of planting material: Budded plantlets should be kept inside a net house to prevent insect vector contamination.

Citrus Nursery Nematode Certification Programme

This program involves three phases: site approval, pit approval, and pre-movement certification of young citrus trees before they are moved from nurseries to orchards. Potential nursery sites are inspected for nematode pests, and certified if found free. Detailed examination and risk analysis of intercepted nematodes are conducted, followed by appropriate disposal or salvaging of plant samples using methods such as fumigation, hot water treatment, or nematicide dips (Inserra *et al.,* 2005).

Steam Sterilization of Media

Steam sterilization is a quick method for potting mixture sterilization, involving passing steam through the mixture for about 20 minutes to kill all pathogens and weeds (Bunt, 2012).

Graft Transmissible Dwarfing of Citrus

This technique, pioneered in Australia, uses citrus viroid to control tree size without adversely affecting cropping efficiency or fruit quality. It is effective for high-density planting, reducing tree size by 33 to 43 per cent in specific rootstock-scion combinations (Broadbent *et al.,* 1988).

Solutions to Citrus Problems

- Use virus-free propagating material through nucellar seedlings.
- Employ tolerant rootstock and compatible stock-scion combinations resistant to viruses, diseases, and nematodes.
- Consider factors such as varietal susceptibility, seed purity, and active insect vectors in the growing area.
- Establish clonal progeny orchards with tested rootstock and scion clones.

Commercially, Citrus species are propagated through various methods: Acid Lime and Rangpur lime by seed, sweet lime by layering and hardwood cuttings, Persian lime by ground or air layering, lemon by cutting, and sweet orange and

Mandarin by budding. In Kinnow Mandarin, wedge and side grafting showed the highest graft union with 90 and 86.27 per cent success rates respectively, while tongue grafting had a 76.67 per cent success rate. Micropropagation of Kinnow Mandarin using nodal segments has been standardized, and standard protocols for Sweet Orange cultivar Blood Red have been established. Veneer grafting trials in Poncirus trifoliata with Mandarin cultivars 'Dhankuta Local' and 'Ota Ponkan' were highly successful in Nepal mid hills. *In vitro* propagation was also successful in Pummelo (Singh *et al.,* 2018).

Budding

In budding, a dormant vegetative bud is attached to an incision made on the rootstock. This method produces uniform and true-to-type plants, unlike seed propagation, which may yield inferior quality seedlings and variations in orchard plants. Compared to grafting, budding forms a stronger union, making the plants more resistant to damage from storms and strong winds. It is also a quicker and more efficient propagation method that does not require professional expertise. The budding process involves several steps, starting with selecting a mother tree and scion wood, and culminating in the delivery of the budded plants to the orchard site (Bordoloi *et al.,* 2022).

Figure 5.3: Budding in Citrus.

Propagation by Cuttings

Propagation by cuttings involves using scions to create clonal plants, identical to their parent trees, making it a vegetative propagation method. While not widely used for citrus nursery production, cuttings have been successful in specific instances. For example, in Sicily, citron (*Citrus medica* L.) has been propagated this way, and in Florida, the technique was employed to produce 'Tahiti' lime (*Citrus latifolia* Tanaka) with over 90 per cent rooting success under nebulization. (Singh *et al.,* 2018)

Rooted cuttings tend to develop a more superficial root system compared to seed-grown plants, potentially making them more susceptible to tip-over after planting. These plants are often smaller and may have poor adaptation to varying soil conditions and variable resistance to abiotic and biotic stress factors. However, cuttings can be beneficial for propagating rootstocks in varieties with low seed counts, like 'Sunki' mandarin and 'Swingle' citrumelo, useful for producing grafted trees or for inarching (Carvalho *et al.,* 2019).

Recent evaluations in Florida have shown that root cuttings can effectively propagate seedless or low-seed materials, accelerating the production of certain rootstocks and budwood for nurseries.

Guava

Guava (*Psidium guajava* L.), part of the Myrtaceae family with over 80 genera and 3,000 species, is widely distributed in the tropics and subtropics, especially in the Americas, Asia, and Australia. The genus *Psidium* has about 150 species, with *P. guajava* being the most well-known globally (Rajan *et al.,* 2019). Guava is among the top 50 edible tropical and subtropical fruits and is commercially important in over 50 countries.

Guava flowers favour self-pollination, but cross-pollination, primarily by bees and insects, occurs at an estimated 35 per cent and varies among genotypes, environmental conditions, and pollinators. Seed propagation results in genetically heterogeneous plants, which is not ideal for high-productivity commercial orchards. Vegetative propagation methods have been extensively studied and their commercial adoption depends on ease, cost, technology transfer, and production chain organization (Pereira *et al.,* 2017). Significant advances in plant cloning over the last century include improved health and disease control, water loss reduction in cuttings, growth regulators, understanding juvenility in propagation, and the use of vegetative methods to avoid pathogen-related diseases.

Guava flowers primarily self-pollinate, with an estimated 35 per cent cross-pollination by bees and other insects, varying by genotype and environment. Seed propagation produces genetically heterogeneous plants, unsuitable for high-yield commercial orchards. Vegetative propagation methods have been extensively studied, with adoption influenced by ease, cost, technology transfer, and production chain organization. Significant advances in plant cloning over the last century include improvements in health and disease control, water loss reduction in cuttings, growth regulators, understanding juvenility in propagation, and the use of vegetative methods to avoid diseases caused by viruses and other pathogens (Omar *et al.,* 2021).

Propagation Techniques

Seeds of guavas are generally considered a waste product and are mainly used in breeding programs or rootstock production for grafting. In countries lacking high technology, guavas are still propagated by seed without grafting. This method, while simple, results in plants with significant variability, leading to low productivity and fruit quality, which is undesirable for commercial orchards. In some guava-producing countries that use grafting, seeds are used for rootstock production. This method has downsides, including genetic variability in root systems and a longer production time for seedlings (1.5 to 2 years) compared to stem cuttings (six to eight months). For seed propagation, ripe fruits are harvested, seeds are cleaned, and dried for 10 days before being sown in nurseries or plastic bags with well-draining substrate. Germination rates can exceed 90 per cent within 15 to 20 days. For grafting, seedlings need a diameter of 12-20 mm and a height of 20 cm (Pereira *et al.,* 2017).

Air layering was once a common method for propagating guava trees but is no longer widely used commercially due to the development of guava wilt disease resistant rootstocks. However, it can still be used effectively. The best time for air layering is from August to February. Select a strong, straight shoot with at least 500 mm length from the tip, remove a 25-40 mm wide ring of bark, and allow the cambium layer to dry for 2 days. Cover the wound with damp sphagnum moss or a 50: 50 mix of peat moss and sterilized decomposed humus, then tie with PVC film (Pereira *et al.,* 2017). Rooting hormones are unnecessary. Roots develop in 2-3 months, after which the layers can be removed and planted in 5-liter plastic bags until ready for orchard planting.

Figure 5.4: Guava Seedlings.

Grafting can be done when the plants reach a height of 200 mm and a diameter of 5-7 mm, typically using the whip grafting technique. Before grafting, plant material should be treated for fungal diseases. The graft wood must be the same thickness as the rootstock or slightly thinner at the grafting point. After cutting the graft wood from the mother plant, leaves should be removed immediately. Both the rootstock and graft wood should have straight, smooth cuts made with a sharp grafting knife for a perfect fit. Once attached, the graft union should be covered with perforated PVC grafting tape, ensuring proper alignment. The entire cutting is then covered with biodegradable Parafilm. Buds start to swell after 2-3 weeks, and when they break through the Parafilm and develop 6-8 nodes with hardened leaves, the grafting can be removed (Garner, 2013). Regularly remove all shoots below the graft union to prevent rootstock growth competition.

Topworking old trees in the orchard can be done effectively. The tree should be cut off at no less than 0.5 m from the ground, with the top treated with fungicide and the rest painted white with diluted PVA paint. Initially, shoots will emerge from the top of the cut trunk; these should be removed to encourage sprouting lower down (about 200 mm). Select the strongest shoots to form the desired structure and whip-graft six to eight of these shoots. Remove all other shoots to reduce nutrient competition. This method is preferred over direct grafting into the main trunk because it produces stronger grafts that are less likely to break in strong winds. Once the desired structure of 4-5 branches is achieved, remove all additional grafts and branches, and take off the grafting tape. The old trunk protruding above the new structure can then be sawn off and treated with a fungicide sealant (Suklabaidya and Mehta, 2019).

Papaya

The significance of papaya to the global economy is evident from its widespread cultivation, substantial production in tropical countries, and high nutritional value. Papaya has been long grown in home gardens in the tropics due to its adaptability to various soil and climate conditions. Although it is largely consumed locally, it holds significant potential for export. Propagation by seed is the most practical method for papaya; however, plants grown from seeds of open-pollinated flowers result in a mix of genotypes with considerable variation in disease susceptibility, fruit quality, and yield. In dioecious papaya cultivars, there is an equal probability of producing male and female plants, leading to the need to remove excess male plants. For gynodioecious cultivars, bisexual types that produce oblong fruits are preferred over the round fruits of female plants due to their higher market value (Senthilkumar *et al.,* 2014). Vegetative propagation can be an alternative to seed propagation to address these issues. Generally, plants produced through vegetative propagation are known to maintain the genotype of the cultivars accurately.

Seedling Variability in Papaya

Commercial papaya production typically relies on seed propagation, which results in significant variability in sex expression and fruit characteristics among plants, even within small plantings. This variability poses challenges for breeding and selection. Open pollination in seed propagation leads to undesirable genetic variations and a mix of genotypes. In addition to germination issues caused by inhibitory substances in the aril, seed extraction and conservation need further research, as seeds can lose their germinative capacity relatively quickly. Conventional production methods, where seeds are often harvested from commercial open-pollinated orchards, do not preserve the original characteristics of parent plants. Papaya plants propagated from seeds exhibit considerable variation in sex types (male, female and hermaphroditic), which leads to undesirable variation in fruit quality and type (Babu *et al.,* 2008).

Asexual Propagation

A key advantage of asexual propagation is that the sex of the plants is predetermined. In dioecious varieties, all productive plants are female, with only a few male plants needed for pollination. Asexual propagation allows for the development of highly productive, virus-resistant papaya plants while preserving the characteristics of the mother plants. Clonal propagation of selected mature female and male papaya plants is particularly beneficial for commercial use in subtropical regions. This method helps overcome the variability in fruit shape, size, flavour, and disease resistance found in dioecious lines. Vegetative propagation thus ensures the maintenance of the parent plant traits and offers benefits such as higher yield, reduced fruiting height, and extended cropping periods.

Figure 5.5: Grafted Papaya Seedlings.

Propagation by grafting in papaya aims to preserve desirable traits. the effectiveness of apical wedge grafting among six papaya species to create periclinal chimeras of *C. papaya* and *C. cauliflora*, resulting in compatible and vigorous plants. Grafting and inarching of promising papaya hybrids onto *C. cauliflora*, a wild species resistant to PRSV-P, has been shown to delay disease symptoms. The terminal wedge grafting was superior to chip budding and side grafting, with success rates of 100, 85 and 75 per cent, respectively, and provided good growth performance. Grafting of Tainung No. 2 and Tainung No. 5 papaya varieties resulted in shorter plants with no scion-rootstock incompatibility and better yields compared to ungrafted seedlings. Top grafting on Papaya resulted in the highest survival rates (83.91 and 75.15%) among six papaya varieties selected as rootstocks (Senthilkumar *et al.,* 2014).

The production of graftable size scion materials (shoots) from the selected genotypes in papaya is a limiting factor due to apical dominance. Hence, many attempts were made to overcome this problem in papaya. Applying a lanolin paste mixture of BA (500 mg/L) and GA (1000 mg/L) to decapitated stems effectively stimulated the production of side shoots. Spraying stems with BA (500 mg/L) and GA (100 mg/L) resulted in cuttings with desirable traits. Additionally, injecting BA and GA into the base of the stem and topping it led to the development of numerous small side shoots. Promalin (BA + GA_4+7) and CPPU + GA_3 were also utilized, showing varied effects on shoot quality. Applying cytokinin and gibberellic acid to 'Solo' and 'Formosa' induced side shoots, with GA (500 ppm) producing the highest number of lateral shoots and GA (1000 ppm) yielding the longest shoots. Treatments with BA + GA_3 improved both shoot number and size. BAP + GA_3 resulted in the most buds on shoots, while removing apical buds promoted growth but not additional bud formation. GA_3 and BA treatments supported the development and growth of side shoots (Senthilkumar *et al.,* 2014).

Grape

The grape (*Vitis vinifera* L.) is part of the Vitaceae family, which includes about 60 species of wild Vitis found in Asia, North America, and Europe, thriving in subtropical, Mediterranean, and temperate continental climates. Native to the Mediterranean region, central Europe, and southwestern Asia-from Morocco and Portugal to southern Germany and northern Iran-grapes are used for various purposes (Hardie, 2000).

Propagation of grapes for commercial vineyards involves techniques such as cuttings, rooting, budding, layering, and grafting. One crucial factor in the success of rooted cuttings is the choice of rooting media, which significantly impacts rooting and growth. The best rooting occurs with cuttings taken from the basal part of the shoots, likely due to higher levels of carbohydrates and root-promoting substances produced in the buds and leaves. The type of rooting medium is essential, as it affects both the quality and percentage of rooting (Jaleta *et al.,* 2019). Effective growth media should provide nutrients, retain moisture, facilitate gas exchange, and ensure good plant anchorage.

The application of growth regulators like IBA and NAA in various concentrations had differing effects on rooting. Specifically, IBA at 100 ppm and NAA at 100 ppm improved survival rates and increased the number of primary roots, while higher concentrations such as IBA at 300 ppm and NAA at 300 ppm produced the longest roots. However, combining IBA and NAA was less effective. 2000 ppm IBA for hardwood cuttings and 500 ppm IBA for softwood cuttings as optimal for achieving high rooting percentages and root lengths in Dogridge and 1613C rootstocks.

Overall, selecting the appropriate rooting medium is critical, as it influences rooting success and plant growth (Singh and Chauhan, 2020).

Hardwood cuttings of the grape rootstock Dog Ridge, each with four internodes, were used for the experiment, with Flame Seedless (a hybrid of Sultanina, Cardinal, Malaga, and Muscat d' Alexandric) serving as the scion variety. The cuttings were sourced from the Fruit Research Farm at the Department of Fruit Science, Punjab Agricultural University, Ludhiana. Vigorous branches from the previous season growth were selected, ensuring they were healthy, well-nourished, and mature, with similar size and appearance. Dog Ridge cuttings, taken from the middle portion of the cane, measured 18-24 cm in length and 0.8-1.4 cm in diameter. The base of each cutting was cut straight across just below a node, while the top was cut at a 45° angle about 2-2.5 cm above a node. The basal 3-4 cm portion of the Dog Ridge hardwood cuttings was treated with 2000 ppm IBA using the quick dip method for 5 minutes, then allowed to dry in the shade for 15 minutes (Singh *et al.*, 2018).

For the scion preparation, Flame Seedless scion sticks were obtained from one-year-old field-grown healthy vines. Dormant vines of matching thickness were selected for grafting. In August and September, the scion shoots were defoliated one week before grafting. All cuttings were dipped in a fungicidal solution containing 2 g of carbendazim per litre for 10 minutes and then dried in the shade for 15 minutes. Scion cuttings with two nodes were then taken for grafting (Kaur, 2018).

Figure 5.6: Grape Cuttings.

Jackfruit

Jackfruit seeds typically produce acceptable fruit, with first-generation seedlings retaining about 90 per cent of the parent's traits and often being more vigorous than grafted plants. For seed propagation, select seeds from trees with consistent bearing, high yields, and desirable traits. After collecting ripe fruits, clean the seeds and dry them in the shade for about an hour. Seeds should be planted promptly to avoid poor germination, though they can be stored in a moist plastic container in the refrigerator for a few weeks. Pre-treating seeds with water or a dilute gibberellic acid solution can enhance germination, which usually starts within 1–3 weeks. Jackfruit seedlings grow quickly, reaching 25 cm in 3–4 months (Rajashekaran *et al.,* 2021).

Figure 5.7: Grafted Jackfruit Saplings.

Grafting is a reliable method for propagating desired cultivars. Scions from mature trees are grafted onto healthy, less-than-one-year-old rootstock seedlings. Ideal grafting conditions include daytime temperatures of 70° to 85°F, nighttime temperatures of 55° to 65°F, and high humidity. Grafted trees typically bear fruit in 2 to 3 years and have a more open canopy compared to seedling trees.

The modified veneer graft is a highly effective method for grafting jackfruit, whether or not the scion retains leaves. This technique involves using scions 10 to

15 cm (4 to 6 inches) long with a swollen terminal bud, and keeping the last fully expanded leaf. Both the rootstock and scion are cut with long, shallow veneer cuts to expose the cambium. The cut on the rootstock leaves a small flap of bark at the base to secure the scion during wrapping. The surfaces of the scion and rootstock are then aligned and wrapped with plastic grafting tape or a rubber band, leaving the terminal bud uncovered (Kumar *et al.,* 2021).

The grafted tree and container are covered with a clear plastic bag and placed in a bright but shaded area, with thorough watering. The terminal bud will unfold and grow, while any rootstock sprouts below the graft should be removed. The bag can be taken off after 2 to 4 weeks when the scion begins to grow. The height of the graft on the rootstock is not critical. This method can also be used without retaining leaves on the scion, which can be stored in a plastic bag at 12°C (54°F) for up to a week. Other successful grafting techniques include chip budding, cleft, and forkert grafts. Key factors for success are the vigor of both the rootstock and scion and the preparation of the bud wood.

REFERENCES

Alagumanian S, Jahirhussain G and Rao MV, 2015. High frequency shoot tip multiplication and exvitro rooting of Centella asiatica (L.) Urb.-a industrial medicinal herb. World *Journal of Pharmaceutical Research,* **4(12)**: 1983-1998.

Al-Amin MD, Karim MR, Amin MR, Rahman SMAN and Mamun ANM, 2009. In vitro micropropagation of banana. *Bangladesh Journal of Agricultural Research,* **34(4)**: 645-659.

Amente G, Gemechu M and Chimdessa I, 2022. Protocol optimization for Micropropagation of Banana Varieties (*Musa* spp.) Using Shoot-Tip Culture. E*nvironment,* **18**: 20.

Angami T, Kalita H, Alone RA, Touthang L, Jini D and Ronya T, 2019. Rapid multiplication of banana through macro propagation techniques. *Agric Food–e Newslett,* **1**: 320-322.

Bhende SS and Kurien S, 2016. Sucker production in banana. *Journal of Tropical Agriculture,* **53(2)**: 97-106.

Bordoloi N, Wangchu L and Nimbolkar PK, 2022. Citrus Budding: Research Perspectives and Recent Trends. In *Biological Forum–An International Journal,* **14(3)**: 841-849.

Broadbent P, Gillings MR and Gollnow BI, 1988. Graft-transmissible dwarfing in Australian citrus. In International Organization of Citrus Virologists Conference Proceedings, Vol. 10, No. 10.

Bunt BR, 2012. Media and mixes for container-grown plants: a manual on the preparation and use of growing media for pot plants. Springer Science and Business Media.

Carvalho SAD, Girardi EA, Mourão Filho FDAA, Ferrarezi RS and Coletta Filho HD, 2019. Advances in citrus propagation in Brazil. *Revista Brasileira de Fruticultura*, **41**: 422.

Chazdon RL and Guariguata MR, 2016. Natural regeneration as a tool for large–scale forest restoration in the tropics: prospects and challenges. *Biotropica*, **48(6)**: 716-730.

Dagnew A, Shibru S, Debebe A, Lemma A, Dessalegn L, Berhanu B, Beyene W and Sierrab YM, 2012. Micropropagation of banana varieties (Musa spp.) using shoot-tip culture. *Ethiopian Journal of Agricultural Sciences*, **22(1)**: 14-25.

Babu Dhines, Patel Ramkishor, Singh DS, Singh Akath, Yadav DS, De L and Deka Bidyut 2010. Seed germination, seedling growth and vigour of papaya under North East Indian condition. *Acta Horticulturae,* **851**: 299-306.

El-Sherif NA, 2019. Impact of plant tissue culture on agricultural sustainability. Sustainability of Agricultural Environment in Egypt: Part II: Soil-Water-Plant Nexus, pp. 93-107.

Galán Saúco V, de Queiroz Pinto AC, Mitra SK, Faleiro FG and Ferreira FR, 2021. Ma ngo Propagation. *The Mango Genome*, pp. 31-44.

Garner RJ, 2013. The grafter's handbook. Chelsea Green Publishing.

Gowen S ed, 2012. Bananas and plantains. Springer Science and Business Media

Hardie WJ, 2000. Grapevine biology and adaptation to viticulture. *Australian Journal of Grape and Wine Research*, **6(2)**: 74-81.

Inserra RN, Stanley JD, O'Bannon JH and Esser RP, 2005. Nematode quarantine and certification programmes implemented in Florida. Nematologia Mediterranea.

Jaleta A and Sulaiman M, 2019. A review on the effect of rooting media on rooting and growth of cutting propagated grape (*Vitis vinifera* L). *World Journal of Agriculture and Soil Science*, **3(4)**: 1-8.

Kamatyanatti M and Singh SK, 2019. Recent advances in citrus propagation and planting. Advances in Horticultural Crop Management and Value Addition, 4-12.

Singh Navpreet and Kaur Gagandeep, 2018. Study on Time and Method of Grafting on the Graft Success in Grape. *J Krishi Vigyan*, **6(2)**: 264-271.

Kaviani B and Kulus D, 2022. Cryopreservation of endangered ornamental plants and fruit crops from tropical and subtropical regions. *Biology*, **11(6)**: 847.

Kumar J, 2012. Evaluation of risks related to the release of biocontrol agents active against plant pathogens. Proceedings of the 26th Training on "Diseases and management of crops under protected cultivation". p. 199-204.

Kumar R, Rai R and Nalini P, 2021. Comparative evaluation of grafting and budding methods in jackfruit. *Indian Journal of Horticulture*, **78(4)**: 398-403.

McKey D, Elias M, Pujol B and Duputié A, 2010. The evolutionary ecology of clonally propagated domesticated plants. *New Phytologist*, **186(2)**: 318-332.

Rajashekaran PE, Rajiv Kumar, Sridhar Gutam, Rohini MR, Linta Vincent and Anushma PL, 2021. E Compendium of invited lectures and abstracts. National Online Training Programme on "Conservation, Management and Utilization of Horticultural Genetic Resources for Livelihood and Nutritional Security", November, 22-26, 2021. Published by Director, ICAR–IIHR, Bengaluru- 560089, pp. 1-133.

Naik BH and Thippesh D, 2014. Fundamentals of horticulture and production technology of fruit crops. *Pomegranate*, **149**: 152.

Nath P, 2014. Horticulture Potential in Fiji Island. *Progressive Horticulture*, **46(2)**: 175.

Norman GG, 1957. Florida State Plant Board program for virus-free budwood. *In International Organization of Citrus Virologists Conference Proceedings* (1957-2010) (Vol. 1, No. 1).

Opata J, 2020. Innovative propagation techniques in banana and plantain (Doctoral dissertation, Dissertation, Hohenheim, Universität Hohenheim, 2020).

Parvin S, 2015. *In vitro* regeneration of gerbera (*Gerbera jamesonii*) from different explants. Ph.D. (Hort.) Thesis. Sher-e-bangla Agri. Univ., Dhaka (Bangladesh).

Pereira FMP, Usman M, Mayer NA, Nachtigal JC, Maphanga OR and Willemse S, 2017. Advances in guava propagation. *Revista Brasileira de Fruticultura*, **39**: e-358.

Prakash J and Singh K, 2015. Epicotyl Grafting: A useful technique of Mango Propagation (An online publication under the category: Agriculture). Biotech Articles.

Rajmohan K, Nazeem PA, Lisamma Joseph, 2004. Plant Biotechnology Research at the Kerala Agricultural University, Kerala Agricultural University, Thrissur. p. 78.

Senthilkumar S, Kumar N, Soorianathasundaram K and Kumar PJ, 2014. Aspects on asexual propagation in papaya (*Carica papaya* L.)-a review. *Agricultural Reviews*, **35(4)**: 307-313.

Singh AK and Kumar A, 2021. *Plant Propagation and Nursery Management*. Bhavya Books (BET).

Sing, HP, Uma S, Selvarajan R and Karihaloo JL, 2011. Micropropagation for production of quality banana planting material in Asia-Pacific. Asia-Pacific Consortium on Agricultural Biotechnology (APCoAB), New Delhi, India, 92.

Singh KK and Chauhan JS, 2020. A review on vegetative propagation of grape (*Vitis vinifera* L.) through cutting. *Global journal of bio-science and biotechnology*, **9(2)**: 50-55.

Singh KK, Krishan C and Singh KKK, 2018. Propagation of citrus species through cutting: A review. *Journal of Medicinal Plants Studies*, **6(1)**: 167-172.

Singh N and Kaur G, 2018. Study on time and method of grafting on the graft success in grape. *J Krishi Vigyan*, **6(2)**: 264-271.

Sivakumar D, Jiang Y and Yahia EM, 2011. Maintaining mango (Mangifera indica L.) fruit quality during the export chain. *Food Research International*, **44(5)**: 1254-1263.

Suklabaidya A and Mehta K, 2019. Rejuvenation of senile horticultural plantations for improved productivity and quality. *Int. J. Hortic. Agric. Food Sci*, **3(4)**: 173-181.

Teskey BJ, 2012. *Tree fruit production*. Springer Science and Business Media.

Vidalakis G, Pagliaccia D, Bash JA and Semancik JS 2010. Effects of mixtures of citrus viroids as transmissible small nuclear RNA on tree dwarfing and commercial scion performance on Carrizo citrange rootstock. *Annals of Applied Biology*, **157(3)**: 415–423.

Vuylsteke DR, 1989. Shoot-tip culture for the propagation, conservation and exchange of Musa germplasm. International Board for Plant Genetic Resources, Rome.

Chapter 6

Orchard and Canopy Management in Tropical Fruit Crops

Prakash Patil and Vani N.U.

ICAR–Indian Institute of Horticultural Research, Bengaluru–560089, Karnataka

Orchard Management in Tropical Fruit Crops

An orchard is characterized by the systematic planting of various fruit crops, managed to ensure consistent yields for economic benefit. Orchards vary in type depending on their intended purpose and management objectives. Managing an orchard for tropical fruits entails meticulous attention to the fruits' specific requirements and the distinct challenges posed by the tropical climate.

Types of Orchard

Commercial Orchard

Orchards managed by fruit growers primarily for commercial gain involve selling the produce for profit. These commercial orchards encompass various types, including low-density, high-density, and dry orchard

Low Density Orchard

- ✰ In low density orchards, standard rootstocks are used, which are vigorous and require wider spacing between plants.
- ✰ Due to fewer trees per unit area, this type of orchard is termed as low density.

High Density Orchard

- High density orchards utilize dwarfing rootstocks, allowing for a greater number of plants per unit area in fruits such as mangoes, apples, citrus, pears, walnuts, and others.
- Various horticultural techniques are employed to dwarf the trees, including the use of dwarfing rootstocks, genetically dwarf scion varieties, specialized training and pruning systems, and plant growth regulators.
- These compact trees offer increased productivity, easier management, and stable income for growers, despite drawbacks such as high initial investment and potential impacts on fruit quality if rootstock selection is inadequate.

Dry Orchard

Dry orcharding involves planting hardy fruit crops like mangoes, pears, sapotas, bael, guavas, custard apples, cashews, and jamuns, employing moisture conservation methods such as contour or terrace planting, crescent bunding, and mulching. These techniques aid in establishing transplants initially and meeting water needs in rain-fed orchards.

Progeny Orchard

Nurserymen maintain progeny orchards primarily for preserving superior fruit tree varieties with desirable traits for propagation. These orchards house mother plants, but management complexities arise from differences in cultural requirements, spacing, fertilization, and irrigation among various fruit varieties.

Home Orchard

Home orchards, found in rural and suburban residences, cater to family fruit needs and are selected based on personal preferences and convenience.

Experimental Orchard

Experimental orchards in colleges and research institutions serve to conduct systematic experimental trials for the benefit of fruit growers and farmers, applying scientific methodologies.

Orchard Management

This optimizes soil for fruit tree yield and quality, ensuring sustainable returns. Decisions should precede planting, with consistent practices. Various systems exist, requiring careful consideration. Integration with pruning, nutrients, and pest control enhances productivity.

Site Selection and Preparation

Choosing the right site is vital for managing a tropical fruit orchard. Key factors include assessing climate conditions such as temperature, rainfall, and humidity, evaluating soil type and fertility, and ensuring adequate water availability and drainage. Preliminary operations involve clearing trees, shrubs, and weeds, followed by deep ploughing to remove large roots. Thorough ploughing, levelling, and manuring of the land are crucial. In hilly areas, creating terraces and ensuring proper levelling within each terrace are essential to prevent erosion.

Establishing an orchard is a substantial, long-term investment that requires meticulous planning. Critical considerations such as selecting the proper location and site, choosing the appropriate planting system and spacing, and carefully selecting varieties and nursery plants are essential to ensure maximum production.

Choosing the location and site selecting a suitable site is crucial and should be based on several key criteria:

- ☆ Opt for a location within a well-established fruit-growing region to benefit from collective grower experience and cooperative marketing opportunities.
- ☆ Ensure proximity to a market area for efficient distribution.
- ☆ Verify that the climate is conducive to the chosen fruit crops.
- ☆ Confirm the availability of consistent water supply throughout the year.

Before finalizing a site for an orchard, growers must assess the following factors:

- ☆ Soil suitability, including fertility, subsoil characteristics, and depth.
- ☆ Adequate drainage to prevent water stagnation, especially during rainy seasons.
- ☆ Quality of irrigation water.
- ☆ Access to transportation via road or rail.
- ☆ Suitability of climatic conditions, considering factors like cyclones, frost, hailstorms, and strong winds.
- ☆ Market demand and potential for seasonal production fluctuations.
- ☆ Competitive landscape and existing orchard operations.
- ☆ Availability of labor.

Preliminary Operations

Once a suitable location and site are chosen, several preliminary operations must be carried out:

- ☆ Clearing of trees without leaving stumps or roots, and removal of shrubs and other vegetation.
- ☆ Deep ploughing to eliminate large roots and preparation of the soil.
- ☆ Thorough ploughing, levelling, and application of manure to enhance soil conditions.

- Levelling of land to optimize irrigation efficiency and prevent soil erosion.
- In hilly terrain, construction of terraces according to land topography, followed by terraced levelling to prevent erosion.
- If the soil is poor, consider planting a green manure crop to improve its physical and chemical properties before undertaking planting operations.

A well-planned orchard ensures efficient and economical management. Consider the following points:

- **Optimum Spacing**: Maximize the number of trees per unit area.
- **Central Buildings**: Place stores and office buildings centrally for better supervision.
- **Well Placement**: Position wells conveniently at a rate of one per 2 to 4 hectares.
- **Fruit Blocks**: Assign each fruit type to a separate block.
- **Ripening Grouping**: Group fruits that ripen simultaneously together.
- **Pollinators**: Include pollinator trees every third tree in every third row for deciduous fruits.
- **Irrigation Channels**: Lay channels along gradients, giving a 7.5 cm slope for every 30m length.
- **Roads**: Minimize road space; clear space between windbreaks and tree rows is advantageous.
- **Tree Height Arrangement**: Place shorter trees in front and taller ones at the back for better supervision and aesthetics.
- **Evergreen and Deciduous Trees**: Evergreen trees should be in front, with deciduous ones behind.
- **Bird and Animal Attraction**: Plant fruits that attract birds and animals near the watchman's shed.
- **Fencing**: Use live fencing with drought-resistant, quick-growing plants like Agave and *Prosopis juliflora*.
- **Windbreaks**: Plant double rows of tall trees around the orchard to reduce wind velocity and protect crops. The trees which are suitable for growing as wind breaks are *Casuarina equisetifolia, Pterospermum acerifolium, Polyalthia longifolia, Eucalyptus globulus, Grevillea robusta, Azadirachta indica etc.*

Planting Distance and Layout

Determining the correct planting distance is crucial and varies depending on the type of tree (*e.g.*, mango, guava, papaya), local rainfall, soil type, rootstock variations, and training systems. Layout options include square, hexagonal, and triangular systems, as well as High Density Planting (HDP) techniques. Calculating the number of trees per hectare for each layout ensures optimal space utilization and growth conditions.

Layout Systems

Aim for maximum tree density, proper development space, and operational convenience. Layouts can be vertical or alternate row patterns:

- ☆ **Square System**: Trees at each corner of a square, allowing intercropping and cultivation in two directions.
- ☆ **Rectangular System**: Trees at each corner of a rectangle, with wider alleys for easier operations.
- ☆ **Hexagonal System**: Trees form a hexagon with a central tree, accommodating 15 per cent more trees than the square system but harder to layout.
- ☆ **Quincunx System**: Square layout with an additional central tree, suitable for larger distances between main trees.
- ☆ **Triangular System**: Trees in alternate rows are staggered, occupying more area and thus fewer trees per hectare. Eg: Amrapali in mango
- ☆ **Contour System**: Used on hills to minimize erosion and conserve soil moisture, planting along contours or terraces.

Planting Distance

The planting distance refers to the minimum vertical space required between any two trees or plants, a critical factor influenced by various considerations. Key principles guiding the determination of planting distances are as follows:

- ☆ **Canopy interaction:** When fully grown, tree canopies should touch each other but without branches interlocking.
- ☆ **Root development:** Tree roots typically spread wider than the canopy, necessitating sufficient space for optimal root growth and nutrient absorption without competition.

Several factors influence the determination of planting distances:

- ☆ **Tree Type:** Different fruit trees require varying planting distances. For instance, mangoes are planted 10m x 10m apart, guavas at 5m x 5m, and papayas at 2m x 2m.
- ☆ **Rainfall:** Areas with lower rainfall require wider spacing compared to higher rainfall regions for the same tree type.
- ☆ **Soil Type and Fertility:** Heavy soils restrict top and root growth, hence require closer spacing, whereas lighter soils allow for wider spacing.
- ☆ **Rootstocks:** Trees grafted on different rootstocks of the same variety grow to different sizes and thus require adjusted planting distances.

- ☆ **Pruning and Training:** Trees trained using the central leader system necessitate closer spacing than other training methods.
- ☆ **Irrigation:** The method of irrigation influences planting density, with irrigated orchards generally allowing closer spacing.

In practice, planting too widely reduces yield per unit area, while overly close planting hampers tree growth, leading to decreased yield and smaller fruits. The choice of planting system—square, hexagonal, or triangular—also impacts the number of trees per hectare:

- ☆ **Mango:** Planted at 10m x 10m, yields 100 trees/hectare in a square system, 115 in hexagonal, and 89 in triangular.
- ☆ **Sapota:** Spaced at 8m x 8m, results in 156 trees/hectare (square), 118 (hexagonal), and 139 (triangular).

High-Density Planting (HDP)

HDP is another approach, where trees are spaced closer than standard recommendations to maximize productivity without compromising quality. Advantages of HDP include early cropping, higher yields, reduced labour costs, and improved fruit quality. Key characteristics include maximizing fruiting branches over structural branches, training trees to minimize shading, and controlling tree size using dwarfing rootstocks, pruning methods, and growth regulators.

Ultimately, choosing the optimal planting distance and system is crucial for balancing yield, fruit quality, and orchard management costs across different horticultural crops and growing conditions.

In mango, Amrapali at 2.5 x 2.5 m in triangular system accommodation of 1600 plants and Dashehari mango at 2.5 m x 3 m (1,333 plants per ha) was raised under HDP with pruning and dehorning after the harvesting followed by paclobutrazol application and yield was secured every year.

In Citrus, Kinnow on Troyer Citrange and Karna khatta rootstocks could be planted at 1.8 x 1.8.m and 3 x 3 m to accommodate 3000 and 1088 plants per ha, respectively.

In pineapple, population density of 63, 758 per ha coupled with improved package of agro-techniques result in increase in yield from 15-20 to 70-80 t/ha.

Pusa Nanha papaya may be planted at a distance of 1.25 x 1.25 m (6,400 plants per ha). Similar observations have been made by various workers in citrus, litchi, banana and pineapple (Mishra and Goswami, 2016).

In an experiment in guava was standardized at CISH, HDP in fruit crops Lucknow. The ultra high density orchard system of guava accommodates 5000 plants per ha with spacing of 2.0 x 1.0 m and managed with regular topping and hedging during initial stages which helped in controlling tree size and getting higher yield (Singh, *et al.,* 2007). In India, high density plantings have successfully been demonstrated in guava (Lal *et al.,* 2007), litchi (Mishra *et al.,* 2014), mango and papaya (Ram, 1996)

Planting Techniques

The ideal planting season, such as during the monsoon or spring, and the planting method are critical. Proper pit preparation, use of planting boards, and ensuring the correct depth and positioning during transplanting are important. Immediate irrigation after planting and protection from wind and sunburn through staking and trunk wrapping are necessary for initial care.

Orchard Cultivation and Maintenance

Effective soil management practices include clean culture, use of cover crops, mulching to conserve moisture and control weeds, and sod and sod mulch methods. Intercropping with compatible crops like legumes and vegetables, and using fillers such as papaya or banana, can enhance productivity. The multitier cropping system combines different crop layers, such as coconut, pepper, and cocoa, to maximize land use.

Nutrient and Water Management

Fertilization involves using organic manures and bio fertilizers, guided by regular soil testing and nutrient management plans. Efficient irrigation systems, particularly drip irrigation, and proper scheduling based on crop requirements and rainfall are crucial for maintaining soil moisture and supporting tree growth.

Insect-Pest and Disease Management

Integrated pest management (IPM) involves regular monitoring for pests and diseases, utilizing biological controls and botanicals, and minimizing chemical pesticide use. Preventive measures include proper plant spacing, pruning, and using disease-resistant varieties and rootstocks to maintain orchard health.

Pruning and Training

Regular pruning maintains tree shape and health by removing dead or diseased branches. Appropriate training methods, such as central leader or open center systems, and techniques for high-density planting (HDP), are essential for maximizing fruit production and tree vigour.

Harvesting and Post-harvest Management

Identifying the optimal harvest time for each fruit type and using proper harvesting tools and techniques ensure high-quality produce. Post-harvest handling includes sorting, grading, packing, and implementing appropriate storage and transportation methods to maintain fruit quality.

Record Keeping and Monitoring

Maintaining thorough records of planting dates, fertilization, irrigation, and pest control activities is crucial. Regular monitoring and periodic assessments of tree health and productivity help in adjusting management practices based on observations and historical data.

Sustainability and Organic Practices

Emphasizing organic farming by avoiding synthetic chemicals, promoting crop rotation, green manures, and composting contributes to sustainable orchard management. Implementing eco-friendly pest and nutrient management strategies and promoting biodiversity and soil health ensure long-term productivity and environmental conservation.

Orchard cultivation involves managing orchard soil to maintain its condition with minimal expenses, ensuring the physical condition, moisture, and nutrient content meet the trees' needs. Key goals include weed control, soil moisture and nutrient conservation, minimal soil disturbance, erosion prevention, and reduced cultivation costs.

Soil Management Practices

Clean Culture

- ✫ Regular ploughing and weed removal.
- ✫ Disadvantages include rapid humus depletion, root injury, nitrogen depletion, hard pan formation, and soil erosion.
- ✫ Minimize these issues by avoiding deep, frequent cultivation and working the soil when it is not too wet.

Clean Culture with Cover Crops

- ✫ After removing weeds, raise a cover crop or green manure.
- ✫ Green manure crops like Sunhemp, Cowpea, and Daincha can be used.
- ✫ Permanent cover crops like *Calapogonium muconoides* and *Centrosema pubescens* help prevent soil erosion.

Mulching

- ✫ Use crop residues like straw, leaves, or materials like polythene films to conserve soil moisture and control weeds.
- ✫ Advantages: keeps soil cool in the day and warm at night, reduces runoff, adds humus, prevents erosion, protects fruits, and reduces irrigation frequency.
- ✫ Disadvantages: risk of fire, potential habitat for rodents, and thick mulches can harm tree trunks.

Sod

- ✫ Raising a permanent grass cover without tillage, useful in sloped areas to prevent erosion.

- ☆ Competes for moisture and nitrogen, requiring increased manuring and water application.

Sod Mulch

- ☆ Similar to sod, but vegetation is cut frequently and left on the ground, reducing moisture loss.

Intercropping

- ☆ Utilizes space between young trees or long-duration crops for additional economic crops, enhancing soil and providing early returns.
- ☆ Should not occupy areas with concentrated tree roots.
- ☆ Selected intercrops should maintain or improve soil fertility and not compete excessively for water or nutrients.

Mixed Cropping

- ☆ Growing certain perennial crops between the main crops for effective land use and increased income.
- ☆ Examples: cocoa, pepper, and nutmeg in coconut and arecanut plantations, showing yield increases due to beneficial crop combinations.

Multitier System of Cropping

- ☆ Combines intercropping and mixed cropping to utilize land, water, and sunlight effectively.
- ☆ Example: coconut and arecanut trees with crops that have different growth heights and rooting habits, maximizing resource use and profitability.

Canopy Management in Fruit Crops

Canopy Definition

The spread of branches and foliage of a plant, including the stem, branches, shoots, and leaves. Canopy density is determined by leaf number and size, while canopy architecture is shaped by the number, length, and orientation of stems, branches, and shoots.

Canopy Management

Definition

Canopy management involves manipulating tree and vine canopies to optimize fruit production and quality. This includes managing the structure, size, shape, branch orientation, and light interception for maximum productivity and quality.

Objectives

- Make the best use of land and climatic factors.
- Optimize light, temperature, and humidity for fruit quality.
- Adjust leaf, shoot, and fruit positions in space to influence plant geometry and leaf orientation.

Components

- Orchard architecture relies on the variety, rootstock, tree spacing, training, and pruning.
- Cultural practices like nutrition, irrigation, planting density, and pruning influence canopy shape, vigour, and microclimate.
- Effective canopy management aims to optimize carbon allocation to fruit sinks while maintaining growth and development.

Techniques and Benefits

Techniques

- Training and pruning
- Use of growth retardants
- Managing light penetration and interception

Essential Features of an Ideal Canopy

- Adequate fruiting units
- Sufficient light and ventilation
- Adequate foliage to protect fruits from sunburn
- Minimized overlapping foliage
- Effective spray coverage
- Reduced risk of pest and disease development

Goals of Canopy Management

- Achieve higher yield with good quality
- Maintain a balance between root and shoot growth
- Form strong crotches
- Remove unwanted, overcrowded, diseased, and pest-affected shoots
- Regulate tree architecture for high-density planting
- Facilitate management practices like spraying and harvesting
- Efficiently utilize air, light, and temperature
- Ensure machinery accessibility between rows

Benefits

- ☆ **Optimized sunlight utilization**: Enhances tree growth, yield, and fruit quality by improving light penetration.
- ☆ **Improved air circulation**: Reduces diseases and pests by enhancing air flow and light penetration.
- ☆ **Reduced unproductive parts**: Eliminates old stems and over mature leaves.
- ☆ **Better phenological understanding**: Improves timing for yield management by understanding leaf cycle, flowering, and fruit bearing habits.
- ☆ **Controlled fruit development**: Manages fruit growth phases to prevent production issues.
- ☆ **Easy fruit load estimation**: Uniform plant stand allows for easy yield estimation.
- ☆ **Induced fruiting in desired seasons**: Techniques like bahar treatment in guava, pomegranate, and ber.
- ☆ **Precise fertilizer and pesticide application**: Tailored application to each tree.
- ☆ **Manageable tree height**: Easier access for maintenance and harvesting.
- ☆ **Mechanization**: Facilitates mechanized spraying, harvesting, and pruning, reducing production costs and labor dependence.
- ☆ **Multi-species orchard management**: Better management in mixed orchards, such as coconut and coffee.
- ☆ **Intercultural operations**: Easier passage for tractors without damaging limbs.

Indirect Advantages

- ☆ Prolonged fruit availability by including various varieties and high-density planting with canopy management.

Principles of Canopy Management

Canopy management is crucial for optimizing tree vigor and effectively harnessing sunlight and temperature to improve quality and productivity while minimizing adverse weather impacts.

Key principles of canopy management include

- ☆ **Maximizing light utilization**: Enhancing yield, productivity, and quality.
- ☆ **Preventing microclimate build-up**: Avoiding conditions favorable to pests and diseases.

- **Facilitating cultural operations**: Ensuring ease of maintenance and operations.

Effective canopy management aims to:

- Develop a strong framework with branches evenly spaced and oriented in all directions.
- Create an open center in the canopy for better sunlight exposure.
- Control plant size for optimal productivity.

The fundamental principles of canopy management are:

Light Interception

- Light is essential for producing quality fruits, influencing both flower induction and fruit development through carbohydrate synthesis in various fruits, including mangoes.
- Yield efficiency in an orchard is proportional to the amount of light intercepted, which depends more on the leaf area index (LAI) than on ground coverage.
- Regions with high light energy and long seasons are ideal for high yields. Orchard design and canopy display play crucial roles in maximizing light interception, which directly impacts total dry matter and fruit yield.
- For apple orchards, light interception is mainly influenced by planting system, tree spacing, shape, height, alley width, row orientation, LAI, and the growing season's length. Wunche and Lakso (2000) found that open canopies with well-exposed fruit-bearing spurs throughout the season enhance fruit numbers, size, colour, and quality.
- Effective canopy management should optimize light interception and distribution, as well as carbohydrate partitioning, to support fruit growth. For regular high-quality fruit production, lower levels of light interception are advisable. For example, pears require a maximum of 65per cent light interception. Ideally, light should be evenly distributed under tree rows and alleys, targeting 60-70 per cent light distribution.

Light Distribution

- Light penetration within the tree is influenced by tree size and shape. Larger trees often experience significant shading, reducing light intensity within the canopy.
- Adequate light distribution is crucial for high fruit quality, as shading can lead to reduced fruit weight and poor colour. Systems like the V-system or Y-trellis can help achieve uniform light distribution.

Photosynthetic Response

- Light is the primary factor affecting photosynthesis, which in turn influences dry matter production and partitioning.
- Canopy photosynthesis depends on leaf photosynthetic rate, whole canopy light interception, and leaf area. These factors are affected by genetic traits, environmental conditions, and cultural practices.

Fruit Quality

- Fruit size and colour are associated with light levels exceeding 70 per cent of full radiation. The best-coloured and largest fruits are typically found on the outer periphery of the canopy.

Studies have shown that using reflective materials can enhance fruit quality:

- In Israel, Moreshet *et al.* (1975) used aluminium plastic film in alleys, significantly improving fruit size, colour, and soluble solids in the lower canopy of Orleans apples and increasing yield by 50 per cent the following year.
- Mika (1980) in Poland used aluminized tar paper under McIntosh, Cortland, and Spartan apple trees, improving the color of lower fruits without carryover effects.
- Doud and Ferree (1980) in Ohio used metallic polyester film under Delicious apple trees, improving fruit colour in the lower canopy.
- Shading within the tree reduces fruit quality, as seen in Satsuma mandarins, sour cherries, and peaches. Shaded leaves are thinner with fewer palisade layers. Apple cv. Fuji requires at least 30 per cent relative light intensity for good fruit colour. Shahak *et al.* (2001) in Israel used reflective metallized polyethylene mulches in mango cv. Keitt, enhancing red coloration.

Flower Bud Differentiation/Flowering and Fruit Set

- Light intensity and distribution significantly influence flower initiation, fruit set, and flower bud production. Canopies receiving less than 30 per cent light produce virtually no flowers, while reducing light from 100 to 70 per cent slightly decreases flowering. Increased shading results in more fruit shedding and fewer flower buds the following season.
- Jackson and Palmer (1977) found a 36 per cent reduction in flowering spurs on Cox's Orange Pippin trees when light was reduced to 37 per cent from early June to leaf fall. This progressive decline in flower bud production varied depending on the previous year's crop load. In mangoes, sufficient light is essential for flower bud formation, with dense canopies impeding flowering, which can be improved by pruning and shoot growth retardation.

Factors affecting Canopy Development

Several factors influence tree canopy development, directly affecting yield:

Layout and Row Orientation

- North-South orientation and rectangular plantings intercept more light than East-West and square plantings.
- Minimum shoot growth observed at a 30° planting angle in 'Royal Delicious.'
- Horizontal branch orientation reduces auxin content, affecting nutrient supply to the apex and spur production.

Planting Density

- High-density planting utilizes more solar radiation and increases bearing surface per unit area.
- Dense canopies can alter the microclimate, increasing humidity and disease incidence.
- Optimal density balances yield and quality, with rapid canopy development in early years.

Planting Geometry

- Closer spacing reduces light intensity, air flow, and photosynthetic activity, while increasing relative humidity.
- Medium spacing (2 × 2 m or 2 × 1 m) provides optimal performance in grapes.

Variety

- Compact/spur varieties, like Red Spur and Red Chief, help in designing efficient canopies.
- Upright cultivars maximize orchard space, while dwarf cultivars allow for close planting in various crops.

Rootstock

- Rootstocks control the size and shape of the canopy.
- In tropical fruits, size-controlling rootstocks are limited.
- Dwarfing rootstocks, like M-9 for apples and Gisela 5 for cherries, are effective in managing canopy size.

Training and Pruning

- Training systems like Central Leader and Open Vase optimize light and ventilation.

- Pruning regulates tree size, improves light distribution, and enhances photosynthesis, leading to higher yields.
- In mangoes, pruning promotes flowering by redistributing hormones and increasing phenolic content.

Soil and Soil Moisture

- Soil moisture management influences canopy vigour, shoot length, and leaf size.
- Moisture stress during key growth periods can reduce shoot vigour and increase productivity.

Wind

- High winds stunt growth, cause plants to tilt, and dehydrate them by increasing water loss.
- Wind reduces photosynthesis by restricting CO_2 diffusion in leaves.

Elevation

- Higher altitudes have thinner vascular pathways, resulting in stunted growth and open canopies.
- Soil fertility is often lower at higher elevations, slowing tree growth.

Sea Coast

- Salt injury from sea spray or dissolved salts reduces stem and leaf growth.
- Salt in soil solution impedes water uptake, while chloride ions in leaves disrupt photosynthesis.

Human Interference

- Canopy is affected by branch breakage during harvesting and root pruning during ploughing.

Methods and Tools for Canopy Management

Methods for canopy management:

- **Tree Thinning**: Remove alternate trees to reduce crowding.
- **Staghorning**: Periodically cut back trees to a bare scaffold to promote regrowth.
- **Individual Limb Removal**: Selectively remove large limbs to increase light penetration and manage tree size.
- **Mechanical Hedgerowing**: Use a mechanical saw to shape trees, typically at an angle.

- **Hedgerowing with PGRs**: Apply plant growth regulators (PGRs) to control vegetative growth.

Tools for canopy management:

- **Training**: Essential in early years to build a strong framework, optimize light and ventilation, and make spraying easier. Common systems include:
  - **Central Leader**: Used for apples, cherries, plums, pears, and pecans.
  - **Modified Leader**: Used for apples.
  - **Open Center**: Used for peaches, nectarines, and plums.
  - **Bower System**: Effective for grapes in tropical regions.
- **Pruning**: Maintains balance among fruit buds, leaf area, and root system. Regular pruning:
  - Regulates tree size and shape.
  - Improves light distribution and photosynthesis.
  - Increases yield and reduces disease incidence.
  - Specific examples:
  - **Mango**: Pruning promotes flowering and reduces malformation.
  - **Citrus**: Pruning upright branches encourages horizontal growth.
  - **Mandarin**: Pruning one-year-old shoots increases yield.
  - **Grapes**: Pruning and topping promote fruiting and ripening.
- **Rootstock**: Influences canopy size and vigour. Ideal rootstocks are precocious, dwarfing, disease-resistant, and adaptable. Examples:
  - **Temperate Fruits**: Extensive use of dwarfing rootstocks like M-9 for apples.
  - **Tropical Fruits**: Limited use, but dwarfing rootstocks like Vellaicolumban for mangoes are effective.
- **Genetically dwarf scion Cultivars**: Suitable for high-density planting. Examples include:
  - **Mango**: Varieties like Amrapali.
  - **Citrus**: Varieties like Meyer lemon and Kumquat.
  - **Apple**: Spur-bearing varieties.
- **Growth regulators**: Chemicals used to control shoot vigour and canopy volume by:
  - **Killing Terminal Buds**: Chemicals like Maleic Hydrazide and ethylene.
  - **Inhibiting Apical Meristematic Activity**: Chemicals like TIBA and Benzyl adenine.
  - **Inhibiting Internode Elongation**: Chemicals like SADH and paclobutrazol.

- Specific treatments:
- **Guava**: 10,000 ppm NAA reduces lateral bud sprouting.
- **Peach**: NAA treatment reduces sprouting without affecting tree growth.
- **Pomegranate**: NAA and MH mixtures inhibit sprout growth.

- **Moisture stress**: Water management is crucial. Moisture stress during specific growth periods can reduce shoot vigour and increase productivity in grapes.
- **Nutrition**: Nutrients significantly influence shoot growth. While nitrogen promotes canopy vigour, potassium reduces it.

Canopy Management Aims to Maximize Light Interception in Plant Models through

a) Training
b) Pruning
c) Planting geometry
d) Tree forms
e) Branch and tree orientation
f) Growth regulating chemicals
g) Planting space and design

Training Systems

Open Centre Training System

- Traditional for peaches, plums, and nectarines.
- Forms a vase-shaped tree with major limbs ('scaffolds') instead of a central leader.
- Facilitates light penetration, reducing shading issues in vigorous trees like peaches.

Central Leader System

- Suitable for areas prone to heavy snow.
- Features a single upright trunk ('leader') with lateral branching.
- Summer pruning enhances interior light interception.
- Divides canopy into well-exposed, marginally exposed, and poorly exposed zones based on sunlight.

California-V-System

- Utilizes two angled scaffolds to mitigate shading.
- Higher density system but poses challenges for equipment manoeuvrability.

Espalier System

- ☆ Trains trees to a central leader with horizontal fruiting branches on trellis wires.
- ☆ Enhances canopy manipulation and uniform sunlight distribution.

Tatura-trellis System

- ☆ Promotes rapid canopy development for mechanical harvesting.
- ☆ Efficient light distribution stimulates fruitwood growth.

Centrifugal System

- ☆ Targets lower canopy and trunk for enhanced fruiting and canopy porosity.
- ☆ Acts as a load-regulating technique.

Palmette Leader Tree

- ☆ Modification of central leader system.
- ☆ Ensures good light exposure with a flat, north-south oriented canopy.

Solen System

- ☆ Low domed form for tip-bearing cultivars like Granny Smith.
- ☆ Restricts tree height and enhances fruiting branch development.

Vertical Tree Walls

- ☆ Trains trees in a narrow leader for maximum light penetration.
- ☆ Supported by trellis for even illumination.

Pruning Systems

- ☆ Removes plant parts to shape canopy and reduce foliage density.
- ☆ Enhances light penetration and photosynthesis.
- ☆ Optimizes fruit coloration and redirects energy to fruit-bearing parts.

Tree Forms

- ☆ Various shapes like round-headed, horizontal, vertical, conical, pyramidal, and tall canopies.
- ☆ Designed to improve light distribution and fruit quality.

Planting Geometry

- ☆ High Density Plantation:
- ☆ Dense planting with dwarfing techniques.
- ☆ Maximizes solar radiation use and yield per unit area.

- Multi-Rows and Bed System:
- Spaced for optimal light distribution.
- Achieves high leaf area index and light interception efficiency.
- Double Hedgerow Planting System:
- East-west orientation with wider inter-row spacing.
- Ensures prolonged daily light incidence and minimal shading.

Types and Structures of Canopies in Fruit Trees

Types of Canopies

Small and Compact Canopy

- Suitable for slow-growing trees controlled by dwarfing rootstock.
- Includes spindle bush, slender spindle, and north Holland spindle in apple.
- Research on tropical and sub-tropical fruits for compact commercial forms ongoing, *e.g.*, modified or Malaysian spindle for carambola.

Columnar Canopy

- Ideal for spur-type branching systems.
- Consists of a single stem with small side branches.
- Natural form in apples via induced mutations.
- Used in super spindle systems for high-density plantings of peaches and apples.

Thin Restricted Plane and Tall Canopy

- Suitable for fast-growing fruit trees.
- Improves light distribution within the canopy.
- Increases fruit-set area due to dense, non-transmitting canopy.
- Examples include T or Ebro trellis systems.

Thin Vertical Canopy

- Receives light exposure from both sides.
- Requires good internal light distribution within the canopy.

Branches and Canopy structures

The canopy structure of fruit trees is influenced by the branching pattern, which varies across genera and families:

- **Erect**: Upward and non-spreading type, *e.g., Polyalthia longifolia.*
- **Semi-erect**: More spreading than erect types, *e.g., Michelia champaca.*

- **Columnar**: Compact upward growth, forming a columnar structure, *e.g.*, in apple trees.
- **Round**: Spreading canopy with a rounded top resembling an umbrella shape, *e.g.*, *Mangifera indica*.
- **Semi-round**: Similar to round but with a less pronounced round shape, *e.g.*, Cassia fistula.

Branching patterns

Branching patterns affect canopy shape and tree performance:

- **Apical dominance**: Controls lateral branch growth; pruning affects lateral shoot development.
- **Canopy types**: Columnar, globular (open-center), conical, open-head irregular, weeping, pyramidal, vase, and horizontal canopies are common types.

Light Interception and Distribution in Tree Canopies

Light Interception

Light interception refers to the process where a canopy captures sunlight, which is essential for various physiological processes such as photosynthesis and fruit development:

- It is calculated as the difference between incident solar radiation above the canopy and the mean radiation beneath it.
- Key factor influencing crop development and quality fruit production.
- Determines energy for photosynthesis and impacts flower induction and fruit carbohydrate synthesis in tropical fruits like mango.
- Orchard yield efficiency correlates more with leaf area index (LAI) than ground coverage index.

Light Distribution in Canopies

- Ideally, orchard light distribution should be uniform, around 60-70 per cent under tree rows and alleys.
- Upper leaves receive direct and diffuse sunlight, while lower leaves intercept more diffuse radiation due to reflection from leaves and soil.
- Visible radiation (400-700 nm) is crucial for photosynthetically active radiation (PAR).
- Only about 50 per cent of incident radiation is used for photosynthesis.

Factors Affecting Light Interception

- Canopy shape and design significantly influence light interception.

- Tree height, canopy width, leaf area density, and alley width affect light penetration.
- Leaf characteristics such as angle, surface properties, thickness, chlorophyll concentration, and leaf arrangement (phyllo taxis) impact light transmission.
- Models show increased light penetration with increased tree height, canopy width, or leaf area density, and narrower alleys.

Optimizing Light Distribution

- Different training systems (*e.g.*, Mikado, HYTEC, Solen) affect light interception and distribution within apple tree canopies.
- Isobilateral and dorsoventrally oriented leaves maximize light absorption efficiency.
- Pruning practices (summer and dormant) can influence light penetration and distribution within canopies.
- Orchard design aims to maximize light interception while ensuring adequate exposure for fruiting sites.

Canopy Management through Rootstock and Scion

Introduction

Achieving optimal light interception and distribution for high-quality fruit production requires strategic canopy management. This involves selecting appropriate rootstock-scion combinations, utilizing dwarfing rootstocks, and implementing effective training and pruning systems to maximize orchard space and maintain overall health.

Rootstock and Canopy Development

Rootstocks play a crucial role in regulating canopy vigour, particularly in temperate crops, although their potential in tropical fruits remains underutilized. The interaction between roots and the canopy is vital: the canopy produces carbohydrates for plant development, while roots provide nutrients and water. Maintaining a balanced root-to-shoot ratio is essential for sustained growth and productivity. Altering this balance through pruning affects natural growth patterns and resource allocation, impacting fruit quality.

Interactions between Rootstocks and Scions

Rootstock and scion combinations influence resource exchange (water, nutrients, carbohydrates) and hormonal signalling between root and shoot systems. Compatibility between rootstock and scion affects these interactions and, consequently, the overall performance and vigour of the tree.

Canopy Management Practices

Various methods are employed to manipulate canopy size and shape:

Control of Tree Vigor

Managing vegetative growth improves light penetration, enhances fruit-bearing potential, and increases yield and fruit quality. Dwarfing rootstocks like M-9 in apples are commonly used for size control. Similarly, GF677 and GF556 in peaches and Colt in cherries provide size control options.

Training Systems and Tree Forms

Different forms such as globular, conical with open framework, vertical tree walls, horizontal canopies, and trellis systems like "Y" and "V" shapes influence light distribution and canopy structure.

Pruning

Pruning practices are essential for controlling canopy size and shape, optimizing light exposure, and improving fruit quality.

Plant Growth Regulators

These regulators can supplement pruning efforts to further manage canopy growth and enhance productivity.

Examples in Fruit Crops

- **Temperate fruits:** Rootstocks like M-9, M27, M26, MM106, and MM111 are extensively used in apples for size control.
- **Stone fruits:** Rootstocks such as GF677 and GF556 are used in peaches, and Colt in cherries for managing tree size.
- **Tropical and sub-tropical fruits:** Limited options exist for size-controlling rootstocks, with some success observed in mango (Vellaikolumban), citrus (Fortunella spp., Rangpur lime), and guava (*Psidium friedrichsthalianum*).

Horticultural Vine Crops

- Dwarfing or vigour-inducing rootstocks like Dogridge (*Vitis champini* clone) have revolutionized the grape industry in India. These rootstocks can moderate vigour under drought and salinity conditions, but excessive use may lead to overgrowth in optimal conditions.

Breeding for Dwarf Scion V`arieties

- Breeding dwarf varieties in perennial fruit crops is challenging due to the polygenic control of traits like low vigour, higher side-branching, spurriness, wide crotch angle, precocity, higher phloem to xylem ratio, higher phenolic content, higher productivity, and smaller leaves.

- ☆ Some varieties, like 'Malaviya Bhog', 'Creeping', and 'Kerala dwarf' in mango are potential gene sources for developing dwarf varieties suitable for high-density planting (HDP) systems.

Canopy Management through Plant Growth Regulators and other Practices

In countries like India, ensuring sustainable development in horticulture poses significant challenges. Strategies must harness natural resources effectively using advanced technologies to meet growing fruit demands. Canopy architecture profoundly influences fruit productivity, a concept emphasized by Halle and Olderman (1970) and Halle *et al.* (1978).

Canopy management stands pivotal in sustaining fruit yield and quality. It's well-established that plant growth and development are regulated by endogenous plant hormones. Synthetic growth regulators play a crucial role in commercial crop management by altering these hormone levels to steer growth and development.

Plant Growth Regulators (PGRs)

Plant growth regulators are organic compounds, excluding nutrients,that modify plant physiological processes and structures when applied in low concentrations. They include natural and synthetic compounds aimed at improving yield, quality, or harvest efficiency.

- ☆ **Auxins** promote cell enlargement, bud formation, and root initiation, enhancing growth.
- ☆ **Gibberellins** stimulate flowering and cell division, contributing to overall plant vigor.
- ☆ **Cytokinins** influence cell division and shoot growth.
- ☆ **Abscisic acid and inhibitors** suppress or delay physiological processes in plants.
- ☆ **Ethylene** regulates fruit ripening and leaf abscission.

Use of Growth Retardants

Growth retardants control plant height and structure by inhibiting gibberellin production, primarily affecting stem elongation without reducing leaf area significantly. Examples include:

- ☆ **Daminozide and Chlorphonium Chloride** are commonly used to reduce stem elongation in ornamental crops.
- ☆ **Ancymidol** prevents shoot elongation effectively.
- ☆ **Mefiquat chloride** inhibits cell elongation and node formation, compacting plant growth.

Methods for Controlling Plant Height

Three main methods are employed:

- ☆ **Internode elongation inhibition:** Using substances like succinic acid-2, 2-dimethyl-hydrazide (SADH) and chlormequat chloride (CCC) to reduce internode length without affecting apical meristems.
- ☆ **Terminal Bud Inhibition:** Substances like maleic hydrazide (MH) and fluorenols inhibit apical meristem function, controlling height effectively.
- ☆ **Promoting Lateral Shoot Growth:** Temporarily reducing apical dominance encourages lateral shoot development, managing overall plant height.

Case Studies

Mango

- ☆ Application of paclobutrazol (PBZ) at 1000 ppm reduces tree height and canopy volume.
- ☆ PBZ at 10 g per tree effectively manages shoot growth and enhances flowering in off-years.
- ☆ Cobalt sulfate foliar spray reduces floral malformation incidence significantly.

Grape

- ☆ Growth retardants like B-9 and Phosphon-D reduce vegetative growth when applied during the 10-leaf stage.
- ☆ Paclobutrazol application before bloom reduces shoot growth and internodal length in Roumi Red cultivars.
- ☆ PP 333 application controls vegetative growth for up to two years.

Fig

- ☆ Foliar PBZ spray or soil drenching with PBZ reduces shoot length and thickness, enhancing tree productivity and fruit quality.

High Density Planting (HDP)

HDP is an innovative approach to enhance fruit productivity by densely planting fruit trees, ensuring efficient resource use and maintaining high fruit quality. It involves planting trees closer together than traditional methods, utilizing techniques like dwarfing rootstocks and pruning to manage tree size and promote earlier fruiting.

a) **Definition and purpose of HDP:** High Density Planting (HDP) is a modern technique aimed at increasing fruit yield per unit area without compromising quality. It involves planting fruit trees closer together than under traditional methods, utilizing strategies such as dwarfing rootstocks and precise pruning to control tree size.

b) **Principles of HDP:** The principles underlying HDP include maximizing vertical and horizontal space efficiency, optimizing resource utilization, enhancing sunlight capture, and achieving a balanced vegetative-reproductive growth ratio.

c) **Advantages of HDP:** HDP offers several advantages, including optimized land and resource use, increased yield of high-quality fruits per unit area, enhanced solar radiation utilization, compatibility with modern agricultural practices (*e.g.*, drip irrigation, mechanization), early economic returns, and improved fruiting precocity and quality.

d) **Challenges of HDP:** Despite its benefits, HDP faces challenges such as high initial establishment costs, potentially shorter orchard lifespan, reduced fruit size and weight, increased difficulty in cultural operations and maintenance, lack of standardized production technologies, and heightened susceptibility to pests and diseases due to increased canopy density.

e) **Key aspects in HDP:** HDP can be applied using mono-species (single type of crop) or multi-species (multiple crops in layered arrangements). It involves manipulating tree growth through techniques like dwarfing rootstocks and precise pruning to enhance fruit-bearing efficiency.

f) **Impact of HDP in fruit crops:** In various fruit crops like Mango, Citrus, Banana, and Pineapple, HDP has shown significant increases in yield per hectare compared to traditional planting methods. For instance, Mango planted at 2.5m x 2.5m spacing achieved a 250 per cent increase in yield.

g) **Plant architecture in HDP:** In HDP, plants are structured to maximize fruit-bearing potential by reducing vegetative growth and ensuring optimal light exposure for photosynthesis. This is achieved through strategic pruning and careful selection of plant forms and rootstocks.

h) **High density planting (HDP):** It is a contemporary fruit cultivation method involving densely planted fruit trees. This technique utilizes small or dwarf tree varieties with modified canopies to optimize light interception, distribution, and facilitate mechanized field operations.

i) **Advantages of HDP:** HDP and meadow orcharding significantly enhance yield and economic returns per unit area. Achieving these benefits requires regular pruning and the use of bioregulators to manage tree size and shape. Modern fruit production increasingly emphasizes maximizing production per unit area.

j) **Plant architecture in HDP:** Ideal plant architecture for HDP involves maximizing fruiting branches while minimizing structural branches. Branches should be arranged and sized to minimize shading between them. Plant architecture is influenced by propagation methods, rootstock selection, and spacing.

k) **Methods to achieve HDP:** HDP employs methods like dwarfing rootstocks, pruning, root pruning, root restriction, growth regulators, and sometimes viral induction to control tree size and enhance fruiting efficiency. Each method plays a crucial role in achieving the desired plant architecture and productivity.

Controlling Tree Size

Methods for controlling tree size include using genetically dwarf scion cultivars, dwarfing rootstocks, training, pruning, root pruning, root restriction, growth regulators, inducing viral infections, and utilizing incompatible rootstocks.

Training and Pruning

Effective training and pruning are essential in HDP and meadow orcharding for shaping trees and controlling size. Properly managed pruning ensures sustained yields without sacrificing tree health.

Root pruning and restriction: Root pruning reduces vegetative growth, promoting balanced tree growth. Root restriction limits root spread, encouraging tree dwarfness and enhancing carbohydrate allocation to flowering and fruiting.

Use of growth regulators: Growth regulators like CCC and Paclobutrazol regulate tree growth, delay flowering, and reduce vegetative growth, enhancing fruiting efficiency in high-density orchards.

Planting systems and density: Various planting systems (square, triangular, hedgerow) and densities (low, medium, high) are employed based on crop type and management goals. Optimal light interception and efficient space utilization are critical in maximizing orchard productivity.

Mechanization and automation: Mechanization plays a crucial role in HDP, particularly in irrigation and fertigation, ensuring consistent orchard management and productivity.

Future Prospects and Research

Ongoing research continues to refine HDP techniques and adapt them to different agro-climatic conditions and fruit varieties. Advances in genetics, horticultural practices, and mechanization are expected to further optimize the benefits of HDP in fruit cultivation.

By embracing High Density Planting, farmers can maximize their orchard's productivity while ensuring sustainable use of resources and maintaining high fruit quality, thus contributing to efficient and profitable fruit cultivation practices.

REFERENCES

Doud DS and Ferree DC, 1980. Influence of altered light levels on growth and fruiting of mature 'Delicious' apple trees. *Journal of the American Society for Horticultural Science*, **105(3)**, 325-328.

Halle F and Oldeman RAA, 1970. *Essai sur l'architecture et la dynamique de croissance des arbres tropicaux*. Masson et Cie.

Halle F, Oldeman, RAA and Tomlinson PB, 1978. Tropical trees and forests: An architectural analysis. *Springer-Verlag*.

Iyer Rejum CPA and Kurian, 2006. High density planting in tropical fruits principle and practices. International books distributing co., Lucknow (UP). pp 48-60.

Jackson ML and Palmer RG, 1977. Soil Chemical Analysis – Advanced Course. University of Wisconsin.

Kumar N, 1997. Introduction to Horticulture. Rajalakshmi Publications, 28/5 – 693, Vepamoodu Junction, Nagercoil. Pp: 15.47- 15.50.

Lal S, Tiwari JP and Mahajan AR, 2007. Studies on planting systems in guava (*Psidium guajava* L.) cv. Sardar. *Acta Hort.*, **735**: 263-266.

Mika A, 1980. Physiological responses of apple trees to pruning. *Horticultural Reviews*, **2**: 337-378.

Mishra DS and Goswami AK, 2016. High density planting in fruit crops. *Hort Flora Res. Spectrum*, **5(3)**: 261-264.

Mishra DS, Lal RL and Chand S, 2014. Precision litchi culture technologies in Uttarakhand. *In: Recent Plasticulture Approaches Towards Precision Horticulture* (Eds.) PFDC, Pantnagar, pp. 73-77.

Moreshet S, Cohen Y and Fuchs M, 1975. Water use and yield relationships of trickle-irrigated peanuts. *Agronomy Journal*, **67(3)**: 274-278.

Ram S, 1996. High density orcharding in mango. *Res. Bull.* 122, DES, GBPUAT, Pantnagar. pp. 1-24.

Singh G, Singh AK and Mishra D, 2007. High density planting in guava. *Acta Hort.*, **735**: 235-241.

Singh KK and Saroj, 2018. Density Concept of Orcharding. *International Journal of Advanced Scientific Research and Management*, **(I)**: 54- *61*.

Wünsche JN and Lakso AN, 2000. The relationship between leaf area and light interception by apple trees. *Journal of Horticultural Science and Biotechnology*, **75(3)**: 321-331.

Chapter 7

Production of Planting Material and Orchard Management of Tropical Minor Fruits

Prakash Chandra Tripathi[1], *G. Karunakaran*[2], *Anuradha Sane*[2], *V. Sankar*[2], *R. Senthil Kumar*[2], *C. Kanupriya*[2] *and Pritee Singh*[2]

[1]*Horticultural Science Division, ICAR, New Delhi–110001*
[2]*ICAR–Indian Institute of Horticulture Research, Bengaluru, Karnataka*

Minor fruit crops refer to fruit species that, while potentially valuable, are not widely cultivated or commercially prominent on a large scale. These fruits are generally consumed in smaller quantities, either because they are less palatable or less available compared to more common fruits. Their market demand may also be lower, and they are cultivated on a limited scale. It is hard to draw a clear distinction between major and minor fruits. A fruit considered major in one region or country may be regarded as minor in another.

Minor fruits can be categorized based on various criteria, such as climate (temperate, tropical, or subtropical) and origin (native or exotic). India, with its rich biodiversity, hosts a variety of habitats, ranging from tropical rainforests and arid zones to temperate forests and coastal wetlands. Several fruit species, including jackfruit, bael, aonla, ber, khejri, jamun, mahua, phalsa, lasora, karonda, wood apple, pilu, bilimbi, and Garcinia, have originated in the Indian subcontinent.

Additionally, many exotic fruits such as rambutan, mangosteen, longan, avocado, water apple, hog plum, macadamia nut, durian, passion fruit, dragon fruit, and carambola were introduced to India over the centuries and have since become naturalized (Tripathi *et al.,* 2017). India is also home to over 100 wild edible fruits that are yet to be fully domesticated but are regularly gathered from forests by rural and tribal communities and sold in local markets (Tripathi *et al.,* 2014a, 2014b, 2017, Tripathi, 2018, 2019).

Tropical minor fruits occupy less than 5 per cent of India's total fruit-growing area, contributing less than 2 per cent to overall fruit production. Jackfruit and passion fruit dominate in terms of cultivation area, while other fruits are mostly grown in Kerala, Tamil Nadu, and Karnataka. These fruits are often grown on less fertile land under rainfed conditions (Anonymous, 2017; Tripathi *et al.,* 2017). Minor fruit crops are hardy and hold significant potential for boosting local and regional productivity and nutritional security. Numerous studies highlight the nutritional and health benefits that indigenous or traditional fruits provide to rural and indigenous communities. Though most minor fruits are lower in carbohydrates, they are rich in fibre, nutrients, vitamins, and nutraceutical properties. While protein content is generally low, certain minor fruits and nuts, such as macadamia nut, chironji, Java almond, Shea butter tree, Brazil nut, Indian almond, marula nut, mongongo nut, breadfruit, jackfruit, durian, longan, rambutan, mangosteen, and passion fruit, offer higher protein or carbohydrate levels (Gopalan *et al.,* 2004; Pareek *et al.,* 1998). Avocado, kokum seeds, Malabar tamarind, and mahua are notable for their fat content. Other fruits, such as peach palm, cape gooseberry, chironji, pine nut, and bael, are excellent sources of vitamins and minerals. Several minor fruits are valued for their medicinal properties. Kokum and Malabar tamarind, for instance, are key ingredients in anti-obesity formulations. Many fruits used by tribal communities in north-east India are employed in treating common ailments like jaundice, diabetes, diarrhoea, and dysentery (Pareek *et al.,* 1998; USDA database, 2017; Tripathi *et al.,* 2020).

One of the major challenges in expanding the cultivation of minor fruits is the lack of quality planting material. New plantations are not being established due to the unavailability of proper planting stock. To address this, the standardization of propagation techniques and rootstocks, as well as the production of high-quality planting material in open conditions, is crucial. The multiplication of quality plants involves several steps, including seedling and rootstock raising, the maintenance of mother blocks for scions and rootstocks, and proper propagation and aftercare techniques.

Propagation

Minor fruits are propagated through seed and vegetative propagation methods.

Seed Propagation

Seed propagation is one of the main methods for multiplying tropical minor fruit crops. Seeds can be categorized into two types: recalcitrant and non-recalcitrant (or orthodox).

Recalcitrant Seeds

Recalcitrant seeds lose their viability quickly and cannot tolerate drying or freezing. These seeds must be stored at ambient temperatures with high humidity to maintain viability. Once dried, they will not germinate and cannot be stored in conventional freezers at -20°C. Therefore, recalcitrant seeds should be sown immediately after extraction from the fruit. Examples of such seeds include rambutan, avocado, and Garcinia species.

Non-Recalcitrant (orthodox) Seeds

Orthodox seeds can be dried to reduce moisture content (below 12%) and stored at low temperatures for longer periods. These seeds may exhibit dormancy, requiring treatments to break it before germination. Examples of such seeds include dragon fruit and passion fruit. Seed treatments using chemicals or growth regulators, like GA3 (gibberellic acid), can enhance germination in certain species.

Propagation Challenges in Minor Fruits

Seed propagation is common for many minor fruits, but challenges exist such as:

- *Slow Growth*: In some crops, such as dragon fruit, seedling growth is very slow, making seed propagation unsuitable for commercial planting.
- *Variability*: Many minor fruits, like rambutan and passion fruit, produce variable seedlings due to cross-pollination, which leads to inconsistent plant performance and fruit quality.
- *Male-Female Ratio*: Certain species, such as rambutan and kokum, produce male and female seedlings in equal proportions, making seed propagation less favorable for commercial cultivation.

Propagation of Specific Minor Fruits

Dragon Fruit

Dragon fruit seeds germinate rapidly if sown soon after extraction. However, seedlings grow slowly and show significant variability, making seed propagation only suitable for breeding programs (Tripathi *et al.,* 2014).

Rambutan

Rambutan seeds are encased in a translucent, whitish aril and are brown in color, measuring 1–1.5 cm in length and weighing 2–4 grams, with a distinctive white basal scar. After removing the aril and thoroughly washing, the seeds should be planted horizontally, with the flattened side facing downward. This orientation helps ensure the seedling grows straight and develops a strong, healthy root system. Under Kodagu conditions, germination typically occurs within 25–28 days after sowing in nursery bags (Tripathi *et al.,* 2014). Rambutan seeds lose viability quickly, so it's crucial to plant them immediately after removing the aril. Fresh seeds (within 2 days of removal) exhibit a germination rate of 87 to 95 per cent, whereas seeds that are one week old see significantly reduced germination rates of 50–60 per cent. Exposure to sunlight for 8 hours or oven drying at 30°C kills the seeds within a week. For temporary storage, fresh seeds can be preserved in moist sawdust, sphagnum moss, or charcoal for 3–4 weeks. Seedlings are generally ready for grafting in 9–10 months. In Indonesia, the local rambutan variety 'Glu' is traditionally considered a superior rootstock. Research on seedlings of the 'Glu' variety showed improved growth traits, including larger leaf area, higher seedling fresh weight, and more root development (Mitras *et al.,* 2011). The same study also found that using farmyard manure compost significantly promoted better root growth compared to other composts. Although propagating rambutan through seeds is easy, it is not recommended for commercial purposes since over 50 per cent of seed-propagated plants produce only male flowers. Therefore, seedlings are typically used for rootstock purposes rather than fruit production.

Passion Fruit

Passion fruit vines can be propagated from seeds, but this method is not ideal for commercial purposes due to the high variability in seed-propagated plants. For seed propagation, fruits should be collected from high-performing vines known for good yield and quality. The seeds are medium-sized, round, brown-black in color, and covered in a mucilaginous pulp. The seeds are typically extracted using the fermentation method, which involves heaping the pulp for 72 hours, extracting the seeds, and drying them in the shade. Sowing is usually done during March-April in a well-prepared seedbed. The seeds begin sprouting about 12–15 days after sowing, with germination generally completed within a month, although in some cases, it can extend up to 50–60 days. Various pre-sowing seed treatments, including chemical treatments with HCl, H2SO4, thiourea, KNO3, sucrose, urea, bleaching powder, and growth regulators like 100 ppm GA_3, have been found to enhance seed germination. Among these, seeds treated with 2 per cent bleaching powder for 24 hours showed the best results in terms of germination and seedling growth. Additionally, treatment with 100 ppm GA_3 for 30 minutes was also found effective (Mehta *et al.,* 2016). Once the seedlings develop four to six leaves, they are transplanted into 10 x 22 cm polythene bags filled with a soil, compost, and sand mixture in a 2: 1: 1 ratio. The seedlings are typically ready for field transplanting after about three months.

Avocado

Avocado seeds are of the recalcitrant type, meaning they retain high moisture content and lose viability quickly if not properly handled. These seeds are almost round, weigh between 25 and 60 g, and have a light dull yellow color. They are located in the fruit's cavity and can be easily removed once the fruit ripens. After cutting the fruit in half, the seeds should be planted vertically with the flattened side down, ensuring that the seedling grows straight and develops a strong root system. Under Kodagu conditions, avocado seeds typically germinate within 1–2 weeks, with a germination rate of almost 90–100 per cent for freshly collected seeds. However, the seeds lose viability quickly when stored at room temperature. The viability period is typically 2 to 3 weeks but can be extended by storing the seeds in dry peat or sand at 5°C. Occasionally, multiple seedlings can emerge from a single avocado seed. Although seed propagation was once a common method for avocado cultivation, grafted plants are now more commonly used for planting due to better uniformity and performance. Seedlings produced through seed propagation are mainly used as rootstocks.

Garcinia Species

Seeds of most Garcinia species, including kokum, Malabar tamarind, mangosteen, and yellow mangosteen, is recalcitrant, meaning they retain moisture and lose viability quickly if not planted immediately. Fresh seeds generally germinate easily, although some species exhibit a brief dormancy period. The germination rate declines rapidly when seeds are stored at room temperature. To ensure good germination, seeds should be extracted from fully ripe fruits. Kokum and Malabar tamarind fruits typically contain 3–5 seeds, while mangosteen and yellow mangosteen usually have 1–2 seeds. It is advisable to select seeds weighing more than 1 gram for optimal germination. Freshly extracted seeds must be sown immediately for a high germination rate. Sowing is typically done in a seedbed, which may be constructed from wood or cement. The sowing medium should consist of a mixture of sand and soil in a 3: 1 ratio by volume, which retains moisture but drains well. Seeds are planted at a depth of 5–10 mm and spaced 2–3 cm apart, then covered with fine sand. Germination usually occurs about 30 days after sowing. Once seedlings emerge, they are transplanted into polyethylene bags using the same sand-soil mixture, while the cotyledons are still attached. Seedlings become ready for field planting within 9 to 18 months. Some Garcinia species produce parthenocarpic seeds, which are useful for multiplying true-to-type plants. For mangosteen and yellow mangosteen, seedlings aged 24 to 36 months are recommended for planting to improve survival rates and reduce maintenance costs, as younger seedlings require extensive care. However, seed propagation is not recommended for the commercial cultivation of kokum and Malabar tamarind, as approximately 50 per cent of seedlings turn out to be male. These seedlings, though, can be used for grafting or budding purposes. Additionally, seedling trees may serve as source of pollen in commercial orchards of elite lines or varieties.

Macadamia Nut

While macadamia nuts can be propagated by seeds, commercial nut production is almost exclusively from grafted trees of known, high-performing varieties. Seedling trees have proven unsatisfactory due to significant variation in productivity, kernel quality, and a longer juvenile phase. Seed-grown trees sometimes produce bitter nuts with low oil content, making them undesirable for commercial cultivation. For raising rootstocks, mature, husked nuts should be used. However, the hard seed coat of macadamia nuts creates a significant barrier to germination. In un scarified seeds, germination rates are as low as 1 to 5 percent. Scarification, a process that involves physically breaking or softening the seed coat, is essential for improving germination rates. Under Bangalore conditions, macadamia seedlings grow slowly, requiring 15 to 18 months to become ready for transplanting.

Jamun

Most fruits of the Syzygium genus, including jamun, malayan apple, rose apple, and wax apple, are generally propagated by seeds. Jamun is primarily multiplied through seed propagation, which is widely practiced for plant production. However, significant variation is observed in plants due to cross-pollination. Jamun seeds weigh between 1 to 2 g and measure 1-1.2 cm in length. They germinate rapidly, with an initial moisture content of 44.2 per cent on a fresh weight basis and an impressive initial germination rate of 89 per cent. The seeds are tolerant to desiccation up to moisture content of 10 per cent. The optimal storage temperature range is 15 to 20°C; storing seeds at 10 per cent moisture content at 20°C can extend their viability for up to 180 days with a germination rate of 92.5 per cent (Anandalakshmi *et al.*, 2005). While fresh seeds do not exhibit dormancy, germination rates decline when stored at room temperature. Seedling trees are vigorous but typically begin flowering only after 8-10 years, making seed propagation unsuitable for commercial plantations. However, it is useful for producing rootstocks. Polyembryony is common in jamun, allowing one seed to produce multiple seedlings, occasionally yielding up to 8-10 seedlings per seed. These seedlings can be utilized for avenue planting and social forestry programs. Vegetative propagation is gaining popularity in jamun due to its advantages, including earlier fruiting, lower vigor of the plants, and more uniform fruit quality.

Custard Apple (*Annona* Species)

Fruits from the Annona family, including custard apple, pond apple, and soursop, are typically propagated by seeds. However, there is a rapid loss of seed viability within six months after harvest. For optimal germination, seeds should be planted as soon as possible after being removed from the fruit. Seeds should be sown in trays or containers, kept moist, and placed in a shaded environment. Fresh seeds generally take 15 to 30 days to germinate, achieving a germination rate of 85-90 per cent.

Pummelo

It can also be propagated by seed; however, seed propagation is not commonly practiced among growers due to the plants' lack of uniformity, inability to come true to type, and the presence of many long spines.

Karonda

Karonda plants can be easily multiplied through seed propagation, which is the most commonly used method for this species. Seeds should be collected immediately after harvesting, as sowing them right after extraction leads to higher germination rates. The seeds are sown in trays, and the resulting seedlings are transplanted into polyethylene bags once they reach the 3-4 leaf stage. The plants become ready for field planting in about 8-10 months. It's important to note that germination rates in seedless or less-seeded varieties tend to be low (Tripathi *et al.,* 2014).

Egg Fruit

It can be propagated through seeds, but the seeds lose their viability very quickly. Therefore, they should be sown immediately after extraction from the fruits. Under Bangalore conditions, seed germination rates were found to be nearly 45 per cent. The seeds typically take about 28 days to germinate, with a final stand of approximately 32 per cent after three months. The seedlings reached an average height of 36 cm during this period.

Wood Apple

Wood apple seeds are small and brown in colour. These seeds exhibit no dormancy and germinate rapidly within 15 to 20 days. However, the growth of seedlings is slow, and the establishment of young seedlings tends to be low. Additionally, seedling trees typically bear fruit only after 10 to 15 years of age and show genetic variability.

Seed propagation is commonly used for the multiplication of various fruits, including karonda, wood apple, noni, alangium, carambola, canistel, bael, egg fruit, star gooseberry, surinam cherry, jackfruit, pummelo, and flacourtia. A study conducted in Bengaluru revealed that fresh seeds of these fruits did not exhibit dormancy, with germination rates exceeding 60 per cent. However, the duration taken for germination varied among these fruits (Tripathi *et al.,* 2021).

Vegetative Propagation of Minor Fruits

Most minor fruits are cross-pollinated and highly heterozygous, resulting in significant variability among seedlings, which makes them unsuitable for commercial plantations. To produce true-to-type plants, various methods of vegetative propagation have been standardized. Techniques such as cutting, grafting, budding, and layering are commonly used for the multiplication of some minor fruit species.

Dragon fruit can be easily propagated using stem cuttings. Typically, 20-25 cm long cuttings are selected for planting. It's advisable to prepare the cuttings one to two days before planting, allowing the latex that oozes from the cut to dry. Cuttings should be taken from elite mother plants during the fruiting season. Either the entire stem segment or individual cuttings of 20-30 cm in length can be used, with the base cut of the cutting slanted to promote better rooting. To prevent diseases, the cuttings should be treated with fungicides. These cuttings are then planted in polyethylene bags measuring 12 x 30 cm, filled with a mixture of soil, farmyard manure, and sand in a 1: 1: 2 ratios. The bags should be kept in a shady location to encourage rooting, while excess moisture must be avoided to prevent rotting. The cuttings will root profusely and become ready for planting in approximately 5 to 6 months.

Rambutan can be vegetatively propagated through several methods, including approach grafting (inarching), air layering, budding, and grafting. Although cuttings have been attempted under mist with the application of growth-promoting hormones, this method has shown low success rates and is not practiced commercially. Among these methods, approach grafting has proven to be the most successful for rambutan propagation under Kodagu conditions. Typically, 8- to 12-month-old seedlings are used as rootstocks. An experiment conducted at the Central Horticultural Experiment Station (CHES), Chettalli, revealed a 42 per cent success rate when using ten-month-old seedlings in November and December. However, approach grafting is labour-intensive and cumbersome, as the number of grafts that can be prepared at one time is limited due to lack of available space and shoots. Other grafting methods, such as cleft and veneer grafting, have also been attempted, but their success rates are very low, ranging from 5 to 10 per cent. In a shade house setting, the success rate for cleft grafting was found to be only 10 per cent (Tripathi *et al.,* 2017). Air layering has been tried using upright branches with a stem diameter of 1-1.5 cm and length of 30-60 cm from well-developed trees. Plant growth regulators, such as IBA at 5,000 ppm, were used in the rooting media (coir pith), resulting in about a 25 per cent success rate (Anonymous, 2007). While these air layers appeared successful initially, many died after being transplanted into field nurseries or poly bags or later in the field, long after separation from the mother tree (Anonymous, 2010). Among the various budding methods, patch budding is preferred due to its higher success rate. Budding is typically performed on well-grown rootstocks aged 8 to 12 months during May or July. In other countries, the success rate for patch budding has been reported as high as 83.6 per cent. T-budding has also shown success. The experiment at CHES, Chettalli, indicated that higher humidity and moderate temperatures are essential for successful budding; success rates decline at temperatures below 20°C. Selecting high-quality budwood is crucial for successful budding. Budwood should be collected from good-bearing trees, with the selected shoots obtained 3 to 4 months post-harvest. Budwood that is one year old and has a diameter of 1.5 to 2 cm is considered most suitable. Removing leaves one week prior to budding can enhance success rates (Tripathi *et al.,* 2017).

Avocado is commercially propagated through cleft grafting, stone grafting, and approach grafting, with cleft grafting being the most successful and popular method among nurserymen. Several experiments have been conducted to standardize propagation methods for avocado under Coorg conditions. Chithiraichelvan *et al.* (2006) found that approach grafting was particularly suitable for avocado multiplication. Experiments were also done to evaluate four methods of propagation: veneer grafting, cleft grafting, T-budding, and patch budding. After 90 days, cleft grafting demonstrated the highest success rate at 32.5 per cent, while lower success rates were observed for veneer grafting (8.3%), T-budding (7.5%), and patch budding (5.8%). Among the various months tested, the highest success was recorded in September (32.5%), followed by March and August with success rates of 22.5 per cent. Notably, the highest overall success (70%) was achieved with cleft grafting performed in September. Additionally, scion growth was superior in grafting methods compared to budding methods. Thus, it was concluded that cleft grafting in September and March is most suitable for vegetative propagation of avocado under humid tropical conditions (Tripathi and Karunkaran, 2019). Khawas and Upadhyay (2022) evaluated the effects of different concentrations of IBA on rooting in cuttings of two avocado cultivars, Hass and Fuerte, under mist chamber conditions. They observed a significant increase in root initiation with the treatment of IBA at 800 ppm combined with the Fuerte cultivar, followed by IBA at 400 ppm with Fuerte. The Fuerte cultivar treated with IBA at 400 ppm exhibited better rooting capacity compared to other treatments. At the Fruit Research Station in Kallar, Tamil Nadu, both layering and approach grafting yielded up to 75 per cent success. In West Bengal, chip budding has also been reported to be successful. Furthermore, softwood grafting achieved success rates of over 90 per cent. Currently, all commercial nurseries in India predominantly use cleft grafting for avocado propagation, with most rootstocks being seedling-based.

Passion fruit can be propagated through stem cuttings, although it is not inherently easy to root. However, satisfactory rooting can be achieved under favourable conditions. Generally, mature portions of the vine, measuring 30-35 cm long and containing 3 to 4 nodes, are selected for cuttings. These cuttings should be planted in a suitable medium, preferably an equal mixture of sand, soil, and farmyard manure. Rooting can be expedited through hormone treatment, with studies showing that treating cuttings with 200 ppm NAA for a very short duration (3-5 seconds) or 80 ppm NAA for 12 hours significantly increases rooting success. Typically, rooting occurs within a month, and the cuttings can be transplanted to the field after about three months (Kumar *et al.*, 2008). Grafting is not a common propagation method for passion fruit, but techniques such as cleft grafting, whip grafting, and side wedge grafting are employed to leverage the benefits of rootstocks. Certain lines of yellow passion fruit that exhibit resistance to nematode infestations and diseases have proven to be effective rootstocks for hybrid passion fruit varieties. Three- to four-month-old seedlings are generally used as rootstocks, and scions from healthy young vines are preferred over those from mature plants. For successful grafting, both the stock and scion should be of pencil thickness, and the diameter of the selected scion should closely match that of the rootstock. Trials conducted at CHES, Chettalli, revealed that cleft grafting of the passion fruit variety Kaveri (as scion) on yellow passion fruit rootstock achieved the highest success rate of 75 per cent in September, while a lower success rate of 30 per cent was recorded in March. Thus, cleft grafting in September is considered the most suitable time for passion fruit grafting under Coorg conditions (Anonymous, 2013). Another effective method for propagating passion fruit is serpentine layering. In this technique, lateral shoots emerging from the main branches are given a partial slanted cut below the nodes, allowing them to root in a medium consisting of soil, sand, and compost (1: 2: 1) with regular irrigation, typically in February. Roots emerge profusely within 45 days, and the plants are separated during April-May. This method has been highly successful, demonstrating a field survival rate of 90-95 per cent and higher (Anonymous, 2012).

Longan can be propagated through various methods, including seed, air layering, budding, grafting, cutting, and inarching. However, propagation by seed is generally not recommended, as seedlings take 7-8 years to bear fruit. Among the different methods of vegetative propagation, air layering is the most popular and widely used technique, offering a high success rate of 80-90 per cent. Air layering is typically conducted during the rainy season. To perform this method, a strong, healthy branch with mature leaves is selected. The branch is girdled, and a 2-2.5 cm wide strip of bark is completely removed down to the cambium layer. The exposed surface is then scraped to remove the phloem and cambial tissue, preventing premature healing. A rooting hormone, such as IBA, is applied to the cut surfaces of the branch to encourage rooting. A root ball is then created using a suitable medium that retains moisture and is well-aerated. Wet peat and sphagnum moss are recommended materials for this root ball. Roots typically begin to form within 2-3 weeks and can be harvested after two to four months, once the roots have changed color from white to creamy brown. It is advisable to prune the rooted air layers to reduce the top in proportion to the roots. Afterward, the rooted air layers are planted in suitable containers and kept in warm, humid, and partially shaded conditions in the nursery to acclimatize before being transplanted to the field. However, trees produced by air layering tend to be more susceptible to wind damage compared to grafted trees. In addition to air layering, longan can also be propagated by grafting, with approach grafting being a common method using seedlings of the same cultivar as the rootstock. Eight- to twelve-month-old seedlings are typically employed as rootstocks. In Thailand, grafting is usually performed in winter when vegetative growth has ceased. Forkert budding and cuttings have also proven successful for longan cultivars introduced into Australia (Menzel *et al.*, 1990). However, instances of graft incompatibility have been reported in China.

Jamun: Several methods of vegetative propagation have been standardized for jamun, including softwood grafting, veneer grafting, budding, and approach grafting. Among these, softwood grafting is the most popular method. For softwood grafting, ten- to twelve-month-old seedlings that have reached a graftable size (approximately pencil thickness) are used as rootstocks. Activated scion sticks, which should be defoliated 7-10 days prior to grafting and measure 4-6 inches in length, are selected for grafting. The best time for grafting is during July and August, which typically yields a maximum success rate of 80 per cent. However, the success of grafting can vary with different climatic conditions, generally showing lower success rates in cooler temperatures and lower humidity. Veneer grafting is also employed in some locations, achieving a success rate of around 60 per cent when performed in February and March, and about 50 per cent in September. Both in situ and ex situ budding techniques on young seedlings of jamun have shown success rates of 60-70 per cent during February. Approach grafting was once the most popular propagation method for jamun, known for its high success rate (>90%) when conducted in June and July. However, due to its cumbersome nature and the limitation on the number of plants that can be produced, it is becoming less popular over time.

Pummelo can be propagated through budding, grafting, and air layering. In India, budding is the most preferred method of propagation. Trials conducted at CHES Chettalli demonstrated that patch budding achieved a high success rate of 80-90 per cent across all months under polyhouse conditions. The Rangapur lime is commonly utilized as rootstock for pummelo in these practices. In other countries, grafting methods are predominantly employed, with Volkamer lemon and calamondin being favored as rootstocks.

Karonda can be multiplied using stem cuttings, air layering, and budding, particularly for elite varieties and lines. Semi-hardwood cuttings are most suitable for plant propagation, typically measuring 25-30 cm in length and around 1 cm in diameter. The optimal time for planting these cuttings is from June to July. Trials conducted at CHES Chettalli revealed that semi-hardwood cuttings achieved higher success rates compared to softwood and hardwood cuttings. Specifically, semi-hardwood cuttings planted during July and August yielded a success rate of 30-40 per cent, outperforming both hardwood and softwood varieties. Additionally, treatments with IBA did not significantly enhance success rates in either hardwood or softwood cuttings. Notably, success rates for table-purpose, bold fruit varieties tend to be lower than those for pickle-type varieties. Air layering has also been successful for Karonda when performed during June to July, with success rates varying between 30-60 per cent in different years. The air layers are typically removed in September and then planted in polythene bags, where they become ready for planting after 6-7 months.

Garcinia species: Various vegetative methods of propagation, including air layering, approach grafting, veneer grafting, softwood grafting, and root cutting, have been explored for Kokum and Malabar tamarind. Studies conducted under Karnataka conditions revealed that among the grafting techniques—cleft, whip, tongue, and epicotyl grafting—the survival rate was lowest for epicotyl grafting and highest for cleft grafting. For successful grafting, scion wood should be sourced from high-yielding trees, utilizing terminal shoots that are greenish-brown in color and approximately 0.5-0.6 cm thick. Healthy and vigorous Kokum seedlings, 10-15 cm in height and 4-5 months old, are selected as rootstocks. The optimal seasons for softwood grafting in Kokum, particularly in the Konkan region of Maharashtra, are from October to November and March to August. Cleft grafting is widely employed as a method of asexual propagation, with grafted plants typically becoming ready for planting 6-7 months post-grafting. Additionally, vegetative propagation through cuttings has been found successful in yellow mangosteen in Assam.

Bael: Various methods of vegetative propagation, including grafting, air layering, patch budding, and root cuttings, have been explored in wood apple, yielding variable success rates across different regions. At the Indian Institute of Horticultural Research (IIHR) in Bangalore, cleft grafting performed in September and October achieved a success rate of 80 per cent, significantly higher than the rates observed in March and July. Grafting was found to be more effective compared to budding. Despite these promising methods, the large-scale adoption of vegetative propagation for wood apple is yet to be realized, with seedlings still predominantly used for planting. In macadamia cultivation, traditionally grafted trees take approximately 24 months from germinating the seedling rootstock to field planting. Mini-grafted trees require about 10-12 months from seed, while the more innovative micro-grafting technique can reduce this timeline to just 8-12 months.

Annona: The multiplication of selected soursop lines can be effectively achieved through cuttings or shield budding. Generally, soursop seedlings serve as the best rootstock for propagation. Successful grafting can also be performed on custard apple (*Annona reticulata*), mountain soursop (*Annona montana*), or pond apple (*Annona glabra*), with the latter being particularly suitable as a dwarfing rootstock. However, grafting on custard apple (*Annona squamosa*) and cherimoya (*Annona cherimola*) has not yielded satisfactory results due to the short lifespan of grafted plants.

Macadamia nut is typically propagated via wedge grafting on seedling rootstocks.

A study conducted at ICAR IIHR focused on identifying ideal propagation techniques for wild and minor fruit crops and revealed that mulberry cuttings showed a positive response with an 80 per cent success rate. Air-layering techniques were successfully employed in Governor's plum (*Flacourtia indica*), karonda (*Carissa carandas*), natal plum (*Carissa macrocarpa*), Citrus hindsi, Surinam cherry (*Eugenia uniflora*), and West Indian cherry (*Malpighia emarginata*), achieving survival rates of 75 per cent and 60 per cent, respectively (Tripathi *et al.*, 2021). Further studies aimed at standardizing vegetative propagation methods and timings for jackfruit, carambola, fig, and star gooseberry under the red and laterite zone conditions of West Bengal showed that inarching performed on July 20 yielded the highest success rate (80%) along with a 70 per cent field establishment rate for detached jackfruit plants. In carambola, grafting on seedling rootstocks resulted in 100 per cent success with robust graft growth during the rainy season, while air layering achieved a 60 per cent success rate and a 50 per cent field establishment rate. In figs, the highest success (50%) and good graft growth were recorded when grafting was conducted on June 15 and 30 using local fig rootstock, with a field establishment rate of 50-60 per cent. Star gooseberry achieved 50 per cent rooting success in air layering with complete field establishment of detached layers when layering occurred on July 30 (Ghosh and Bera, 2015). Cleft grafting proved to be the most successful method for jackfruit and jamun, achieving over 80 per cent success when performed during the rainy season under Bangalore conditions.

Orchard Management: Planting Distance

The planting distance for fruit trees is influenced by the vigor and size of the mature tree. Seedling trees, which tend to exhibit vigorous growth and larger sizes, are typically planted at wider distances. In contrast, vegetatively propagated trees, which are usually less vigorous, are planted closer together. Additionally, climatic conditions and soil fertility play significant roles in determining the optimal planting distance. Research indicates that the growth rates of certain tropical fruits, such as rambutan, durian, and pulasan, are slower in Indian conditions compared to their growth in tropical humid countries. Consequently, grafted plants of rambutan, mangosteen, avocado, and jackfruit are commonly spaced at 6 x 6 m rather than the broader 8 x 8 m. Similarly, grafted plants of kokum and Malabar tamarind are also planted at 6-meter intervals. For crops like passion fruit, karonda, and dragon fruit, which require specific training and pruning systems, a closer planting distance is adopted.

Seedling trees of **rambutan** are particularly vigorous and spread out, necessitating more space for their growth. Therefore, these trees should be planted at greater distances to prevent overlapping after 10 to 15 years of growth.

Dragon fruit is typically planted at a spacing of 4 x 4 meters, supported by poles that are 1.5 to 2 meters long. Depending on climatic conditions, each pole can support two to four plants. In contrast, seedling trees of rambutan are generally spaced between 10 x 10 to 12 x 12 meters. In the Philippines, it is recommended to plant rambutan trees at least 10 meters apart in all directions; however, in rich soil, a distance of 12 meters is also advisable. Planting trees too closely can lead to overcrowding within a few years, significantly affecting production. For vegetatively propagated trees, which are smaller in size, a spacing of 8 x 8 or 8 x 6 meters is appropriate. When planting at 6 x 6 meters, pruning may be necessary later to control tree size. In orchards with mechanization, the row spacing should accommodate machinery, with an optimal planting density of 200-250 plants per hectare. The square or triangular planting system is commonly used for rambutan (Tripathi, 2021).

In relatively new **avocado** plantations, careful selection of varieties is crucial. Both A and B type varieties should be chosen, ensuring their flowering periods overlap. A common proportion for planting is 1: 1 or 2: 1 for A and B group varieties. Avocados should be planted at distances ranging from 6 to 12 meters, depending on the vigor and growth habit of the variety. For spreading varieties like Fuerte, wider spacing is recommended. In areas susceptible to excess water, avocados should be planted on mounds, as they cannot withstand waterlogging. In Sikkim, a planting distance of 10 x 10 meters is preferred on hillside slopes. In South India, when planted alongside coffee, the spacing can vary between 6 to 12 meters. Pits measuring 1 cubic meter are dug during April-May and filled with a 1:

1 ratio of farmyard manure and topsoil before planting. Planting typically occurs in June-July or sometimes in September. In Coorg, Karnataka, avocado trees are also planted as part of a mixed cropping system primarily based on coffee (Tripathi *et al.*, 2014, 2022).

Longan is typically planted at the onset of the rainy season. Prior to planting, pits measuring 60 x 60 x 60 cm should be dug about two weeks in advance. To enrich the soil, organic manure and phosphate fertilizer can be incorporated into the pit filling. Tree spacing for longan ranges from 6 x 6 meters to 12 x 12 meters, resulting in plant densities between 70 to 300 trees per hectare. In mature orchards in Thailand, the tree density averages around 50 trees per hectare. Conversely, in Taiwan, a spacing of 4 to 6 meters is common, with densities reaching 300 to 600 trees per hectare in mature orchards. In Fujian province, China, plant densities range from 195 to 300 trees per hectare, with an average density of approximately 270 trees per hectare. Under normal spacing conditions, well-established trees typically fill their designated spacing within 6 to 7 years. At the Yehai Litchi Science and Technology Garden in Dongguang, Guangdong province, the longan variety 'Shixialongan' is planted at a high density of about 1,500 to 1,800 trees per hectare, maintained as dwarf trees through heavy pruning. Additionally, it is essential to stake and tie the plants securely to prevent them from being uprooted by strong winds.

Soursop and **custard apple** are planted with a spacing of 4 to 6 meters apart, depending on local climate and soil type. Pits of the same dimensions (60 x 60 x 60 cm) are dug and filled with a mixture of three parts topsoil and one part compost. It is preferable to plant during cloudy days in June or July, shortly after the onset of the monsoon, to ensure that the plants become well-established by the end of the rainy season. The plants selected for transplanting should be 9 to 12 months old and have a height of 60 to 100 cm additionally, plants should be irrigated as soon as possible after transplanting to promote healthy growth.

Jamun can be cultivated under various cropping systems, including as a pure crop, a component of horti-silvi-pastoral models, or as a hedgerow. It serves as an excellent windbreak when established around the periphery of an orchard. Typical spacing for jamun trees is 6 x 6 meters or 8 x 8 meters in a square system. However, there is no standard spacing recommended for exclusive commercial plantings. In some trials, jamun was planted at distances of 10 x 5 meters (200 plants/ha) and 5 x 5 meters (400 plants/ha). In high-density planting systems, a spacing of 2.5 x 2.5 meters (1,600 plants/ha) is used, accompanied by canopy management techniques to enhance productivity. It is important to note that closer planting requires careful canopy management through training and pruning, and such practices are not yet recommended for commercial planting. Pits measuring 90 x 90 x 90 cm are typically dug during the summer months, with planting occurring during the rainy season, ideally in July or August. In the initial 2-3 years of orchard establishment, it is advisable to protect the plants from adverse conditions.

Pummelo planting distances vary from 6 to 8 meters by 8 to 10 meters, depending on cultivar vigor and soil fertility. While planting is best done at the onset of the rainy season, it can also occur year-round if irrigation is available. Young pummelo plants require shade and frequent watering until they are well established.

The **table purpose variety of Karonda** should be planted at a spacing of 3 x 3 meters in a square layout. Pits measuring 3 x 3 feet should be prepared at least one month before planting and filled with an equal mixture of farmyard manure (FYM) and soil. The recommended planting time is June-July, and the land should be cleaned and levelled, sloping away from the water source. For hedge planting, karonda can be planted at 2-foot distances. The planting trench should measure 1 x 1 feet, with the option to create individual pits of the same size. For orchards, karonda is planted at 3 x 3 meters with pits of 2 x 2 feet dug before the rainy season. In rocky soil, pits of 3 x 3 feet should be opened.

In **macadamia nut** orchards, the design typically favors long rows to optimize land use and facilitate machinery operation. Planting distances can vary, ranging from high density (6 x 3 meters) to lower density (10 x 4 meters), depending on tree variety, soil conditions, and topography. Some growers prefer staggered planting within rows to form equilateral triangles, though trees commonly form long hedgerows.

Egg fruit should be planted at a spacing of 6 x 6 meters, with the rainy season being the best time for planting. The plants typically begin flowering 6 to 7 years after planting.

The planting distance for wood apple depends on soil type, land slope, and whether seedlings or grafts are used. Seedlings are usually planted at 8 x 8 meters, while grafted plants can be spaced at 6 x 6 meters. Planting is best done in July-August, with one to one and a half-year-old plants being optimal for planting. Older plants (2-3 years) have lower establishment rates. Mulching with dry leaves is recommended to conserve moisture. Planting is conducted in pits measuring 1 x 1 x 1 meter, which are dug in April or May, or 1-2 months prior to planting. The size of the pits may vary depending on the vigor of the tree, soil types, and topography (Table 7.1). The pits are filled with a mixture consisting of three parts topsoil and one part compost. It is preferable to plant during June or July, following the onset of the monsoon, to ensure that the plants are well established by the end of the rainy season. Grafted plants should be carefully removed from their containers, ensuring that the soil ball remains intact. They should then be placed in the prepared pits, with their stems straight and properly aligned with adjacent plants. After positioning the plants, cover the pits with topsoil and press gently. It is crucial to ensure that sufficient water is available to the plants for an extended period. If rain is not expected, irrigate the plants immediately after planting. Subsequently, regular irrigation should continue until the plants are fully established.

Table 7.1: Planting Methods and Timing for Humid Tropical Underutilized Fruits

Fruit	Planting Distance	Pit Size	Planting Time
Passion fruit	2 x 3 m	45 x 45 x 45cm	June -July
Soar sop	4 x 4 to 6 x 6 m	60 x 60 x 60cm	June -July
Rambutan (grafted	6 x 6 m	1 m X 1m X1m	June -July
Jack fruit (grafted)	6 x 6 m	1 m X 1m X1m	June -July
Avocado (grafted	6 x 6 m	1 m x 1m x1m	June -July
Karonda	3 x 3m0.6 x 6m (hedge row)	60 x 60 x 60 cm	June -July
Mangosteen	6 x 6 m	1 m x 1m X1m	June -July
Kokum (grafted)	6 x 6 m	1 m x 1m x1m	June –July
Malabar tamarind	6 x 6 m	1 m x 1m x1m	June –July
Dragon Fruit	5 x 5 m	60 x 60 x 60 cm	June –July
Wood apple	6 x 6 m	1 m x 1m x1m	June –July

Training and Pruning of Tropical Underutilized Fruits

Most tropical underutilized fruits are evergreen and generally require less pruning compared to temperate fruits. However, it is essential to train these plants to manage their height effectively, facilitating better management practices.

Training Methods

Open Centre System

Most of these plants are trained in an open center system, which allows for maximum light penetration and air circulation. The process of training branches and establishing the framework should be completed within the first 3-4 years of growth.

Pruning

After the initial training period, pruning is conducted primarily to remove dried, diseased, or damaged branches.

Maintenance Pruning

Additionally, water suckers, crowded branches, and crisscrossing branches should be removed to promote better growth and increased yields.

Some fruits, such as **passion fruit**, **karonda**, and **dragon fruit**, require specialized training and pruning techniques to optimize yield and overall plant health. By implementing these training and pruning practices, growers can enhance the productivity and longevity of tropical underutilized fruit crops.

Dragon fruit plants are fast-growing vines that initially produce a dense network of branches. To support their growth, sturdy pillars with rings at the top are typically used. Each pillar can accommodate two to three main shoots, with the total number of branches limited to around 500 to ensure optimal growth and management. A round or solid metal rod is placed at the upper part of the pole to help control plant growth. The growth rate of dragon fruit stems is rapid, reaching approximately 8 to 10 cm per week. After about eight months post-planting, a dense cluster forms at the top of the dragon fruit stem, which may bend towards the ground. Lateral buds and branches should be pruned to encourage growth towards the support stands. Once the vines reach the top of the stands, branches are allowed to grow freely. Trimming the tips of the stems induces lateral branching, enhancing overall plant structure. A well-grown vine can produce between 30 to 50 branches within the first year and may exceed 100 branches by the fourth year. To ensure better aeration, facilitate cultural operations, and prevent disease and pest infestations, regular trimming of these branches is essential.

Rambutan trees exhibit strong apical dominance, often leading to long, upright growth. To manage this growth effectively, early pruning and training to form an open center structure are recommended. After harvesting, fruited twigs should be pruned back to stimulate new growth, allowing up to four new side shoots. Research indicates that approximately 22 per cent of these shoots will bear fruit in the following season. Regular removal of dead branches and water suckers is essential to maintain tree health and vigor. While mature rambutan trees generally do not require extensive pruning, they can develop a dense central crown if left unpruned. In the early stages of growth, it is crucial to establish a framework that will promote optimum yields. This can be achieved by training young trees to develop a strong structure with 2 or 3 wide-angled branches originating from the main trunk. After establishing the main branches, lateral branches should be allowed to develop from these primary supports.

Passion fruit is a woody vine that requires support for optimal growth and fruiting. For commercial cultivation, the vines are trained on a frame of wires and poles that are positioned 1.5 to 2 meters above the soil surface.

Trellising System

Among the various trellising methods, the Kniffin system is the most economical. In this system, 2.5-meter-long poles are erected six meters apart, with a wire fixed at the top. The trellis should run across the slope or in the North-South direction to ensure maximum and even sunlight exposure for the vines. To support the weight of the vines and fruit, it is essential to use eight or ten-gauge wire, turnbuckles, and strong supports such as stone pillars or cement or wooden poles (Anonymous, 2009). This construction is vital for maintaining yield and withstanding the weight of the vines and fruits in various weather conditions for at least five years. If wooden poles are used, they should be treated with tar up to the portion that will be buried in the soil to prevent deterioration and protect against white ants. Young vines are supported with hardwood stakes or gunny twine, which should be stretched vertically from the base of the plant to the top of the wire.

Pruning and Training Techniques

After establishment, the plants grow rapidly, producing several branches from the base. As new shoots appear, all but two vigorous shoots should be removed. The main shoots are bound to the stake or twine, and all lateral growth in these shoots should be removed until they reach the wire. Once the main vines reach the wire, the tips are pinched to encourage leader formation. Two leaders are directed on either side of the wire and tied with loops until they meet the tips of the adjoining plants, at which point the tips are cut. This encourages lateral growth that hangs down from the wire.

Passion fruit vines bear fruit only on current season's growth. Systematic pruning encourages new growth, resulting in a regular and higher yield. Lateral branches from leader branches are allowed to grow and bear fruit. After fruiting, these laterals should be cut back to four to six buds to induce regular bearing. Pruning is typically done twice a year: first in March and April, and again in October and November, depending on the harvest cycle. Pruning should focus on cutting back laterals or the buds of those laterals that have fruited.

In the case of older laterals, cutting back should be limited to the nearest active bud; otherwise, the basal buds may become dormant or sterile. Indiscriminate or drastic pruning of inactive or dormant vines can lead to setbacks in growth and reduced yield.

The **karonda** plant initially has a comparatively weaker stem, making it essential to provide support after planting. As the plant tends to produce multiple branches near the ground surface, these emerging branches should be removed to facilitate cultural operations. In the early stages, training is crucial to establish the desired framework for the plant. Unwanted branches should be pruned to provide a defined shape and to promote the growth of the trunk and crown. Three to four branches should be allowed to develop 30-45 cm above the ground, spaced opposite each other, to create a proper frame for the tree. The branches should be trained in an open center system to enhance light penetration and air circulation, ultimately leading to better production.

Training is primarily performed during the first two years of growth. Once the plant matures, ongoing pruning becomes necessary to maintain its size and shape. In arid areas, the karonda plant grows slowly and requires less frequent pruning. However, in humid and tropical regions, the plant grows vigorously, necessitating heavy annual pruning to manage size and ensure regular yield. Regularly remove water suckers, crowded branches, crisscross branches, and those with narrow angles to promote healthier growth. Additionally, dried and diseased branches should be periodically removed to maintain plant health. In regions like Coorg, pruning is typically performed in October to align with the plant's growth cycle.

Hedge Planting: When karonda is planted for hedge purposes, it does not require the same level of training. Instead, 3-4 branches can be allowed to grow near ground level to form a compact hedge. Hedges should be trimmed after harvesting the fruits to maintain their shape and density.

Immature **longan** trees, typically in their first 3-4 years after planting, require formative pruning to develop an open canopy that facilitates good light penetration. This pruning process is crucial for establishing the desired tree structure. Only a limited number of main branches are retained to achieve the desired structure. Surplus branches and all water shoots should be removed to promote healthy growth. The skirt of the tree is maintained at a height of 1-2 meters above the ground, which helps in managing the overall shape and encourages better airflow. After each growth flush, one strong branch should be retained to form a natural, rounded crown consisting of 6 to 10 main branches. This structure optimizes light exposure and air circulation within the canopy.

Benefits of Pruning

Canopy development: Early pruning produces a well-formed tree canopy, strengthening fruit-bearing branches and ensuring annual cropping.

Pest and disease management: An open canopy helps limit the incidence of insect pests and diseases by improving air circulation and reducing humidity within the tree.

Height control: Pruning also helps reduce the height of the tree to a manageable level, making it easier for maintenance and harvesting.

Additional Considerations

Flower spike removal: In regions like China and Thailand, the flower spikes of young trees under five years of age are often removed to encourage vegetative growth and crown expansion during the summer months.

Soursop

Soursop trees typically develop a symmetrically conical shape and are well-suited for the central leader training system. To optimize fruit production and facilitate harvesting, these trees can be trained into a mushroom shape by topping them at a height of 2.0 to 2.5 meters.

Central leader system: This method encourages a strong main trunk that supports lateral branches, which bear the fruit. The fruit hangs down, making it easier to harvest.

Mushroom shape: Topping the tree creates a mushroom-like canopy, enhancing light exposure and air circulation. After initial training, little pruning is typically required. In later stages, pruning should focus on thinning out diseased, interlocking, and weak branches to maintain tree health and shape. To manage tree size within a specific space, the longest branches—both horizontal and vertical—should be pruned annually, ideally right after the harvest.

Impact of severe pruning: It's important to note that excessive pruning can negatively affect subsequent fruiting, so care should be taken to avoid over-pruning.

Garcinia and Yellow Mangosteen

Garcinia plants are evergreen and tend to grow slender and tall, which can pose challenges for harvesting as the trees mature. To mitigate these issues and enhance fruit accessibility, proper training and pruning practices are essential.

- ☆ **Height restriction**: In the initial year, it is beneficial to restrict the tree size through training. Pinching the main branch in subsequent years helps keep the plants shorter and ensures the fruiting zone remains within easy reach for harvesting.
- ☆ **Open centre system**: For seedling plants that naturally grow upright, trimming the apical branches in the early years can help establish an open centre system, promoting better air circulation and light penetration.
- ☆ **Annual maintenance**: Each year, remove drooping, interlocking, diseased, and dried branches to improve fruit yield and quality.
- ☆ **Branch direction**: Grafted plants often exhibit directional growth with only one branch developing. Regular pinching should be practiced to encourage branching in all directions, enhancing overall tree structure.
- ☆ **Sucker removal**: Regularly remove suckers emerging from the rootstock below the graft union to prevent nutrient competition and maintain tree health.

Pruning for Bearing Trees

- ☆ **Minimal pruning**: For mature bearing trees, minimal pruning is required, focusing primarily on the removal of broken or dead branches.
- ☆ **Stimulating growth**: For trees around 8 years old or older, pruning small inner branches from unproductive plants can stimulate new growth and fruit bearing.

Yellow Mangosteen

Yellow mangosteen plants tend to grow compact and dense. Similar pruning practices apply. Remove suckers, drooping branches, and interlocking branches annually to enhance yield and fruit quality.

Macadamia Nut

To develop a well-structured macadamia nut tree, it is recommended to prune for a single main stem and a framework of horizontal branches. The following practices should be followed:

Initial pruning: Begin forming the tree by selecting one of the three upper buds found in the vertical row of leaf axils at each node after pruning. Allow this chosen bud to grow while clipping off the other two. This method encourages the buds below to grow horizontally.

Branch maintenance: Regularly remove weak and damaged branches to promote overall tree health and vigour.

Wood Apple

Wood apple plants grow slowly and produce numerous thorny branches near the main stem during the initial years. In the first 2-3 years, it is crucial to prune these thorny branches to develop a single main stem. Once the plant establishes apical dominance, training the main stem is essential to reduce excessive upright growth.

Kaith

Kaith trees can be trained using a modified central leader system. Once the tree matures, minimal pruning is required, mainly for the removal of dead or diseased twigs and branches. This pruning should be conducted annually after fruit harvesting and before the onset of flowering, ideally in February or March.

Wax Apple

Wax apple plants typically adopt a symmetrically conical shape and are well-suited to the central leader training system.

Mushroom shape training: The main stem can be topped at a height of 2.0 to 2.5 meters to train the tree into a mushroom shape. This allows flowering on lateral branches, which hang down for easier harvesting. Little pruning is necessary after initial training. Later, pruning should focus on thinning, disease, interlocking, and weak branches. To regulate tree size, prune the longest horizontal and vertical branches annually, preferably right after harvest. However, it is essential to avoid severe pruning, as this can negatively affect subsequent fruiting.

Water Management

Tropical humid fruits thrive in high humidity conditions year-round, necessitating effective water management practices, particularly during dry spells. The following guidelines are essential for optimal water management:

Irrigation Requirements

During dry seasons, irrigation is crucial to maintain adequate moisture levels. Newly planted fruits require consistent watering to establish strong root systems.

For young plants, irrigation should occur every 10-15 days in winter and every 6-7 days in summer to ensure proper growth and development.

Water Manipulation for Flower Induction

Certain fruits, such as rambutan and longan, require a dry period of 21-30 days to induce flowering. After this dry period, rambutan plants typically begin to bloom within 10-15 days of reintroducing water.

Irrigation Methods

The basin or flood method of irrigation is commonly practiced to provide adequate water to the plants. However, drip irrigation has shown significant advantages in optimizing water use and promoting enhanced growth by delivering water directly to the root zone.

Mulching Practices

Applying mulch around the base of trees is highly beneficial. It helps reduce water loss from the soil, increase soil organic matter, mitigate extremes of soil temperature, encourage the growth of feeder roots. When done correctly, under-tree mulching also aids in weed control, minimizing competition for nutrients and water (Menzel *et al.,* 1990).

Dragon Fruit

Dragon fruit (*Hylocereus* spp.), belonging to the cactus family, is adapted to arid conditions and requires less water compared to other fruit crops. Effective water management is crucial for achieving optimal growth and fruit production. Dragon fruit requires irrigation primarily during dry seasons. Regular watering during this period is essential to support healthy growth and development. To maximize fruit production, it is important to avoid water shortages during the flowering and fruiting stages, which occur from June to October.

Water Requirements

In the first and second years, each dragon fruit plant requires approximately 1,000 to 1,500 liters of water per year. This water should be distributed appropriately throughout the growing season. In India, the flowering and fruiting process coincides with the rainy season, which can lead to excessive humidity. It is vital to monitor and control humidity levels to prevent early germination of flowers and ensure optimal conditions for fruit set.

Passion Fruit

Passion fruit (*Passiflora edulis*) exhibits sensitivity to water deficits, impacting both leaf production and flowering. Effective water management is crucial for maximizing yield and ensuring healthy plant development. Water stress significantly reduces leaf and floral bud initiation. While flower bud development and fruit set are less affected by water deficits, the overall health of the vine can be compromised. Prolonged dry spells can adversely affect the development of flowering laterals, potentially leading to reduced crop yield.

Water Requirements

On average, passion fruit requires approximately 12-15 liters per vine per day during the summer and 6-8 liters per vine per day in winter. If rainfall is insufficient during dry months, supplementary irrigation becomes essential to maintain plant health. The crop coefficient increases throughout the growth stages: it starts at about 0.6 during apical vegetative growth and rises to around 1.25 during flowering and fruiting. This indicates a higher water demand during these critical stages. Research by Rao *et al.* (2014) indicates that irrigating every 12 days yields higher growth and fruit production compared to an 18-day interval, suggesting that more frequent irrigation can be beneficial during peak growth periods.

Irrigation Methods

Micro-sprinklers and drip irrigation are the most effective methods for delivering water precisely to passion fruit vines. These systems help ensure optimal moisture levels while minimizing water wastage. Passion fruit vines respond significantly to fertigation. A study on potassium fertigation demonstrated increased commercial production, highlighting the importance of integrating nutrient management with irrigation practices.

Longan

An ideal annual rainfall for optimal longan (*Dimocarpus longan*) production ranges from 1,200 to 1,400 mm, distributed over 100 to 150 days during the flowering and fruiting periods. This rainfall pattern is crucial for ensuring healthy growth and high yields.

In the event of prolonged dry spells, supplemental irrigation is necessary to maintain plant health and productivity. Any rainfall that deviates from the ideal pattern requires careful water management strategies, particularly in the form of irrigation. In Thailand, it is widely recognized that irrigation is a critical factor influencing longan yields. Trees growing near rivers typically exhibit higher yields due to better access to water. In the longan-producing provinces of Chiang Mai and Lamphun, irrigation practices are particularly emphasized, especially during the initial four years after planting.

Irrigation Practices

Drip irrigation system: A drip irrigation system is commonly installed in longan orchards, providing efficient water delivery directly to the roots. Critical Growth Stages: For bearing trees, irrigation is essential during key growth stages, including panicle emergence, flowering, fruit set, fruit development, post-harvest growth flush (up until maturation). Irrigation is typically withdrawn before the next flowering season to induce a water stress condition. This controlled stress is beneficial as it promotes flower initiation for the subsequent season's crop, ensuring consistent yields year after year.

Soursop

Soursop (*Annona muricata*) plants are relatively drought-tolerant and thrive in drier conditions. However, it is crucial to prevent water stress during critical growth phases, including flowering, fruit set, and fruit development, as the fruit is more sensitive to water deficiencies compared to the leaves. Soursop has a shallow, fibrous root system, making it more susceptible to water stress. To enhance moisture retention and promote healthy root growth, mulching is highly beneficial. Organic mulch can help reduce evaporation and improve soil organic matter. Prolonged dry spells, particularly from January to March, can significantly reduce the main summer crop and adversely affect the development of flowering laterals. This highlights the importance of timely water management during these months. If rainfall is insufficient during the dry season, supplementary irrigation should be provided at fortnightly intervals to ensure adequate moisture levels. This practice helps maintain plant health and supports optimal fruit development, leading to improved yields.

Karonda

Karonda (*Carissa carandas*) is a hardy plant that requires careful water management, especially during its early growth stages. Newly planted karonda trees should be adequately irrigated to promote establishment. Young plants should receive water at intervals of 10-15 days during winter and 6-7 days during summer to ensure healthy growth. The basin or flood method of irrigation is commonly practiced for karonda cultivation. However, the adoption of drip irrigation has proven to be an effective strategy, allowing for economical water use while promoting enhanced growth and development. Once established, adult karonda orchards typically do not require regular irrigation, as the plants are resilient and can withstand drier conditions.

Mulching with dry leaves or agricultural residues around the base of the plants is highly recommended. This practice aids in moisture conservation, helping to maintain soil humidity and reduce evaporation, ultimately supporting the plant's health and productivity.

Garcinia

Irrigation is crucial for young *Garcinia* plants, particularly during their first three years of growth, especially in dry months. In the Coorg region, irrigation is typically required from January to March to enhance fruit quality. In other areas, ensuring a regular supply of water from October to November is key to achieving higher yields and better fruit quality.

The *Garcinia* root system lacks root hairs, necessitating a constant and abundant supply of water in the soil. A continuous dry period of at least one month, followed by heavy rainfall, can stimulate the plants to induce flowering. Regular watering is essential during fruit growth and development, as insufficient water can significantly slow down fruit development.

Macadamia Nut

For Macadamia nut trees, maintaining soil moisture is critical. The soil should be kept moist, but not saturated, to a depth of at least 1 meter. This moisture balance helps support healthy growth and optimal nut production.

Wood apple plants can tolerate drought for long period and does not require irrigation. But young plants need to be watered 10-15 days interval in summer and one-month interval in winter for their establishment and rapid vegetative growth.

Weed Management in Orchards

Weeds pose a significant challenge in orchards as they compete with young trees for essential nutrients and moisture in the soil. Major weeds affecting orchards include:

- ✰ **Parthenium** (*Parthenium hysterophorus L.*)
- ✰ **Red sprangletop** (*Leptochloa chinensis (L.) Nees*)
- ✰ **European water clover** (*Marsilea quadrifolia L.*)
- ✰ **Purple nutsedge** (*Cyperus rotundus L.*)
- ✰ **Bermuda grass** (*Cynodon dactylon (L.) Pers.*)
- ✰ **Lantana** (*Lantana camara*)

Controlling weeds in the initial years is crucial for the optimal growth of young trees. Creeping weeds can cover young trees during the rainy season, potentially harming their development.

Weed Control Methods

Mechanical Control

Utilize tools such as mechanical grass cutters, grass hooks, sickles, or slashing bolos to effectively underbrush the orchard.

Regular ring-weeding helps maintain tree sanitation and reduces weed competition.

Chemical Control

Herbicides provide a quick solution for weed management. However, they may affect beneficial microorganisms in the soil.

Commonly used herbicides in rambutan cultivation include **Paraquat** (0.5 mL/L) and **Glyphosate** (1.8-2.1 kg/L).

Cultural Practices

Cover cropping: Planting cover crops can help control weeds and enhance soil fertility.

Mulching: Applying a **2 to 4-inch (5–10 cm)** thick layer of mulch around the base of trees suppresses weed growth and retains soil moisture.

Soil Preparation

Before planting trees, remove sod from the planting area and ensure that grass and weeds are kept away from the tree trunk.

Intercropping

Implementing intercropping strategies can effectively reduce weed populations in both young and mature orchards.

By adopting these weed management practices, orchardists can enhance tree growth and improve overall productivity.

Intercropping in Orchards

During the initial establishment period, the space between orchard plants can be effectively utilized for intercropping, providing additional income without competing with the main crop. Suitable intercrops include **cowpea**, **French bean**, and other regionally appropriate crops. In mature orchards, green manure crops can be grown and incorporated into the soil to enhance its fertility, moisture retention, and overall physical condition.

In the Kodagu and Kodaikanal regions of South India, crops like **avocado** are successfully cultivated as mixed crops alongside coffee and mandarin orchards. For instance, in Coorg, Karnataka, avocado trees are integrated into primarily coffee-based cropping systems (Tripathi and Karunakaran, 2013).

Rambutan

In newly established rambutan orchards, cultivating the spaces between trees with annual food crops can yield profitable returns before the rambutan trees bear fruit. Early in the orchard's life, food crops such as beans, leafy vegetables, and root crops can be planted without competing with the young rambutan trees.

The benefits of intercropping extend beyond added income; these crops also provide shade to young trees. Rambutan trees can be profitably intercropped with **coconut**, **coffee**, and **cacao**.

As the tree canopies develop, growing food crops may become unfeasible. At this stage, planting leguminous cover crops becomes advantageous. Low-growing legumes such as **Canavalia**, **Crotalaria**, and **Vigna** can be effectively intercropped with rambutan. These cover crops help prevent soil erosion, retain soil moisture, and enrich the soil with nitrogenous compounds due to the activity of nitrogen-fixing **Rhizobium** and other bacteria.

Additionally, cover crops contribute to reduced orchard maintenance costs, suppress weed growth, and minimize pest infestation, making them a beneficial component of sustainable orchard management.

Mulching in Rambutan Orchards

The application of dried weeds, grasses, and other organic residues as mulch around the base of both young and mature rambutan trees is a common practice that offers multiple benefits. This method effectively retains soil moisture, prevents weed growth, and reduces soil temperature. Additionally, as the organic residues decompose, they contribute to maintaining soil fertility.

Mulching should be applied around the base of the tree to a depth of 10-15 cm, ensuring that the area immediately surrounding the trunk remains free of mulch. This approach not only supports the health of the rambutan trees but also enhances the overall productivity of the orchard.

Passion Fruit

The plants are shallow-rooted, with most feeder roots located within 15 cm of the soil surface, necessitating light digging for soil maintenance. Deep digging is avoided to protect these roots, while weed growth is managed through surface weeding, scraping, and scuffling. Mulching with dried leaves or grass is also practiced to conserve moisture during the summer months.

Effective weed management is crucial for the healthy growth of passion fruit vines. A clean space of about 60-90 cm radius around the base of each vine should be maintained, while the rest of the plantation can have an undergrowth of grass. However, caution must be exercised regarding the choice of undergrowth, as passion fruit can infect legume crops, making it inadvisable to plant leguminous cover crops. Additionally, some weeds serve as hosts for aphid vectors, which are responsible for transmitting passion fruit mottle and ring spot viruses; therefore, these weeds should be promptly eliminated.

The use of herbicides is not recommended, as passion fruit vines are highly susceptible to chemical treatments. Instead, mechanical weed control measures should be employed to ensure the health and productivity of the plants.

Longan

Intercropping longan trees with legumes such as *Wisteria chinensis, Tephrosia candida, Sphenostylis dolichoris, Macroptilium lathyroides,* and *Lablab purpureus* has shown several benefits. These legumes exhibit good shading tolerance, regeneration ability, and high palatability, along with significant crude protein content. Additionally, they promote the growth of longan trees by inhibiting weed growth and enhancing soil fertility, thus creating a more conducive environment for fruit tree cultivation (Xian-Feng, 2010).Intercropping with crops such as papaya, banana, and soybean after sweet corn, as well as shallot after pumpkin and lemongrass in young longan orchards, was conducted in Ban Mae Hor Phra, Phrao district of Thailand. The study indicated that net returns from all intercrops were positive in the second year, with notably higher returns from banana and papaya intercrops. The economic justification for using lemongrass as an intercrop in the early stages of young longan orchards was also supported. Additionally, intercropping experiments led to an increase in the pH level of the area, making macronutrients such as nitrogen (N), phosphorus (P), potassium (K), calcium (Ca), and magnesium (Mg) more available (Saw Hto *et al.,* 2011).

Kokum and Mangosteen

In the Konkan region and coastal areas of Karnataka, kokum is often planted as a mixed crop in coconut and areca nut plantations. In several parts of Karnataka, kokum plants are also grown along the borders of plantations, where the planting distance may vary. Intercropping with short-duration crops, such as banana, is common to provide necessary shade to mangosteen seedlings. Mangosteen trees undergo a juvenile phase that can last between 12 to 20 years; however, with proper care and good agricultural practices, trees can begin fruiting as early as 5-7 years after planting.

Karonda

Karonda (*Carissa carandas*) typically grows in dry soil, often in regions where water facilities are limited. However, during the rainy season, certain vegetables can be grown as intercrops. During the initial period of establishment, the space between the plants can be effectively utilized for growing intercrops. This practice provides additional income in the early stages of orchard development without competing with the main crop. Suitable intercrops include cowpea, French bean, okra, brinjal, and other region-specific crops.

In mature karonda orchards, green manure crops can be grown and later incorporated into the soil. This practice not only improves soil fertility but also enhances moisture retention and improves the physical structure of the soil, contributing to better orchard management and yield.

Conclusion

Underutilized fruits are gaining increasing popularity due to their remarkable health benefits and their potential contribution to crop diversification, particularly in tropical humid regions. Many of these fruits have tremendous potential for the pharmaceutical and processing industries. Despite some research efforts in identifying superior lines, there remains a significant gap in the development of comprehensive production technologies, especially in India. There is an urgent need to focus on the production, protection, and post-harvest technologies for these crops to fully harness their economic and health potential. Enhanced research and attention to these fruits will not only boost the agricultural economy but also promote sustainable farming practices and improved nutritional outcomes.

REFERENCES

Akali Sema and Maiti CS, 2006. Status and prospects of Passion Fruit industry in Northeast India. Extension Bulletin Central Institute of Horticulture, Medziphema, Nagaland.

Anandalakshmi R, Sivakumar V, Warrier RR, Parimalam R, Vijayachandran SN, Singh NB, 2005. Seed storage studies in *Syzigium cuminii*. *J. Tropical For. Sci.*, **17(4)**: 566-573.

Anonymous, 2010. Final report of National Network Project on Underutilized Fruits. CHES Chettalli.

Anonymous, 2009. Diamond Jubilee souvenir of CHES Chettalli.

Anu Ann Augustine, Jyothi Bhaskar and Suma A, 2021.Morphological diversity with respect to fruit characters of ***avocado*** (*Persea americana* Mill.) in High Ranges of Kerala. *Journal of Tropical Agriculture,* **59(1)**: 83-85.

APAARI, 2012. Jackfruit Improvement in the Asia-Pacific Region – A Status Report. Asia-Pacific Association of Agricultural Research Institutions, Bangkok, Thailand. 182 p.

Chithiraichelvan R, Sudhir Kumar, Karunakaran G and Sakthivel T, 2006. Standardization of propagation techniques in Avocado (*Persia americana* Mill) Abstracts of Poster papers in National Symposium on underutilized horticultural crops during June 8 -9 at IIHR, Bangalore, pp: 68.

Geethu M and Saji Gomez, 2022.Biochemical and morphological characteristics of *avocado* genotypes in Kerala, **24(3)**: 369-374.

Ghosh SN and Bera B, 2015. Studies on standardization of propagation methods of some minor fruit crops in India. *International Journal of Minor Fruits, Medicinal and Aromatic Plants,* **1(1)**: 31-36.

Ghosh SP, 2000. Avocado Production in India. In Avocado production in Asia and the Pacific. FAO Corporate Document Repository.

Gopalan C, Ramashastri BV, Balasubramanian SC, Rao NBS, Deosthale YG and Pant K C, 2004. *Nutritive Value of Indian Foods,* ICMR Hyderabad: National Institute of Nutrition.

Joy PP, 2010. Status and prospects of passion fruit cultivation in Kerala. Pineapple Research Station (Kerala Agricultural University), Vazhakulam, Kerala, India.

Kader AA, Veazie PP and Lester GE, 2005. Nutritional Quality of Fruits, Nuts, and Vegetables and Their Importance in Human Health, Department of Pomology, University of California, Davis, CA.

Karunakaran G, Arivalgan M, Sakthivel T, Tripathi PC and Muralidhara BM, 2022. Production technologies of emerging fruit crops of humid tropics. National Symposium on Horticultural crops of humid tropics for nutrition and livelihood security (NSHCHT)- 2022 organized by CHES, Chettalli at Madikeri (Kodagu) during Dec. 2-3, pp.12-18

Karunakaran G, Thirugananvel A, Rajindran S, Arivalgan M, Sakthivel T and Tripathi PC, 2022. Orchard management *In*: Jack fruit: Botany, Production and uses, Eds(Dr. S.K. Mitra, ISHS), CABI International. Pp.296.

Karunakaran G, Tripathi PC, Arivalgan M, Sakthivel T, Sankar, Puspa Chetan kumar, Narayana CK, Sriram S, Venugopalan R, Sreenivasmurthy D, Rajindran SR, Jayanthimala BR and Ramchandran R, 2022. Bharat Me Kamal phal : Vartaman sthiti and sambhvanayein. (Dragon fruit cultivation in India: Present status and prospectus) (in Hindi), ICAR–IIHR, Bangalore Tech. Bulletin. No.-TB21/2022. Pp-13.

Karunakaran G, Tripathi PC, Sakthivel T, Ravishankar H, Sankar V and Senthil Kumar R, 2016. Rambutan is a potential future fruit. *Indian Horticulture,* **61(2)**: 3-5.

Khawas E and Upadhyay S, 2022. Effect of plant growth regulators on rooting of *avocado* (*Persea americana* Mill.) cuttings in Sikkim. *Bhartiya Krishi Anusandhan Patrika,* **37(2)**: 157-162.

Kumar Sudhir, Chithiraichelvan R, Karunakaran G, Sakthivel T, 2008. Studies on Propagation of *Passion Fruit* Cv. Kaveri by Cuttings under Coorg Conditions 2008. *Indian Journal of Horticulture,* **65(1)**.

Mehta Sangita , Prasad Birendra and Das Ajay Kumar, 2016. Pre-sowing seed treatments with different chemicals and growth regulator on seed germination and growth of passion fruit. *Advance Research Journal of Crop Improvement,* **7(2)**: 201-206.

Pareek OP, Sharma S and Arora RK, 1998. Underutilised edible fruits and nuts-An inventory of genetic resources of their regions of diversity. IBPGR office of South Asia. New Delhi, India, 235p.

Sankar V, Tripathi PC, Karunakaran G and Senthil Kumar R, 2014. Pummelo, Technical Bulletin8/2014, ICAR–IIHR, Central Horticultural Experiment Station, Chettalli, Kodagu, Karnataka. PP.10.

Senthil Kumar R, Tripathi PC, Sankar V and Karunakaran G, 2014. Kokum – A potential fruits of Western Ghats. Technical Bulletin 6-/2014, ICAR–IIHR, Central Horticultural Experiment Station, Chettalli, Kodagu, Karnataka.

Tripathi PC, Karunakaran G, Sankar V and Senthil kumar R, 2015. Improved production technologies of Rambutan. In souvenir and Abstract of Seminar cum field day on Rambutan held at CHES Chettalli on October 10, P. 40-44.

Tripathi PC, Karunakaran G, Sakthivel T, Sankar V, Senthil kumar R, Jayanthi Mala BR, Priti Sonavane, Ravisankar H and Chithraichelvan R, 2015. Avocado Research at CHES Chettalli. In souvenir and Abstract of Brain Storming session on avocado held at CHES Chettalli on May 27. P.7-17.

Tripathi PC and Karunakaran G, 2019. Standardization of time and method of propagation in avocado. *Journal of Applied Hortic.* **21(1)**: 67-69.

Tripathi PC, 2023. Status and prospects of Avocado cultivation in India. *Brazilian Journal of Development,* **10(2)**: 1-25.

Tripathi PC, Karunakaran G, Senthil kumar R, Sakthivel T, Sankar V, Muralidhara BM, Rajindran R, Venkataravanappa V, Madhu GS and Nesara Begane, 2022. Advances in avocado production in India, ICAR–IIHR, Bangalore Tech. Bulletin. No.-TB18/2022. Pp-30.

Tripathi PC, Karunakaran G, Senthil kumar R, Sakthivel T, Sankar V, Muralidhara BM, Rajindran S, Venkataravanappa V, Madhu GS, Nesara Begane and Jha AK, 2022. Bharat Me Navneet phal kee kheti. (Advances in avocado production in India) (in Hindi), ICAR–IIHR, Bangalore Tech. Bulletin.No.-TB19/2022. Pp-27.

Tripathi PC, 2023. Longan cultivation. Agro India Jan, 28-31.

Tripathi PC, Shetti DL and Rupa TR, 2019. Nutrient content of some underutilized fruits *Progressive Horticulture*. **50(2)**: 115-22.

Tripathi PC, 2020. Lasoda (Indian Cherry or glue berry) : An unexploited fruit of medicinal importance, Agro India July, pp. 16-19.

Tripathi PC, Yogeesha HS and Shetti DL, 2022. Standardization of propagation methods in minor wild fruit crops. *Current Horticulture,* **10(1)**: 32-36.

Tripathi PC, 2021. Kokum- an underutilized fruit with medicinal importance. *Agro India*, **26(6)**: 10-13.

Tripathi PC, Karunakaran G and Sakthivel T, 2012. Production technologies of Rambutan. Souvenir, Rambutan Seminar Nov.23 Central Horticultural Experiment Station (IIHR), Chettalli, pp-7-13.

Tripathi PC, Karunakaran G, Sankar V and Kanupriya. 2019. Genetic Resources of wild and underutilized fruits in Western Ghats In Shaping Future of Indian Horticulture (Ed Chaha *et al*). Astral Publication, New Delhi. pp. 143-158.

Tripathi PC, 2021. Canistel (egg fruit): An underutilized nutritious fruit, Agro India Oct., 2021: 14-16. Tripathi, P. C. 2022. Jamun cultivation. *Agro India* April, pp.13-17.

Tripathi PC, 2021. Carambola (Star fruit) at a glance. *Indian Hortic.*: May June Cover III-.

Tripathi PC, 2021. Malabar tamarind at a glance. *Indian Hortic.*: March- April Cover III-.

Tripathi PC, 2021. Medicinal and therapeutic properties of minor fruits - A Review. *International Journal of Minor Fruits, Medicinal and Aromatic Plants,* **7(2)**: 01- 28.

Tripathi PC, 2021. Some unique fruits for home garden. *Indian Hortic.* **66(3)**: I and 34-39.

Tripathi PC, 2021. *Garcinia*: A genus of nutraceutical fruits. *Indian Hortic.*, **66(6)**: II and 3-8.

Tripathi PC and Karunakaran G, 2013. Production technologies of minor fruits with special emphasis on Kodagu held at Madikeri, Karnataka during November 27- 29, Souvenir and Abstracts, Pp. 97-105.

Tripathi PC, Karunakaran G, Sankar V and Senthil Kumar R, 2014. Karonda - A Potential fresh fruit of future. Technical Bulletin 7/2014, ICAR–IIHR, Central Horticultural Experiment Station, Chettalli, Kodagu, Karnataka. pp. 14.

Tripathi PC, Karunakaran G, Sankar V and Senthil Kumar R, 2014. Dragon fruit – Nutritive and Ruminative fruit. Technical Bulletin 11/2014, ICAR–IIHR, Central Horticultural Experiment Station, Chettalli,Kodagu, Karnataka. pp. 10.

Tripathi PC, Sankar V and Senthil Kumar R, 2014. Sour Sop – An emerging fruit of future. Technical Bulletin, 5/2014, ICAR–IIHR, Central Horticultural Experiment Station, Chettalli, Kodagu, Karnataka. PP14.

Tripathi PC, 2021. Yellow mangosteen-a future fruit with medicinal value, *Agro India* **26(6)**: 20-22.

Tripathi PC, 2018. Minor fruits for Nutritional security. In Zero Hunger India Ed. Dr. K. V. Peter, Brillion Publishing, New Delhi pp706.

Tripathi PC, 2021. Longan in Sub-Tropical Fruit Crops: Theory to Practical (Eds. Dr. Ghosh and Dr Sharma).Jaya Publishing Company, New Delhi. pp. 384- 412.

Tripathi PC, 2021. Rambutan In Tropical fruits (Eds. Dr. Ghosh and Dr. Sharma). Jaya Publishing Company, New Delhi. pp. 542-575

Tripathi PC, Karunakaran G, Sakthivel T, Sankar V and Senthil Kumar R, 2014. Passion fruit cultivation in India. Technical Bulletin 3/2014, ICAR–IIHR, Central Horticultural Experiment Station, Chettalli, Kodagu, Karnataka, Kodagu, Karnataka. pp. 18.

Tripathi PC, Karunakaran G, Sakthivel T, Sankar V and Senthil Kumar R, 2014. Rambutan cultivation in India. Technical Bulletin 1/2014, ICAR–IIHR, Central Horticultural Experiment Station, Chettalli, Kodagu, Karnataka, Kodagu, Karnataka pp. 18.

Tripathi PC, Karunakaran G, Sankar V and Senthilkumar R, 2017. Improved production technology of Rambutan. Souvenir and Abstracts. Nat. Con. on Horticultural Crops of Humid Tropics - Diversification for Sustainability" at Madikeri, May 20-21, pp. 55-58.

Tripathi PC, 2018. Passion Fruit. (In) Peter,K V (Ed).Horticultural Crops of high nutraceutical values. Brillion Publishing New Delhi p. 245-270.

Tripathi PC, 2019. Advances in production technologies of major underutilised fruit crops. Workshop on Exploring prospects of under-utilized fruits crops and their processing and value addition. College of Hortic. (UHS) Kolar, Karnataka, Sept 13, P. 29-42.

Tripathi PC, 2020 Kaith, *Phal Phool*, **41(4)**: 51-53.

Tripathi PC, Karunakaran G, Sakthivel T, Sankar V and Senthil Kumar R, 2020. Status and prospects of rambutan cultivation in India. *Acta Hortic.*, **1293**: 33-40

Tripathi PC, 2021. Malabar Tamarind : A potential fruit, condiment and medicine. *Spice India*, **34(7)**: 4-10.

Tripathi PC, Anuradha Sane, Kanupriya Pritee Singh and Muralidhara BM, 2022. Mangosteen: A fruit with splendid taste. *Agro India*, December, pp. 10-13.

Tripathi PC, Karunakaran G, Sakthivel T, Sankar V and Senthil Kumar R, 2022. Advances in Avocado cultivation in India. Technical Bulletin2/2022, ICAR–IIHR,Bangalore. pp. 32.

Tripathi PC and Anuradha Sane, 2023. Sour Sop. *Agro India* October 9-14.

Tripathi PC, Anuradha Sane, Kanupriya and Pritee Singh, 2023. Macadomia nut : A fruit with splendid taste. *Agro India* March, pp. 24-27.

Tripathi PC, Anuradha Sane, Kanupriya, 2024. Wax apple cultivation. *Agro India*, Feb 25-28.

Chapter 8

Production Technology of Acid Lime, Sweet Orange, Tamarind in Dry Tropical India

***V. Krishnamoorthy*[1] *and M. Tamil Selvan*[2]**

[1]*Agricultural Research Station, TNAU, Pattukkottai–614906, Tamil Nadu*
[2]*ICAR–Krishi Vigyan Kendra, Pudukkottai–622303, Tamil Nadu*

The nutrients present in acid lime and sweet oranges are plentiful and diverse. The fruit is low in calories, contains no saturated fats or cholesterol, but is rich in dietary fiber, *pectin.* Pectin, by its virtue as a bulk laxative, helps to protect the mucous membrane of the colon by decreasing its exposure time to toxic substances as well as by binding to cancer-causing chemicals in the colon. Pectin has also been shown to reduce blood cholesterol levels by decreasing its re-absorption in the colon by binding to *bile acids* in the colon. Citrus fruits, are an excellent source of vitamin C. Vitamin C is a powerful natural antioxidant. Orange fruit contains a variety of phytochemicals *viz.,*Hesperidins, naringin, and naringenin are flavonoids found in citrus fruits. Naringenin is found to have a bio-active effect on human health as antioxidant, free radical scavenger, anti-inflammatory, and immune system modulator. This substance has also been shown to reduce oxidant injury to DNA in-vitro studies. Total antioxidant strength (ORAC) of oranges (navel variety) is 1819 µmol TE/100 g. It is also a very good source of B-complex vitamins such as thiamin, pyridoxine, and folates. These vitamins are essential in the sense that body requires them from external sources to replenish. Orange fruit also contains a very good amount of minerals like potassium and calcium. Potassium is an important component of cell and body fluids that helps control heart rate and blood pressure through countering sodium actions. Citrus fruits, as such, have long been valued for their wholesome nutritious and antioxidant properties. It is now beginning to be appreciated that the other biologically active, non-nutrient compounds in the citrus fruits such as *phyto-chemical antioxidants, soluble and insoluble dietary fiber* helps in cutting risk for cancers, chronic diseases like arthritis, obesity, and coronary heart diseases.

In India, citrus is grown in 1.45 million ha with a production of 3.79 million tones. The most important commercial citrus cultivars in India are the mandarin (*Citrus reticulata* Blanco), following by sweet orange (*Citrus sinensis* Osbeck) and acid lime (*Citrus aurantifolia* Swingle) sharing 41, 23 and 23 per cent, respectively of all citrus fruits produced in the country. Commercially, Kinnow mandarin is grown in Punjab, Haryana, Himachal Pradesh, western part of Rajashtan and Uttar Pradesh; Khasi mandarin in north eastern region; Darjeeling mandarin in Darjeeling; Acid lime in khera district of Gujarat, Akola in Maharashtra, Periyakuulam in Tamil Nadu; Nagpur mandarin in Vidharbha region of Maharashtra and adjoining areas of Madhya Pradesh; Mosambi in Marathawada region of Maharashtra; Sathgudi in Andhra Pradesh and Coorg mandarin in Coorg area of Karnataka.

Area, production and productivity trend under citrus in each citrus growing state provides an idea about the extent to which the cultivation is successful and sustainable. In terms of area, Maharashtra tops followed by Andhra Pradesh Punjab, Karnataka, Uttar Pradesh and Bihar. But the perusal on productivity trend gives an entirely different scenario. Highest productivity is obtained in Tamil Nadu followed by Madhya Pradesh and Gujarat, Andhra Pradesh and Bihar. During the last 35 years, the Indian citrus industry attained a quantum expansion in area from 0.091 million ha in 1961 to 0.45 million ha in 1996 (about 5 times), while production increased from 0.824 to 3.79 million tones.

Citrus consists of a group of fruits belong to the Family Rutaceae. Special feature is presence of juice sacs, leaves posses transparent oil glands, flower contain an annular disc. Fruits are stalked with spiller shape; leaves are persistent and simple except *Poncirus* deciduous and trifoliate.

Origin: It is believed that most of the species under genus citrus are native to tropical and sub-tropical regions of Southeast Asia, Particularly, India, and China and in the region between these two countries. The north eastern region of India is considered as one of the natural home of citrus. The major citrus producing countries of the world are USA, Spain, India, Italy, Japan, Mexico, Brazil, Morocco, South Africa, Australia and Israel. The U.S.A. is considered to be the largest producer of citrus fruits. Covering about 35-40 per cent of the world total production.

Citrus fruits include oranges, lemons, limes, pummelo and grape fruit. Being a native of tropical and subtropical region of South East Asia, these have been under cultivation from time immemorial in South China, Malaya and sub-Himalayan parts of Assam. From here, they spread to other tropical and subtropical parts of the world. Next to mango and banana, citrus represents the third most important group of fruits in India. The botanical classification of the genus is highly confusing since more and more interspecific and inter generic hybrids are going on added to the list each on deserving a separate species status. All the edible fruits of citrus come under subgenus *Eucitrus* which can be divided into 5 horticultural groups.

Acid Group

1. Acid lime : *Citrus aurantifolia*
2. Tahiti or Persean lime: *Citrus latifolia*
3. Rangpur lime: *C. limonia*
4. Lemon: *Citrus limon*
5. Rough lemon: *C. jambhiri*
6. Citron: *C. medica* (Kidarankai in Tamil, used for pickling)
7. Sweet lime: *Citrus limettoides*

Orange Group

1. Sweet orange: *Citrus sinensis*
2. Sour orange: *Citrus aurantium* (Narthankaai in Tamil, used for pickling)
3. Multiple leaf orange: *C. multifolia*
4. Japanese summer grape fruit: *C. natsudaidai*

Mandarin Group (loose jacket)

1. Coorg mandarin, Nagpur Santra and Kodai orange: C. *reticulata*
2. Japanese Satsuma mandarin: C. *unshiu*
3. Willow leaf mandarin: *C.deliciosa*
4. King mandarin: *C. nobilis*
5. Kinnow mandarin: King x willow leaf
6. Tangerine orange var Dancy (Trifoliate x mandarins): *Citrus tangerina*

Pummelo and Grape Fruit Group

1. Pummelo: *C. grandis*
2. Grape fruit: *C. paradisi*
3. Kumquat: *Fortunella sp.*

The fifth group consists of mainly hybrids of different citrus fruits with trifoliate orange (*Poncirus trifoliata*):

Mainly used as rootstock.*e.g.* Citrange (*Poncirus trifoliata* x *C. sinensis*) var. Troyer, var. Carrizo

1. Citrangor (Citrange x *C. sinensis*)

2. Tangelo (Tangerine x grape fruit)
3. Citrangequat (Citrange x kumquat)

Acid Lime (*Citrus aurantifolia*)

It is otherwise called as sour lime (or) Kagzi nimboo. The fruit juice is rich in citric acid and ascorbic acid.

Climatic and Soil Requirements

Acid lime is tropical in its climatic requirement and sensitive to frost than any other citrus fruits. It thrives well in a dry climate of Bombay-Deccan, North Karnataka as well as central Tamilnadu. The optimum temperature is 20° to 30°C. Right from sea level upto even 1500 M, acid lime can be grown. A well drained soil with a pH of 6.5 to 7.0 will be ideal for better growth and yield.

Varieties

1. Kagzi Nimboo: It is susceptible to Tristeza virus and Bacterial canker disease.
2. PKM – 1 (Jaya Devi): Less susceptible to leaf miner and citrus butterfly.

Released by Mahatma Phule Krishi Vidyapeeth, Rahuri

3. Sai-Sharbati : (MPKV): yield : 30kg/Tree, Tolerant to Canker and Tristeza.
4. Phule Sharbati: (MPKV): 52.19 t/ha.

Released from Marathwada Krishi Vidyapeeth, Parbhani.

5. Pramalini: Canker tolerant.
6. Vikram - The fruits are borne in cluster.
7. Balaji : Released from regional citrus experimental station – Tirupati.
8. Kasipentla, Mungalipattu, Petlur selection:

Released from citrus experimental station, Petlur (Andra Pradesh).

9. Rasraj (IIHR): Hybridization *C.aurantifolia* X *C. Limon*
   - ☆ It is resistance to bacterial canker.
10. Seedless: Selection from Himachal Pradesh.
11. Chakradhar lime: It is a thornless and seedless variety.
12. Kagzi Kalan (IARI): 7-10 kg per tree

Released from ICAR–Central Citrus Research Institute, Nagpur

13. NRCC Acid Lime-7: Yield 54kg/tree, fruit 45g, acidity 7.05 per cent, Juice 50 per cent
14. NRCC Acid Lime-8: 55kg/tree, fruit weight 50g, Juice 51. per cent, TSS 7.85

Tahiti Lime or Persian Lime

- It is related to lime but *Citrus latifolia*.
- It is triploid and does not possess seeds.
- Multiplied by ground layering and air layering.

Rangpur Lime

- Related to lime but *citrus limonia*.
- It is mainly used as root stock
- Adaptability to a wide range of soil conditions especially for heavy soil.
- It is tolerant to tristeza virus and salt.

Propagation

- Acid lime is mainly propagated through seeds.
- Seeds should be sown in raised nursery bed immediately after extraction (atleast within 3 days) as they lose their viability rapidly.
- Since it produces polyembryonic seedlings which are nucellar origin, there is not much variation among the seedling progenies.
- GA 25 ppm or Thiourea 1 per cent soaking improves seed germination.
- As the plants are susceptible to a virus disease called Tristeza.
- Root stock: Rangpur lime and Rough lemon.
- Can also be propagated by ground and airlayering.
- Tahiti lime is seedless and hence propagated by air layering.
- Acid lime can also be propagated by ground and air layering.
- Acid limes can also be budded by patch budding on Rangpur lime and Rough lemon.

Pre-Immunised Acid Lime Seedlings

As the plants are susceptible to a virus disease called 'tristeza', pre-immunisation by inoculating a mild strain of this virus through patch budding of a bark piece on healthy seedlings is done to induce resistance to virulent form of virus. Such seedlings are called **Pre-Immunized Acid Lime Seedlings (PIALS)**.

Field Preparation and Planting

Field is ploughed 2-3 times and pits of 75 cubic centimeters are dug at a spacing of 5 to 6m depending on the fertility and type of soil. After filling the pits with top soil and 20kg of FYM, the seedlings (10-12 months old or airlayers a can be planted in the centre of the pit and supported by staking with a bamboo stick to prevent any possible break by wind. The pits are irrigated immediately after planting.

Manures and Manuring

A bearing lime tree is to be fertilized with 600g of N, 200g of P and 300g of K besides 50kg of Farm Yard Manure. Normally these doses would be required from 4th year of planting. One fourth of the above dose will be sufficient during first year, 1/2 will be during second year, and 3/4 during third year. The total nutrients have to be split into two equal parts and applied once during June-July and other during November- December. The manures and fertilizer are mixed and applied as a band application 60cm away from the trunk. NPK @ 600: 200: 300 g + 15kg neem cake/ plant/year significantly increased the fruit yield and quality in acid lime.

Spray Zinc sulphate @ 0.5 per cent (500 g/100 lit of water) thrice in a year (March, July and October) after the emergence of new flushes.

Interculture and Irrigation

Under normal planting system, pulses like blackgram, cowpea or vegetable like cucurbits, (beans and carrots during winter season) can be grown as intercrops. Under tropical condition water requirement of acid lime has been found to be 875 mm per year. The daily water requirement would be 50-60 liters per plant during humid months and 90-100 liters during summer. The chloride content of the irrigation water should not exceed 100 ppm. Soil moisture stress will be detrimental to plant growth and also during the critical phases like flowering, fruitset, fruit enlargement and juice formation. Weed management can be done by a pre-emergent application of 3 kg diuron per ha followed by gramaxone 1.5 lit as a post emergent application in the interspaces.

Paclobutrazol @ 18g/tree as soil application enhanced the flowering, fruit set and produced 1600 fruits/tree/year in acid lime. Flowering can be regulated by with holding irrigation for 1-2 months and again stress is alleviated which will result in flowering. But excessive stress will lead to over flowering, poor fruit set and plant health will also be affected adversely. Foliar application of 500 ppm of paclobutrazol (PP^{333}) at new flush will encourage profuse flowering. To increase fruit set, spray 2, 4 – D @ 20 ppm (200 mg/10 l) during flowering. Application of 2, 4, 5-T 20 ppm at pepper stage or 2, 4 – D @ 20 ppm or NAA 30 ppm (300 mg/l) marble size stage prevent fruit drop and increase fruit retention.

Harvest and Yield

Though limes can be harvested throughout the year, the main crop is harvested during different periods in different parts of the country. The average yield is 20-25 kg/tree/year. Limes can be stored at 18°C. At Horticultural College and Research Institute, Periyakulam a low cost storage tank has been developed with double layer brick work, the inter space filled with sand which is kept wet by periodical watering.

Sweet Orange (*Citrus sinensis*)

Introduction

Due to sweet taste, pleasant flavour and refreshing qualities, the demand for orange as a dessert fruit is very great. It is a good source of vitamin C (64 mg/100 g of juice) besides vitamin A and B.

Climatic and Soil Requirements

Sweet orange thrives better in a subtropical dry climate (low humidity) having well defined summer and winter seasons, coupled with low rainfall. Rainfall seems to be unimportant if irrigation is provided. Even in a tropical climate it can be grown. But atmospheric humidity exerts a greater influence. Though it can be grown in a wide temperature range of 10°C to 38°C, the optimum temperature would be 20-30°C. Sweet oranges do not bear fruits successfully in regions of heavy rainfall and high atmospheric humidity like Konkan, Kerala, Mysore, East Coast of Tamil Nadu, W. Bengal, Bihar and Assam. Here trees flower irregularly at different times and fruits do not develop colour and ripen properly.

Dry climate of Pune, Ahmedabad and Hyderabad (RF 500 mm to 1750 mm/ annum restricted to June-September) favors the cultivation of an excellent variety Mosambi. In Cuddapah, Kurnool, Chithoor and Rayalaseema of Andra Pradesh, a variety 'Sathgudi' is extensively cultivated. 'Malta Blood Red' is a popular variety grown in plains of Punjab, Rajasthan and Uttar Pradesh where summer and winter are marked and severe.

Ideal soil for sweet orange cultivation would be medium or light loam. The deep alluvial loam of Indo-Gangetic plains and coarse friable black soils of Madhya Pradesh are considered excellent for sweet orange. It is very sensitive to high concentration of salt and cannot withstand water-logged condition. The best pH would be 4.5-6.5.

Varieties

- ☆ **Sathgudi:** Sweet in taste. Granulation is a problem.
- ☆ **Mosambi:** Apex is marked with circular ring. Pulp is pale yellow.
- ☆ **Pineapple:** Pulp is orange yellow in colour.
- ☆ **Jaffa:** Pulp is light orange in colour.
- ☆ **Malta Blood Red:** Excellent variety, red pulp. Grown in Punjab, U.P RJ.
- ☆ **Valencia: A variety from USA.**
- ☆ **Washington Navel Orange:** USA variety, It is a seedless budsport of the Brazilian orange variety "Laranja Selecta".
- ☆ **Shamouti:** Fruit is ellipsoidal in shape, medium sweet in taste with rich in flavour.
- ☆ **Hamlin:** Florida origin, fruits orange red.
- ☆ **Batauion:** Yellowish green, axis is hollow, mainly grown in coastal areas of A.P.
- ☆ **Phule Mosambi:** released from MPKV, Rahuri

Planting Material

- ☆ Budded plants (Root Stock: Rangpur lime is best, now Rough lemon is also preferred).

Rootstocks for Sweet Orange

1. **Rangpur lime** (*C limonia*)*:* Tolerant to salt and tristeza virus
2. **Sour orange** (*C. aurantium*)*:* High yield with quality fruit but susceptible to tristeza virus and nematode.
3. **Citrange**: (Trifoliate orange x sweet orange)

   e.g. Troyer citrange, Carrizo citrange are tolerant to tristeza and resistant to burrowing nematode.
4. **Cleopatra mandarin** (*C.reticulata*) *:* Tolerant to tristeza and suitable for sodic soil.
5. **Rough lemon or Jambhiri** (*C. jambhiri*)*:* Tolerant to Tristeza and susceptible to root rot and blight.
6. **Sweet orange** (*C.sinensis*) : Plants are cold hardy and have wide range of soil adaptation.
7. **Trifoliate orange**: A dwarfing rootstock, good yield and fruit quality, susceptible to exocortis virus, but tolerant to most other virus diseases.
8. **Wood apple (*Feronia limonia*)**: Dwarf and bushy growth habit.

Among these Rootstocks

- ☆ Rangpur lime, Jatti Khatti (*C.jambhiri*), Troyer citrange is the commonly used best root stock.
- ☆ Buds are taken from the trees which are free from virus and diseases.
- ☆ Budded plants are ready for planting in the main field in 12 months.

Preparation of Field

- ☆ Dig pits at 75 cm x 75 cm x 75 cm in size at 7 x 7 m spacing. Fill up the pits with top soil and 10 kg of FYM. Plant the budded plants in the centre of the pits and stake it.

High Density Planting System

- ☆ Kinnow mandarin orchards were now converted to high density types where the most suitable rootstock identified was Troyer citrange (1.75 x 1.75 m – 2990 pl ha^{-1}) and Kharna khatta (2.25 x 2.25m - 1682 pl ha^{-1}). Such orchards are precocious and at the same time gave double returns to growers compared to conventional system of planting (Goswami *et al.,* 1993).

Irrigation

- ☆ Immediately after planting irrigate copiously. Irrigations may be given once in 10 days. Avoid water stagnation near the plant.

Manures and Fertilizers per Plant

- ☆ N to be applied in two doses during March and October.
- ☆ FYM, P2O5 and K2O are to be applied in October.

Manures and Fertilizers	*I Year (kg)*	*Annual Increase (kg)*	*From 6th Year (kg)*
FYM	10.00	5.00	30.00
N	0.200	0.100	0.600
P	0.100	0.020	0.200
K	0.100	0.040	0.300

- ☆ Manures are applied in the basin 70 cm away from the trunk and incorporated in the soil.
- ☆ Foliar praying of Zinc Sulphate (0.5%), Manganese (0.05%), Iron (0.25%), Magnesium (0.5%), Boron (0.1%) and Molybdenum (0.003%) once in 3 months at the time of new flush production. In addition to that, apply 50 g in each of Zinc Sulphate, Manganese and Iron per tree per year. To control fruit drop, 2, 4-D @ 10ppm (10mg/lit) should be sprayed two times, once in the first week of May and second in the first week of September.

Harvest, Yield and Storage

For sweet oranges the main harvesting season in North India is from December to February while in South, it extends from October to March. The fruits remain fresh for several weeks without any deterioration after attaining full maturity which facilities harvesting according to one's choice and demand in the market. Application of 10 ppm of GA by spraying on the trees just before fruits turn yellow will not only help to extend harvesting season but also improves resistance to fruit files, transport tolerance and resistance to post harvest moulds and mechanical injury by keeping the peel tough. The yield is 200-300 fruits/year/tree. Oranges can be stored for 12 weeks at 50C (If stored at 0°C, there will be chilling injury).

Physiological Disorders

Compared to any other fruits, citrus fruits exhibit a number of physiological disorders which are listed below.

Name	Features
Granulation (Selerocystosis, corkiness, stem end granulation)	Noticed in mature fruits on stem end. There will be decrease in soluble sugars, acids and accumulation of starch, pectin and cellulose. Juice sacs enlarged, develop thicker cell walls, juice vesicles lose colour, fruit volume increases. Temperature, humidity, light, water logging are other factors responsible. Control – reduce frequency of irrigation, foliar spray with zinc sulphate and copper sulphate + lime and 2, 4-D (10 ppm)
Xerocystosis (Blossom end granulation)	Occurs in the distal end of fruit. Juice vesicles loose water, become needle like and collapse.
Endoxerosis (Blossom end decay)	Common in lemons. Tissue disorder starts at stylar end. Juice vesicles near stylar end lose water. Fruits develop dull colour, immature fruits shed. At drought condition, severe incidence is noticed, so provide adequate irrigation.
Oleocellosis (rind oil splitting)	Appear as slightly depressed areas on fruit and oil glands become prominent. Care of fruits after picking, packing methods and time of packing are factors responsible.
Creasing of puffiness	Defect of the peel. Seen on mature fruits – appearance of random grooves on fruit surface. Caused by climatic and nutritional factors, higher proteolytic activity and higher content of soluble pectin. Application of GA (20 ppm) at early stages of fruit development check this disorder.
Splitting	Splitting of fruits and premature drop caused by fluctuation in environmental factors. Frequent light irrigation in drought season, early picking (just at maturity) are the control measures.
Degreening	Seen in lemon. Colour of mature fruits remain green – do not fetch good price. Dip fruits in ethrel (2500 ppm) and store in polythene bag for 3 days – fruits turn to yellow without affecting quality.
Sun burn	Sweet oranges are the most affected when grown in hot and arid regions.
Frost injury	Common to citrus plants grown in temperate region. Root stock like trifoliate orange is tolerant.

Major Production Problems

Citrus Decline (Citrus chlorosis, Die back, Frenching, Neglectosis)

A serious problem prevalent in all citrus growing tracts in the world is citrus decline. In India also citrus decline is seen in all species but it badly affects sweet oranges and mandarins. Affected trees become unhealthy and uneconomic, slowly it shows dieback symptoms and eventually dies off. This is a complex malady caused by a number of factors. Major factors responsible are:

- ☆ Climate – too high/too low temperature, frost, heavy rainfall.
- ☆ Soil – high and low pH, high CEC, high water table, hard pan in the soil.
- ☆ Nutritional factors – deficiency and imbalance of Zn, Cu, B, Fe and Mn and improper manuring.
- ☆ Rootstocks – Incompatible rootstocks, susceptible rootstocks to salinity, water logging, cold and virus.
- ☆ Improper orchard management practices.
- ☆ Entomological – Aphids, leaf miner, psylla *etc.*
- ☆ Pathological – Viral, fungal, bacterial and mycoplasma.
- ☆ Nematodes

Major control measures include evolving planting materials by following bud wood certification programme, regular manuring which includes 1/3rd in organic form and 2/3rd in inorganic form. Microelements like zinc, copper, manganese and boron are to be applied as foliar sprays. Soil application of dolomite can be done once in 2-3 years for improving soil reaction and availability of calcium and magnesium. Regular weeding, inter cultivation and mulching during drought periods will reduce the incidence. Pruning dried and diseased branches, application of Bordeaux paste in tree trunk, spraying of insecticides and fungicides, providing proper drainage facilities, use of appropriate root stocks (Eg: Rangpur lime and trifoliate orange for heavy soils, rough lemon and sweet lime for light-soils) are the other control measures.

Crop Regulation in Citrus Fruits

Flowering/fruiting cycle management is another term for crop regulation. Crop regulation's major goal is to force the tree to rest and produce abundant blossoms and fruits during one of the two or three flushes. Only by limiting the crop to a single season can a good yield be achieved (bahar). Crop regulation provides the foundation for a consistent and high-quality harvest. It is also known as "vegetative-reproductive balance management." The availability of water, the occurrence of diseases and pests and the market position all play a crucial role in bahar selection. Typically, 30-40 leaves per fruit are adequate to produce high-quality fruit. Many fruit crops bloom all year, with no rest period in between and produce 2-3 bahars per year, with poor quality fruits and reduced output. To determine which bahar will produce the best harvest, it is vital to study the flowering and fruiting behaviour of crops while taking into account all relevant parameters.

Bahar	*Month*	
	Flowering	*Fruiting*
Ambe bahar	February - March	July – August
Mrig bahar	June - July	November - December
Hasth bahar	October - November	February – March

Principle of Crop Regulation

Crop regulation's main premise is to regulate the fruit plant's natural flowering in the appropriate season to maximize fruit output, quality and profitability. The majority of crops only bear blooms on young, succulent and vigorously emerging vegetative growth which supports this theory. These fresh growth flushes might appear as new lateral bud emergence on older stems or as expansions of already established terminals of varying size and vigour.

Why the Need?

The bahar determines the yield and fruit quality. Many crops that bloom more than once a year do not produce a consistent yield and fruit quality throughout the year. This necessitates many ways for controlling blooming (from ambe bahar to mrig bahar) in order to achieve the most profitable harvest. Depending on meteorological parameters, cropping pattern, cultivar and other considerations, different methods of crop management are used in different regions. This crop regulation of flowering from ambe to mrig bahar to obtain most profitable crop is done using several methods:

With respect to fruit crops, crop regulation aim at

- Optimal crop annually.
- Improve fruit quality.
- Modify the balance between growth and fruiting to obtain higher yield and reduce management costs.
- Improvement in flowering and fruit set, especially where the yields are suboptimal.
- Reduce the crop load in order to avoid competition.
- Uniform flowering.
- Maximize production.
- Profit to the grower.

Methods of Crop Regulation

In order to get only appropriate season crop it is necessary to manipulate the flowering. The following practices can be adopted

- De-blossoming or Thinning.
- Withholding of irrigation.
- Root exposure or Root pruning.
- Shoot Pruning.
- Chemical/PGRs application.
- Nutrients application.

De-Blossoming or Thinning

One way for crop management is to deblossom rainy season crop. Hand thinning of flowers and tiny fruits, done twice in April and May at fortnight interval has been proven to be highly successful in this regard. However, the method is expensive and inconvenient. This approach also reduces the plant's total yield for the year. Under various agro-climatic conditions, growth regulators (NAA, NAD, 2, 4-D and ethrel) and specific chemicals (carbaryl) have been found to be very effective in flower thinning and manipulating the cropping season. Flower thinning has been attempted with varied degrees of success using Naphthalene Acetic Acid (NAA), Naphthalene Acetamide (NAD), 2,4- Dichlorophenoxy acetic acid (2,4-D), potassium iodide (KI), 2-chloroethyl phosphonic acid (ethephon), 4,6-Dinitro-OCresol (DNOC) and urea. Chemical thinning, according to the majority of workers, is cost-effective and boosts winter yield while also improving fruit quality.

Withholding of Irrigation

During the months of February and March, irrigation is turned off for a couple of months. Trees go into dormancy when their growth cease, leaves turn yellow and fall off owing to water stress. At the end of May or early June, the orchard soil is ploughed and individual tree basin is manured with farm yard manure and afterwards irrigation is performed. The first two irrigations should be spaced three days apart, with successive irrigations spaced 10-15 days apart until the monsoon arrives. The plants subsequently continue growth and blossom profusely in August–September, yielding a large number of high-quality fruits over the winter season.

Root Exposure and Root Pruning

Root trimming of various degrees has been found to be effective in reducing the rainy season crop. In areas with a high atmospheric humidity and water table, root exposure and pruning is practiced. After removing the top 8 cm of soil, the method comprises exposing the tree's higher roots up to 1.5 ft. radius around the trunk. The primary root system is not harmed, but the fibrous roots on them are pruned away with a pruning shear. Hence, the tree sheds its leaves. The exposed roots are now covered with dirt, mixed with manures, and irrigated promptly. Root pruning, on the other hand, should be avoided as it shortens the life of the plant. The plant's roots are exposed to the sun by removing up to 7-10 cm of soil from a 40-60 cm radius around the tree trunk. Before blossoming, the water is withheld for a month or two. Leaves wilt and fall to the ground as a result of water stress. Roots are again covered with a mixture of soil and FYM and irrigated right before one month of targeted bahar. Irrigations are administered at appropriate intervals after that. As a result, plants produce fresh vegetative growth, as well as abundant flowering and fruiting.

Shoot Pruning

Crop management in fruits like guava has been proven to be particularly effective when new shoot (current season) growth is pruned in the first week of May.

This procedure entails removing half to three-quarters of the shoot development. As a result, of this pruning, the flowers and flower buds of the spring flowering season are automatically removed, and the rainy season crop can be lowered. This approach has been determined to be effective and cost-efficient, with little impact on the plant's annual total production. When citrus plants were clipped up to 45 cm in May, highest TSS, sugar: acid ratio and yield were observed

Chemical PGRs/Application

The application of growth regulators to manipulate the flowering season has been discovered to be particularly successful for crop regulation. Growth regulators including NAD, NAA, and 2, 4-D has been proven to be particularly effective in flower thinning and cropping season manipulation. Summer flower thinning can be accomplished with chemical treatments of NAD at 30 and 50 ppm, NAA at 100 and 125 ppm and 2, 4-D at 15 and 30 ppm.

Nutrition Application

Nutrients also have a role in crop regulation. Two treatments with 15 per cent urea at

15 day intervals in April-May were more successful in the summer season proved economically feasible for crop regulation

Crop Regulation Methods in Citrus Fruits

Acid Lime

- ✰ Acid lime plants may be trained to modified central leader system, with a smooth trunk up to 75-100 cm height from the ground level and 4-5 well spaced and well spread branches, as scaffolding branches.
- ✰ All sprouts appearing on the trunk up to a height of 75-100 cm should be removed. Similarly on grown up trees, the water suckers appearing on main trunk and scaffolding branches should be removed promptly.
- ✰ Once a young plant is trained to a desired shape, it requires very little pruning. Light pruning may be given during later years.
- ✰ Light pruned trees make more development of roots and shoots, producing fruit earlier that those pruned heavily Pruning of bearing trees though differ with variety, chiefly consists of removal of dead, dried, diseased, broken and cris-cross branches, whose existence is detrimental to the health of trees. Removal of water suckers is also essential.
- ✰ Pruning may be done just after harvesting. Soon after pruning, the cut ends may be smeared with Bordeaux paste or Blitox.
- ✰ Untrained or unpruned guava trees become huge and unmanageable after a few years of growth. The bearing area is reduced and the interior of plants become entirely without fruits.

- ☆ Trees are topped to a uniform height of 60-70 cm from the ground level, 2-3 months after planting to induce the emergence of new growth below the cut points.
- ☆ Three to four equally spaced shoots are retained around the stem to form the main
- ☆ Scaffold limbs of tree. These shoots are allowed to grow for 4-5 months after topping until they attain a length of 40-50 cm.
- ☆ The selected shoots are further pruned to 50 per cent of their length for inducing multiple shoots from the buds below the cut end. Newly emerged shoots are allowed to grow up to 40-50 cm and pruned once again for emergence of new shoots. This is chiefly done to obtain the desired shape.
- ☆ The pruning operations continue during the second year after planting. After two years, short branches within the tree canopy produce a compact and strong structure. All the plants are confirmed to a hedge shape of 2 m inter row width and 2.5 m height for which pruning is performed in January and May-June every year.

Sweet Orange

- ☆ The trees at planting time are headed back more severely to a height of 70-80 cm from the ground level.
- ☆ Pruning or cutting back of one year old shoots to half length (50 per cent of the total) or to full length is recommended for obtaining proper yield of high quality fruit. The pruning, therefore, is done to keep the balance between fruiting and vegetative growth.
- ☆ Pruning of some of the shoot certainly removes a part of fruiting area and helps
- ☆ Maintain regular cropping. The dried up branches found in the lower part of the plant too are removed.

Tamarind (*Tamarindus indicus*, Caesalpiniaceae)

Varieties

PKM 1, Urigam, Hasanur,Tumkur Prathisthan, DTS 1, Yogeshwari.

Soil and Climate

Grown on variety of soils ranging from poor degraded, eroded, gravelly, saline and alkaline soils. Productivity is higher in red loamy, deep well drained soils. The absolute maximum temperature varies from 36-47.50 C and the absolute minimum temperature varies from 0-17.50 C. Rainfall requirement – 750-1900 mm. Altitude – up to 100 m above MSL.

Season

June – December is found to be optimum.

Propagation

Seeds/Grafts

Nursery

Fresh seeds are sown in nursery beds in March – April. Soaking of seeds in 10 per cent cow urine or in cow dung solution (500 g in 10 l of water) for 24 hours. Two year old seedlings are transplanted to the main field.

Vegetative Propagation

- ✰ Softwood grafting: March- April
- ✰ Air Layering: Shoots treated with IBA 4000 ppm.

Spacing: 8-10m x 8-10m is adopted.

Planting

The grafts should be planted in the pits of 1 m x 1 m x 1 m filled with FYM and top soil. Add 50 g of Methyl parathion 1.3 per cent dust in the pit. Immediately after planting, support the graft with stakes.

Irrigation

Regular watering should be given once in seven days.

Fertilizers

Apply 200: 150: 250 g of NPK per tree per year along with 25 kg of FYM and 2 kg of Neem cake.

After Cultivation

- ✰ Remove the rootstock sprouts.
- ✰ Remove the dried and diseased parts.
- ✰ Intercrops like leguminous crops, short duration vegetables, annual drumstick, sesamum and sorghum may be raised in the alley spaces up to four years.

Training

Early training is necessary to form a high head and uniform scaffold branches in all directions.

Pruning

Removal of dried, diseased and criss cross branches.

Harvest

Plants starts bearing from 4th year onwards and the economic yield will be achieved from 9th year onwards. Pods are harvested in March - April every year

REFERENCES

Bose TK, VA, Parthasarathy, Mitra SK, Chakraborty B, Ghosh and D Sanyal, 2022. *Fruits: Tropical and Subtropical Vol 4. 4th Revised and Illustrated edn.* ISBN 10: 9390435463/ISBN 13: 9789390435463

Karan Kumar and Sharma, 2020. Canopy Management in Fruit Crops. Dr. Y.S. Parmar U.H.F., Nauni, Solan (HP) *Just Agriculture.* **1**, Article Id: 046.

Krajewski A and Krajewski S, 2011. "Canopy Management of Sweet Orange, Grapefruit, Lemon, Lime and Mandarin Trees in the Tropics: Principles, Practices and Commercial Experiences." *Acta Horticulturae,* **894**: 65–76.

Krajewski A and Pittaway T, 2000. "Manipulation of Citrus Flowering and Fruiting by Pruning." *Proceedings of the International Society of Citriculture,* **1**: 357–360.

Mahala K, Bhateshwar MC and Solanki SK, 2022. Advances In Canopy Management in Fruit Crops. *Recent Innovative Approaches in Agricultural Science*: 155.

Nanhe Lal Saroj, Karobi Handique, Sunil Ram, Ajay Patel and Jahnabi Hazarika, 2024. Canopy Management in fruit crops. *Just Agriculture,* **4(9)**: P.49-57.

Practical Manual on Canopy Management in Fruit Crops-Dr. Gorakh Singh (2010). Department of Agriculture and Cooperation, Ministry of Agriculture, Govt. of India, New Delhi-110001.

Chapter 9

Propagation Techniques for Humid Tropical Fruits

Pooja Bohra[1], Ajit Arun Waman[1] and Gourish R. Karanjalker[2]

[1]Division of Horticulture and Crop Improvement, ICAR–Central Island Agricultural Research Institute, Port Blair- 744 105, Andaman and Nicobar Islands
[2]Goa College of Agriculture, Ela, Old Goa–403 402, Goa

India with its huge geographical area has diverse agroclimatic regions. Each of these regions is characterized by distinct flora of horticultural significance including fruits, vegetables, ornamental plants, plantation crops, spices and medicinal and aromatic plants. Some of these crops are restricted in their distribution, while others may occur in more than one agro-climatic zones of the country. As the crop husbandry in horticulture is known to get influenced by a number of region-specific soil and climatic factors, agro-techniques suitable for each region are required to be developed.

Propagation is one of the first and prime operations in the cultivation of any crop and it is also one of the most sensitive operations that is highly influenced by the area/region in which it is being carried out. Even though the microclimate prevalent in the protected structures could be easily manipulated using routine as well as high-end techniques, the propagation activities are influenced by a number of external factors. For example, the appropriate growth stage of mother plant for scion collection may not coincide in two different agro-climatic zones of its cultivation. Hence, the propagation activities for the same crop may not necessarily remain the same for different agro-climatic conditions. In this chapter, efforts have been made to provide an overview about propagation techniques in various fruit crops grown in humid tropics of India. In crops which are grown in multiple regions, only those techniques which have been standardized under humid tropics have been covered.

Humid Tropical Region of India

Humid tropical region (HTR) of India is typically rich in the horticultural diversity, wherein diverse varieties of major, minor and underexploited fruit species are distributed and cultivated. This HTR with warm or high temperature is broadly divided into two sub-groups (Tripathi, 2021). The first sub-group covers area with *tropical monsoon climate* and covers region of South-western lowlands of Malabar Coast, the Western Ghats, southern Assam, Lakshadweep and the Andaman and Nicobar Islands. These regions usually have high humidity, moderate to high temperature and high rainfall (above 2000 mm per year) in most parts of the year. The second sub-group is with *tropical wet and dry climate* and covers regions of inland peninsular India except a semi-arid rain shadow area towards Eastern Western Ghats, parts of West Bengal, North Eastern States, Odisha and Coastal Andhra Pradesh. This region experiences very high temperature during summer and receives annual rainfall of about 750–1,500 mm per year. The crops grown in both the areas generally have higher tolerance to moisture in the air as well as soil.

Diversity of Fruit Crops in the HTR of India

A number of cash crops such as mango, pineapple, banana *etc.* are commercially grown in the HTR of India. Improved varieties suitable for this region have been developed by various regional research stations of the State Agricultural Universities/ICAR Institutes located in these regions. Further, a number of exotic crops such as mangosteen, rambutan, durian *etc.* were successfully introduced in these areas and their area is increasing in the country and hence, these have been identified as potential crops for the country. Additionally, as the HTR covers diversity-rich areas (hotspots) such as the Western Ghats, Andaman and Nicobar Islands and parts of northeast India, wide range of underexploited species are known to exist in these regions, that have been regionally valued for various food purposes. For example, crops such as Andaman Kokum (*Garcinia dhanikhariensis*) in the Andaman Islands and Blood fruit (*Haematocarpus validus*) in the north-eastern India and Andaman Islands have emerged as potential sources of livelihood to the native farmers in these regions.

Propagation Techniques for Fruits of HTR

Plant propagation is multiplication of any plant for the benefit of mankind. Of the two major methods of propagation, namely sexual and asexual (vegetative), the use of appropriate technique depends on the crop/species in question, purpose of multiplication, season of operation, availability of resources including skills *etc.*

Sexual Propagation in Fruits of HTR

In sexual method of propagation, seeds obtained from fusion of the male (pollen) and female (ovule) germ cells are utilized for regeneration. This method is mainly used in the breeding programmes for developing hybrids. Further, in fruit crops like papaya and mangosteen, wherein no other feasible methods of multiplication are available, seed is the only viable option for multiplication. Another important application of using seeds for multiplication is for raising rootstocks, which could further be used for grafting or budding operations. Unlike in most of the temperate fruit crops, wherein the concept of clonal rootstocks is widely adopted, the research efforts in case of humid tropical fruits are limited. Hence, seed-derived rootstocks are to be utilized till such options of clonal rootstocks are explored in these fruits.

As emphasized earlier, the HTR also covers a number of region-specific underutilized fruit species, a number of which are commonly harvested from the natural stands and in some cases, the populations are dwindling in wild. In such cases also, seed propagation serves the purpose of conserving the diversity of the species apart from providing propagules for area expansion in home/commercial gardens, habitat enrichment activities and other societal development programmes.

The sexual propagation, though seems easy, requires standardization of a number of factors as each species generally pose typical germination requirement. These factors could be seed related such as nature of seed (orthodox/recalcitrant/intermediate), selection of mother plant/tree source, maturity of seed, size of seed, presence of any form of seed dormancy, seed moisture content *etc.* Similarly, external factors such as seed germination treatment, type of substrate, container size, the microclimatic requirement, aftercare, age of transplanting, incidence of nursery disease/pests *etc.* are known to govern the success in this method.

Seed propagation is regarded as a rather easy method for obtaining planting material and progenies with limited requirement of special skills and techniques. Seedling derived plants in a number of cases have unique features such as high yield (due to higher canopy spread than grafted/budded counterparts) and longer life of the plantations. Seedlings are hardy with deep root system, which helps in better transportation of nutrients and water within the plant. Their competing ability is much higher than that of vegetatively propagated plants for anchorage, absorption of the nutrients, water, fertiliser, allelopathy, stress tolerance and mitigating the environmental effects. Hence, the water, nutritional and other requirements are generally lower in seedling derived plants. A number of rootstocks with tolerance to abiotic/biotic stresses and canopy manipulation ability have been identified for vegetative propagation, which are usually raised by seeds *e.g.* Vellaikolumban in mango and Rayan/Khirni (*Manilkara hexandra*) in Sapota. Polyembryony is the phenomenon of production of more than one embryo from a single seed that produces true to type/nucellar embryonic seedlings. This is found in some fruit crops that could be used as rootstock for their better performance *e.g.* Mango *var.*

Olour and Bappakai. Moreover, seed propagation is necessary when vegetative propagation is unsuccessful or expensive.

Considering the disadvantages, seed derived progenies are neither true to type nor uniform in yield and quality, which are the desirable attributes in commercial fruit orchards. Seedlings have a long juvenile period and normally take longer time (more than 6-12 years) to come to bearing. Seedlings require more spacing as they have vigorous growth habit and canopy coverage. Unlike various vegetables and flower crops, a number of fruit species have recalcitrant seeds and hence, seeds must be sown immediately after extraction. Seed related problems such as dormancy, low germination rate, low vigour and vitality have to be counteracted with proper treatments. Many crops or varieties are seedless and seed propagation is difficult in fruit crops like banana.

Asexual Propagation Method

Vegetative propagation has been considered as the boon for perennial horticultural crops, mainly fruit crops, for mass multiplication of elite mother plants. For establishment of commercial orchards for fresh fruit purposes (mainly export markets) as well as processing purposes, high uniformity is expected in the produce, which could be effectively addressed when asexual propagation methods are developed and adopted. Vegetative propagation has a number of advantages to offer. The plants are generally true-to-type, uniform in growth, yielding capacity and fruit quality. They have short juvenile phase and thus, come into bearing earlier than seedling plants. The advantages of rootstocks can be obtained by budding or grafting susceptible varieties on resistant/tolerant rootstocks. Plants have restricted growth so cultural practices and harvesting are easy. In some fruit plants like banana, which do not bear seeds, this is the only method of propagation. However, new varieties cannot be evolved by this method and plants are not so vigorous and long-lived as compared to the seedling trees.

A number of methods are relevant for propagation of fruits of HTR. Of these, cuttings, air layering, budding and grafting are notable ones. Propagation by cuttings can be practiced in some tropical fruit species. In fruit crops, generally hardwood or semi-hardwood cuttings are prepared wherein, adventitious roots are induced in the cuttings by exogenous application of rooting hormones. However, in some recent studies, leaf cuttings have been tried in guava with good success percentage (Neelavathi *et al.,* 2021). Factors governing success of rooting in cuttings include season of cutting, length of cutting, concentration of growth regulator, substrate and growing environment. In a study by Mng'omba (2023), none of the cuttings collected from fruiting trees could root as carbohydrate reserves are depleted in stem portion of the plant during fruiting phase. Ryadin *et al.* (2014) studied five varieties of Malay apple for their performance in propagation by cuttings using two substrates and two auxin sources. Use of naphthaleneacetic acid was found to be better in induction of rooting and shoot growth whereas combination of top soil and sand as substrate was found to be better than sawdust and sand combination.

Similarly, air layering is a popular method in difficult-to-root species of horticultural significance. In this case also, optimization is required to identify suitable season, kind and concentration of growth regulators, substrates, age/thickness of branches *etc.* to get better rooting. Thus, a number of factors are required to be studied to identify the most suitable combination for obtaining quality planting material in large numbers.

In case of grafting and budding, importance of rootstock is paramount. In case of newly developed hybrids or fruit species with less seed production, use of a commonly available and compatible species as a rootstock facilitates mass multiplication of the improved genotypes. In lai-durian, a natural interspecific hybrid of *Durio zibethinus* and *D. kutejensis*, use of *D. zibethinus* or *D. kutejensis* seeds for raising rootstocks was found to be advantageous as the hybrid suffers from an issue of lesser number of fully developed seeds per fruit (Sunaryo *et al.*, 2019). Success of cleft grafting and chip budding was statistically similar on these rootstocks as that on its own rootstock. In case of jackfruit, five vegetative propagation methods were tried at monthly interval under Tarai conditions of Uttarakhand (Kumar *et al.*, 2021). Wedge grafting was found to be the best method of propagation, if performed during December to April with the highest success (74.38 per cent survival) during March. Thus, in grafting as well as budding, identification of suitable rootstock, age of the rootstock and scion, season of operation, growing environment, method of grafting/budding *etc.* are required to be standardized.

Propagation Techniques in Humid Tropical Fruits

Soursop (*Annona muricata*)

Soursop is a member of the custard apple family, which has spine like projections all over the fruit surface. The pulp is relished in fresh form in most of the coastal regions of India. Popularity of soursop has been on rise in recent years due to identification of medicinally valued compounds in the fruit. Seed germination has been attempted under Karnataka conditions. For improving seed germination, fresh seeds should be extracted from the fully ripe fruits and soaked in water for 96 h before sowing. Use of red soil: sand: farmyard manure (2: 1: 1) substrate resulted in 82.85 per cent germination in soursop (Nadukeri *et al.*, 2018). Among the vegetative propagation methods, use of 12 cm long softwood cuttings with halving of top two leaves was recommended and use of auxin *viz.* 2,000 ppm IBA exhibited 70 per cent rooting success (Santos *et al.*, 2011). Under Tamil Nadu conditions, propagation using soft wood cuttings of 15 cm length with 2-4 nodes collected from young branches of 5 years old trees of soursop was attempted. About 4 to 6 leaves were retained on the cuttings and lower slant portion was dipped in 1000 ppm NAA for 5 minutes prior to placing them in into the polybag containing sand + coir pith+ FYM (3: 1: 1) under mist. Within 20 days, sprouting was noticed and higher number of roots (9.12) with better plant growth parameters could be obtained (Maheswari *et al.*, 2021). Iwalesin *et al.* (2023) reported 100 per cent success in air layering of soursop when the girdled portion was treated with 2,000 ppm IBA and coconut peat was used as substrate.

Breadfruit (*Artocarpus altilis*)

Breadfruit is a staple food crop of several tropical countries as the fruit is processed into flour for preparing breads, cakes *etc.* The fruit is among the most relished commodities by the Nicobarese tribal population in the far-flung Nicobar Islands of the country. In a number of coastal states, the shallow fried breadfruit slices are considered as a delicacy. It has been reported to be propagated by stem cutting, root cutting, layering, budding, grafting and *in vitro* techniques; however, only a few methods are considered commercially feasible for propagating it.

The species is generally seedless and hence, other species are required for carrying out grafting in it. *Artocarpus camansi* (breadnut) is a closely related taxon that commonly produces abundant seeds, which germinate naturally with more than 95 per cent germination success. Further, the growth of seedlings of breadnut is very vigorous. Grafting of breadfruit on 45 days old seedlings of breadnut rootstock was found to be successful with 83 per cent graft take and thus, could be a commercially viable method of its propagation (Medagoda and Chandrarathana, 2007). Three months old grafted plants could be transferred to the field condition and start fruiting within 3 years of planting. Use of root suckers or root cuttings can give rise to limited number of plants. The root cuttings can be separated from identified elite mother plant and placed in polythene bags with dimension of 10 cm × 15 cm prefilled with growing media (soil and compost). The cuttings root well when dipped in 1,500 ppm IBA solution for 5 min before placing in the polybags. The cuttings are ready for planting in the main field after six months of hardening in nursery. Approach grafting has also been reported to be successful in breadfruit (Nandwani and Kuniyuki, 2005).

Jackfruit (*Artocarpus heterophyllus*)

Jackfruit is a traditional crop that has gained wide popularity in the recent past. A number of institutes and organizations have ventured in to research on this crop. The popularity of the crop is also evident by the large number of value-added products reaching the markets and increasing number of improved varieties developed in the country including those identified by the farmers. The crop has been a prominent component in the homesteads of west coast of the country and commercial scale plantations are being set up in most of these places.

Jackfruit is a large sized tree requiring long gestation of about 10 years and the existing plant populations show huge variations in terms of morphological, yield and quality parameters. Thus, use of vegetatively propagated material is desirable to boost the production as well as productivity. Commercially, jackfruit is propagated by epicotyl or softwood grafting methods in humid tropical climate. The seeds are extracted manually without delay as they lose viability drastically after extraction. Use of bolder seeds of 7 g was found to give better germination and seedling vigour parameters (Nivetha and Maheshwari, 2015). However, in case of non-availability of seeds, minimum 3 g sized seeds should be used. The seeds can be grown in 15 cm × 20 cm polybag filled with equal proportion of soil, vermicompost, cocopeat and rice husk for the production of rootstocks. About 93 per cent germination could be obtained within 24 days of sowing using this substrate (Gawankar *et al.,* 2021).

Efforts have been made to extend the viability of seeds. For this, seeds should be washed thoroughly to remove the pulp and aril, and only those weighing more than 3 g should be selected. Seeds should then be treated with Carbendazim (0.2 per cent for 5 min) and air dried before storing them in the screw cap bottles. This could extend the viability of seeds to about 150 days (Gawankar *et al.,* 2020). The study also suggested that germination process in jackfruit was under the control of phytochrome action and for promotion of germination, dark and humid conditions are required. The germplasm/variety used for grafting also influences success of a protocol (Nivetha and Maheshwari, 2015) and hence, genotype specific optimization is required.

In softwood grafting, use of seedlings of 60 days has been recommended for getting better results (Baskaran *et al.,* 2008). Epicotyl grafting performed well under Kerala conditions by using 3 to 4 months-old scions and 5-days-old rootstocks with highest sprouting (29.50%) and survival (21.67%) (Jose and Velsalakumari, 1991). Time of grafting plays an important role and hence, standardization is required based on the regional agro-climatic situation. For inarching in jackfruit, July 20 was found to be the optimum time with 80 per cent success under West Bengal conditions (Ghosh and Bera, 2015). Epicotyl grafting was successful in June on 14 days old rootstock with 53.71 per cent graft take under Bangladesh conditions (Islam *et al.,* 2003). The grafts should be re-bagged in polybags of 103 ×143 size filled with soil: vermicompost (2: 1) for better growth and field establishment (More *et al.,* 2023). Under Konkan conditions of Maharashtra, for improving the growth of grafted var. Konkan Prolific, use of soil + sand + FYM (1: 1: 1) and spraying of vermiwash (10%) was recommended (Varadkar *et al.,* 2020).

Bilimbi (*Averrhoa bilimbi*)

Bilimbi is an underutilized tropical fruit, which is used to prepare various traditional value-added products in coastal parts of India and abroad. It has been a common fruit used for pickling in Andhra Pradesh, while the fruit is common is backyards of Goa, West Bengal and Andaman Islands, wherein the local people relish fresh fruits with salt and spices. The presence of high content of oxalic acid; however, deters the masses to consume it in large quantity. Seed propagation is the common method of propagation in bilimbi. The ripened fruits are harvested and seeds are extracted from the fruits by fermentation or manually. Seeds are then air dried for at least one or two days and sown in black biodegradable polybags (10 × 15 cm) of 200 micron. Germination generally takes place within 10-15 days and about one year old seedlings can be planted in main field during onset of monsoon

(Karanjalker and Karanjalker, 2021). Scientific studies on vegetative propagation are not available and require attention of researchers.

Figure 9.1: Cauliflorus Bearing in Bilimbi.

Carambola (*Averrhoa carambola*)

Carambola or five-cornered fruit is a minor yet well-known fruit of the tropics; wherein it is popularly known as *kamrakh*, *kamrang* or star fruit. Both, sweet and sour fruited types are reported in carambola of which sweet types are suitable for fresh consumption as well as processing into value added products such as beverages; while the sour types are used in some culinary preparations. The species bears fruits multiple times a year and three main crops could be obtained under the climatic conditions of Andaman Islands. The branches are not strong and hence, at times get damaged in the cyclonic conditions in the growing region. However, carambola exhibits high regeneration capacity and the damaged plants give rise to new shoots in no time. This peculiar character makes it a suitable crop for the wind-damage prone regions of humid tropics.

Carambola has been reported to be propagated by seeds, grafting, budding and air layering methods. Fully ripe fruits, ideally fallen naturally, are used for manual extraction of the seeds. Seeds are air dried for 2-3 days and sown in polybags or root trainers in nursery. Storage of seeds is possible as higher germination speed index was reported in seeds dried at 38 °C and stored for 6 months at 15 °C (Oliveira *et al.,* 2008). Seedlings of about 45 to 60 cm could be used for field planting. As the seedling derived plants exhibit variability, vegetative propagation is of advantage, especially when targeting multiplication of sweet fruited types. Further, seedling derived plants usually take about 4 years for fruiting, whereas vegetatively propagated plants come to bearing within 1-2 years of planting (Radha and Mathew, 2007).

Figure 9.2: Seedlings of Carambola.

Softwood grafting has been found to be an efficient method using own rootstock. Effect of different grafting seasons suggested that July-August was the ideal time for grafting with 90 per cent success (Sonawane *et al.,* 2012). Under West Bengal conditions, grafting on one year old rootstock during June 30 to August 15 gave 100 per cent success with better graft growth during hardening (Ghosh and Bera, 2015). Air layering has also been found to be successful with 60 per cent success without use of auxins, when layering was performed during June 15. However, field establishment in this case was mere 50 per cent with poor root development (Ghosh and Bera, 2015). Under Arunachal Pradesh conditions, collections of sweet carambola were used for 'I' budding on one-year-old rootstock and June to August was found to be the most suitable season for budding with 51 to 55 per cent success (Singh and Devachandra, 2021).

Figure 9.3: Cut Section of Carambola Fruit.

Karonda (*Carissa carandas*)

Karonda is a regionally popular underutilized fruit that has been mainly sourced from the non-commercial plantations, wild stands and homestead gardens. The bushes being thorny, karonda is an excellent bio fence material apart from yielding nutritious fruits mainly rich in iron. The species is very hardy and is known to survive in marginal and wastelands including saline-sodic soils with high pH level. Karonda could be propagated by cutting, air layering and seeds.

Fully ripe fruits should be soaked in water overnight and pulp should be removed by rubbing on hard surface. Such seeds should be shade dried and used for sowing immediately since they lose viability soon. Seeds should be sown at 1.0 to 1.5 cm depth in black polythene bags of 15 cm × 22 cm size and 300-gauge thickness filled with red soil: farmyard manure: soil (1: 1: 1) and 1 g/m^3 carbendazim as prophylactic measure (Deepika *et al.*, 2014). Freshly sown seeds germinated quickly (9 days) and showed highest germination of 67 per cent, which dropped to 57 per cent after 10 days of storage at room temperature. Viability reduced drastically with time and mere 20 per cent germination was obtained after 60 days of storage (Deepika *et al.*, 2014). Seedlings of about one year age are suitable for field planting (Banik *et al.*, 2012).

Hardwood cutting propagation in karonda is comparatively easier method for mass multiplication of superior genotypes. For this, from selected mother plants, about 1 cm thick hardwood cuttings with 4-5 nodes *i.e. ca.* 20-25 cm length should be collected. Cuttings should then be disinfected with fungicide for a few minutes, washed and kept in shade for 10 minutes. Under West Bengal conditions, quick dip treatment of cuttings with IBA (8000 ppm) solution and planting in polythene bags filled with sand: soil: farmyard manure (1: 2: 1) gave highest rooting of 39 per cent. Dipping of roots in 4 per cent sucrose solution could also induce rooting in 37 per cent cuttings (Dey *et al.*, 2017). Under Andhra Pradesh conditions also, similar results were obtained with 34 and 33 per cent rooting success (Deepika and Vanajalatha, 2015).

Velvet Apple (*Diospyros blancoi*)

Velvet apple or *Bilayati Gaab* is a dioecious fruit tree of the tropics, which has velvet like appearance of its peel and firm pulp. Trees grow very tall and bear fruits of variable sizes *i.e.* 100-220 g weight with 75-90 per cent pulp content. Seed germination has been reported to be more than 75 per cent when seeds were soaked for 12 h in water, coconut water or a combination of both (Origenes and Lapitan, 2020). Vegetative propagation of elite types of velvet apple was attempted in Bangladesh, in which scion sticks were cleft grafted on one-year-old seedling rootstock of the same species. Success rate varied among the genotypes and highest success rate was 60 per cent after 12 weeks of grafting (Robbani and Sarker, 2022).

Surinam Cherry (*Eugenia uniflora*)

Surinam cherry or Brazilian cherry or Pitanga is one of the popular fruit crops in the tropics of South America. It has been successfully introduced in southern parts of our country and Andaman Islands, wherein it is being popularized. Based on fruit colour, it could be orange red or dark purple, the latter one having higher antioxidant content. Though commercial orchards of Surinam cherry are not present in the country, it has been popular among the gardening enthusiasts for its attractive fruits and aromatic leaves. It can be propagated using seeds, which should be sown immediately after extraction from the fruits. Though immature and mature fruits have distinct colour variations, seeds from both the stages have been reported to have similar moisture content and hence, the degree of germination under laboratory conditions did not vary drastically in seeds from these two stages; however, mature seeds exhibited higher germination percentage (Teixeira and Barbedo, 2012).

Figure 9.4: Ripe Fruits of Surinam Cherry.

Figure 9.5: Field Ready Seedlings of Surinam Cherry.

As the fruits are naturally predated by birds and other herbivores, use of seed fragments has also been explored to regenerate the seedlings from halved or quarter sized seeds successfully. It was also possible to regenerate seedlings from the pieces obtained from the germinating portion with roots intact or unrooted portion of the seeds (Teixeira and Barbedo, 2012). Cipriano *et al.* (2022) conducted a series of experiments to standardize vegetative propagation using cuttings; however, little success was obtained even with IBA concentration of up to 3,000 ppm. Under Hawaii conditions, veneer grafting of two days stored scion stick on more than one-year-old rootstock was found to be successful and 87 per cent graftage success was achieved after six months of operation (Manners *et al.*, 2011).

Andaman Kokum (*Garcinia dhanikhariensis*)

Figure 9.6: Cut Section of Carambola Fruit.

Andaman Kokum is an endemic fruit from Andaman Islands and has been identified for its commercial potential in the island and coastal states of India. The species yields fruits which have attractive dark red colour and are novel sources of biologically active anthocyanins, hydroxycitric acid and phenolic acids (Bohra *et al.*, 2021a). Fruits are highly suitable for preparation of value-added products, with high potential in the tourism dominated Andaman Islands. For propagation using seeds, seeds are extracted from the ripe fruits and washed thoroughly to remove the adhering pulp. The seeds could be soaked in 0.1 per cent potassium nitrate or 500 ppm GA_3 solution for 24 h before sowing in portrays filled with soil: vermicompost (1: 1) to improve the germination and promote seedling growth (Bohra *et al.*, 2021b). Selection of mother plant is also important in endemic species for enhancing the success of propagation as inter-plant variations for germination have been reported in Andaman Kokum (Bohra *et al.*, 2021). Removal of seed coat and use of large seeds (more than 0.5 g weight) have been reported to enhance germination speed and overall germination percentage (Horo *et al.*, 2022).

Figure 9.7: Seedlings of Andaman Kokum Ready for Field Planting.

Mangosteen (*Garcinia mangostana*)

Mangosteen, also known as "the Queen of Tropical fruits", is being promoted in India as a prioritized exotic crop by the Government of India. Despite its commercial importance, the crop faces issues in propagation including poor seed set and slow growth of seedlings owing to poor development of root system (Harahap *et al.*, 2014). Pulp segments within a fruit are unequal in size and bigger ones generally have chances of possessing well developed seeds. Under Andaman conditions, fruits produced 0.9 to 1.6 seeds per fruit (Bohra and Waman, 2019). Porous, nutrient-lacking substrate like sand has been found to be suitable for its germination. Even though mangosteen has been regarded as an important exotic crop, only limited studies on its propagation are available (Othman *et al.*, 2019).

Figure 9.8: Developing Fruits of Mangosteen.

Figure 9.9: Mangosteen Seedlings with First Pair of True Leaves.

Attempts have been made for improving the seedling growth through the use of Arbuscular Mycorrhizal Fungi and *Azospirillum* (Yusuf *et al.,* 2014), CO_2 supplementation in growing environment (Jawal *et al.,* 2002); however, the techniques have not been commercially adopted so far. Mangosteen seedlings of 26 months age have been reported to develop first lateral branch, which is one of the prerequisites for seedlings to be field ready (Mansyah *et al.,* 2013).

Blood Fruit (*Haematocarpus validus*)

Blood fruit is a tropical liana found growing naturally in the forests of Andaman group of islands and north eastern parts of our country besides some neighbouring countries. The fruits are valued for their blood red juicy pulp, which is rich in anthocyanins and hence, has potential as a novel fresh fruit cum natural food colourant for cultivation in the backyards (Bohra *et al.,* 2020b). As the fruits are generally harvested from the wild, natural populations are in constant danger (Bohra *et al.,* 2022). Use of standardized propagation technology could increase the availability of planting material for habitat enrichment and domestication. Under Andaman conditions, seed pre-treatment with 0.1 per cent thiourea and sowing in coir pith could significantly improve the germination to 85.4 per cent (Bohra *et al.,* 2018). In another experiment, vermicompost was found to be better substrate for germination of blood fruit over sand and soil + vermicompost (Bohra *et al.,* 2020a).

Figure 9.10: Blood Fruit Bunch (Left) and Seedlings of Blood Fruit Ready for Field Planting (Right)

Barbados Cherry (*Malpighia glabra* and *M. emarginata*)

Barbados cherry or acerola is one of the richest sources of vitamin C in the plant kingdom. The name has been used for both the aforementioned species. It grows as a large bush to small tree, and fruits multiple times in a year under humid tropical conditions. Though seed germination is possible, the success is low and the regenerated plants are variable in performance (Ritzinger *et al.,* 2017). It can be successfully propagated by hardwood cuttings. Under island conditions of Andaman, 0.5-1.0 cm thick hardwood cuttings taken from 1-year-old shoots and treated with 1,500 ppm IBA rooted successfully (90%) (Singh and Attri, 2000). In another study conducted in hill zone of Karnataka, maximum rooting (80%) could be achieved, when hardwood cuttings were treated with 5,000 ppm IBA (Samim *et al.,* 2018). Bharanidharan *et al.* (2023) achieved 90 per cent success in air layering of acerola during rainy season under Tamil Nadu conditions. Wedge grafting has also been reported to be highly successful method of propagation (90 per cent success) for West Indian cherry (Ritzinger *et al.,* 2017). One season old plants are considered ready for field planting.

Figure 9.11: Various Developmental Stages of Barbados Cherry Fruits.

Figure 9.12: Rooted Cutting of Barbados Cherry (Left) and Hardened Plants (Right).

Mango (*Mangifera indica*)

Mango needs no introduction, especially for Indians as mango is a part of their life in fresh, dried, pickled or value-added product forms. Though, mango is being cultivated in various agroecological zones of India, a number of popular varieties including Alphonso are being cultivated in the HTRs of the coastal states.

Propagation of mango can be done by sexual and vegetative methods. However, due to cross pollinated nature and segregation of traits, the vegetative method is mostly preferred to obtain the true to type saplings. Commercial method of propagation adopted in humid tropical climate is stone grafting/epicotyl and softwood grafting depending on its high success rate (85-95%).

Seeds should be collected from fully ripe fruits and raised on the nursery bed of 100 cm width raised to the height of 20-30 cm under the shade net conditions with proper water management and other care. Seeds should be sown within 15-20 days of extraction (Reshma and Simi, 2019). Alternatively, stones available from processing industries could be utilized for raising of rootstocks. Seeds require about 10 to 20 days from sowing to germinate. Soaking of seeds in cow dung slurry for 24 h could be practiced for enhanced germination and performance of the seedlings (Chaudhari and Patel, 2012). Soaking of seeds in 100-200 ppm GA_3 solution enhanced germination within 13 days with 50 per cent germination in 23 days (Reshma and Simi, 2019; Kolekar *et al.,* 2017).

Fifteen to twenty days old, copper red coloured seedlings can be chosen for epicotyl grafting. The scion should be selected from the elite mother tree with good yield and quality traits. Grafting can be performed during July to August. Stones could also be sown directly in black polythene bags containing a mixture of farm soil, sand and FYM for raising rootstocks for softwood grafting. Under Bangladesh condition, epicotyl grafting was observed to be a promising method of propagation in mango. Variety Mollika showed better grafting compatibility among the varieties tried and use of 30 days old rootstocks was suggested (Bose *et al.,* 2019).

Softwood grafting can be practiced in the months of October-November on the rootstocks of one season or year old. The stionic-relationship is often observed in many experiments conducted in the Konkan region. The plant height, plant volume and yield decreased when Vellaikolamban was used as the rootstock with scion of 'Alphonso' and the hybrid 'Ratna' (Gawankar *et al.,* 2010).

Sapota (*Manilkara achras*)

Sapota is a major fruit crop that is popularly cultivated in diverse agroclimatic conditions including humid tropical areas in the country. Apart from use as a table fruit, sapota is used in preparation of value-added products and latex is an industrial product from the fruits. Sapota is commercially propagated by soft-wood grafting on the *Rayan* or *Khirni* (*Manilkara hexandra* Roxb.) rootstock with 3-4 months old scion during October-November. *Khirni* is used as a rootstock due to its strong root system and better performance. Seeds of *khirni* should be sown in polythene bags in nursery in June and raised to at least 30-45 cm height for using as rootstock. However, this species is not compatible with all the genotypes. Under West Bengal condition, of the ten cultivars studied, high compatibility of *khirni* in softwood grafting was observed using CO-2, Cricket Ball and DSH-2, whereas no grafting was possible in cultivars CO-1, DSH-1 and Guthi. Hence, testing the compatibility of a new variety is desirable before mass production. Optimum season for carrying out softwood grafting was found to be July 1 (72%) and August 15 (70%) (Ghosh *et al.,* 2010). Under Odisha condition, July 20 was found to be optimum for soft wood grafting in sapota on one year old *khirni* rootstock (Ghritlahare and Ashutosh, 2018).

Chrysophyllum lanceolatum, an indigenous species of the Sapotaceae family, was identified as a new rootstock for propagation of sapota under Kerala condition. For this, shade dried seeds were sown in fine river sand and germinated seedlings of 30 days were transplanted into polybags filled with equal proportion of sand: top soil: cow dung. Approach grafting was carried out on CO-2 variety using ten-month-old seedlings as rootstock for which a slice of bark (5 cm) was removed from scion and rootstock and the branches were held together using 100- guage polythene strip. April-May was identified as the optimum period for grafting. Grafts were detached after 75 days and under field conditions, no incidence of root rotting was observed. The grafts fruited well indicating their compatibility (Kalesh *et al.,* 2005).

The technique of veneer grafting under South Bengal condition has been developed. For this, 15 cm long cured scions of six-months maturity and 2-years old rootstocks showed highest success. Grafting should be done on rootstock at 20 cm above the collar region and September-November was the ideal season for the operation (Sandhu, 1992).

Rambutan (*Nephelium lappaceum*)

Rambutan is an exotic tropical species which was introduced in Kerala and the cultivation has spread to about 1000 acres in the states of Kerala, Tamil Nadu and Karnataka. Even though good scope for expansion exists in other humid tropical belts of the country, non-availability of quality planting material is one of the important factors hindering it (Tripathi *et al.,* 2020). Rambutan is propagated by seeds, air-layering, approach grafting and chip budding.

Figure 9.13: Mature Fruits of Rambutan.

Seed germination is better when seeds are sown at temperature of 25 °C (Andrade *et al.,* 2017). Under Coorg conditions of Karnataka, patch budding was performed using seedlings of 25 -30 cm height and pencil size thickness in November using Arka Coorg Arun as budwood. The method gave success rate of 70 per cent with 37 days taken for sprouting and low rate of mortality (6.28%) (Bhattacharjee *et al.,* 2022). Further, approach grafting was also found to give 73 per cent success under the same set of conditions (Bhattacharjee *et al.,* 2022).

Passion Fruit (*Passiflora edulis*)

Passion fruit is a popular crop that is gaining importance in recent past due to its ease of growing and processing potential. Two basic types of passion fruit *viz.* yellow and purple are under cultivation in the country. It could be easily propagated through seeds for which fully ripe/naturally fallen fruits are used for extraction of seeds. Seeds should be washed to remove the pulp and air dried before sowing in pro-trays or polybags. For vegetative propagation of superior types, hardwood cuttings of 30-35 cm length with 3 to 4 nodes should be treated with 200 ppm NAA and raised on a media containing sand, soil and farmyard manure. A study suggested use of semi-hardwood cuttings of 20 cm length without rooting hormone and perlite as a substrate. About 90 per cent success could be obtained using this method (Atlinkaya *et al.,* 2024). Grafting has also been found to be successful in passion fruit. Lo *et al.* (2023) developed a holistic procedure to develop passion fruit 'Yellow' grafts on passion flower rootstock. Germination temperature of rootstock seeds was optimized resulting in 100 per cent germination. Using these rootstocks, grafting was done using single node scion and successful union was observed in less than two months' time (Lo *et al.,* 2023).

Star Gooseberry (*Phyllanthus acidus*)

Star gooseberry is a small tree of Phyllanthaceae family that finds place in a number of home gardens. Both fruits and leaves are edible and fruits could be consumed in fresh as well as processed forms. Star gooseberry is mainly propagated by seeds and cuttings (Kishore *et al.,* 2007). Layering has been identified as a promising method of multiplication in it. Under West Bengal conditions, air layering was successful with 100 per cent rooting using 8 months old branches when performed during firs half of July and September 1; however, 100 per cent field establishment was obtained from the propagules layered on September 1 (Ghosh and Bera, 2015).

Egg Fruit (*Pouteria campechiana*)

Canistel or yellow sapote is a fruit of Sapotaceae family that has been introduced in India and is being promoted for cultivation in the southern parts of our country. Fruit pulp is very rich in carotenes that resembles boiled egg yolk and hence is popularly known as egg fruit. However, the strong aroma of the pulp is not liked by everyone and only selected people could relish it.

Figure 9.14: Seedlings of Passion Fruit in Nursery.

The species can be propagated commercially by veneer grafting, cleft grafting, patch budding and air layering (Radha and Mathew, 2007); however, limited research efforts have been made under Indian conditions. Freshly harvested seeds should be used as the seeds lose viability very quickly. Seed sowing could be done in polythene bags during June –July with 45 per cent germination success under Bengaluru condition (Tripathi, 2021).

Under Mexican conditions, 100 per cent seed germination with improved plant growth was achieved, when seeds were sown in horizontal position (Perez-Barcena *et al.*, 2021) and scarification treatment marginally reduced the germination process. The fruit size has shown no influence on seed size (Perez-Barcena *et al.*, 2021) and hence, even small sized fruits which could not fetch higher prices in the market could be used for nursery production. This would be helping the farmers in better disposal of their produce with minimized economic losses. As the crop is new in India, efforts on research on its vegetative propagation are warranted.

Hog Plum and Golden Apple (*Spondias* species)

The genus *Spondias* of Anacardiaceae family is known for production of fruits which are regionally popular for fresh consumption as well as processing. While *S. mombin* and *S. pinnata* are commonly called as hog plum, *S. cytherea* (syn. *S. dulcis*) is also known as June plum or golden apple. These species have largely been found in the semi-wild state or are sometimes planted in the home gardens. Propagation of these species is difficult and some efforts have been made for standardizing their multiplication techniques.

Figure 9.15: Mature Fruits of Hog Plum (*Spondias pinnata*).

To overcome the seed dormancy in *S. dulcis*, use of mechanic scarification in the opposite region of the embryonic axis, followed by the soaking in gibberellin + cytokinin solution (350 ppm) for 12 hours was found to be useful (de Souza *et al.*, 2020). Under West Bengal conditions, propagation of *S. pinnata* using cuttings was standardized and the highest success (66.67%) was attained when 15 cm long cuttings were treated with 1500 ppm IAA before planting (Dey *et al.*, 2017). Air layering trials conducted in *S. pinnata* revealed that rainy season (July) was the best time for preparing air layers. Use of IBA (500 ppm) on the girdled portion was effective in promoting root formation (Tomar, 2016). In an experiment by de Souza *et al.* (2012), *S. dulcis* cuttings of different lengths were tested with different concentrations (0-45 g/L) of refined sugar, and pure sucrose as a control. Cuttings of 15 cm length were found to be optimum, while there was no significant difference among the carbohydrate concentrations tested and hence, refined sugar was recommended for the purpose.

Water Apple (*Syzygium aqueum*)

Watery rose apple is generally found cultivated in the home gardens of tropical Asian countries including India. Fruits are mildly sweet and texture resembles as that of watermelon. If propagated through seeds, seeds must be shown fresh as they are recalcitrant. It can be successfully propagated using air layering technique. Optimum time for air layering was identified for red laterite zone of

West Bengal. For layering, about one year old shoots were used for air layering. Air layering performed during June and July showed 100 per cent rooting and field establishment. Rooted layers were ready for separation from the mother plant after about 80 days during September-October, while longer time of 110 days was required when layering was performed during June-July (Ghosh *et al.,* 2008). Bohra and Waman (2020) developed a novel approach for round the year propagation of watery rose apple through air layering in protected conditions. Use of coir pith compost as a substrate and IBA (2,000 ppm) resulted in 100 per cent rooting success. Use of coir pith compost reduced the time taken for layering by more than 70 per cent. The technique that has been developed at ICAR– Central Island Agricultural Research Institute, Port Blair, has been christened as 'Dweep Goottee-365'. The technique holds promise for use in multiplication of rare germplasm of other horticultural species as well.

Jamun (*Syzygium cumini*)

Jamun is a highly perishable fruit that has been valued for medicinal and nutraceutical properties. Due to the presence of anthocyanins in the fruits and anti-diabetic properties of seeds, the crop is gaining rapid popularity in the recent past in the country. The species has wider adaptability and thus, its cultivation has spread across arid to subtropical to tropical regions. However, the crop is highly heterozygous and hence, use of vegetative propagation techniques is desirable. Further, several improved varieties in the crop have been released for cultivation from various ICAR institutes and State Agricultural Universities, which demand vegetative propagation for maintaining the true-to type nature of the varieties (Singh *et al.,* 2019).

Soft wood grafting and patch budding are commercial methods of propagation in humid tropical region. Soft wood grafting can be performed during the months of June to October on one season or one year old rootstock. Under Tamil Nadu conditions in var. Konkan Bahadoli, softwood grafting performed during July-August was found to give highest survival success of 94.72 per cent. Further, use of whip and tongue grafting during the same season showed about 89 per cent success. Limited success of 65, 46 and 43 per cent were obtained in other methods such as patch budding, hardwood cutting and semihard wood cuttings, respectively (Muniyapan *et al.,* 2019).

For softwood grafting, seedlings of about 8-10 months, with vigorous growth and pencil size thickness should be used as rootstocks; while scions should be collected from one season old twigs. About one week prior to separation of scions, removal of leaves from scion branches is recommended. Separation of scions should be done during morning time. Collected scions should be wrapped in the moist cloth covered with polythene (Chander *et al.,* 2016). Under Junagadh condition, first week of March was optimum for grafting on 6 months old rootstocks as it gave quicker sprouting (18 days), higher success rate (58%) and highest survival

(49%) after 120 days. Further, use of polyhouse supported grafting operation and improved the survival percentage (Makavana *et al.*, 2022).

Apart from the nursery techniques developed in the aforesaid fruit crops, a few studies have been conducted in the minor underutilized fruits of HTR and the key details of the propagation techniques have been presented in Table 9.1.

Table 9.1: Propagation Techniques Standardized in other Minor Fruit Crops of HTR

Species	*Propagation technique*	*References*
Dillenia indica	Stem cuttings from 1.5-2 years old shoots (15-20 cm, 10 mm) treated with 2,000 ppm IBA showed 43 per cent success.	Angami and Das (2011)
Flacourtia indica	Double node stem cuttings with 2 leaves intact when treated with 4,000 ppm IBA showed rooting in 81 per cent cuttings.	Kabir *et al.* (2017)
Flacourtia jangomas	Double node cuttings with 2 leaves trimmed to half collected from 2 years old stock plants when treated with 4,000 ppm IBA showed 100 per cent rooting.	Hossain *et al.* (2011)
Garcinia morella	10 cm long cuttings with 2 top leaves intact collected from 6 months old seedlings when treated with 4,000 ppm IBA showed rooting in 40 per cent cuttings.	Bhuyan *et al.* (2017)
Garcinia pedunculata	Treatment of stem cuttings with 2,000 ppm IBA resulted in root induction in 80 per cent cuttings.	Bhuyan *et al.* (2017)
Blighia sapida	Air layering on previous season branches (1-1.5 cm diameter) with 3 cm bark removal and treatment with 3,500 pm IBA showed 100 per cent rooting.	Maurya *et al.* (2013)
Baccaurea sapida	Soaking of freshly harvested seeds in 2 per cent NaH_2PO_4 for 12 h followed by sowing in equal ratio of sand, soil and FYM resulted in 87 per cent germination.	Dey *et al.* (2022)
Flacourtia jangomas	Soaking in seeds in water (25 ± 2°C) for 48 h followed by sowing in soil showed germination in 81 per cent seeds.	Hossain *et al.* (2011)
Dimocarpus longan	Cold water soaking of seeds for 24 h and use of soil: FYM (3: 1) as substrate resulted in 80 per cent germination.	Hazarika *et al.* (2023)
Spondias mombin	Soaking of seeds in 60 per cent concentrated sulphuric acid for 25 minutes and sowing in loamy soil showed 75 per cent germination.	Fadimu *et al.* (2014)
Garcinia indica	Seedlings can be produced by soaking in water with up to 100 per cent germination. Seedlings can be sown in polybags of 10 × 15 cm size containing soil and FYM (3: 1). Softwood grafting using 5 to 6 months old scion and 22 weeks old rootstock is recommended.	Haldankar *et al.* (1992)

Conclusion

The humid tropical regions of India have been the areas of cultivation of a number of traditional fruit crops contributing to the nutritional and economic security of the farmers here. Further, considering the similarity of the agro-climatic conditions in this region with other tropical countries, a number of exotic fruit species have been introduced in this region and are being popularized. Further, the underutilized fruits of regional significance and having good potential in the livelihood of native dwellers are naturally distributed in these regions. In order to promote scientific cultivation of fruit species of all three categories, standardization of nursery protocols is of prime importance. Even though a good number of techniques have been standardized for various crops, considerable scope exists for fine tuning them to make them commercially viable. Standardised propagation techniques in such crops will ensure uniform and quality saplings available for the production and conservation of these crops. Knowledge of various standardised methods of propagation applicable to the humid tropical fruits will be helpful in an efficient production and nursery management in this region.

REFERENCES

Andrade RAD, Barreto LF, Nacata G and Sauco VG, 2017. Advances in the propagation of rambutan tree. *Rev. Bras. Frutic.* **39(5)**: (e-690).

Angami T and Das RP, 2011. Standardization of IBA concentration for rooting of cuttings of some indigenous fruit crops of Assam. *J. Hortl. Sci.* **6(2)**: 109-113.

Atlinkaya L, Gubbuk H and Biner B 2024. Effect of microbial fertilizers and growth media on the rooting of passion fruit (cv. Possum Purple) cuttings. *Indian J. Hortic.* **81(1)**: 110-114.

Banik BC, Ghosh SN and Singh SR, 2012. Research and development in karonda (*Carissa carandas*), a semi wild fruit in India. *Acta Hort.* **948**: 61-65.

Baskaran A, Saraswathy S and Parthiba S, 2008. Standardisation of propagation methods for jack (*Artocarpus heterophyllus* Lam.). *Asian J. Hort.* **3(2)**: 361-363.

Bharanidharan A, Ranchana P, Gopi V, Patricia KAJ, Dinesh KP, Irine C, Giriprasath RS, Vishal S, Krishna PY and Pavethra A, 2023. Effect of IBA concentrations and time of air layering of West Indian cherry (*Malpighia glabra* L.) in Coimbatore region. *The Pharma Innov. J.* **12(5)**: 971-975.

Bhattacharjee P, Sakthivel T, Chander S and Nimbolkar PK, 2022. Standardization of asexual propagation techniques in rambutan (*Nephelium lappaceum* L.) for humid tropical region of India. *Environ. Conser. J.* **23(1 and 2)**: 31-38.

Bhuyan L, Sharma CL, Singh B and Sharma MB, 2017. Vegetative propagation of two important *Garcinia* species of Assam, NE India. *J. Med. Plants Stud.* **5(3)**: 273-277.

Bohra P, Waman AA and Devi RK, 2021b. Seed Fatty Acid Composition and Germination Studies in *Garcinia dhanikhariensis* S.K. Srivastava (Clusiaceae): a Novel Tropical Fruit Species from Bay Islands, India. *Intl. J. Fruit Sci.* **21(1)**: 970-978.

Bohra P and Waman AA, 2019. Mangosteen (*Garcinia mangostana* L.): a potential crop for increasing incomes of island farmers. *J. Andaman Sci. Assoc.* **24(1)**: 9-13.

Bohra P and Waman AA, 2020. Air layering under protected condition: a new technique for year-round propagation. *J. Andaman Sci. Assoc.* **25(2)**: 147-150.

Bohra P, Waman AA and Devi RK, 2020. Seed germination in blood fruit (*Haematocarpus validus* (Miers.) Bakh. f. ex Forman) as influenced by substrates. *Curr. Agric. Res. J.* **8(2)**: 104-107.

Bohra P, Waman AA and Devi RK, 2021. *Garcinia dhanikhariensis* S.K. Srivastava. In: *Perennial Underutilized Horticultural Species of India* Eds. Ajit A. Waman and Pooja Bohra). Jaya Publishing House, Delhi. pp. 144 - 151.

Bohra P, Waman AA, Basantia D, Devi HL and Reang E, 2018. Domestication and conservation efforts in *Haematocarpus validus* (Miers.) Bakh. f. ex Forman: an underutilized fruit species and natural colourant. *Curr. Sci.* **115(6)**: 1098-1105.

Bohra P, Waman AA, Kumar SS and Devi HL, 2021. Morphological and physicochemical studies in blood fruit (*Haematocarpus validus* Bakh. f. ex Forman): A tropical fruit and natural colorant. *S. Afr. J. Bot.* **145**: 48-55

Bohra P, Waman AA, Roy TK and Shivashankara KS, 2020. Blood fruit (*Haematocarpus validus* (Miers.) Bakh. f. ex Forman): a novel source of natural food colourant. *J. Food Sci. Technol.* **57**(1): 381-389. https: //doi.org/10.1007/s13197-019-04064-2

Bose SK, Ghosh K and Howlader P, 2019. Effect of variety and rootstock age on the success and survivability of epicotyl grafting in mango. *J. Agri. Res. Adv.* **01(01)**: 13-25.

Chander S., Kumar S, Kavino M and Bora L, 2023. Effect of seasonal variation on softwood grafting under different environmental conditions in jamun (*Syzygium cumini* Skeels.). Res. on Crops **17(3)**: 524-528.

Chaudhari PM and Patel BN, 2012. Effect of pre-sowing treatments, sowing position and duration on germination of mango. *Bioinfolet*, **9**: 277-279.

Cipriano RR, Tomazzoli MM, Tofanelli MBD, Sant'Anna-Santos BF, Ribas KCZ and Deschamps C, 2022. Rooting of *Eugenia uniflora* cuttings: substrate, seasonality, auxine and reinvigoration methods. *Communicata Scientiae*, https: //doi. org/10.14295/CS.v13.3739.

de Souza Pinto JR, de Freitas RMO, Araújo PC, Praxedes SC, 2012. Propagation of *Spondias* species by cuttings of different lengths treated with sugars. *Genet. Plant Physiol.*, **2(3–4)**: 171–176.

de Souza PHM, Ragagnin ALSL, Ribeiro RC, da Silva GZ, Rocha DI, da Silva DFP, 2020. Dormancy overcoming in seeds of cajá-manga (*Spondias dulcis*). *Communicata Scientiae Hortic. J.* **11**: e3341.

Deepika N and Vanajalatha K, 2015. Growth regulators and sucrose influence on rooting and sprouting of Karonda (*Carissa carandas* L) cuttings. *International Journal of Tropical Agriculture*, **33(4)**: 2691-2694.

Deepika Vanajalatha and Ashok Yadav, 2014. Effect of seed storage on seed viability, germinability and morphological characteristics of karonda (*Carissa carandas*) seedlings. *Asian J. of Adv. Basic Sci.*, **2(3)**: 1-6

Dey K, Ghosh A, Mani A, Bauri FK and Dey AK, 2017. Root generation of Karonda (*Carissa carandas* L.) cuttings in response of sucrose and IBA. *Journal of Pharmacognosy and Phytochemistry*, **6(6)**: 803-806.

Dey AN, Panda MR, Sharma B and Ghosh A, 2022. Effect of pre-treatments on germination of Latka (*Baccaurea sapida* Muell. Arg.). *Intl. J. Minor Fruits Med. Arom. Plants* **8(1)**: 30-34.

Dey K, Ghosh A, Dey AN and Bhowmick N, 2017. Efficacy of IAA concentration and cutting length on rooting of stem cuttings in *Spondias pinnata* Linn. *Indian J. Hort.* **74(1)**: 127-131.

Fadimu OY, Idowu OTH and Ipinlaye SJ, 2014. Studies on the dormancy and germination of stony fruits of hog plum (*Spondias mombin*) in response to different pre-soaking seed treatments. *Int. Res. J. Biological Sci.* **3(6)**: 57-62.

Gawankar MS, Haldankar PM, Haldavanekar PC, Salvi BR, Burondkar MM, Pawar CD and Jamadagni BM, 2021. Effect of Potting Media on Seed Germination and Seedling Vigour of Jackfruit (*Artocarpus heterophyllus* Lam.) under the Konkan Region of Coastal Maharashtra. *J. Indian Soc. Coast. Agric. Res.* **39(1)**: 61-65.

Gawankar MS, Haldankar PM, Salvi BR, Haldavanekar PC, Burondkar MM, Pawar CD, Dhekale JS and Jamadagni BM, 2020. Influence of storage methods and phytochrome activities on seed germination in jackfruit seeds (Artocarpus heterophyllus Lam.). *Intl. J. Chem. Studies* **8(1)**: 776-786.

Gawankar MS, Salvi BR, Chavan SA and Dalvi NV, 2010. Comparative performance of mango varieties grafted on Vellaikolamban and mixed rootstock. *J. Hortl. Sci.* **5(2)**: 114-116.

Ghosh SN, 2008. Effect of season on success of air layering in water apple in red laterite zone of West Bengal. *J. Hort. Sci.* **3(2)**: 164-165.

Ghosh SN and Bera B, 2015. Studies on standardization of propagation methods of some minor fruit crops in India. *Intl. J. Minor Fruits Med. Arom. Plants* **1(1)**.

Ghosh SN, Bera B, Roy S and Banik BC, 2010. Effect of cultivars and season on grafting success in sapota under Paschim Midnapur conditions of West Bengal. *J. Hortl. Sci.* **5(2)**: 138-139.

Ghritlahare S and Ashutosh, 2018. Performance of sapota (*Manilkara achras*, Mill.) softwood grafts on precuring and different season. *International Journal of Chemical Studies* **6(6)**: 1768-1772.

Haldankar PM, Salvi MJ, Joshi GD and Patil JL, 1992. Standardization of softwood grafting in kokam. *J. Maharashtra Agric. Univ.* **17(2)**: 227-229.

Harahap F, Poerwanto R, Suharsono SC and Rahayu S, 2014. In vitro growth and rooting of mangosteen (*Garcinia mangostana* L.) on medium with different concentrations of plant growth regulator. *HAYATI J. Biosciences* **21(4)**: 151-158.

Hazarika P, Kumar A and Das D, 2023. Evaluating pre-sowing seed treatment on the germination of important forest fruit plant species. *Crop Res.* **58(5-6)**: 251-259.

Horo S, Roopa Ranjani Ajit Arun Waman, Santosh K Jha and Pooja Bohra, 2022. Seed Germination Studies in Andaman Kokum (*Garcinia dhanikhariensis* S.K. Srivastava): An Endemic Species from Bay Islands, India. *J. Andaman Sci. Assoc.* **27(SI)**: 95-103.

Hossain MA, Sen M, Jewel M/I.U, Kabir MA, 2011. Propagation of *Flacourtia jangomas*: an approach towards the domestication of a wild fruit species in Bangladesh. *Dendrobiology* **65**: 63-71.

Islam MM, Haque MA and Hossain MM, 2003. Effect of age of rootstock and time of grafting on the success of epicotyl grafting in jackfruit (*Artocarpus heterophyllus* L.). *Asian J. Plant Sci.* **2(14)**: 1047-1051.

Iwalesin BA, Oyun MB and Oke DO, 2023. Domestication of *Annona muricata* L. in (soursop) by marcotting. *J. Res. For. Wildlife Environ.* **15(3)**: 106-115.

Jawal MAS, Mansyah E and Usman TPFF, 2002. Pengaruh sungkup plastik dan sistem perakaran terhadappertumbuhan semai manggis. *Jurnal Hortikultura* **12(3)**: 158-164.

Jose M and Valsalakumari PK, 1991. Standardization of technique of epicotyl and softwood grafting in jack. *South Indian Hort.* **39(5)**: 264 – 267.

Kabir MA, Khan MHM, Rahman MM, Alam ATMR, Jahan SMH, Zaman AKMM and Mamun MZA, 2017. Clonal propagation of *Flacourtia indica* for ensuring quality planting materials and sustainable supply of edible fruits. *Intl. J. Agril. Sys.* **5(1)**: 42-52.

Kalesh KS, Shareef SM, Mathew SP and Chemburkar MS, 2005. *Chrysophyllum lanceolatum* — A new rootstock for sapota (*Achras zapota* L.). *J. Appl. Hort.* **7(1)**: 23-24.

Karanjalker GR and Karanjalker AG, 2021. *Averrhoa bilimbi* Linn. In: *Perennial Underutilized Horticultural Species of India* Eds. Ajit A. Waman and Pooja Bohra). Jaya Publishing House, Delhi. pp. 10-18.

Kishore K, Pathak KA and Bharali R, 2007. Star gooseberry an underutilized fruit of north-eastern region. *Indian Hort.* **52(3)**: 24.

Kolekar SN, Kadam AS and Gend DG, 2017. Effect of different organics and chemicals treatments on germination, growth and success of softwood grafting in mango during nursery stage. *Intl. J. Chem. Studies* **5**: 880-884.

Kumar R, Pratibha Rai R and Nalini P, 2021. Comparative evaluation of grafting and budding methods in jackfruit. *Indian J. Hortic.* **78(4)**: 398-403.

Lo H, Lou L, Huang T, 2023. Establishment of integrated propagation system on grafting plantlets of passion fruit (*Passiflora edulis*). *Hort Science* **58(2)**: 170–177.

Maheswari TU, Sinduja A, Rajavel A, Padmanaban J and Sudhagar R, 2021. Effect of plant growth regulators in adventitious rooting of cutting in soursop (*Annona muricata* L.). *Plant Arch.* **21(1)**: 2013-2015.

Makavana PC, Patel KD, Kavad MH, Kanzaria DR and Hathi HS, 2022. Standardization of grafting time and various environmental conditions in Jamun (*Syzygium cumini* Skeels). *Pharma Innov. J.* **11(12)**: 2085-2088.

Manners MM, Griffis JL and McDonald TG, 2011. Time after scion harvest and grafting method influence graft success rate for purple-fruited pitanga (*Eugenia uniflora* L.). *Proc. Fla. State Hort. Soc.* **124**: 28–30.

Mansyah E, Muas I, Jawal MAS and Affandi, 2013. The research for supporting sustainable mangosteen (*Garcinia mangostana* L.) production. *Intl. J. Adv. Sci. Eng. Info. Tech.* **3(1)**: 16-22.

Maurya RP, Lewis DM and Chandler JStA, 2013. Studies on the propagation of Jamaican Ackee (*Blighia sapida* L.) by air-layering. *Hort Science* **48(10)**: 1298–1300.

Medagoda I and Chandrarathana WMCJK, 2007. Grafting of breadfruit (*Artocarpus altilis*) using breadnut (*Artocarpus camansi*) as root stock. *Acta Hortic.* **757**: 10.17660/ActaHortic.2007.757.20.

Mng'omba SA, 2023. Delayed rooting hormone application and stem cutting collection time improved *Flacourtia indica* rooting ability and root growth. *Southern For.* **85(3-4)**: 194-200.

More VV, Haldankar PM, Salvi BR, Kasture MC and Haldavnekar PC, 2023. Standardization of media and bag size for re-bagging jackfruit (*Artocarpus heterophyllus*) grafts under Konkan agroclimatic conditions. *Pharma Innov. J.* **12(9)**: 2361-2364.

Muniyappan C, Rajangam J, Kumar CSR and Venkatesan K, 2019. The Standardization of method and time of propagation in jamun (*Syzygium cuminii* Skeels) var. Konkan Bahadoli. *J. Pharmacog. Phytochem.* **8(3)**: 467-471.

Nadukeri S, Thanuja PC, Kolakar S, Hanumanthappa M and Rakshith KR, 2018. Effect of pre-sowing treatments on seed germination and seedling growth of *Annona reticulata* L. *J. Pharmacogn. Phytochem.* **SP3**: 457-460.

Nandwani D and Kuniyuki AH, 2005. Grafting and improvement of breadfruit production in Micronesia. *Acta Hortic.* **694**: 10.17660/ActaHortic.2005.694.50.

Neelavathi R, Sridhar P, Rani CI, Pugalendhi L, 2021. Studies on novel method of propagation in guava through leaves. *Indian J. Hortic.* **78(4)**: 423-427.

Nivetha K and Maheswari TU, 2015. Influence of genotype and seed weight on the performance of rootstock and the success of softwood grafting in jack (*Artocarpus heterophyllus* Lam.). *Plant Arch.* **15(2)**: 875-878.

Oliveira MT, Berbert PA, Vieira HD, Thiébaut JT, Carlesso VD and Pereira RD, 2009. Vigour of carambola seeds as affected by drying and storage. *Rev. Bras. Eng. Agricola e Ambient.* **13:** 477-482.

Origenes MG and Lapitan RL, 2020. effect of coconut water on pre-sowing treatments additive on seed germination and initial seedlings growth performance of Kamagong (*Diospyros discolor*). *Asian J. Res. Agric. For.* **6(4)**: 58-71.

Othman M, Aizat WM, Zuddin NSA, Baharum SN and Noor NM, 2019. Metabolite profiling of mangosteen seed germination highlights metabolic changes related to carbon utilization and seed protection. *Sci. Hortic.* **243**: 226-234.

Perez-Barcena JF, Cruz-Castillo JG, Jesus-Sanchez AD, Jimenez-Aparico AR and Evangelista-Lozano S, 2021. Germination and developmental conditions of *Pouteria campechiana* (Sapotaceae) plants. *Bot. Sci.* **99(2)**: 377-387.

Radha T and Mathew L, 2007. Under and un-utilized fruits of tropics and subtropics. In: (K.V. Peter ed.) *Fruit Crops, Horticulture Science Series-3*. pp. 343-392.

Reshma UR and Simi S, 2019. Effect of pre-sowing treatments, sowing positions and age of stones after extraction on germination of mango. *Int. J. Curr. Microbiol. App. Sci.* **8(4)**: 2565-2573.

Ritzinger R, Ritzinger CHSP, Fonseca N and Machado CF, 2017. Advances in the propagation of acerola. *Rev. Bras. Frutic. Jaboticabal* **40(3)**: (e-928).

Robbani M and Sarker CR, 2022. Growth performance of cleft grafted velvet apple (*Diospyros discolor* Willd.). *South Asian J. Agric.* **9(1 and 2)**: 1-10.

Ryadin AR, Ranamukaarachchi SL, Soni P and Shrestha RP, 2014. Vegetative propagation of five local cultivars of Malay Apple (*Syzygium malaccense* spp.) in Ternate Island. *Intl. J. Adv. Sci. Engg. Info. Technol.* **4(2)**: 35-39.

Samim AK, Shivakumar BS, Yallesh HSK and Ganapathi M, 2018. Study on rooting of stem cutting in Barbados cherry (*Malpighia glabra* L.) under hill zone of Karnataka. *J. Pharmacogn. Phytochem.* **SP3**: 418-421.

Sandhu MK, 1992. Standardization of grafting techniques in sapota (*Achras zapota* L.). *Acta Hortic.* **321**: 610-615.

Santos MQC, Lemos EEP, Salvador TL, Rezende LP, Salvador TL, Silva JW, Barros PG, Campos RS, 2011. Rooting soft cuttings of soursop (*Annona muricata*) 'Giant of Alagoas'. *Acta Hortic.* **923**: 10.17660/ActaHortic.2011.923.36.

Singh DB and Attri BL, 2000. Effect of IBA on rooting in West Indian cherry (*Malpighia galbra* L.) cuttings. *J. Appl. Hort.* **2(2)**: 134-135.

Singh S, Singh AK, Saroj PL and Mishra S, 2019. Research status for technological development of jamun (*Syzygium cumini*) in India: A review. *Indian J. Agric. Sci.* **89(12)**: 1991–1998.

Singh SR and Devachandra N, 2021. Effect of seasonality on budding and air layering propagation techniques in carambola (*Averrhoa carambola* L.) under foothills condition of Arunachal Pradesh. *Ind. J. Hill Farming*. **34**: 264-267.

Sonawane GR, Khandekar RG, Korake GN, Haldankar PM and Mali PC, 2012. Effect of season on softwood grafting in carambola (*Averrhoa carambola* Linn.). *Asian J. Hort.* **7(2)**: 412-415.

Sunaryo W, Pranoto H, Nurhasanah and Rahman, 2019. Interspecific grafting to solve the rootstock shortage in vegetative propagation of Lai-durian (*'Durio zibethinus* x *kutejensis'*) originated from East Kalimantan. *Australian J. Crop Sci.* **13(4)**: 642-648.

Teixeira CC and Barbedo CJ, 2012. The development of seedlings from fragments of monoembryonic seeds as an important survival strategy for *Eugenia* (Myrtaceae) tree species. *Trees* **26**: 1069–1077.

Tomar A, 2016. Impact of seasonal changes on air layering and rooting hormone in *Spondias pinnata* (J. Koenig ex L. f.) Kurz. *Trop. Plant Res.* **3(1)**: 131–135

Tripathi PC, 2021. Canistel (egg fruit): An underutilized nutritious fruits, *Agro India*, p. 14-16.

Tripathi PC, Karunakaran G, Sakthivel T, Sankar V and Senthil Kumar R, 2020. Status and prospects of rambutan cultivation in India. *Acta Hortic.* **1293**: 33-40.

Varadkar R, Salvi BR, Kulkarni MM and Haldavnekar PC, 2020. Effect of potting media on per cent germination and growth of Jackfruit (*Artocarpus heterophyllus* L.) grafts cv. Konkan prolific. *Int. J. Chem. Stud.* **8(6)**: 668-670.

Yusuf LM, Kurien S, Surendragopal K and Augustin A, 2014. Growth studies in mangosteen (*Garcinia mangostana* L.). II. activation of seedling growth in Mangosteen using Arbuscular Mycorrhizal Fungi and *Azospirillum*. *Adv. Hort. Sci.* **28(3)**: 153-163.

Chapter 10

Orchard and Canopy Management in Humid Zone Fruit Crops

V. Arunachalam

ICAR–Central Coastal Agricultural Research Institute, Ela, Old Goa–403402

Orchard and canopy management are key issues in increasing productivity of fruit crops. The current chapter deals with the spacing and high density planting techniques for humid zone fruit crops along with canopy management tools including training, pruning techniques. The chapter covers mainly humid zone fruit crops such as avocado, citrus, passion fruit, and pineapple. Trellis system of training in passion fruit, high density planting in pineapple, dwarfing rootstocks in citrus, high density and pruning in avocado are discussed in detail. Role of paclobutrazol as a chemical growth regulator for regulating vegetative growth is discussed in all these fruit crops.

Effective land use management for obtaining maximal fruit yield and quality from a unit area forms the core of fruit tree planting and orchard management. Canopy management covers the aspects of reducing excessive canopy shade and regulating the plant canopy and size to ensure optimal light penetration to ensure high photosynthetic efficiency, leading to better fruit set, yield, and quality. Because short statured compact canopy trees with optimal canopy architecture are efficient at capturing and converting light than tall and spread trees to increase the productivity of orchards.

Canopy management is an essential tool to ensure optimal fruit quality and size by reducing fruit-to-fruit competition during peak fruiting. Fruit growers are able to obtain financial gains based due to better shape and design of the canopy, facilitating sunlight penetration and distribution. Management of canopy of short statured fruit tree gives improved yield and quality. It is achieved by several strategies including appropriate choice of location, judicious use of planting grafts of combination of right rootstock and scion varieties at appropriate density, training the juvenile trees, followed by appropriate pruning, trimming, thinning and use of required plant growth regulators.

Planting Density

Seedlings require wide spacing whereas grafted plants with compact canopy are amenable for high density planting at narrow spacing. About 2 m distance between both row to row and plant to plant could be reduced for spacing of grafted plants of several humid zone fruit trees compared to seedlings planted at 8 m x 8 m spacing (Tripathi, 2021). A modern fruit production method is high density planting, or HDP where the plants are accommodated at double to triple densities compared to normal wide conventional planting densities. It is achieved either by dwarfing rootstocks or short compact canopy varieties, to obtain high yield of quality fruits per unit area (Mitra, 2019). A review of high density planting and pruning techniques reported in avocado (Menzel and Lagadec, 2023) indicates various strategies available for orchard and canopy management. Avocado trees can reach tall stature of heights ranging from 15 to 18 m. Köhne and Kremer-Köhne, 1991) evaluated 400 and 800 trees per hectare and found high fruit yield per unit area in high density planting. A range of six different planting densities from 204 to 816 trees per hectare (Cano-Gallego *et al.,* 2023) were evaluated for Avacado where two intermediate densities of 333 and 400 plants per hectare gave the highest yield of quality fruits with high mesocarp proportion.

Of the two spacing (narrow at 1.2 m x 3 m spacing and wide at spacing 1.8 m x 3 m spacing) tried in passion fruit, narrow spacing was found better than wide spacing (Gachanja, 1977). Of three densities of pineapple planting (35700, 47600, 55500 plants per ha) for three cultivars Golden, Smooth Cayenne, and Santa Rosa, highest yields above 100 tonnes per ha was found at the highest density of planting at 55500 plants per ha. (Neri *et al.,* 2021). Of the six densities from 43,036 to 63,758 plants per hectare, highest yield of 106 tons per ha was found at high density of 63,758 plants per ha (Chadha *et al.,* 1973). In general in tropical fruit crops fruit size and quality is shown to decrease with increase in planting density whereas total yield per unit area increases with increase in planting density (Haque and Sakimin, 2022). High density of 2,020 plants per ha over the conventional 350-900 plants per ha gave good performance in citrus (Wheaton *et al.,* 1991).

Dwarf Varieties and Dwarfing Rootstocks

Genetic manipulation of stature, growth habit and compactness of canopy traits available with suitable dwarf cultivars or dwarfing rootstocks play crucial role in management of canopy and orchard without much costs. Of the sapota varieties, PKM 1 is dwarf cultivar, whereas PKM 3 variety shows a vertical growth habit, and Co-3 variety grows upright with a compact canopy (Gopu and Balamohan, 2016). All three sapota varieties listed above are useful in cost effective canopy management due to their genetic make-up. Askel and Chiquitita are two olive varieties with suitable plant architecture for high density planting. Chiquitita expressed low levels of cellulose synthase genes and high levels of WUS-Clavata genes (González-Plaza *et al.,* 2016). Two new cultivars of Kokum (*Garcinia indica*) Goa Kokum 2 and Goa Kokum 3 evolved at Goa, India possess bush habit and amenable for high density planting.

Of the two dwarfing rootstocks attempted for oranges, Forner-Alcaide FA 517 showed better performance for Salustiana orange scions than FA 418 for growth, yield and quality (Hervalejo *et al.,* 2024). Flying Dragon variety of *Poncirus trifoliate* performed well as a dwarfing rootostock for citrus where the dwarf habit was controlled by monogenic inheritance (Roose *et al.,* 33).

Training

Training system appropriate for the crop and its growth pattern is a crucial aspect of canopy management to shape the canopy in the initial years. Rambutan fruit trees grow straight with a high degree of apical dominance and hence a open centre system is required to be established by early years of pruning and training (Tripathi *et al.,* 2019). In cherimoya fruit trees two systems of training were compared open centre vase system with 6 m x 4 m spacing and slender spindle system (4m x 1 m) at high density. The high density slender spindle system gave the favorable environmental parameters of high relative humidity and low temperature for natural pollination and better fruit set (Cautin *et al.,* 2010). Central leader system of training is recommended for sapota trees (Ausari *et al.,* 2023) where the young trees are cut at 60-70 cm from ground level and only 3-4 spaced new shoots are allowed to grow to form canopy.

Passion fruit is a vine and need to be trained on support for optimal, growth, flowering, and fruit set. Kniffin system of training was found better for passion fruit than Bower, tatura, systems in terms of economic returns over the costs involved (Sulladmath *et al.,* 2012). The passion fruit vines trained on vertical trellis gave high fruit yield and least fungal infection than horizontal trellis (Gachanja, 1977). A simple trellis provided higher yield of 30.5 tonnes per hectare than a T or horizontal trellis for yellow passion fruit (Cleves-Leguizamo, 2024).

Shoot Manipulation by Branch Bending or Girdling

Stem bending is a physical method to manage growth and flowering of fruit trees especially in mandarins and wax apple. Branches with intact leaves were bent with the help of a rope to a certain known angle to achieve canopy development, encourage flowering and obtain quality fruits. Sometimes trees tend to have dense arrangement of vertical branches with acute angle (Budiarto *et al.,*

2018) where one-fourth of leaves are forced to get shaded by other leaves. Hence bending of branches here allow the canopy to open up from ellipsoid to obloid and allow sun light and airflow leading to better physiological state due to increased photosynthesis (Budiarto *et al.,* 2018). When the branch is bent, there is decrease in phloem formation leading to slow movement of sap flow leading to changes in the ratio of C: N ratio which breaks dormant buds and increases flowering or fruit set (Lal *et al.,* 2017). Of the different branch angles bending (5°, 20°, 45°, 65° and 85° evaluated, 65° and 85° bending angles gave good results of flowering, fruit development and quality in wax apple trees (Khandaker *et al.,* 2020). In Citrus fruit trees of Mandarin group cultivar Batu 55, branch bending followed by nutrient application increased flowering percentage by 100 per cent and harvested fruit number by 1.6 fold (Devy *et al.,* 11). In another mandarin cultivar Borneo Prima, branch bending induced flowering and increased fruit set and reduced fruit drop (Budiarto *et al.,* 2018).

Girdling is a technique of removal of bark of woody perennial plants using a sharp tool all around the stem to reduce the growth by reducing the nutrient flow from the foliage and the roots of the woody plant due to splitting of the vascular cambium. In olive trees, fruit yield was increased by girdling at a width of 1 to 1.5 cm during December to January followed by PVC tape healing (Lavee *et al.,* 1983). Girdling the stem followed by a plastic strip is found to restrict the sap flow of vessel elements thereby the growth is reduced in olive trees (Tombesi and Farinelli, 2016). Citrus fruit trees of mandarin group Murcott cultivar, when girdled during October showed increased foliar starch content and number of flowers than ungirdled trees (Goldschmidt *et al.,* 1985).

Canopy Management

Canopy management is the art and skill of shaping perennial fruit trees to achieve a specific size and structure to allow sufficient sunlight, air flow and to reduce juvenile duration. It helps to attain a proper ratio of vegetative to floral buds, and to get optimal yields of high-quality fruit over years (Dhillon and Thakur, 2014). Canopy management in fruit crops aims at opening the canopy for maximum light penetration and is achieved by one or more of physical methods including training, pruning, bending the branches, girdling, use of appropriate rootstock, or chemicals like growth regulators, fertilizer, by exposing to water stress (Wani *et al.,* 2021).

Use of cycocel at the time of differentiation of fruit bud has shown to reduce shoot length in sapota (Agrawal and Dikshit, 2008). Avocado trees produce short lived leaves continuously leading to shading of the photosynthetic leaves by young leaves for nearly 40 days (Schaffer and Whiley, 1987). Avocado trees are planted at rectangular plantings and trained to central leader system with a pyramidal shape. Later they are pruned regularly, cinctured (Feb to May) treated with paclobutrzal 0.4 per cent during full bloom, with appropriate nitrogen management in soil to obtain optimal yield and quality (Stassen *et al.,* 2022).

Interception of light in tree canopies depends on total leaf area index hence canopy management needs to be devised to get optimal shape of tree by appropriate planting density, training, shoot arrangements and pruning strategies (Nath *et al.*, 2019). Citrus canopy management involved the knowledge of growth habit and strong and weak bearing branching units as pruning removed the weak bearing branching units with appropriate nutrient, pest and disease, and floor management in orchards (Krajewski and Krajewski, 2010).

Pruning intensity to the tune of 30-50 per cent is recommended for Citrus grown in Mediterranean climate (Matias *et al.*, 2023). Pruning followed by application of urea 5 per cent gave early sprouting, generated several new long shoots in acid lime (Pavani *et al.*, 2018). Mangosteen (*Garcinia mangostana*) trees bear fruits on small twigs of new growth of the terminals of the stem and large branches. Hence, such terminal shoots and twigs should be promoted by appropriate pruning (Salakpetch, 2005).

Chemical Regulation of Canopy by Paclobutrazol

Paclobutrazol is a triazole compound also known as cultar or pp333 is a potent growth retardant and known to inhibit gibberellins at kaurene oxidation step in plants. Gibberellins are involved in cell elongation and internodal length. Hence, the growth retardant paclobutrazol has a major role in cost effective canopy management strategy. Paclobutrazol is applied through soil or foliar spray and has shown promising results in wide range of horticultural crops (Yadav *et al.*, 2005) in reducing shoot growth, intermodal length and ensuring regular bearing in fruit crops. Role of paclobutrazol in reducing the vegetative growth and ensuring regular bearing in different fruit crops is reviewed (Bazurto *et al.*, 2022). However caution needs to be exercised while using this stable compound which persists in soil for more than a year to reduce residual toxicity to aquatic life (Kishore *et al.*, 2015).

Paclobutrazol applied on leaves of avocado trees at 2.5 to 5 g active gradient per liter of water could reduce the shoot length by nearly 40 per cent (Wolstenholme *et al.*, 1990). A decrease in the canopy area and increase in trunk circumference in avocado trees by the application of paclobutrazol at three weeks after full bloom was found (Symons and Wolstenholme, 1990). Paclobutrazol at 500 or 1000 mg per liter spray before summer flush on Minnelo tangelo fruit trees reduced the growth of shoots, and intermodal length (Monselise, 1985; Aron *et al.*, 1985). Soil drenching of paclobutrazol to seedlings of lime fruit trees decreased branches, internodal length and increased early flowering (Snowball *et al.*, 1994). Soil application of paclobutrazol at 500 mg per plant reduced the plant dry weight by 61 per cent and height of stem by 74 per cent in sweet orange (Vu and Yelenosky, 1992). According to authors paclobutrazol reduced the rate of photosynthesis process and the activity of carboxylase leading to reduction of biomass in Citrus trees.

Paclobutrazol when used at 50 ppm concentration along with thiourea at 500 ppm and pendimethalin at 375 ppm induced sucker production in pineapple (Suwunnamek *et al.*, 1992). Tebuconazole another triazole plant growth regulator at 2 per cent applied after 65 days of flower initiation reduced length and weight of crown on fruit but did not reduce the leaves on crown (Bello-Amez *et al.,* 2024).

Paclobutrazol applied by immersion to roots of 45 days old passion fruit seedlings affected stem, leaf, and root traits and especially enhanced lignifications at 87.6 ppm concentration (Teixeira *et al.,* 2019). Paclobutrazol reduced the stem growth of seedlings of yellow passion fruit but did not reduce the root mass (Teixeira *et al.,* 2021). Paclobutrazol enhanced starch accumulation on the stem due to decrease in gibberellins and promoted early flowering in passion fruit (Teixeira *et al.,* 2023). Soil application of paclobutrazol 8 ml per liter of water to olive trees cultivar Ascolana decreased growth of vegetative shoots (Moreira *et al.,* 2016).

Conclusion

Humid zone fruit trees are constantly exposed to high relative humidity in air during most part of the year. Humidity below 55 per cent leads to low water stress and lead to decreased growth (Grange and Hand, 1987) whereas high humidifies (90 per cent and above) especially in leaf crowded canopies are prone to diseases and is sometimes required to have sufficient root pressure to avoid calcium uptake. Choosing a dwarf cultivar of a dwarfing rootstock is an important step in canopy manangement of humid zone fruit crops. High density planting appropriate to the canopy development is next step. Training the plant in initial years to central leader or open centre system depending on the species and pruning the branches as per the bearing nature of position of reproductive buds, bending the branches, girdling and other physical operations alone or in combination with chemical regulation agents such as paclobutrazol forms the successful canopy management.

REFERENCES

Agrawal S and Dikshit SN, 2008. Studies on the effect of plant growth regulators on growth and yield of sapota (Achras sapota l.) cv. cricket ball. *Indian J. Agric. Res.,* **42(3)**: 207 -11.

Aron Y, Monselise SP, Goren R and Costo J, 1985. Chemical control of vegetative growth in citrus trees by paclobutrazol. *HortScience,* **20(1)**: 96-98.

Ausari PK, Ninama N and Bagul HB, 2023. Canopy Management of Horticultural Crops. *Revolutionizing Horticulture The Green Path,* 118-136.

Bazurto FP, Celi A, Corozo L and Solís L, 2022. Importance of paclobutrazol in out-of-season citrus production. *Manglar,* **19(1)**: 117-127.

Bello-Amez S, Roman-Aquino A, Elera-Anicama M, Borjas-Ventura R, Alvarado-Huamán L, Castro-Cepero V and Julca-Otiniano A, 2024. Effect of two growth regulators on the fruit crown of Ananas comosus L. Merr. cv'MD-2'. *Siembra,* **11(2)**: e6284-e6284.

Budiarto R, Poerwanto R, Santosa E and Efendi D, 2018. Shoot manipulations improve flushing and flowering of mandarin citrus in Indonesia. *J. Appl. Hortic,* **20**: 112-118.

Cano-Gallego LE, Bedoya-Ramírez SI, Bernal-Estrada JA, Barrera-Sánchez CF and Córdoba-Gaona ODJ, 2023. Yield and fruit quality of avocado grown at different planting densities in Colombia. Pesquisa Agropecuária Brasileira, **58**: e03146.

Cautin R, Reig C, Mesejo C, Martinez-Fuentes A and Agustí M, 2010. Flowering and fruit set in cherimoya (Annona cherimola Mill.) as affected by the tree-training system. *The Journal of Horticultural Science and Biotechnology*, **85(6)**: 511-515.

Chadha KL, Melanta KR and Shikhamany SD, 1973. Effect of Planting Density on Growth, Yield and Fruit Quality in Kew Pineapple [*Ananas comosus* (L.) Merr.]. *Indian J Horti.* **30(3 and 4)**: 461-466.

Cleves-Leguizamo JA, 2024. Fruit productivity and quality in yellow passion fruit orchards with different trellis systems and planting density in the Colombian low tropics. *Revista Colombiana de Ciencias Hortícolas*, **18(1)**: In press.

Devy NF, Sugiyatno A and Mufidah L, 2023. Manipulation of flowering and fruiting induction through branch bending and fertilizer application in mandarin citrus cv. Batu 55. In *IOP Conference Series: Earth and Environmental Science.* **1287(1)**: 012006.

Dhillon WS and Thakur A, 2014. Canopy management and effects of pruning on flowering tendencies in fruit trees. In: *National Seminar-cum-Workshop on physiology of flowering in perennial fruit crops. Lucknow, India* (pp. 182-201).

Gachanja SP, 1977. The effects of trellising, spacing and pruning on fruit yields and quality, disease incidences and production life of purple passion fruit at Thika, Kenya (Doctoral dissertation, University of Nairobi).

Goldschmidt EE, Aschkenazi N, Herzano Y, Schaffer AA and Monselise SP, 1985. A role for carbohydrate levels in the control of flowering in citrus. *Sci Hort.* **26(2)**: 159-166.

González-Plaza JJ, Ortiz-Martín I, Muñoz-Mérida A, García-López C, Sánchez-Sevilla JF, Luque F and Beuzón CR, 2016. Transcriptomic analysis using olive varieties and breeding progenies identifies candidate genes involved in plant architecture. *Frontiers in plant science*, **7**, 240.

Gopu B and Balamohan TN, 2016. Canopy Management in Sapota (*Achras sapota* L.). *Global Journal For Research Analysis*, **4(7)**.

Grange RI and Hand DW, 1987. A review of the effects of atmospheric humidity on the growth of horticultural crops. *Journal of Horticultural Science*, **62(2)**: 125-134.

Haque MA and Sakimin SZ, 2022. Planting arrangement and effects of planting density on tropical fruit crops—A Review. *Horticulturae.* **8(6)**: 485.

Hervalejo A, Arjona-López JM, Romero-Rodríguez E and Arenas-Arenas FJ, 2024. Suitability of two dwarfing citrus rootstocks for 'Salustiana'orange trees grown under super-high-density conditions with mechanical harvesting. *New Zealand Journal of Crop and Horticultural Science*, **52(1)**: 64-75.

Khandaker MM, Ismail NH, Abdulrahman MD, Yusoff N, Mohd KS, Badaluddin NA and Saifuddin M, 2020. Effects of branch bending angle on growth and flowering of wax apple (*Syzygium samarangense*). *Plant Archives*, **20(2)**: 5907-5913.

Kishore K, Singh HS and Kurian RM, 2015. Paclobutrazol use in perennial fruit crops and its residual effects: A review. *Indian J Agril. Sci.* **85(7)**: 863-872.

Köhne JS and Kremer-Köhne S, 1991. Avocado high density planting–a progress report. *South African Avocado Growers' Association Yearbook.* **14:** 42-43.

Krajewski AJ and Krajewski SA, 2010. Canopy management of sweet orange, grapefruit, lemon, lime and mandarin trees in the tropics: Principles, practices and commercial experiences. *Acta Hortic.* **894**: 65-76 DOI: 10.17660/ActaHortic.2011.894.5

Lal N, Sahu N, Marboh ES, Gupta AK and Patel RK, 2017. A review on crop regulation in fruit crops. *International Journal of Current Microbiology and Applied Sciences*, **6(7)**: 4032-4043.

Lavee S, Haskal A and Ben Tal Y, 1983. Girdling olive trees, a partial solution to biennial bearing. I. Methods, timing and direct tree response. *J Hort Sci.* **58(2)**: 209-218.

Matias P, Barrote I, Azinheira G, Continella A and Duarte A, 2023. Citrus pruning in the Mediterranean climate: A Review. *Plants.* **12(19)**: 3360.

Mitra SK, 2019. High density orcharding and canopy management of some tropical and subtropical fruits. *Acta Hortic.* **1255**: 127-138

Monselise SP, 1985. Growth retardation of shoot and peel growth in citrus by paclobutrazol. *Acta Hortic.* **179**: 529-536.

Moreira RA, Fernandes DR, da Cruz M DCM, Lima JE and de Oliveira AF, 2016. Water restriction, girdling and paclobutrazol on flowering and production of olive cultivars. *Sci Hort.* **200**: 197-204.

Nath V, Marboh ES, Gupta AK, and Lal N, 2019. Canopy management for sustainable fruit production. *International Journal of Innovative Horticulture*, **8(2)**: 115-126.

Neri JC, Meléndez Mori JB, Vilca Valqui NC, Huaman Huaman E, Collazos Silva R and Oliva M, 2021. Effect of planting density on the agronomic performance and fruit quality of three pineapple cultivars (*Ananas comosus* L. Merr.). *International Journal of Agronomy*, **2021(1)**: 5559564.

Pavani K, Mishra S, Dash DK and Vaka DV, 2018. Effects of pruning and chemicals on growth behaviour of acid lime (*Citrus aurantifolia* Swingle) cv. Kuliana lime. *IJCS*, **6(6)**: 1959-1963.

Roose ML, Cheng FS and Federici CT, 1994. Origin, inheritance, and effects of a dwarfing gene from the citrus rootstock *Poncirus trifoliata* Flying Dragon. *HortScience*, **29(5)**: 482e-482.

Salakpetch S, 2005. Mangosteen (*Garcinia mangostana* L.) juvenility and pruning in Thailand. *Hawaii Tropical Fruit Growers*, **27**: 203-207.

Snowball AM, Warrington IJ, Halligan EA and Mullins MG, 1994. Phase change in citrus: the effects of main stem node number, branch habit and paclobutrazol application on flowering in citrus seedlings. *J Hort Sci.* **69(1)**: 149-160.

Sulladmath VV, Srinivas K and Laxman RH, 2012. Yield and quality of passion fruit in relation to training systems. *J Hort Sciences.* **7(1)**: 46-50.

Suwunnamek U, 1992. Effect of paclobutrazol, thiourea, and pendimethalin alone or in combination on the induction of suckering in pineapple. Acta Hortic. **34**: 93-100.

Symons PRR and Wolstenholme BN, 1990. Field trial using paclobutrazol foliar sprays on Hass avocado trees. *South African Avocado Growers. Association Yearbook*, **13**: 35-36.

Teixeira EC, Matsumoto SN, Silva DDC, Pereira LF, Viana AES and Arantes ADM, 2019. Morphology of yellow passion fruit seedlings submitted to triazole induced growth inhibition. *Ciência e Agrotecnologia*, **43**: e020319.

Teixeira EC, Matsumoto SN, Ribeiro AFF, Viana AES, Tagliaferre C, Carvalho FD and Silva V A, 2021. Morphophysiology and quality of yellow passion fruit seedlings submitted to inhibition of gibberellin biosynthesis. *Acta Scientiarum. Agronomy*, **43**: e51541.

Teixeira EC, Matsumoto SN, Pereira LF, Silva DDC, Viana AES and Gonçalves ANS, 2023. Paclobutrazol as a strategy to induce reproductive precocity in sour passion fruit. *Revista Brasileira de Fruticultura*, **45:** e-501.

Tombesi S and Farinelli D, 2016. Trunk constriction effects on vegetative vigour and yield efficiency in Olive tree (*Olea europaea* L.). *Journal of Agricultural Science and Technology*. **18(6)**: 1667-1680.

Tripathi PC, 2021. Production Technologies of Humid Tropical Underutilized Fruits in India. In: Current Horticulture- Improvement, Production, Plant Health Management and Value-Addition Vol-II, (Eds.) Balraj Singh, A. K. Singh, B. S. Tomar, J. K. Ranjan, Som Dutt. Brillion Publishers,New Delhi, 315-334.

Tripathi PC, Karunakaran G, Sakthivel T, Sankar V and Senthil Kumar R, 2019. Status and prospects of rambutan cultivation in India. *Acta Hortic*. **1293**: 33-40.

Vu JC and Yelenosky G, 1992. Growth and photosynthesis of sweet orange plants treated with paclobutrazol. *Journal of Plant Growth Regulation*, **11**: 85-89.

Wani RA, Din S, Khan M, Hakeem SA, Jahan N, Lone RA and Khan J, 2021. Canopy management in fruit crops for maximizing productivity. *IJCS*, **9(3)**: 160-165.

Wheaton TA, Castle WS, Whitney JD and Tucker DPH, 1991. Performance of citrus scion cultivars and rootstock in a high-density planting. *HortScience*, **26(7)**: 837-840.

Wolstenholme BN, Whiley AW and Saranah JB, 1990. Manipulating vegetative: reproductive growth in avocado (Persea americana Mill.) with paclobutrazol foliar sprays. *Sci Hort.* **41(4)**: 315-327.

Yadav RK, Rai N, Yadav DS and Asati BS, 2005. Use of paclobutrazol in horticultural crops–a review. *Agricultural Reviews*, **26(2):** 124-132.

Chapter 11

Epicotyl and Stone Grafting in Mango

D.P. Waskar, V.S. Khandare and B.M. Kalalbandi

Vasantrao Naik Marathwada Krishi Vidyapeeth, Parbhani–431402, Maharashtra

Mango (*Mangifera indica* L.) belongs to the Anacardiaceae family; it is the choicest, traded, popular and widely cultivated fruit in the world (Naik and Kumar 2020). Due to its appealing color, delicious flavour, and high nutritional value, mango has long been regarded as the "King of fruits" and is currently thought to be the most valuable tropical fruit (Ullah *et al.,* 2017). It has got its origin from the southern part of Asia, especially from eastern India, Burma and the Andaman Islands (Honja, 2014). In India, mango is cultivated in almost all the states, while UP, MP, Bihar and Andhra Pradesh are the leading states in area and production. Other states where mango cultivation exists include Orissa, West Bengal, Karnataka, Gujarat, Maharashtra, Madhya Pradesh, Tamil Nadu, Kerala and Punjab. Districts of Uttar Pradesh like Lucknow and Amroha (City of Mango) are very famous in mango production and producing lots of varieties of mango.

Mango is highly cross-pollinated heterozygous in nature that's why sexually raised plants usually acquire huge genetic variation and fails to produce identical true to type plants. Stone-propagated plants also have a long juvenile phase and poor performance (Bobade *et al.,* 2018). Therefore, it must be propagated asexually to produce true-to-type off springs. Moreover, most commercial varieties of mango are mono-embryonic. Thus, the grafting technique is needed to produce genetically identical copies of desired cultivars (Minja *et al.,* 2017). Among the several produce genetically identical copies of desired cultivars (Minja *et al.,* 2017). Among the several asexual methods, grafting is proposed as the most appropriate technique to preserve the fidelity of a given variety, allowing the production of offspring with similar characteristics to the mother plant. It involves joining a scion and a rootstock together, growing as a single plant with the scion becoming the upper fruiting part of the grafted plant and the rootstock developing into a root system. (Prasanth *et al.,* 2009). According to Hartmann *et al.* (2010) rootstocks can influence a variety of physical and pomological characteristics of grafted plants, including yield, tree growth habit, fruit maturity time, and size. Grafting has several benefits including early fruiting, reduced stature and bushy canopy. The majority of improved cultivars are mono-embryonic, necessitating grafting to produce trees that are true to type. Additionally, some perform poorly because they are unsuited to tropical conditions. In Pakistan, the most commonly used techniques for successful grafting are veneer and cleft grafting that are being practiced commercially from ages. However, establishing a grafted mango orchard of better tree architecture is somewhat difficult when done by veneer grafted plants because this technique failed to produce erect and spout plants of uniform canopy which is a pre-requisite for successful modern mango orcharding (Soleimani *et al.,* 2010).

Currently, among several grafting methods, epicotyl grafting is well-documented as a quick, affordable and effective technique. Numerous researchers have reported success with epicotyls grafting (Jaganath *et al.,* 2014; Simon *et al.,* 2010). This technique has been widely adopted in hot and humid climates where temperature reaches up to 40-45°C. But in Central and Northern conditions, this method has yet not been popularized. Further studies in different climate zones are needed before recommending it to nurseries. The success of epicotyl grafting is dependent on a number of variables, including rootstock age, temperature, relative humidity, light, soil moisture, time, method of grafting, and the grafter's skill. Epicotyl grafting involves grafting the scion on the stem of the young seedlings just above the embryonic leaves and allowing the scion to grow out of the seedling. The newly germinated seedlings under juvenile stage are usually taken for epicotyl grafting which may result into greater success of grafts. In addition, the type and age of rootstock are also crucial factors affecting the success rate, survival, and growth of mango grafts. Bhan *et al.* (1969) standardized the epicotyl grafting in mango at the Horticulture Research Station, Krishnanagar and they claimed about 75-80 per cent success by using germinating seed as a rootstock and semi-mature terminal shoot as a scion. In epicotyls grafting, young seedlings are used as rootstock for grafting. Epicotyl grafting was successfully used as an efficient, economic and rapid method for the propagation of mango.

Season of Grafting

Singh and Shrivastava (1982) reported that the best results (84 per cent take) were obtained with inarching and soft wood grafting both in July. Khalil *et al.* (1983) reported that the best month for grafting were June, July and August, during which 100 per cent take could be expected. He obtained in trials with the cv. Dashehari, softwood grafting on 20th August gave the highest grafting success (90%) compared with 67 in July and 70 per cent in late September.

Figure 11.1: Step by Step Procedure for Epicotyl Grafting in Mango.

Patil *et al.* (1991) observed that when stone grafting in mango done during first week of July using 7 to 10-day old seedling as rootstock, Mulgoa gave 100 per cent successful graft while Alphonso, Pairi and Totapari, recorded 40.00, 53.33 and 53.33 per cent grafting success, respectively. Further observed that the highest average sprout length was recorded under June grafting and survival after 180 days was highest in August grafting.

Kumar *et al.* (2000) conducted an experiment at Ranchi to study the effect of stone grafting in mango and observed the best growth and survival (79%) of grafted plants in the treatment with used scions which had been defoliated for 8 days and stored for 3 days before grafting in the rainy season. Nayak and Sen [17] reported that the percentage of success was also greater 78.8 per cent when it was done in July- August compared with 75 per cent in January- March. However, overall rate of growth was greater in winter grafted plants. Jacob *et al.* (2014) reported that the higher graft success (90%) was obtained during the months of July-August, September and October. Prasanth *et al.* (2006) studied epicotyl grafting in mango and reported that grafting in the first fortnight of July resulted earliest sprouting (28.71 days), while grafting in the first fortnight of August recorded late sprouting (35.10 days) under North- Eastern dry zone of Karnataka.

Effect of Variety

Singh *et al.* (1992) observed that the Dashehari gave significantly higher average sprouting success (65-90%), average number of leaves (8.27) and survival after 6 months (65.35%) than Langra in July. Radha *et al.* (2000) reported that the height of the plants varied from 26.3 cm in Chandrakaran to 36.5 cm in Bangalora at 6 months. Karim *et al.* (2018) reported that the minimum time required for first flush (15.69 days), highest percentage of success (93.58%) and survival (83.56%) were achieved in Sindhu, at grafting above the 2nd node. Maximum scion length and stionic height were achieved in Langra at grafting above 2nd node (30.65 cm and 70.38 cm, respectively). The highest numbers of leaves per graft and canopy volumes were recorded at 240 DAG in Mallika and Amrapali (27.89 leaves and 55468.33 cm3, respectively) while grafting above the 2nd node. So, grafting above the 2nd node is suitable for the highest success, survivability and stionic growth for all varieties of mango. Bobade *et al.* (2018) reported that the maximum values of growth parameters like height of sprout (6.02 cm), length of shoot (27.97 cm), length of secondary roots (23.73 cm) were also observed in grafts of variety Mallika and it was closely followed by grafts of variety Kesar. The maximum value of stionic ratio (0.93) was recorded in grafts of variety Pairi and it was at par with variety Amrapali (0.91).

Muzamil *et al.*, [16] reported that epicotyls grafting resulted in minimum days for bud break (8.33) days required to first flush in grafts (15.66) produced more rootstock length (18.57 cm), rootstock diameter (5.34 cm),scion length (8.14 cm),scion diameter (4.34 cm), stionic height of graft (38.57 cm), number of leaves per graft (8.57), success per cent (68.84) and survival per cent (65.53) in variety Chausa compared with Langra and Sindhri.

Scion Characteristics

Prasad *et al.* (1990) studied certain aspects of veneer grafting in mango (*Mangifera indica* L.) cv. Banganapally in Tirupati and they recorded that the Pre cured scions gave higher success (78.96%) than non-precured scions (52.75%). Success was highest (71.08%) with 100 days old scions and lowest (62.67%) with 120 days old scions. The mean graft survival after potting was 75.39 per cent with precured scions and 52.28 per cent with non-precured scions that high mortality often occurs when material grafted in the nursery is lifted and potted after grafting. Kumar *et al.* (2000) reported that the high success in veneer and cleft grafting (>85%) of mango. Alam *et al.* (2006) reported that the minimum success (10.0%) was recorded in BARI Aam-3 grafted onto 5 and 30days old seedlings. The tallest shoots (25.07 and 24.73 cm consecutively) were produced by Langra grafted on 15 and 20 days old seedlings. Maximum final success (76.67%) was recorded in 10.0 cm long scions followed by 7.5 cm long ones (70.00%) in variety Langra. Singh and Srivastava (1979) stated that the best results were obtained from 6 months old scions, grafted on 2 years old rootstocks in July/August. The bud sticks were defoliated and kept in moss wrapped with polythene for 3-5 days before grafting. Among the several tying materials, the polythene strips gave the best results. Alam [1] concluded that Langra variety grafted on 10 cm long scion shoots secured the highest final survibility (76.6%) on 50 days old rootstock seedling.

Rootstock Attributes

Muzamil *et al.* (2024) reported that epicotyls grafting using different age rootstock significantly affected the traits of grafts. Results revealed that 50 days rootstock resulted in minimum days for bud break (13.92), days required to first flush in grafts (23.11), produced more rootstock length (19.96 cm), rootstock diameter (6.03 cm), scion length (9.64 cm), scion diameter (5.04 cm), stionic height of graft (37.57 cm), number of leaves per graft (8.33), success (70.25%) and survival (68.82%) in comparison with 20, 30 and 40 days rootstock seedling. Mude *et al.* (2020) reported the grafting on Kurukkan rootstock with various cultivars of scions such as Amrapali, Dashehari, Mallika, and Langra at Raipur, Chhattisgarh conditions. In the Dashehari variety, followed by Langra, the most successful percentage (93.33%) was recorded. Tadda, *et al.* (2018) concluded that epicotyl grafting was found to be feasible using the available local rootstock and desired scion. The 10 cm scion was identified to be the most suitable. Alphonso grafted onto Gwaiwarago was the best combination as these produce the highest success of grafts and longer grafted saplings. Hence, epicotyl grafting presents quickest means of rapid production of low cost seedlings in mango. The maximum number of sprouted grafts (14.83), maximum sprouting percentage (74.17), minimum days for leaf emergence, maximum number of leaves per graft (15.65), girth (above the union) (1.82 cm), minimum mortality and maximum survival (19.17%) of grafts when grafts were made on 6 cm height of rootstock. Consequently, maximum growth in terms of height and girth (28.94 cm) (below the union) were recorded in grafts made on 10 cm height of rootstock of mango cv. Kesar (Nalage *et al.* (2010). Among the growing media used in the study the soil + sawdust was found to be the best growing mixture in respect of sprouting, survival and overall performance of stone grafts over other growing mixtures (Prabhat *et al.,* 2012).

Bhuiyan *et al.* (2018) the highest height of the rootstock was recorded (36.50cm) and highest canopy volume was produced by Amrapali scion grafted onto the Amrapali rootstock followed by the plants of Gopalbhog scion onto Gopalbhog rootstock combination plant.

Conclusion

After studying various practices by researchers and their noted records, it can be concluded that factors like varieties, time of grafting, method, growing conditions, defoliation period of scion, age of the scion, leaf and node retention on rootstock *etc.* influenced the success and survivability of mango grafts. Rainy season grafting in mango shows best results of growth and survival percentage compared to other seasons of grafting. Inarching, veneer grafting, side and wedge grafting methods are most common, but the stone grafting method is easy, cheap and rapid method to produce high quality and quantity in comparatively less time and area. Higher graft success was obtained during the months of July-August, September and October. The success of grafting depends on the season, age of rootstock (50 days old) and scion, preparation of scion before grafting, cultivars, height of grafting and media used in stone grafting methods. Epicotyl grafting technique plays a vital role in the success of graft union and the advantages is that the germinating seedlings are in juvenile condition and the cells have the potentiality of quick differentiation.

REFERENCES

Alam A, 2004. Effect of variety, scion length and age of seedling on stone grafting in mango. M.Sc theses submitted to Bhangabandhu Sheikh Mujabir Rehman Agrixulture University, Salana,Gazipur.

Alam MA, Mortuza MG, Uddin MZ, Sarker D and Barman JC, 2006. Effect of length and variety of scion in stone grafting of mango. *International Journal of Sustainable Crop Production.* **1(2)**: 7-11.

Bhan KC, Samander HN and Yadav PS, 1969. Chip budding and stone grafting of mangoes in India. *Tropical Agric Trinidad.* **46(3)**: 247-253.

Bhuiyan MFA, Rahim MA and Alam MS, 2010. Study on the growth of plants produced by epicotyl (stone) grafting with different rootstock-scion combinations in mango. *J. Agrofor. Environ.* **3(2)**: 163-166.

Bobade DH, Ingole RH and Kadam AS, 2018. Effect of different scion varieties of mango on growth and biomass production performance of stone grafts (*Mangifera indica* L.). *Int J Curr Microbiol App Sci.* Special Issue, **6**: 1642-1648.

Jacob S, Ray DP, Sahu GS and Chandra A, 2001. Studies on the success of soft wood grafting in some commercial hybrid mango (*Mangifera indica L.*). *Orissa Journal of Horticulture*. **29(2)**: 6-9.

Jaganath B, Subramanyam K, Mayavan S, Karthik S, Elayaraja D, Udayakumar R and Ganapathi A, 2014. An efficient in planta transformation of *Jatropha curcas* (L.) and multiplication of transformed plants through in vivo grafting. *Protoplasma*, **251(3)**: 591-601.

Hartmann HT, Dale EK, Davies FT and Geneve RL, 2010. Plant Propagation, Principles and Practice. 7 edition Pp 145.

Honja T, 2014. Review of mango value chain in Ethiopia. *Journal of Biology, Agriculture and Healthcare*. **4(25)**: 230-239.

Karim MR, RahimMA and Alam MS, 2018. Effects of leaf and node retention on rootstock of epicotyl grafting in mango. *International Journal of Natural and Social Sciences*. **5(1)**: 69-77.

Khalil FA, Hamoouda AM, Bondik AZ, El-Din lS, 1983. Ann Agric Sci Ain Shams Univ **28**: 1677-1688.

Karim MR, 2004. Effect of leaf and node retention in rootstock on grafting in mango. M.Sc. (Ag.) thesis, Hort Dept BAU Mymensingh 30- 53.

Kumar S, Sant R and Singh CP, 2000. Success of veneer and cleft grafting different grafting heights of seedling rootstocks in Dashehari mango. *Indian Journal of Horticulture*. **57(3)**: 212-214.

Minja R, Kimaro A and Nyalusi G, 2017. Effects of rootstock type and scion cultivar on grafting, success and growth of mango (*Mangifera indica* L.) seedlings. *J. Hortic. Sci.* **32(2)**: 179-189.

Mude A, Naik K and Nangunuri VK, 2020. A review on recent developments on propagating mango (*Mangifera indica* L.) through grafting. *Int.J. Curr. Microbiol. App. Sci* **9(11)**: 3481-3487.

Muzamil Farooque Jamali1, Noor-un-Nisa Memon, Niaz Ahmed Wahocho, Afifa Talpur, Waqas Ahmad, Fayaz Ali Jamali, Ali Raza Jamali, Hub e Ali Mangsi, Asghar Ali, 2024. Comparative effects of varieties and rootstock age on success and survival rate of epicotyl grafting in mango (*Mangifera indica* L). *Pak. J. Biotechnol.* **21(1)**: 219-227.

Nayak G and Sen SK, 2000. Evaluation of vegetative propagation methods of mango. *Environment and Ecology*. **18(1)**: 243-245.

Naik MAK and Kumar NVA, 2020. Review on Recent Developments on Propagating Mango (*Mangifera indica* L.) through grafting. *International Journal of Current Microbiology and Applied Science*. **9(11)**: 3481-3487.

Nalage NA, Magar SS, Bhosale SS and Mhetre DA, 2010. Effect of height of rootstock on success of epicotyl grafting in mango (*Mangifera indica* L.) cv. Kesar. *Internation J.of Agricultral Scien*ce. **6(1)**: 124-128.

Simon A, Mngomba Akinnifesi FK, Sileshi G and Ajayi OC, 2010. Rootstock growth and development for increased graft success of mango (*Mangifera indica*) in the nursery. *African Journal of Agriculture* **9(9)**: 1317–132.

Singh AR, Pandey SP, Sing RK and Sing ND, 1992. Influence of cultivar and period of operation on the success of veneer grafting in mango. *Adv Hort Forestry.* **19(2)**: 17-23.

Singh NP, Srivastava RP, 1982. Studies on various factors involved in soft wood grafting in mango. *Progressive Hort* **14(2 and 3)**: 117-120.

Singh NP and Srivastava RP, 1979. Studies on the different aspects involved in veneer grafting in mango. *Prog Hort* **11(1)**: 67-74.

Soleimani A, Rabiei V and Hassani D, 2010. Effect of different techniques on walnut (*J. regia* L.) grafting. *Journal of Food, Agriculture and Environment.* **8(29)**: 544-546.

Patil AA, Cadigeri BG, Nalawadi UG, 1991. Response of mango varieties to stone grafting. *Curr Res Agril Sci* **20(2)**: 135-136.

Prabhat K and Nidhika K, 2012. Effect of age of rootstock and different growing media on the success of stone grafting in mango. *International Journal of Farm Science,* **6(1)**: 255-261.

Prasad PV, Suryanarayana V and Naidu N, 1990. Studies on certain aspects of veneer grafting in mango (*Mangifera indica L.*) cv. Banganapally. *South Indian Horticulture.* **38(1):** 1-7.

Prasanth J, Mukherjee SK and Litz RE, 2009. Introduction: botany and importance. in *The mango: Botany, Production and Uses* (pp. 1-18). Wallingford UK: CABI.

Prasanth JM, Reddy PN, Pampanagouda B and Patil SR, 2006. Effect of cultivars and time of grafting on percent success and survival percentage of veneer grafting in mango (*Mangifera indica L.*). *Internat J Agric Sci.* **2(1)**: 19-21.

Radha T and Aravindakshan K, 2000. Differential response of mango varieties to epicotyl grafting on commercial scale. *Acta-Horticulturae,* **509**: 265-268.

Tadda SA, Muhammad M, Manga AA and Yahaya SU, 2018. Epicotyl grafting in mango (*Mangifera indica* L.) as influenced by length of scion, scion type and rootstock. *Bayero Journal of Pure and Applied Sciences,* **11(2)**: 99–106.

Ullah SS, Malik S, Kumar R and Kumar M, 2017. Effect of Time and Technique of Grafting for Quality Production of Nursery Plants of Dashehari Mango (*Mangifera indica* L.). *International Journal Current Microbiology and Applied Sciences,* **6(10)**: 685-90.

Chapter 12

Nursery Production, Orchard and Canopy Management in Grapevine in Tropical Climate

N.A. Deshmukh, Shradha Takale, Shalaka Sinhasane, Sudarshan Gat and P.H. Nikumbhe

ICAR–National Research Centre for Grapes, Pune–412307, Maharashtra

Grapes (*Vitis vinifera*), esteemed for their cultural and commercial significance, have a storied history intertwined with human civilization. Originating from Western Asia and Europe, these fruits have adapted to diverse climates, including temperate, tropical, and subtropical regions. Historically, grapes spread from the Caucasus region to various parts of the world, reaching North India around 1300 AD via Muslim invaders and South India later through French Christian missionaries.

Grapes are a non-climacteric fruit, woody, perennial vines belonging to the genus *Vitis* (Hiwasa Tanase and Ezura, 2014). Grapes are cross-pollinated plants and are highly valued for their rich nutritional content, including vitamins B1 and B2, calcium, phosphorus, and iron. They exhibit remarkable versatility, being consumed fresh or processed into various products such as jam, juice, jelly, vinegar, wine, and raisins (Adsule *et al.,* 2012; Sharma *et al.,* 2020).

In India, grape cultivation spans an area of 171 thousand hectares with a production of 3781 thousand metric tons, as per the NHB's 2023-24 first advance estimate. During the 2022-23 period, India exported 267,950.39 metric tons of fresh grapes, generating revenue of Rs. 2,543.42 crores (313.70 million USD). Major grape-growing states include Maharashtra, Karnataka, Tamil Nadu, and Mizoram, with Maharashtra leading in both production and productivity. (APEDA, 2024)

Grape productivity and quality are influenced by various factors, including planting material, climate conditions, training systems, pruning, irrigation management, and canopy management (Keller, 2015). Despite the expansion of grape cultivation areas, average production remains low in some regions due to the lack of quality planting material and limited adoption of advanced cultural practices. However, modern agricultural practices and technologies have significantly boosted grape production, particularly in regions that have adopted scientific methods. Efforts to optimize grape cultivation continue to evolve, with a focus on advancements in rooting media selection, propagation techniques, rootstock choices, and orchard management strategies. These innovations aim to enhance productivity, improve fruit quality, and sustainably meet domestic demand while strengthening India's position as a significant player in the global grape market.

This chapter will explore the optimization of grape cultivation through detailed discussions on nursery production techniques like rooting media, propagation, rootstock choices, orchard and canopy management strategies. These insights aim to benefit viticulturists, researchers, and enthusiasts, offering valuable knowledge for enhancing grape production and quality.

Nursery Production

Planting Material Availability and Requirement

Grape vines are propagated through vegetative methods, with rootstocks grown from cuttings and scions grafted onto these rootstocks (Singh and Chauhan 2020). Typically, nurseries maintain nucleus and mother blocks to produce true-to-type, high-quality planting material. In India, growers often prefer in-situ grafting, purchasing rootstock material and grafting scions directly in the field. Certified nurseries ensure the establishment of virus-free vineyards, but there is a shortage of such nurseries in India (Naik *et al.,* 2023). While some private nurseries supply rootstock material, certification and traceability are often lacking.

Establishing certified nurseries is crucial to meet the high demand for virus-free grape planting material. The National Horticulture Board has initiated programs to certify and recognize nurseries. Investing in certified nurseries ensures healthier vines, higher yields, and better-quality fruit, reducing disease losses. Research and development in grapevine propagation, disease resistance, and cultivar improvement can further support the industry's growth. Certified nurseries should also adopt sustainable practices, like using organic fertilizers and pest management techniques, to minimize the environmental impact of grape cultivation. Grapevines grown for table purposes are typically planted at a spacing of 2.75 m x 1.5 m, resulting in a density of 2,400 plants per hectare. With an annual compound growth rate (ACGR) of 6.35 per cent for vineyard acreage, it is estimated that approximately 24.7 million vines will be needed to plant 10,287 hectares of new vineyards annually. This demand for clean planting material is expected to increase over time. Additionally, 25.9 million new vines will be required to replant old and aging vineyards that are more than 15 years old. Therefore, there is a demand for more than 50 million new quality vines annually to meet the needs of both existing and new vineyard acreage. (Naik *et al.,* 2023)

Rooting Media

The choice of rooting media significantly influences root growth parameters such as length, diameter, and density in grape cuttings (Table 12.1). Media with optimal aeration and water retention, like peat-based or vermiculite-perlite mixtures, promote longer roots, while those rich in organic matter support thicker roots, aiding in nutrient absorption and stress resistance (Dewayne *et al.,* 2003; Shah *et al.,* 2021; Khan *et al.,* 2017). For seedling development, media composition plays a crucial role in either facilitating or impeding soil-borne pathogens (Sahin *et al.,* 2002). Various media blends are used, including sand, soil, and FYM, or alternatives like vermiculite and perlite, which optimize moisture and oxygen for root growth (Meissner *et al.,* 2019). However, challenges such as the high costs of certain media necessitate economical alternatives like sugarcane bagasse and farmyard manure, which have shown promising results in grape seedling growth (Alexander *et al.,* 2008; Muhammad Saqib *et al.,* 2021).

Table 12.1: Rooting Media for Grape Cutting Propagation

Rooting Media	*Description*	*Advantages*	*Disadvantages*	*References*
Soil-based media	The mixture of garden soil, sand, and organic matter	Nutrient-rich composition promotes root development	May be heavy, limited drainage, variable nutrient availability	Dewayne *et al.,* 2003
Peat-based media	Peat moss or sphagnum moss is known for excellent water retention properties	High rooting percentage, robust root development	Expensive, sustainability concerns (peat moss)	Shah *et al.,* 2021
Vermiculite-perlite	Lightweight, sterile, inert mixtures	Enhances aeration, and drainage; higher rooting success compared to soil-based media	Requires additional nutrients, the initial cost of vermiculite and perlite	Meissner *et al.,* 2019
Sand-based media	Coarse sand primarily, offers excellent drainage and stability	Good drainage, stable environment for cuttings	Lacks organic nutrients, may restrict nutrient availability, may require supplemental fertilization	Song *et al.,* 2001; Muha-mmad Ami *et al.,* 2021

Propagation

Cuttings

Cutting is a traditional practice widely used for the vegetative propagation of plants and is commercially utilized in viticulture for multiplying rootstock and own-rooted vines (Patil *et al.,* 2001; Gambetta *et al.,* 2012; Gokbayrak, 2010). Among various vegetative techniques, hardwood and softwood cuttings are the most popular methods for grape propagation. Hardwood cuttings are particularly preferred due to their high success rate, cost-effectiveness, and simplicity, as they do not require special rooting techniques. The success of propagation (Table 12.2) is significantly influenced by the type, position, and timing of the cuttings (Somkuwar *et al.,* 2017). Research indicates that cuttings taken from the basal portion of the shoot are optimal due to the higher accumulation of carbohydrates and root-promoting substances in buds. Prepare cuttings that are 30-45 cm long with a diameter of 8-10 mm and have four to five nodes. Using basal 4-5 node hardwood cuttings ensures optimal reserve food material, rooting percentage, sprouting, and shoot production (Somkuwar *et al.,* 2004).

Table 12.2: Rooting of Grapes at Various Substrates, Cutting Types, and Auxin Treatments

Rootstock	*Substrate*	*Cutting Type*	*Cutting Length (cm)*	*Auxin Dose (mg L^{-1})*	*Rooting (%)*	*Reference*
Paulsen 1103	Commercial	Hardwood	10	IBA 2000	100	(Amaral *et al.,* 2009)
	Soil + sawdust + cattle manure	Semi-hardwood	15	IBA	84	(Coradini *et al.,* 2014)
Dog Ridge	Soil + sand + FYM	Herbaceous	2	IBA	92.5	Somkuwar *et al.,* 2011)
IAC 572 'Jales'	Commercial	Herbaceous	10	IBA 2000	95	Souza *et al.,* 2015)
5BB	Perlite	Hardwood	35	IBA2000	87.5	Dogan, Uyak, and Kazankaya (2016)
Crimson (own root)	Sandy soil	Hardwood	10	IAA 1000	100	Ahmed *et al.* (2017)
SO4	Vermiculite	Herbaceous	5-8	IBA1000	75	Nascimento *et al.* (2019)

Rao (2004) noted that the ideal months for taking hardwood cuttings for Dogridge and 1613C rootstocks are February and March, while March is optimal for softwood cuttings. Studies have also indicated that applying plant growth regulators such as IBA and NAA can influence the number of sprouts and the vegetative growth of grapevine cuttings (Maninderdeep and Singh 2022; Ausari *et al.*, 2023). The optimum IBA concentrations for rooting are 2000 ppm for hardwood cuttings and 500 ppm for softwood cuttings in Dogridge and 1613C rootstocks (Rao, 2004; Ramesh, 2023). Maximum success rates have been reported in *Vitis longyi*, SO-4, and 99 R rootstocks at 1000 ppm of IBA; in Dog Ridge, Salt Creek, 110 R, and 1103 P rootstocks at 750–1000 ppm of IBA; and in Thompson Seedless rootstocks at 500 ppm of IBA (Satisha and Adsule, 2006). Additionally, cuttings treated with these IBA concentrations showed early sprouting, with the highest success rates and root lengths observed in cuttings from mother vines with a moderate carbon-to-nitrogen (C/N) ratio and reduced phenol content (Ausari 2021). At the age of 3-4 months, rooted cuttings are ready to transplant during February-March. This is the ideal time to transplant in the tropical climate to get the desired shoot thickness for grafting in September. (Somkuwar and Adsule, 2004).

Grafting

Selection of Rootstock

All of the rootstock shoots ought to grow for up to 50 days in India's semi-arid tropical environment before being trained to keep two shoots on each plant and cut other shoots. To knot the shoots, grow them straight, and prevent dead wood forming on the stem throughout the vine-growing process. To maintain the food reserve and grow straight growth, 2 to 3 leaves should be plucked from the main shoots at intervals of 15 days. When grafting, the rootstock plant's leaves and side shoots were cut off up to 45 cm above the ground (Somkuwar and Ramteke, 2006; Somkuwar *et al.,* 2015).

Selection of Scion Shoot

Selecting the appropriate scion cuttings is pivotal for achieving optimal success and subsequent plant performance. The ideal scion cuttings should be one season old, fully matured with a thickness of 6-8 mm, sourced from elite, disease-free, healthy, and high-yielding mother vines, as emphasized by Somkuwar and Ramteke (2006). These scion shoots must possess at least two healthy buds and exhibit short internodes to ensure vigorous growth post-grafting. Additionally, the scion should be round and contain adequate food reserves to support the initial growth phase following grafting, thereby enhancing the overall success rate of the grafting process and the subsequent development of the grafted vine.

Process of Grafting

The optimal time for in-situ wedge grafting in India's semi-arid tropical climate is from August to September. During this period, the most successful grafts are achieved due to optimal environmental conditions, with temperatures ranging from 30°C to 40°C and relative humidity levels between 80 per cent and 90 per cent. These conditions activate alpha-amylase enzymes, leading to the highest success rate for grafting (Jogaiah, 2009). In situ wedge grafting procedure, selected shoots were cut at a height of 45 to 50 cm above the ground, and a vertical slit cut of 4 to 5 cm was made on the rootstock. To produce a 3 to 4 cm wedge-shaped cut without harming the cambium layer at the basal end of the scion, the bark on two opposite sides should be removed (Somkuwar *et al.,* 2006). Understanding rootstock-scion interactions (Table 12.3) is important for selecting the best graft combinations for various settings and producing high-quality fruit. (Rasool *et al.,* 2020).

Table 12.3: Graft Success Rates and Stock: Scion Ratio of Various Grapevine Cultivars

Name of Cultivar	*Graft Success (%)*	*Stock : Scion Ratio*
Tas-A-Ganesh	90.0	0.81
Thompson Seedless	91.56	1.12
Flame Seedless	83.3	0.79
Sharad Seedless	87.50	0.86

Source: Somkuwar and Ramteke, 2006.

Rootstock Selection

Selection of plant material is critical for successful propagation, with mother vines chosen for cuttings, budding, or grafting requiring vigor, health, and freedom from diseases and pests. Rootstocks play a significant role in affecting scion behavior by regulating vigor, and physiological and biochemical processes such as bud burst, photosynthesis rate, and fruit quality, thus increasing grape yield and quality(Verma *et al.,* 2010). Studies have shown that grafting grapevines to specific rootstocks influences various scion characteristics, including mineral content (Migicovsky *et al.,* 2019; Gautier *et al.,* 2020). Ghule *et al.* (2019) reported superior performance of Dogridge and 1103P rootstocks under semi-arid conditions. Marin *et al.* (2019) observed that cv. Syrah and Tempranillo grafted onto the 3309C rootstock demonstrated the highest vigor. Rootstocks also influence biotic resistance through the accumulation of phytoalexins and polyphenols and enhance drought tolerance by increasing water usage efficiency and osmolytes. The following table shows varieties and suitable rootstock for various conditions (Table 12.4).

Orchard Establishment

Since grapevines remain in the field for 12-14 years, careful consideration should be given to proper planting, especially concerning the spacing between vines and rows. The layout of vineyards is crucial and is typically planned based on several factors including the soil type, climate conditions, and training system.

Table 12.4: Suitable Rootstock for different Varieties

Sl.No.	Variety	Suitable Rootstock	Reference
1.	Anab-e-Shahi	1616 and Dogridge	(Prakash and Reddy, 1990; Verma *et al.*, 2023)
2.	Tas-A-Ganesh	110 R	(Upadhyay, 2006; Somkuwar, Satisha and Ramteke, 2010)
3.	Pusa Urvashi	Dogridge	(Verma *et al.*, 2010)
4.	Sonaka	Dogridge	(Kathe *et al.*, 2011; Shelke *et al.*, 2019)
5.	Thompson seedless	110R and Dogridge	(Upadhyay *et al.*, 2013; Jogaiah *et al.*, 2014)
6.	Fantasy seedless	110R and Dogridge	(Somkuwar *et al.*, 2015;)
7.	Nana saheb purpule	1103P and Dogridge	(Ghule *et al.*, 2019)
8.	Manjari Naveen, Crimson seedless	1103P and Dogridge	(Ghule *et al.*, 2019; Thutte *et al.*, 2024)
9.	Superior seedless	Salt Creek,1103P	(Elaidy *et al.*, 2019)
10.	Manjari Medika, Manjari Kishmish	110R and 140 Ru	(Ghule *et al.*, 2019; Karagatiya *et al.*, 2023)
11.	Cabernet Sauvignon	Dogridge and 420A	(Gautier *et al.*, 2020)
12.	Red Globe	Salt Creek and Dogridge	(Ghule *et al.*, 2021), (Taha and Arafat, 2024)

Climate and Soil

Grapevines thrive in sunny conditions, requiring an optimal light intensity of around 2,400-foot candles for ideal growth. However, low light levels during fruit bud development stages can negatively impact both yield and quality. Conversely, excessive humidity can lead to rampant vegetative growth, reducing yield and quality while inviting increased pest and disease pressure (Radha and Mathew, 2007). Untimely rains, particularly during blooming and ripening stages, can hinder fruit set and lead to crop spoilage. A temperature range of 15°C to 35°C, coupled with annual rainfall up to 900 mm, is conducive to grapevine development.

Regarding soil requirements, the ideal substrate for vine growth is sandy loam, offering adequate water retention capacity and maintaining a pH level between 6.5 and 8.0 (Satisha *et al.*, 2006 Sood, 2008). Sustainable viticulture also necessitates soil with an electrical conductivity not exceeding 4 mhos/cm and an exchangeable sodium content lower than 15 per cent. Exceeding a free calcium concentration value of 12 per cent can result in severe iron deficiency (Chadha, 2019).

Land Preparation

For vineyard establishment, land should be leveled using a tractor, and major weeds like Cyanadon dactilon and nutgrass (*Cyperus rotundus*) must be eradicated using herbicides (Ramteke *et al.,* 2012). Trench opening should begin from December to January to utilize rainless days for easier operations. Trenches should be approximately 75 cm deep and wide, and 200 to 250 feet long, allowing for better root penetration and soil aeration (Somkuwar and Adsul, 2004). The top 30 cm of soil should be set aside and the trench exposed to sunlight for 10-15 days before filling. Fill the trench with topsoil up to 2 feet, then add green manure crops like Sunhemp and Dhaincha, followed by well-decomposed farmyard manure (20 kg) and a nutrient mixture per running foot of the trench. Finally, cover the trench with the remaining soil and level the surface (Somkuwar *et al.,* 2015).

Planting

The optimum time for planting rooted rootstock cuttings is February to March, during this time temperature starts rising above 15°C, which is congenial for vegetative growth (Somkuwar *et al.,* 2015) and cuttings get desired shoot thickness for grafting during September. For planting, the pits of 30 cm x 30 cm x 30 cm should be dug at suitable spacing, and plant root cuttings in the center of the pit. In the case of polybag plants, care should be taken while removing the plant from the polybags so that the roots are not cut with a sharp blade. The new growth starts after 10 to 15 days of planting and plants should be irrigated at suitable intervals.

Spacing

Maximum vineyard productivity is attained when most of the available sunlight is intercepted by grapevine leaves. The vigour in heavy black cotton soil is more than the in light soil. Hence, to accommodate the vigour, 10 feet distance between the rows and 6 feet between the vines is considered optimum. However, considering the moderate vigour in light soils, the distance maintained as 9-feet between the rows and 5-feet between the vines is sufficient (Somkuwar, 2008).

Table 12.5: Spacing for Planting is Based on the Purpose of Grapes, Soil Type, and Local Environment

Soil Type	*Humid Climate*		*Dry Climate*	
	Table Variety	*Wine Variety*	*Table Variety*	*Wine Variety*
Heavy	10'x6'	8'x5'	9'x5'	8'x4'
Medium	9'x6'	8'x4'	8'x5'	8'x3'
Light	8'x5'	6'x4'	8'x4'	6'x3'

To determine the number of plants required per acre based on spacing, it is essential to consider the total area and the specified planting distances. Here's how to calculate it:

- ☆ One acre equals 4,000 square meters or 43,560 square feet.

For a spacing of 10 feet by 6 feet, the calculation is as follows:

For example: One acre = 4000 sq. mtr or 43560 sq. ft.

The planting material requirement for a spacing of 10' X 6' in a one-acre area will be

43560/60 = 726 plants

Therefore, for a spacing of 10 feet by 6 feet, you will need 726 plants per acre

Training Systems in Vineyards

Training grapevines serve various purposes. Firstly, it allows for the arrangement of perennial wood and canes to manipulate the exposure of leaf area, maximizing light interception. This manipulation leads to higher yield potential, optimized leaf area-to-fruit ratio, improved quality, and better disease control (Ausari *et al.,* 2024). Secondly, distributing bearing units on a trellis facilitates the movement of equipment through the vineyard and enables the mechanization of vineyard operations. Thirdly, proper disposal of trunks and canes prevents competition for light between vines (Wolf *et al.,* 2008). Fourthly, effective training can establish a renewal zone, ensuring the perpetuation of vine form and maintenance of yield (Geller *et al.,* 2018). The choice of training system should adequately meet all these objectives, considering the specific site, cultivar, growth habit, winter hardiness, fruitfulness of base buds, and adaptability to mechanization (Table 12.6).

Table 12.6: Different Training Systems Used in Grape Cultivation

Sl.No.	*Training System*	*Remark*
1.	Bower/Pergola system	Suited for vigorous cultivars
2.	Kniffin system	Suited for moderately and less vigorous cultivars with less apical dominance, closer planting adopted
3.	Telephone Trellis system	Suited for moderately vigorous with more apical dominance
4.	Head system	Suited for cultivars producing fruitful shoots from the basal bud, very less vigorous
5.	Y Trellis system	The foliage and bunches are protected from sunburn due to good light interception and good fruit bud growth. Commercially use table grape varieties like Thompson Seedless and its mutants, such as Tas-A-Ganesh, Sonaka, and Clone 2A.
6.	Flat Roof Gable system	Best suited for vigorously developing vines where the bunches hang below the canopy and are protected from direct sunlight while the shoots receive sunlight exposure for the production of fruit buds

Source: Somkuwar, 2008; Mallick and Sahoo, 2022.

Grapes are cultivated for various purposes such as table grapes, raisins, and wine (Somkuwar and Gawande, 2022). The genetic makeup of the vine determines its vigor, which in turn influences the training system used. Commercially grown table grape varieties like Thompson Seedless and its mutants, such as Tas-A-Ganesh, Sonaka, and Clone 2A, are generally vigorous, especially when grown in heavy soil (Somkuwar *et al.,* 2018; Naik, 2018). To manage this vigor, training systems like the Y-trellis or flat roof gable system can be employed. Conversely, wine grape varieties tend to exhibit less vigor. To achieve an economical yield, these vines should be spaced more closely and trained using systems like the T-trellis, single wire, or Kniffin system.

Sunlight is crucial for photosynthesis until the veraison stage, influencing canopy strength. After veraison, bunches are shaded to prevent sun damage. Row direction is vital; an East-West orientation leads to increased berry scorching (Jogaiah, 2022). For flat roof gable training, North-South rows are ideal, while East-West rows suit bower-trained vines. Y trellises should be North-South for, optimizing sunlight exposure and reducing bunch sunburn in hot areas. (Minnaar *et al.,* 2022)

Presently vines are generally trained on extended Y (Figure 12.1); flat roof gable and bower systems (Figure 12.2) while, the kniffin and telephone systems have also been used in the past (Mallick and Sahoo, 2022; Sharma *et al.,* 2020).

Figure 12.1: Y Trellis System.

Figure 12.2: Bower System.

Training in New Vine

Training young grapevines to establish a vine compatible with the chosen trellis system is the primary goal (Jogaiah, 2023). In the first year, the focus is on developing a strong root system and trunk, which can be achieved through either the untrained or trained method on the other hand, involves a more proactive approach to trunk development. Here, the best shoot is carefully selected and supported, while other shoots are pruned away, concentrating the vine's energy on a single, strong trunk. Post-planting, only the two best basal buds are retained to further concentrate growth. This approach to training in the first year aims to establish a solid foundation for the vine, whether through allowing natural growth or directing it through the trained method in the second year, the selected shoots are trained along the wire to form the desired cordon or cane structures (Reynolds and Heuvel, 2009). The third year involves continued training and shoot management to properly establish the fruit-bearing structures, ensuring that the cordons or canes are positioned as intended (Keller, 2015)

Canopy Management

A well-trained grapevine should meet several key criteria to ensure optimal growth and yield. (Sharma *et al.,* 2020). Firstly, it should have a desirable shape and support the crop load effectively. Secondly, it should facilitate convenient field operations and mechanization. Additionally, the vine should have an adequate number of fruitful canes. During the growing season, sufficient light and ventilation should penetrate the canopy. Maximizing the proportion of quality grapes per unit area is essential (Somkuwar *et al.,* 2022) Adequate foliar coverage is necessary to nourish and protect the bunches during the fruiting season, which spans from November to March. Overlapping foliage should be avoided to ensure efficient photosynthesis for every leaf. The vine should also allow for effective spray coverage with pesticides and growth regulators. Finally, it should prevent the development of a micro-climate conducive to disease.

Pruning

Pruning is a critical aspect of grape production (Gutierrez-Gamboa *et al.,* 2021) requiring standardization of severity for each grape variety to ensure optimal quality and quantity (Champa, 2015). In the tropical climate of India, particularly in regions such as Maharashtra, Andhra Pradesh, and North Karnataka, grapevines undergo two annual pruning sessions. Foundation pruning occurs in April, and fruit pruning takes place in October (Kumar *et al.,* 2017). These pruning practices involve the careful removal of canes, shoots, leaves, and other vegetative components (Adsule *et al.,* 2013).

Foundation Pruning (Back Pruning)

Foundation pruning in grapevine cultivation is crucially timed for the first week of April to optimize uniform and early bud sprouting, taking advantage of ideal temperature and humidity conditions. The following practices are implemented before and after foundation pruning:

Bud Break

During the summer, foundation pruning is carried out when temperatures exceed 30-40°C and humidity drops below 30 per cent. These conditions often result in erratic bud break or delayed bud sprouting. To ensure uniform and early bud sprouting, Hydrogen Cyanamide is recommended (Kokare *et al.,* 2024). Bud sprouting is initiated after 12-14 days. The application rates for Hydrogen Cyanamide based on vineyard age and temperature are given in Table 12.7.

Table 12.7: Application Rates for Hydrogen Cyanamide based on Vineyard Age and Temperature

Type of Vineyard	*Temperature in the Vineyard*	*Hydrogen Cyanamide Application Rate (mL/L)*
New vineyard (1-3 years)	30-40°C	30 mL/L
	More than 40°C	25 mL/L
Old vineyard (4-8 years	30-40°C	30 mL/L
	More than 40°C	25 mL/L
Old vineyard (more than 8 years	30-40°C	25 mL/L
	More than 40°C	20 mL/L

Somkuwar and Ramteke, 2007.

Vegetative Growth (1- 40 days)

In the vegetative growth stage, spanning from 1 to 40 days post-pruning. Excess, double, and downward-growing shoots are removed to regulate growth.

Shoot Thinning

This removal of excess shoots is a crucial canopy management practice, as it regulates vegetative growth and ensures a well-structured vine. Each vine typically produces more than 70 shoots of varying diameters; retaining all these shoots would result in a dense canopy and create shading effects, reducing photosynthesis activity. Thus, careful shoot thinning is essential to maintain vine health and optimize grape quality. Shoot thinning should be done at the 6-7 leaf stage to prevent nutrient loss from excessive shoots (Somkuwar, 2008). The number of shoots retained should match the vine's spacing, typically 0.5 to 0.6 shoots per square foot (Somkuwar and Satisha, 2013). Priority should be given to removing highly vigorous, weak, and downward-growing shoots, ensuring only uniform, healthy, and straight-growing shoots are retained to optimize vine growth.

Sub-cane Development

Sub-cane development is encouraged under conditions of high vine vigor. When new growth reaches the 9 leaf stage, it is pinched back to 7 leaves. This practice promotes the development of one or two lateral shoots, depending on the main shoot's vigor (Chougule, 2004). These lateral shoots are further pinched or topped, leaving only 5 leaves, resulting is known as a single sub-cane. Each axillary shoot's basal bud, stimulated by pinching, becomes fruitful. This method not only enhances bud fruitfulness but also provides clear guidance for accurate forward pruning, reducing uncertainty about flower cluster emergence from sprouting buds. However, vine vigor can sometimes be compromised due to factors like inadequate irrigation, nutrient deficiency, high temperatures during vegetative growth. In such cases, instead of sub-cane development, a straight cane approach is adopted. The main shoot is tipped when it reaches the 10-11 leaf stage.(Mallick and Sahoo, 2022; Somkuwar, 2008).

Fruit Bud Differentiation (40-70 days)

Fruit bud differentiation in grapevines typically occurs between 40 to 70 days after pruning. Pruning in April is particularly beneficial as it enhances fruit bud formation and reduces susceptibility to diseases (Somkuwar *et al.,* 2022). Factors such as shoot vigor, light exposure, and temperature play significant roles in this process. Higher shoot vigor, characterized by longer internodal lengths and greater weight, often leads to reduced productivity. Adequate sunlight and optimal temperatures are crucial for maximizing bud fruitfulness. In addition, the use of plant growth regulators (PGRs) like 6-BA and Uracil is recommended during this stage. Specifically, a foliar spray of 6-BA at 10 ppm on the 40th and 50th days after foundation pruning helps increase the cytokinin to gibberellin ratio. Similarly, applying a foliar spray of uracil at 50 ppm on the 45th day after foundation pruning aids in enhancing nucleic acid levels necessary for effective fruit bud development (Yadav *et al.,* 2022). To manage vine vigor at this stage, phosphorus application is encouraged while nitrogenous fertilizer is completely withdrawn. Stress is induced in the vine by reducing irrigation water by approximately 50 per cent compared to the amount applied during vegetative growth (Somkuwar, 2023).

Cane Maturity (61-90 days)

The stage of cane maturity typically begins 60 days after foundation pruning and completes within 90 days, influenced by factors such as soil type, weather conditions, and irrigation. During this period, the cane transitions externally from purple to milky white, and finally to brown, while the pith darkens, indicating full maturity. Immature canes fail to produce quality bunches and often convert to foliage (Thokeet *et al.,* 2024) This structured approach to managing grapevine pruning and growth stages ensures healthy vine development and optimal fruit production. To promote cane maturity, it's important to control vine growth during this stage. Reducing irrigation and applying potassic fertilizer help manage growth. Additionally, tying shoots onto foliage wires prevents microclimate buildup in the canopy, thereby reducing the risk of downy mildew. A spray of Bordeaux mixture at 1.0 per cent during this stage can control diseases and aid in advancing cane maturity. In conditions of continuous rainfall, anthracnose disease becomes active. This fungus infects leaves, causing pinpricks on new tender leaves and shoots. In severe cases, the fungus can enter growing shoots. Therefore, priority should be given to removing new shoots to prevent disease spread. (Abdel Mohsen, 2013)

Fruit Pruning

The optimal period for fruit pruning in tropical regions is from October 10th to November 10th, as this timing helps ensure a dry period during the berry maturation phase (Sharma *et al.,* 2016; Somkuwar *et al.,* 2022). Pruning during October helps mitigate the risk of berry cracking and rotting. The various practices carried out before and after fruit pruning are detailed in the subsequent sections.

Before Fruit Pruning

Grapevines take approximately 140-150 days to reach cane maturity after foundation pruning. Following cane maturity, defoliation should occur approximately 15 days before fruit pruning to induce leaf drop and stress the vine. This stress is further intensified by stopping the water supply in the vineyard, restricting vine growth, and aiding the maturation of remaining green canes (Somkuwar and Satisha, 2013). As canes mature, they accumulate lignin, which enhances nutrient storage. Depending on climatic conditions, leaf removal in grafted vines should occur 12-14 days before fruit pruning and 6-8 days prior in own-rooted vines, facilitating early and uniform bud sprouting (Elmer and Michailides, 2007; Reynier, 2012).

Fruit Pruning Practice

Applying Hydrogen Cyanamide directly to the apical 2-3 buds has been demonstrated to effectively stimulate early and uniform bud sprouting (Kokare *et al.,* 2024). Swabbing the buds once with 1.5 to 4.0 per cent (active ingredient) of Hydrogen Cyanamide within 24-48 hours after forward pruning promotes early and uniform bud break, considering cane thickness and current temperature conditions.

Increasing Distance between Rachis

Bunches typically begin to appear at the 3-leaf stage after sprouting. Achieving loose bunches is crucial, and one method to facilitate this is by increasing the distance between the rachises. This approach not only helps in reducing the need for berry thinning after fruit set but also enhances overall grape quality (Somkuwar and Ramteke, 2007). Application of gibberellic acid (GA) during the pre-bloom stage, when the flower panicle is at a parrot green color, promotes cell elongation, thereby increasing the distance between rachises.

Berry Thinning

A typical grape bunch contains approximately 600 to 700 berries. Retaining all these berries can lead to intense competition for resources such as fertilizers, water, and light. This competition often results in smaller berry sizes that may not meet the standards required by the importing market (Khalil *et al.,* 2024). Therefore, berry thinning is essential to achieve optimal bunch size and quality. Generally, to achieve a bunch weighing 450-500 g with berries of 18 mm diameter and weighing 3.5-4.5 g each, about 100-120 berries are sufficient. Thinning excess berries when they reach 4-6 mm in size helps in effectively utilizing available resources and promoting better berry development. (Shikhamany *et al.,* 2016).

Crop Load per Vine

To optimize grape quality, the crop load per vine after forward pruning should align with grape utilization goals, whether for local markets, export, raisin production, or juice processing (Table 12.8). Vine capacity, defined as producing maximum clusters without delaying maturity or compromising grape quality, depends on factors such as variety, canopy structure, climate, and resource availability (Dry *et al.,* 2004; Somkuwar *et al.,* 2024). Bunch thinning should only occur during the pre-bloom stage. Bunch retention for high-quality fruit production is primarily determined by vine spacing (Somkuwar and Ramteke, 2006). For optimal growth, table grapes require 1 bunch per 1.0-1.5 square feet of spacing, raisin grapes need 2.0-2.25 bunches per square foot, and juice grapes should have 1.3-1.5 bunches per square foot.

Table 12.8: Training System and Pruning Intensity in different Cultivars

Name of Cultivar	*Planting Distance (m)*	*Training System*	*No. of Canes*	*No of Buds per Cane*	*Reference*
Thompson seedless	3×3	Telephone	16-20 cane/vine	8	(Bhujbal, 1972)
Perlette	3×3	Head, Kniffin	30-40 cane/vine	3-4	(Anonymous, 2002)
Himrod	3×3	Kniffin	15-20 cane/vine	6-8	(Anonymous, 2002)
Pusa Narang	-	Head system	12 cane/ vine	4	(Palanichamy, *et al.,* 2004)
Red wine grape (Cabernet Sauvignon, Cabernet Franc, Pinot Noir, Pinot Meunier, Syrah and Grenache	3×1.5	-	25 cane/ vine	4	(Chalak *et al.,* 2011)
H-516 (Catawba x Beauty seedless)	3×3	Bower system	-	4	(Singh *et al.,* 2014)
Red Globe	3×2.5	Bower system	15 cane/ vine	6	(Porik *et al.,* 2017)
Beauty seedless	3×3	Bower system	60-80 cane/vine	4	(Meena and Sarolia, 2017)
Anab-e-Shahi	3×6	Arbour	60-80 cane/vine	5-6	(Meena and Sarolia, 2017)
Superior	2×3	Y system	-	8	(Khamis *et al.,* 2017)

REFERENCES

Adsule PG, Yadav DS, Upadhyay A, Satisha J and Sharma AK (Eds), 2013. Good Agricultural Practices for Production of Quality Table Grapes, pp 57. Pub. by Director, ICAR–National Research Centre for Grapes, Pune.

Ahmed S, Jaffar MA, Ali N, Ramzan M and Habib Q, 2017. Effect of naphthalene acetic acid on sprouting and rooting of stem cuttings of grapes. *Science Letters* **5(3)**: 225–32.

Alexander PD, Bragg NC, Meade R, Padelopoulos G and Watts O, 2008. Peat in horticulture and conservation: the UK response to a changing world. *Mires and Peat* **3**: 08.

Amaral U, Bini DA and Martins CR, 2009. Rapid propagation of rootstock of grapevine semi hardwood cuttings in Uuguaiana-RS. *Revista da FZVA* **15(2)**: 85–93.

APEDA (Agricultural and Processed Food Products Export Development Authority). 2023. https: //apeda.gov.in/apedawebsite/SubHead_Products/Grapes.html.

Ausari PK, Soni N, Kanpure RN, Ninama N and Bhandari J, 2023. Effect of indole-3-butyric acid (IBA) on hardwood cutting of grapes (*Vitis vinifera* L.) cv. Pusa Navrang. *International Journal of Environment and Climate Change* **13(12)**: 61–69.

Ausari P, Somkuwar R, Soni N, Gurjar P, Nale R and Thutte A, 2024. Bunch load management in relation to the training system in producing quality grapes. *Journal of Experimental Farming and Agriculture* **19**: 344–50.

Bhujbal BG, 1977. 'Studies on nutrient status of Thompson Seedless variety of grapes (*Vitis vinifera* L.) in western Maharashtra'. Ph.D. Thesis, Mahatma Phule Krishi Vidyapeeth, Rahuri, Maharashtra.

Chadha KL 2006. Handbook of Horticulture, pp 365–8. Indian Council of Agricultural Research (ICAR), New Delhi.

Chalak S, Kulkarni SS, Kolse SV and Garad BV, 2011. Effect of pruning levels in some red wine grape varieties for yield and quality parameters under western Maharashtra conditions. *Ecology, Environment and Conservation* **17**: 87–93.

Coradini DM, da Silva ESB and Korte KP, 2014. Rooting of semi hardwood cuttings of grapevine rootstock treated with indolebutyric acid. *Revista Campo Digital* **9(2)**: 80–85.

Dogan A, Uyak C and Kazankaya A, 2016. Effects of indole-butyric acid doses, different rooting media and cutting thicknesses on rooting ratios and root qualities of 41B, 5 BB and 420A American grapevine rootstocks. *Journal of Applied Biological Sciences* **10**(2): 8–15.

Elaidy AA, Abo-Ogiala AM and Khalf IR, 2019. Effect of different grape rootstocks on the growth, yield and quality of Superior grape under salt stress. *Middle East Journal* **8(1)**: 167–75.

Elmer PAG and Michailides TJ, 2007. Epidemiology of Botrytis cinerea in orchard and vine crops. *Botrytis: Biology, Pathology and Control*, pp. 243–272. Elad Y, Williamson B, Tudzynski P and Delen N (Eds), Springer, Netherlands

Gautier A, Cookson SJ, Lagalle L, Ollat N and Marguerit E, 2020. Influence of the three main genetic backgrounds of grapevine rootstocks on petiolar nutrient concentrations of the scion, with a focus on phosphorus. *Oeno One* **54(1)**: 1–13.

Ghule VS, Ranpise SA, Somkuwar RG, Kulkarni SS, Wagh RS, Naik RM and Nimbalkar CA, 2021. Effect of rootstocks on growth parameters of Red Globe grapevines (*Vitis vinifera* L.). *International Journal of Chemical Studies* **9(1)**: 3483–87.

Ghule VS, Zagade P, Bhor V and Somkuwar R, 2019. Rootstock affects graft success, growth and physiological parameters of grape varieties (*Vitis vinifera* L.). *International Journal of Current Microbiology and Applied Sciences* **8(1)**: 799–805.

Gutierrez-Gamboa G, Zheng W and de Toda FM, 2021. Current viticultural techniques to mitigate the effects of global warming on grape and wine quality: A comprehensive review. *Food Research International* **139**: 109946.

Hiwasa-Tanase K and Ezura H, 2014. Climacteric and non-climacteric ripening. *Fruit Ripening: Physiology, Signalling and Genomics*, pp. 1–14. Nath P, Bouzayen M, Mattoo A K and Pech J C (Eds), CAB International, Wallingford, UK

Jogaiah S, Maske SR and Upadhyay A, 2014. Rootstock induced changes in enzyme activity and biochemical constituents during budbreak in 'Thompson Seedless' grapevine. *Vitis* **53(2)**: 57–64.

Jogaiah S, 2023. Vineyard management strategies in scenario of climate change–A review. *Grape Insight* **1**: 11–22.

Karagatiya FP, Patel S, Parasana JS, Vasava HV, Chaudhari TM, Kanzaria DR and Paramar V, 2023. Adapting fruit crops to climate change: Strengthening resilience and implementing adaptation measures in fruit crops. *Pharma Innovation Journal* **12(7)**: 3159–64.

Kathe DD, Barhate BG and Bade SJ, 2011. Reactions of different grape varieties to rust caused by *Physopella ampelopsidis*. *Journal of Plant Disease Sciences* **6(1)**: 35–38.

Keller M, 2015. The Science of Grapevines: Anatomy and Physiology, 2nd edn, pp. 1–509. Academic Press, USA.

Khalil U, Rajwana IA, Razzaq K, Brecht JK and Sarkhosh A, 2024. The impact of fruit thinning on size and quality of fresh–market muscadine berries. *Journal of the Science of Food and Agriculture* **104(4)**: 2198–2203.

Kishan Rao K, 2004. 'Studies on the propagation of grape rootstocks through hardwood and soft wood cuttings'. MSc. Thesis, Acharya N.G. Ranga Agricultural University, Rajendranagar, Hyderabad.

Kokare HD, Ramteke SD, Varma S, Sawant NV, Mapari AR and Banerjee D, 2024. Enhancing uniform sprouting, yield, and quality in grapevines: The impact of hydrogen cyanamide (50 per cent S.L) treatment. *Journal of Experimental Agriculture International*, **46(5)**: 56–67.

Kose B, Karabulut B and Ceylan K, 2014. Effect of rootstock on grafted grapevine quality. *European Journal of Horticultural Science* **79(4)**: 197–202.

Kumar AR, Parthiban S, Subbiah A and Sangeetha V, 2017. Effect of severity of pruning on yield and quality characters of grapes (*Vitis vinifera* L.): A review. *International Journal of Current Microbiology and Applied Sciences* **6(4)**: 818–35.

Maninderdeep GS and Singh G, 2022. Study of IBA containing rooting powder on root induction behavior of hardwood cutting of grape (*Vitis vinifera* L.). *Indian Journal of Agricultural Research* **1**: 7.

Marin D, García R, Eraso J, Urrestarazu J, Miranda C, Royo JB, Abad FJ and Santesteban LG, 2019. Evaluation of the agronomic performance of 'Syrah' and 'Tempranillo' when grafted on 12 rootstocks. *Vitis* **58**: 1–8.

Meissner G, Athmann ME, Fritz J, Kauer R, Stoll M and Schultz HR, 2019. Conversion to organic and biodynamic viticultural practices: impact on soil, grapevine development and grape quality. *Oeno One* **53(4)**. https: //doi.org/10.20870/oeno-one.2019.53.4.2470

Michel JC, 2009. Physical properties of growing media: State of the art and future challenges. *Acta Horticulturae* **819**: 65–72.

Migicovsky Z, Harris ZN, Klein LL, Li M, Mc Dermaid A, Chitwood DH, Fennell A, Kovacs LG, Kwasniewski M, Londo JP and Ma Q, 2019. Rootstock effects on scion phenotypes in a 'Chambourcin' experimental vineyard. *Horticulture Research* **6**: 1–12.

Naik S, Banerjee K and Dhekney SA, 2023. Quality planting material of grape: need to develop plant certification standards for the Indian grape and wine industry. *International Journal of Agricultural Innovation and Research* **12(1)**: 15–16.

Nascimento DC, Dini M, Moreira RM, Sampaio NV and Schuch MW, 2019. Mini cutting of Chardonnay grapevine and SO4 grapevine rootstock. *Revista Científica Rural* **21(3)**: 270–81.

Palanichamy V, Jindal PC and Singh R, 2004. Studies on severity of pruning in grapes (*Vitis vinifera* L.) cv. Pusa Navrang a teinturier hybrid. *Agricultural Science Digest* **24(2)**: 145–47.

Patil VA and Naik VN, 2018. Texture profile analysis of Sonaka and Thompson Seedless raisins. *Journal of Postharvest Technology* **6(4)**: 75–81.

Porika, H, Jagadeesha M and Suchithra M, 2015. Effect of pruning severity on quality of grapes cv. Red Globe for Summer Season. *Advances in Crop Science and Technology*, S1: 004. DOI: 10.4172/2329-8863.1000S1-004

Possingham JV, 2008. Developments in the production of table grapes, wine and raisins in tropical regions of the world. *Acta Horticulturae* **785**: 45–50.

Prakash GS and Reddy NN, 1990. Effect of different rootstocks on budbreak in grape cv. Anab-e-Shahi. *Crop Research* **3**: 51–55.

Radha T and Mathew L, 2007. Fruit crops. *Horticulture Science Series*–3, pp. 236–270. Peter K V (Ed.), New India Publishing Agency, New Delhi.

Ramteke SDR, Bhange M and Kor R, 2012. Chemical management of broad-leaved weeds in grapes. *Journal of Weed Science and Technology* **44**: 198–202.

Rasool A, Mansoor S, Bhat KM, Hassan GI, Baba TR, Alyemeni MN, Alsahli AA, El-Serehy HA, Paray BA and Ahmad P, 2020. Mechanisms underlying graft union formation and rootstock scion interaction in horticultural plants. *Frontiers in Plant Science* **11**: 590847.

Reynolds AG and Wolf TK, 2008. Grapevine canopy management. *Wine Grape Production Guide for Eastern North America*, pp. 124–134. Wolf T K (Ed.), Natural Resource, Agriculture, and Engineering Service (NRAES-145), Ithaca, New York.

Sahin F, Yildirim E and Sezgin A, 2002. Utilization of sugarcane bagasse and coconut fiber as low-cost planting media. *Agricultural Science Journal* **16(4)**: 215–23.

Satisha J and Adsule PG, 2006. Rooting behavior of grape rootstocks in relation to IBA concentration and biochemical constituents of mother vines. In: International Symposium on Grape Production and Processing, *Acta Horticulturae* **785**: 121–26.

Satisha J, Prakash GS, Murti GSR and Upreti KK, 2006. Response of grape rootstocks to soil moisture stress. *Journal of Horticultural Sciences* **1(1)**: 19–23.

Shah SU, Ayub Q, Hussain I, Khan SM, Ali S, Khan MA, Haq N, Mehmood A, Khan T and Brahmi NC, 2021. Effect of different growing media on survival and growth of grape (*Vitis vinifera*) cuttings. *Journal of Advances in Nutrition Science and Technology* **1**: 117–24.

Sharma AK, Upadhyay AK and Somkuwar RG, 2020. Grape growing: Opportunities for better returns. *Progressive Horticulture*, **52(2)**: 134–43.

Shelke TS, Shikhamany SD, Kalbhor JN and Mungare TS, 2019. Performance of Thompson Seedless grape and its clones on Dogridge rootstock. *International Journal of Horticulture*, **9(1)**: 1-9

Shikhamany SD, Borade SV, Jeughale SK and Patil SY, 2016. Effect of cane regulation and GA_3 spray on berry thinning in 'Thompson Seedless' grape (*Vitis vinifera* L.). *Journal of Horticultural Sciences* **11(2)**: 131–42.

Singh K and Chauhan J, 2020. A review on vegetative propagation of grape (*Vitis vinifera* L.) through cutting. *Global Journal of Bioscience and Biotechnology*, **9(2)**: 50-55.

Somkuwar RG and Adsule PG, 2004. Grape cultivation on rootstock. Technical Bulletin No. 5, Pub. By Director ICAR–National Research Centre for Grapes, Pune, pp. 1-11.

Somkuwar RG and Ramteke SD, 2006. Wedge grafting compatibility in different grape varieties raised on Dogridge rootstock. *Annals of Plant Physiology* **20**: 22–25.

Somkuwar RG and Ramteke SD, 2007. Bunch development through canopy management and growth regulators. Technical Bulletin No. 9, ICAR–NRC Grapes, Pune pp.1-6.

Somkuwar RG and Satisha J, 2013. Foundation pruning in grapes. Technical Bulletin No. 14, ICAR–NRC Grapes, Pune, pp. 1-11.

Somkuwar RG, Bhange M, Sharma J, Upadhyay AK and Khan I, 2017. Interaction of biochemical and nutritional status of nodal sections with rooting success in grape rootstocks. *Journal of Environmental Biology* **38(1)**: 115–21.

Somkuwar RG, Bondage DD, Surange MS and Ramteke SD, 2011. Rooting behaviour, polyphenol oxidase activity, and biochemical changes in grape rootstocks at different growth stages. *Turkish Journal of Agriculture and Forestry* **35(3)**: 281–87.

Somkuwar RG, Sharma AK, Deshmukh NA, Saha S and Rao BV, 2022. Innovative Approaches to Develop Entrepreneurship in Grapes, E-book, Agri Business Incubation Centre of ICAR– National Research Centre for Grapes, Pune and National Institute of Agricultural Extension Management (MANAGE), Hyderabad: 100.

Somkuwar RG, Taware PB, Bhange MA, Sharma J and Khan I, 2015. Influence of different rootstocks on growth, photosynthesis, biochemical composition, and nutrient contents in 'Fantasy Seedless' grapes. *International Journal of Fruit Science*, **15(3)**: 251–66.

Somkuwar RG, 2008. Training, pruning and canopy management. Technical Bulletin No. 9, ICAR–NRC for Grapes, Pune, pp. 1-15.

Somkuwar RG, Ramteke SD, Sawant S and Takawale P, 2018. Canopy modification influences growth, yield, quality, and powdery mildew incidence in Tas-A-Ganesh grapevine. *International Journal of Fruit Science*, **19**: 437–451.

Song YG, Lu WP, Wang J, Shen YJ, Wu ZS and Liu WD, 2001. Study on promoting the rooting ability of hardwood cuttings of Amurien grape varieties. *China Fruits*, **1**: 4–7.

Souza ER, Lenk FL, Ono EO and Rodrigues JD, 2015. Carbohydrate content in vine cuttings of rootstock cv. IAC 572. *Brazilian Journal of Applied Technology for Agricultural Science*, **8(2)**: 7–15.

Taha NA and Arafat LA, 2024. Analysis of the 'Superior Seedless' vines' berries shattering during their shelf-life after being grafted onto several rootstocks. *Journal of Agricultural Chemistry and Biotechnology*, **15(3)**: 41–45.

Thoke S, Patil S, Gollagi S, Hipparagi K, Satyanarayana C, Hiremath V, Peerajade D and Kore D. 2024. Evaluation of different table grape varieties in response to growth and physiological traits during foundation bud pruning. *Journal of Experimental Agriculture International*, **46**: 126–40.

Thutte A, Somkuwar R, Dhemre JK, Gharate PS, Ausari P and Kakade P, 2024. Rootstock impact on quality attributes and shelf life of Crimson Seedless grapes. *International Journal of Advanced Biochemistry Research*, **8**: 556–560.

Upadhyay AK, Sharma J and Satisha J, 2013. Influence of rootstocks on salinity tolerance of Thompson Seedless grapevines. *Journal of Applied Horticulture* **15**(3): 173–77.

Sharma J and Upadhyay AK, 2008. Rootstock effect on Tas-A-Ganesh (*Vitis vinifera* L.) for sodium and chloride uptake. *Acta Horticulturae* **785**: 113–16.

Verma SK, Singh SK and Krishna H, 2010. The effect of certain rootstocks on the grape cultivar 'Pusa Urvashi' (*Vitis vinifera* L.). *International Journal of Fruit Science*, **10(1)**: 16–28.

Chapter 13

Nursery Techniques, Canopy and Orchard Management in Banana

M. Tamil Selvan

ICAR–Krishi Vigyan Kendra, Vamban, Pudukkottai–622303, Tamil Nadu

Banana is one of the oldest fruit known to mankind and also important food for man. It is the cheapest and plentiful fruit. This fruit is the food of sages since ancient time. Banana is the staple food in Uganda, Bukoba and Tanzania. It is the most important world traded tropical fruit. Global export is 15.5 million tones which include the Cavendish varieties like Dwarf Cavendish, Grand Naine, Robusta and Lacatan.

It is also called as Kalpatharu means a plant of Virture, Antique fruit crop, Tree of Wisdom, Tree of Paradise, Adams fig, Apple of paradise. It has the somatic chromosome number 2n=2x=22, 3x=33, 4x=44. Origin: *M. acuminate* = Malaysia, *M.Balbisiana* = Burma (Indo Malayan-South East Asia). Plantains are the cooking banana, humid tropical herb. Banana leaf is considered as a biological plate. India is the largest producer of banana in the world. It is Monocotyledonous, Monocarpic, herbaceous perennial herb, fruit is produced through vegetative parthenocarpy.

India is the largest producer of banana in the world. Even though India is the largest producer, India's share is negligible in the export market. Total area in India is about 0.47 million ha with an annual production of about 15.7 million tonnes (2004). Tamil Nadu, Kerala, Maharashtra, Andhra Pradesh, Assam, Bihar, Gujarat, Karnataka, Orissa, Madhya Pradesh, West Bengal and Himachal Pradesh are the major banana growing states. Rich source of energy (137 K. Ca/100g). It is a good laxative. Banana is the most nourishing fruit, contains vitamins and minerals, and has several medicinal properties. It is a rich source of energy, low in protein and fat contents and provides a balanced diet. The composition of the fruit is given below (Simmonds, 1966). Moisture (70%), Carbohydrate (27%), Crude fibre (0.5%), Protein (1.2%), Fat (0.3%), Ash (0.9%) Phosphorus (290ppm), Calcium (80ppm), Iron (0.6ppm), β-carotene (0.5ppm), Riboflavin (0.5ppm), Niacin (7ppm), Ascorbic acid (120ppm), Energy (Cals/100 g). Edible cultivars of banana are derived from interspecific hybridization between *M.acuminata* (AA) and *M.balbisiana* (BB).

Propagation and Nursery Management

In India, use of conventional suckers cannot be completely avoided because of some regional special varieties tissue culture technique is yet to be perfected. Further for resource poor small scale growers, the cost of tissue culture plants stills a hurdle in adoption. There are three types of planting material used for propagation namely, Suckers, Peepers, Rhizomes.

- ☆ Natural regeneration of cultivated bananas through suckers is very slow due to hormone-mediated apical dominance of the mother plant.
- ☆ A plant produces only 5-20 suckers during its life time of 12-14 months. For accelerating the propagation rate, suckers with growing buds or cut rhizomes called 'bits' and peepers' are used.
- ☆ Several good bits, each with a centrally placed germinating eye can be cut from an un branched rhizome after trimming the roots.
- ☆ Selection of appropriate mother plant for raising new propagules either through *in vivo* or *in vitro* methods is important.
- ☆ About one kilogram uniformly sized rhizomes or bits, well-trimmed around the growing sprout are the best starting material.
- ☆ In some parts of India rhizomes are sun dried for 2-3 days after paring and pralinage treatment (trimmed of all roots, dipped in mud slurry and sprinkled with nematicide) and stored in shade for a week before planting.

Selection of Mother Plants

- ☆ Mother plant should be healthy, true to type and free from diseases and pests,especially virus diseases.
- ☆ The male flowers buds should be retained to check the presence of virus diseases (male flower buds exhibit symptoms of late infection of viruses.
- ☆ Mother plants should be raised under roofless insect proof shade net with sufficient height.
- ☆ Mother nursery must be located away from other banana plantations with an isolation distance of 500 m to maintain purity and to avoid spread of virus diseases.

- ☆ Mother plants should be grown under very good management conditions so as to facilitate the true expression of traits.
- ☆ Individual plants should be tagged with a master code number so that the plantlets developed could be traced back to the mother plant.
- ☆ Pedigree record and source of each mother plant should be maintained and catalogued.
- ☆ Once indexed, the mother suckers can be maintained in field or concrete rings with frequent decapitation to facilitate production of more auxiliary buds. They also serve as explants for culture initiation.

Propagation Material

Suckers

Sword sucker: A sucker which arises from the mother rhizome with sword like (narrow) leaves at an early stage which produce superior fruit (Healthy banana clump) and are recommended for commercial cultivation.

Water sucker: A sucker next to but only superficially attach to the mother rhizome with broad leaves at an early stage. Water suckers produce inferior fruit (not healthy banana clump) and are therefore not recommended.

Peepers: Peepers (very young suckers) produce late and poor crop. Four month old suckers and split rhizomes(each having about 2.0 Kg) produced heavier bunches compared with those obtained from peepers.

Rhizomes: Whole or split rhizomes can also be used when suckers are not available. Bits of rhizome of 2.0 Kg or more may be planted in the nursery for sprouting or directly sown in the main field. For quick multiplication of a variety rhizome bits may be used; though the plants will require little longer time to fruit.

Seed

- ☆ Seed propagation is common in wild species which are diploid and undergo normal meiosis, fertilization and seed set.
- ☆ The extent of seed set, germinability and dormancy depends on the species.
- ☆ In *Ensete*, the only other genus of Musaceae, seed propagation is the only means of perpetuation since sucker production is absent.
- ☆ Seeds are generally brown to black in colour, 2-6 cm in diameter, round or triangular in shape, and mostly compressed in appearance.
- ☆ Fruits of wild species are inedible, being full of seeds that are enveloped in thin mucilaginous pulp.

☆ In India, this propagation system is not followed because all cultivated commercial bananas are parthenocarpic.

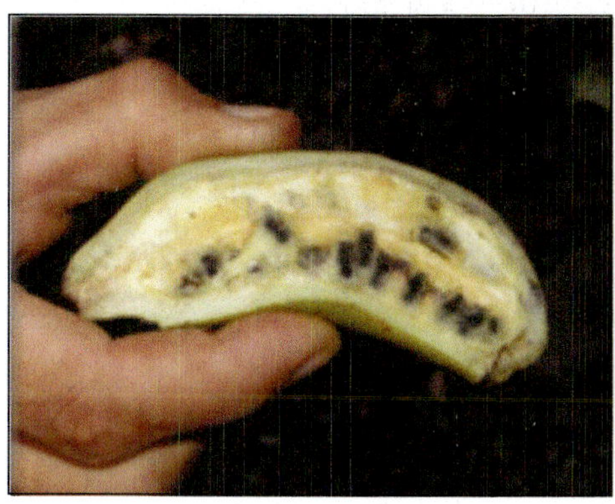

Wild Banana Fruit

Seed

Micro-propagated Plants

Micro-propagation is the practice of rapidly multiplying stock plant material to produce a large number of progeny plants under aseptic conditions, using modern plant tissue culture methods.

Steps involved in the micro-propagation of banana:

1. Selection of pedigree mother plant.
2. Establishment and maintenance of mother block nursery.
3. Selection of superior planting material for in vitro initiation.
4. Rouging at various stages of proliferation.
5. Rooting, primary hardening and rouging.
6. Secondary hardening and rouging off-types.
7. Virus indexing and genetic fidelity testing at various stages of micro propagation.

Propagation Using Suckers

Conventionally bananas are propagated through suckers. There are two types of suckers *viz.*, sword sucker and water sucker. The sword suckers have a well developed base and a pointed tip with narrow sword shaped leaves while the water suckers produce broader leaves at an early stage.

Water Sucker

Sword Sucker

Tissue Culture Banana

Selection of Planting Material

- Sword suckers should be vigorously growing and weighing 1.5 to 2.0 kg
- The suckers should be swollen at the base and tapering towards the top.
- Free from pests, diseases and nematodes.
- They are prepared by trimming off the roots and decayed portion of the corm and cutting the pseudostem leaving 20 cm from the corm.
- The suckers can be graded and grouped based on their size before planting.

Pre-Treatment of Suckers (Paring and Pralinage)

The rhizomes selected from mother plant should be free from pests, diseases and nematodes. The rhizomes are trimmed and then dipped for 5 minutes in 0.2 per cent Carbendazim solution (2g in 1 lit of water). This treatment will help to control Panama wilt. Then the sucker is dipped in clay slurry solution of 4 parts clay plus 5 parts water to have a thick coating of slurry over which 40g of carbofuran 3G granules is sprinkled. This treatment will control nematodes. Alternatively, the corms can also be dipped in with 0.2 per cent Monocrotophos (2ml Monocrotophos 36 WSC/l of water), shade dried for at least 24 hours. The sucker is then planted at the centre of the pit and the soil is pressed around.

Orchard and Canopy Management in Banana

Clean Culture with Cover Crops

This type of soil management involves raising of a cover crop or green manure after removing the weeds. If clean cultivation is attempted during the rains, considerable erosion is almost sure to occur. It is probably best to plant a green manure crop between the trees early in the rains and plough it into the soil towards the end of monsoon season. In India, green manure crops like Sunhemp, Cowpea, Daincha, Lupins *etc.* are more commonly used. Legume cover cropping in grape, mango, guava and other fruit crops is becoming a common practice in the management of orchards. Cowpea and French beans grow well under guava and sapota tree.

In some places to prevent soil erosion, certain permanent cover crops like *Calapogonium muconoides, Centrosema pubescens* and *Peuraria phaseoloides* are raised in the alley spaces. They are leguminous crops, establish in a short period, dry up during summer to conserve moisture. With summer showers they come up again because of their profuse seeding habit and spread themselves as a vegetative mat by the time the heavy monsoon starts pouring in. Such permanent cover cropping is a common feature in rubber plantations of Kerala and Kanyakumari district.

Mulching

This is one of the important soil management practices adopted in certain countries. Crop residues like straw, cotton stalks, leaves, saw dust, pine needles, coir dust arid other materials like polythene films or certain special kinds of paper are spread in the tree basins and in inter spaces between trees. Main objective of mulching is to conserve soil moisture and to control the weed growth.

The other advantages of mulching are:

1. Keeps soil cool in day; warm at night hours
2. Reduces surface run-off
3. Adds humus to the soil
4. Prevents soil erosion

5. Fruits are protected and kept clean since they fall on the mulches
6. It allows the absorption of more rain water and
7. It reduces irrigation frequency.

The following are some of the disadvantages:

1. Dry materials used as mulches encourage the risk of fire and consequent damage to trees.
2. Thick mulches may act as places for mice and rodents to live and multiply. They may cause damage to tree trunks and roots by eating the bark and burrowing to the land. The mulching materials should be placed too close to the tree trunk and it should be spread in such a way that they give a good cover to the root system of the trees.

Sod

In this method, permanent cover of grass is raised in the orchard and no tillage is given. This type of orchard cultivation is followed in USA and Europe. This may be useful in slopy lands for preventing soil erosion. But they compete for soil moisture and available nitrogen. The drawbacks of this system are the need for increased manuring and water application. They are harmful to shallow rooted trees. Hence sod may be useful with deep rooted trees because soil moisture will be very low on the top layers.

Sod Mulch

This is similar to sod with the only difference is that the vegetation is cut frequently and the cut material is allowed to remain on the ground. This is slightly better than the previous one, as the moisture loss is not so great as in sod. In both sod and sod mulch, more nitrogen should be applied to the fruit trees than usual application because the vegetation utilizes more soil nitrogen.

Intercropping

In young orchards, the question of how best one can use the soil between the trees arises. If the trees are properly spaced there is considerable land which will not be used by the permanent trees for several years. Similarly in the case of other long duration horticultural crops like tapioca, turmeric, ginger and banana some area between adjacent plants will be remaining unoccupied by the main crop for few months. It naturally appeals to the grower to get some return from this vacant land especially when he is getting no return in the early periods. The practice of growing any economic crop in alley spaces of the fruit trees in the first few years or in the' unoccupied spaces of the long duration crop in the early periods is referred as intercropping. They also act as a cover crop and the land benefits by the cultivation, irrigation, manuring given to the intercrops.

Intercropping **Intercropping** **Intercropping**

The following important principles should be observed while growing intercrops.

1. Intercrops should not occupy the area where the roots of the fruit trees are concentrated.
2. Soil fertility should be maintained or improved when intercrops are grown.

   Water requirements of the intercrops should not clash with those of the main fruit trees.
3. The intercrop may require an irrigation at a time when it would be detrimental to the trees.
4. Intercrops should be selected with reference to their effect on soil moisture. Grain crops remove excessive moisture to the detriment of fruit trees.
5. The intercrops selected should not exhaust the soil water and nutrients and should not demand more water than is allowed for fruit trees.

Vegetables are the best inter crops when compared to millets. But whatever may be the intercrop grown, it should be kept well away from the main fruit trees and irrigated independently. The intercropping should be stopped when trees occupy the entire orchard space. Thereafter, green manuring or cover cropping should be only practiced.

Many growers prefer some quick growing fruit trees to grow as intercrops. A satisfactory fruits are available for this purpose. In temperate regions peaches are often grown between apple trees. Similarly, in properly spaced mango orchard, guava trees can be planted to bear in two or three years and will produce a number of crops before it is necessary to remove them. Such short-lived trees are known as 'fillers'. Papayas, bananas or phalsa may be well grown as fillers in orchards. The danger in using fillers is when they are allowed to remain in the orchard for too long periods. As normally root system makes a faster growth than the branches, the roots *of* the permanent trees come in contact with the roots *of* the fillers before there is any crowding above ground. Therefore, the fillers should be removed after a few years usually immediately after the main fruit trees have commenced bearing.

The recommended intercrops for some important horticultural crops are given.

Crop	*Age*	*Intercrop*
Mango	Upto 7 years	Leguminous vegetables, Papaya (filler)
Grapes	Upto 8 months	Snake gourd or bitter gourd in pandal
Apple, pears	Upto 5 years	Potato, Cabbage
Banana	Upto 4 months	Sunhemp, onion
Tapioca	Upto 3 months	Onion, beans, lab-lab, black gram
Turmeric	Upto 3 months	Small onion, coriander
Arecanut	Upto 10 years	Pineapple
Coconut	Upto 3 years	Banana, tapioca, vegetables

Mixed Cropping

It refers to the practice *of* growing certain perennial crops in the alley spaces of the main perennial crops. The main advantage is the effective utilization of available area and increase in the net income of the farm per unit area. Extensive research conducted by CPCRI, Kassargode on mixed cropping in coconut and arecanut plantations showed that cocoa, pepper, cinnamon, clove and nutmeg can be grown as mixed crops in coconuts while nutmeg and clove as mixed crops in between four arecanut palms on alternate rows. In all the above cases, increase in yield (upto 10%) is obtained in the main crop due to the synergistic effect of the crop combinations arising out of beneficial micro- organisms in the rhizosphere and the more availability of major nutrients in the active root zone of the crop mix as compared to the pure stand.

Mixed Cropping

Multitier System of Cropping

Certain horticultural plants like coconut and arecanut are grown for about 50 years in a particular land. It takes nearly 4 to 7 years for the above trees to reach the bearing stage. Adequate alley spaces (nearly 75%) are available in between these trees and being the palm trees, their root system will not also spread beyond one metre in diameter.

Hence, these vacant spaces can be profitably used for raising other crops, thereby increasing the employment opportunities and profit. This is the chief objective of the multitier system of cropping. Intercropping and mixed cropping involve jointly multitier system of cropping and is defined as a compatible companion of crops having varying morphological frames and rooting habits, grown together in such a manner that their canopies intetercept solar energy at varying heights and their roots forage the soil at different zones. The main principle here is that the land, water and sunlight should be effectively used.

An ideal combination of crops for multitier cropping in coconut and arecanut plantations is as follows.

Tier	*Crop*
First (Top)	Coconut or arecanut
Second	Pepper trained over the trunk of coconut or arecanut trees
Third	Cocoa or cloves planted at the centre of four arecanut or coconut
Fourth (ground)	Pineapple, ginger and dwarf coffee

Multistorey System

Crop Rotation

- Crop rotation is the growing of two or more crops one after the other in the same piece of land.

- It is advantageous that the succeeding crop belongs to a family different from that of the previous crop.
- The period of crop rotation may be for two to three years or longer.
- Farm income significantly increased
- Better control of weeds
- Better control of pests and diseases
- It improves soil structure, soil fertility and organic matter content
- Banana may be rotated with paddy, sugarcane, pulses, vegetables *etc.*

Special Cultural Practices

Bunch covers: During the development stages, the surface of the fingers may get bruised or damaged due to their exposure. Use of transparent polythene sleeves provided with 2 per cent (during cool season) and 4 per cent (during summer season) ventilation to cover the bunches can be helpful to avoid such damages. Besides, reducing the blemishes, it improves bunch appeal, advances the maturity by 7-10 days in banana. The right stage of bunch covering is when the last hand has opened and the male bud has to be removed at that time. The top end of the sleeves should alone be tied with the peduncle leaving the bottom open for better aeration.

Removal of male bud (Denavelling): The male bud along with the removal of withered styles 8-11 days after bunch emergence along with the removal of withered styles is recommended to favour fruit development.

Removal of withered styles and perianth: In some varieties the styles are persistent and hence they remain in the bunch till maturity. They can be easily removed by a light brushing movement of the hand a few days after flowering and if it is delayed, it is difficult to remove and later, they become brown and shriveled. This operation eliminates the removal of infection by saprophyte fungi especially to avoid the fingertip disease.

Application of growth regulator: To improve the grade of bunch -2, 4-D at 25 ppm (25 mg/lit) may be sprayed after the last hand has opened. This also helps to remove the seediness in Poovan variety. Spray CCC 1000 ppm of 4th and 6th month after planting. Spray plantozyme @ 2ml/lit at 6th and 8th month after planting to get higher yield.

Micronutrients: Spray micronutrients *viz.*, $ZnSO_4$ (0.5%), $FeSO_4$ (0.2%), $CuSO_4$ (0.2%) and H_3BO_3 (0.1%) at 3rd, 5th and 7th MAP to increase yield and quality of banana.

Mattacking: After harvesting, the pseudostem should be cut leaving a stump of about 0.6 m height. This practice is called 'mattacking'. The food material stored in the left out stump continues to nourish the daughter sucker till it withers and dries up.

Ratoon Crop

Bananas are perennials which produce succeeding generations of crops. The first cycle after planting is called the plant crop. The ratoon is the sucker (also called the follower) succeeding the harvested plant. The second cycle is called the first ratoon crop. The third cycle is the second ratoon crop, and so on.

Ratoon Management

Keeping the field and neighborhood free of weeds generally is advised to avoid spread of infestations. Though the TC plants, which are supplied are healthy and virus free, field contamination can be avoided only by keeping the generally field free of weeds. Ratoon management is important for sustained income. Some places in Tamil Nadu, three ratoons have been done in 24-25 months. Once the flowers emerge and all the fruits have fully opened, the bunches are covered using a skirting bag which prevents the tender fruits from insect attack, besides maintaining uniform temperature inside. This is important for colour and fruit development.

Plant Canopy and Light Intersection

- ✰ Unlike other fruits, the vegetative growth, flowering and fruit growth is not seasonal in banana and are largely influenced by time of planting, type and size of planting material and prevailing temperature.
- ✰ Planting density and their interception. Reduced light intensity at ground level with increase in size energy conversion efficiency was maximum in 1.2 x 1.2m spacing and minimum in 2.1 x 2.1m spacing.

Light Intersection.

Plant Height and Girth

- ✰ Height of pseudostem is invariably increased with reduction in spacing.
- ✰ Plant height is adversely influenced, depending upon cultivators.

- ✰ Stem girth is reduced with height on increase in density.
- ✰ Girth of Robusta banana was not influenced by varying plant density although tallest plants were produced under closer spacing of 1.2 x 1.2 m.
- ✰ In poovan cultivar, plant height was significantly increased and girth reduced when spacing was brought down from 2.1 x 2.2 m to 1.5 x 1.8 m.

Plant Height and Girth

Leaf Number and Leaf Area Index

- ✰ Canopy characteristics such as leaf area index (LAI) and transmission of photosyntheically Active radiation (PAR) can be used to correlate with optimum density.
- ✰ Leaf emergence is reduced under very close planting owing to lower temperature inside the canopy since temperature had significant influence on rate of leaf emergence.

Leaf Number

Sucker Production and Root Growth

- ✰ The number of suckers per plant was more in wider spacing compared to that in closer to that in closer spacing.
- ✰ The suckers per plant decreased with reduction in planting distance with reduction in planting distance in Robusta and Poovan.
- ✰ The length of fine roots increased with increase in plant density to shooting and declined thereafter.

Sucker Production and Root Growth

Weed Growth

- ✰ High density planting (4440 to 6950 plants/ha) coupled with initial check on weed population through intercropping of cowpea caused reduction in weed growth.
- ✰ Under high density planting, at all stages of plant growth, weed growth was minimized and cowpea intercrop also improved the soil fertility besides suppressing weed growth.

REFERENCES

Baker S, 1959. Bananas in East Africa. *Journal Experiment Agriculture*. **20**: 66-76.

Bhande S and Kurien S, 2016. Sucker production in banana. *Journal in Tropical Agriculture* **53(2)**: 97-106.

Bose TK, Parthasarathy VA, Mitra SK, Chakraborty Ghosh B and Sanyal D, 2022. Fruits: Tropical and Subtropical Vol 4. 4th Revised and Illustrated edn. ISBN 10: 9390435463/ISBN 13: 9789390435463

Bose TK, Mitra SK, Sadhu MK, Das P, D Sanyal D and Parthasarathy VA, 2005. Propagation of Tropical and Subtropical Horticultural Crops Vol: I: *Advances in Propagation Techniques of Banana.* ISBN – 10: 8185971366. Naya Udyog Publications.

Manzur MD, 2001. In situ mass propagation of the FHIA-20 banana hybrid using benzylaminopurine, *Info Musa.* **10**: 3-4.

Singh HP, Uma S, Selvarajan R and Karihaloo JL, 2011. Micro-propagation for Production of quality banana planting material in Asia-Pacific, *Asia-Pacific Consortium on Agricultural Biotechnology.* New Delhi, India.

Vuylsteke DR and Ortiz R, 1996. Field performance of conventional vs. in vitro propagules of plantain (Musa spp., AAB Group), *Horticultural Science.* **31**: 862–865.

Uma S, Lakshmi S, Saraswati MS, Akbar A, Mustaffa MM, 2011. Embryo rescue and plant regeneration in banana (Musa spp.). *Plant cell, Tissue and Organ Culture (PCTOC).* **105(1)**: 105-111.

Chapter 14

Orchard and Canopy Management in Pomegranate

R.A. Marathe and Chandrakant Awachare

ICAR–National Research Centre on Pomegranate, Kegaon, Solapur–413255, Maharashtra

Pomegranate, *Punica granatum* L., is one of the ancient and important cultivated fruit crop and has re-emerged in contemporary culture because of its countless health benefits. It is native to dry hot regions of Persia and surrounding areas of Afghanistan and Baluchistan (Holland, 2009). Globally, India is the *numero uno* country in pomegranate acreage (2.76 Lakh ha.) and production (30.78 Lakh tones), contributing to more than 40 per cent global share followed by China (1.6 million tonnes), Iran (0.7 million tonnes), and Turkey (0.5 million tonnes) (Ferrara *et al.,* 2021). In India, Maharashtra state is considered as Pomegranate Basket of India, contributing more 65 per cent of total area and production followed by Karnataka and Andhra Pradesh and is fast picking up in Gujarat, Rajasthan, Himachal Pradesh and Madhya Pradesh. Pomegranate has gained unique credits in last few years due to its versatile adaptability, immense medicinal and nutritional properties, built-in-ability to tolerate heat and drought, low resource input demanding nature and alluring monetary returns per unit area, which has resulted remarkable increase in area and production (Singh *et al.,* 2019). This fruit has growing consumer demand both in the domestic as well as international market owing to its attractive, juicy and refreshing arils and immense nutritional profile.

The present pomegranate production is confronted with several biotic and abiotic factors, however, besides these pre-orchard planting information about the crop, orchard planning, area suitability, resource availability, agro-ecological factors like climatic conditions and soil types and other factors like canopy management are important considerations for successful pomegranate production.

Nursery Management

Pomegranate is considered as highly remunerative fruit crop for rocky areas due to its marvellous yield even under drought and edaphic stresses. This crop is facing huge losses both in terms of production and quality due poor quality and infected planting material. Use of unhealthy planting material (having latent/symptomless blight infection) results in spread of bacterial blight in new areas where blight was not prevalent earlier resulting into declining of economic longevity of orchards. The intensification of production system and introduction of improved varieties have led to the occurrence of diseases especially, bacterial blight in serious proportions resulting in 30-50 per cent yield losses, sometimes even goes up to the range of 80-100 per cent under favorable environmental condition. Presently, the disease has become a threat to pomegranate productivity, badly affecting production in all the three major pomegranate growing states *viz.* Maharashtra, Karnataka and Telangana. Though, the disease is an air borne but infected planting material (carrying latent infection) is a major source of spread of this disease in many parts of the India. This situation demands an urgent need to produce healthy planting material by hygienically maintaining the mother plants under protected structures.

Moreover, average annual requirement of planting materials are increasing day by day which demands an urgent need for utilizing conventional (layering and cutting) and non - conventional methods (tissue culture, budding and grafting) to meet the requisite demand of healthy planting materials.

Preparation of Quality Planting Material

Pomegranate is commercially propagated through stem cutting, air layering and recently by *in vitro* propagation. Air-layering is commonly used for propagation of pomegranate in the Deccan Plateau and hardwood cutting in the rest of India. Few Indian firms have come up with the tissue culture raised pomegranate planting material but endeavour is still on to produce biohardened and better field performing pomegranate plants using tissue culture technique in an economic manner. Recently, systematic efforts for standardization of grafting technique and identification of suitable rootstocks to overcome problems of biotic and abiotic stresses have been initiated by this Centre.

In vitro Propagation

There is a growing demand of pomegranate planting material in the country which may not be fulfilled only by conventional methods of propagation, thus,

Figure 14.1: ***In Vitro*** **Propagation in Pomegranate.**

mass multiplication of pomegranate through tissue culture is the need of hour. These *in vitro* raised plants are initially free from any kind of infection. The *in vitro* raised plants of pomegranate should be exactly similar to the mother plant from where explants had been excised provided multiplication cycles are judiciously standardized with optimum level of growth regulators and there is non-occurrence of somaclonal variations which is a rare phenomenon in case where explant is nodal segment or shoot tip. It is well established that the tissue cultured plants have synchronized flowering and are highly amenable for mechanized cultivation. Besides, these plants have precocity in bearing, higher uniformity and better yield and fruit quality compared with conventionally propagated plants.

Recommendations for Propagation through Tissue Culture

- ☆ Mother plant with proven horticultural trait, obtained from reliable source and in healthy state should be used for excision of explants
- ☆ It should be ensured that the mother plants should be kept under protected structures/insect proof shade net and monitored regularly for any visible symptoms of infection.
- ☆ All the plant protection and phytosanitory measures should be regularly followed to keep mother plants healthy.
- ☆ Shoot tips or meristem portion or axillary bud with nodal segments should be used as explants and other explants which follow indirect regeneration pathway should be avoided to reduce the chances of occurrence of somaclonal variations.
- ☆ Number of multiplication cycles should be limited to 5-6 so as to reduce the chances of somaclonal variations
- ☆ Bio-hardening/biopriming of tissue culture raised plants should be carried out using various plant beneficial microbes like *arbuscular mycorrhizal* fungi to improve the field performance of in vitro raised plants. At least 3-4 months of hardening should be done to reduce field mortality.
- ☆ Clonal fidelity testing of in vitro raised plants should be carried out using molecular tools to detect variability if any.
- ☆ As there is no reported economical damage due to viruses in pomegranate so virus indexing may be kept optional.

Orchard Management

Climate, Soil and Orchard Establishment

Pomegranate now-a-days is extensively cultivated in Mediterranean, semi-arid and sub-tropical regions having marginal agro-climate at an altitude up to 1800 – 3330 m above mean sea level (Singh *et al.,* 2020). Pomegranate is light loving plant and thrives well in hot and dry climate. It is a hardy, evergreen or partially deciduous crop and can withstand up to 45 - 48 °C temperatures coupled with dry hot winds. However, the areas experiencing hot and dry summers and mild winters have witnessed best quality fruits. Further, the optimum temperature for development of attractive aril colour and sweetness in aril is 35 - 38 °C. The high temperatures and low humidity adversely affects the fruit quality.

Although, pomegranate trees have adaptability to wide range of soil conditions but it is highly sensitive to waterlogging due to development of soil borne diseases like *Ceratocystis fimbriata* and also to soil moisture fluctuation particularly during fruit developmental stages. Sandy or alluvial loam soils with optimum water holding capacity having pH range of 5-7 are well suited. It can grow well in slightly saline soils as it can tolerate salinity up to 6.0 dsm^{-1} where the other fruits fail to grow (Hasanpour *et al.,* 2015).

Planting Methods and Density/Spacing

The rectangular system of planting at 4.5 m x 3.0 m spacing, accommodating 740 plants per ha. is recommended commercially for 'Bhagwa' and 'Solapur Lal' pomegranate. However, the hedge row system of planting at 4.5m x 2m (1110 trees/ha) spacing may also be practiced as it that improves orchard productivity and facilitates the farm mechanization but closer spacing is not recommended due to the prevalence of diseases and insect pests. Planting should be preferably done on raised beds of 1.5m - 2.0 m wide bed with 45 cm high at the middle and sloping down towards both ends with about 30 cm height at ends and are separated by dead furrows oriented in the direction of land slope (Chandra *et al.,* 2011). This bedding system of planting helps to develops favourable conditions in the rhizosphere, enhances nutrient uptake and improves the yield and fruit quality of pomegranate on the bedding system of plantation (Marathe *et al.,* 2017).

The prevailing agro-ecological conditions are important considerations for pomegranate plantation. Planting should be generally practised during July - September till spring season and it should be avoided during harsh and extreme weather and unfavourable soil moisture conditions, as it leads poor orchard establishment (Singh *et al.,* 2021). The raised beds or trenches should be made during summer months (April – May) or may be solarized by covering with white transparent polyethylene (50 µ) for six weeks. This will help to solarize the soil and reduces the intensity different types of pathogens, insects, their eggs and control of weeds. The top 30 cm of each pit/bed should be mixed with well decomposed FYM (10 kg), SSP (1 kg), Neem cake (0.5 kg), Trichoderma mixture (25 g), PSB (15 g) and Azotobacter/Azospirillium (25 g) (Marathe *et al.,* 2017).

Canopy Management

Pomegranate tree is a shrub or small tree, grows up to 8 m height with many spreading and erect branches which tend to collapse and break (Chandra *et al.,* 2011). Therefore, it is highly necessary to train the trunk of tree and its branches, especially when it is in fruit bearing stage. Pomegranate is a light loving plant thus there should be proper light interception within the canopy for quality fruit production (Holland *et al.,* 2009). The proper canopy architecture enhances the photo synthetically active area of the plant by exposing the green leaves to trap sunlight and produce carbohydrates which are then transported to the sites (buds, flowers, fruit *etc.*) (Saroj and Kumar, 2019). Developing an ideal canopy architecture is of paramount important to restrain the exuberant vegetative growth and to have improved fruit quality and yields. The convenience in carrying out the cultural operations and restricting the build-up of congenial micro-climate congenial for the development of diseases and insect pests are also important considerations in ideal canopy management. The common method used in orchards is to support the trunk and tie its branches during the initial years of establishment. The ties and support require a lot of hard work. The growing fruits cause the branches to lean earthwards and to stick to each other. In this situation, fruit is exposed to the sun and its quality is degraded (sunburns, scratches and pest attack in closed canopy) reducing the profitability. For effective flowering, it is desirable to have vigorous shoots with sylleptic branches which may give about 30 per cent flowering on current season and contribute to 70 per cent flowering in the ensuing season (Reddy and Reddy, 2000). The canopy management is mainly aimed to manipulate the tree vigour and use maximum available light and temperature to increase productivity, fruit quality and also to minimize the adverse effects of the weather.

Canopy Management through Training Systems

Conventionally, pomegranates were trained to multi-trunk system, wherein, 3-5 stems that originates from the main stem at ground level were maintained followed by heading back of these stems to 45-50 cm (Singh *et al.,* 2021). Further, the new shoots originated from these stems were reduced from 1/3rd to 50 per cent of annual growth during next winter season to achieve the compact canopy framework. Apart from that single stem system of training also been widely practiced, wherein 3-4 strong scaffold branches are trained on single trunk to grow as an open vase with optimum interception of incident solar light. In general, 3-4 growth flushes are observed in pomegranate during a year.

Canopy Management through Pruning

In general, two pruning are practised under Deccan Plateau conditions, main pruning just after harvest and Light pruning (upper 10-15 cm) just before bud break (After stress) for flower regulation. During the pre-bearing stage, pruning should be aimed to develop and maintain 3-4 main scaffold branches, either as main stem for multi-trunk system as bush form or as scaffold branches arising from main trunk to from a single stem system. For compact and strong framework regular shoot tipping and removal of unwanted shoots/suckers/water sprouts should be done periodically, owing to its suckering tendency which diverts growth from fruiting into non-productive vegetative growth. Various forms of pruning practised in pomegranate are appended below.

Tip Pruning Shoot Pinching

Generally practiced in pre bearing trees (1-2 years old) wherein only top 10-15 cm growth is cut off to promote rapid but temporary accumulation of carbohydrates and flowering hormones near the uppermost bud of the shoots.

Skirting: Practiced for removal of the lower twigs at the ground level to avoid the build-up of microclimate congenial for pest and disease incidence.

Topping

Practiced in senile, low productive and old orchards (12-15 years) in the form of rejuvenation/top working. Even a blight affected orchard with stem cankers can be rejuvenated by topping the main stems from 8-10 cm below the canker to encourage new growth below cut ends.

Canopy Management through Use of Plant Growth Regulators

Flowering and fruiting in pomegranate can be remarkably regulated a by the application of different chemicals, PGRs and PGR antagonists. Reddy and Reddy (2000) noted that GA and IAA at higher concentrations inhibited flowers per shoot, however, GA at 25 ppm promoted the flowering by 25 ppm increasing the number of flowers per shoot. Further, this incremental effect was not countered even by presence of NAA at higher concentration in the cultivar Bassein Seedless. An early flowering (30 days) was observed by Chaudhari and Desai, 1993 with Ethrel @ 250 ppm in Ganesh, similarly, Ahire *et al.* (1994) also reported highest percentage of hermaphrodite flowers with the application of Ethrel at 250 ppm while working with Bhagwa. GA at 50 ppm concentration increased fruit set as well as average fruit size (Supe and Marbhal, 2008). Furthermore, Venkateshwarulu (2006) observed 75-90 per cent of leaf fall and heavy flowering with the application of ammonium sulphate at 5000 ppm and Ethrel at 500 while, ammonium sulphate at 7500 ppm could defoliated only to an extent of 68-84 per cent.

Bahar Regulation

Synchronized and systematic flower regulation is of paramount important for optimum production. Pomegranate behaves as a deciduous shrub under temperate climate and evergreen or partially deciduous in sub-tropical and tropical climate. Under regular irrigation conditions Deccan Plateau, pomegranate flowers uninterrupted with unsynchronized fruit maturity. However, continuous blossoms would produce light crops over the whole year which requires high cost on watch and ward and marketing. It is a hardy crop which needs a considerable period of water stress to flower in a synchronized manner. The duration of withholding irrigation depends upon type of soil, in general 30-45 days in light textured soil and about two months in heavy textured soil is sufficient to create stress resulting into 50-60 per cent leaf drop. This is followed by spray of Ethrel® (Ethepon 39 SL) @ 1.0 -2.0 ml per litre (depending on intensity of natural stress) + 5 g 0: 52: 34 per litre to facilitate drop of remaining leaves. Light (upper 10-20 cm from tip) pruning at flower regulation, this pruning is immediately done after stress period, before or after chemical defoliation to promote flower initiation. The recommended doses of fertilizers (inorganic source of nitrogen should not be given to the plats till completion of flowering) are applied immediately after pruning and irrigation is resumed. It is also recommended to remove water shoots at the time of flowering and carry out shoot pinching which lead to rapid but temporary accumulation of carbohydrates and flower bud forming hormones near the uppermost bud on the shoot. Flowering is regulated to encourage complete harvesting in one or two pickings. In Deccan Plateau region of India, there are three flowering seasons, *viz.*, January-February (*ambe bahar*) June-July (*mrig bahar*) and September-October (*hasta bahar*).

The flowering/fruiting of choice is regulated taking into consideration the availability of irrigation, fruit quality, incidence of pests and diseases, price and market demand. The pomegranate plants are allowed to undergo different phases *viz.*, rest or storage, stress and flowering phases during the cropping cycle. In rest or storage period, which is immediately after harvest, the plants are provided with basal application of nutrition coupled with limited irrigation, it may last for about 3 months. During rest period, deep pruning (main pruning just after harvest-pruning up to 60-75 cm, removal of *criss - cross* and diseased branches followed by application of basal dose of nutrients is practiced. Application of recommended basal dose of organic and inorganic fertilizers is required for better accumulation of carbon reserves for optimum flowering and fruiting in the coming crop season. During stress period of 4-6 weeks, irrigation is withheld and the plants are allowed to undergo moisture stress. Once the stress requirement is fulfilled, the plants undergo defoliation (shedding of leaves), new flush initiation and flower bud differentiation to result in flowering.

REFERENCES

Ahire GZ, Desai YT, Chaudhari SM, Masalkar SD and Kale PN, 1994. Effect of growth regulators on flower induction, sex and drop in pomegranate. *J. Maharashtra Agril. Univ.,* **19(1)**: 113-154.

Chandra R, Suroshe S, Sharma J, Marathe RA, Meshram DT, 2011. Pomegranate growing manual. ICAR–National Research Centre on Pomegranate, Solapur, Maharashtra, p. 69.

Choudhari SM and Desai YT, 1993. Effect of plant growth regulators on flower sex in pomegranate. *Ind. J. Agri. Sci.*, **63(1)**: 34-35.

Ferrara G, Selahvarzi Y, Ahmadpourmir H, Mazzea A and Giancaspro A, 2021. Production and Growing area. In: *The Pomegranate: Botany, Production and Uses*, eds Sarkhosh A., Yavari A. M., Zamani Z., Wallingford, UK: CABI International, pp. 59–93.

Hasanpour Z, Karimi HR and Mirdehghan SH, 2015. Effects of salinity and water stress on echo-physiological parameters and micronutrients concentration of pomegranate (L.). *Journal of Plant Nutrition*, **38**: 795–807.

Holland D, Hatib K and Bar-Ya'akov, 2009. Pomegranate: Botany, Horticulture, Breeding (Ed. J. Janick). *Hort. Rev.* **35**: 127-191.

Kumar A, Singh SK, Pandey SD, Patel RK and Nath V, 2017. Effect of Foliar Spray of Chemicals on Flowering and Fruiting in Litchi. *Int. J. Curr.Microbiol. App. Sci*, **6(5)**: 1337-1343.

Marathe RA, Sharma J, Babu KD and Murkute A, 2017. Bedding system: a unique plantation method of pomegranate in arid and semi-arid region. *National Academy of Science Letters,* **40(4)**: 249-251.

Reddy SS and Reddy YN, 2000. Studies on basipetal thickening of fruit bearing shoots and subtending previous growth in pomegranate. 4th International Symposium on tree, Montreal, Canada, p. 1137.

Reddy SS and Reddy YN, 2000. Studies on basipetal thickening of fruit bearing shoots and subtending previous growth in pomegranate.*4th International Symposium on tree*, Montreal, Canada.

Saroj PL and Kumar R, 2019. Recent advances in Pomegranate production in India- A review. *Annals of Horticulture* **12(1):** 1-10.

Singh NV, Babu KD and Sharma J, 2021. Pomegranate. In: Subtropical fruit crops: theory to practical (Ghosh, S.N. and Sharma, R.R. eds.), Jaya Publishing House, New Delhi, pp. 454-523.

Singh NV, Karimi HR, Babu KD and Sharma J, 2019. Propagation Techniques and Nursery Management. In: The Pomegranate: Botany, Production and Uses (Eds. Z. Zamani, A. Sarkhosh and A. M. Yavari). CABI- Wallingford OX10 8DE, United Kingdom, pp. 196-224.

Singh NV, Shilpa P, Roopasowjanya P, Babu KD, Mundewadikar DM, Sangnure VR, Sharma J and Pal RK, 2020. Performance evaluation of plants raised through different propagation methods in pomegranate (*Punica granatum* L.). *Indian Journal of Agricultural Sciences,* **90(4)**: 285-288.

Supe VS and Marbhal SK, 2008. Effect of plant growth substances on fruit size of pomegranate cv. Mridula, *Asian J. Hort.* **3(1)**: 18-20.

Venkateshwarlu B, 2006. Studies on the evaluation of precursor inoculum interactions in pomegranate (*Punica granatum* L.) orchard establishment and induction of flowering. M.Sc. (Hort) thesis submitted to ANGRAU, Hyderabad.

Chapter 15

Propagation Techniques for Edible Palms

Maheswarappa H.P. and Mahendra Belagumpi

Directorate of Research, University of Horticultural Sciences, Bagalkot–587104, Karnataka

Palms are monocotyledonous angiosperm that belongs to Arecaceae family represent the third most important plant family with respect to human use (Johnson, 1998) and numerous edible and non edible products are obtained from palms. Major edible palms include coconut, arecanut and Palmyra. Propagation is crucial for ensuring the long-term viability and productivity of edible palms, which are key sources of food, oil and economic value globally (Abubakar *et al.,* 2023). Through methods like vegetative propagation and advanced techniques such as tissue culture are being studied, still majority of the planting material is produced through seed propagation to ensure a steady supply of healthy palm plants to growers which is particularly important for commercial cultivation, where consistent quality and yield are necessary to meet global demands and support economies that rely heavily on palm products, such as the coconut and oil palm industries. Effective propagation techniques not only ensure a steady and sustainable supply of these vital crops but also preserves genetic diversity, enabling palms to adapt to changing environmental conditions and resist diseases. As the demand for palm products continues to grow and as challenges like climate change intensify, the role of propagation becomes even more crucial. It underpins the ability of these plants to thrive in various conditions, supports global food security, and contributes to economic stability, while also playing a significant role in environmental conservation efforts.

Coconut

Coconut is the most useful palm among the cultivated palms. It is a primary source of food, drink and shelter. It is popularly known as Kalpavriksha/tree of divine life/tree of heaven/tree of abundance. It is the only plantation crop which accommodates maximum biodiversity in the inter space and it is grown in more than 93 countries. It is a crop of humid tropics, confined to coastal regions between 20° N and S of equator (Thampan, 1981). Coconut industry provides livelihood to more than 10 million people and value added products from copra, shell, coir, coir products and handicraft items from coconut fetch considerable foreign exchange to the country (Chowdappa *et al.,* 2017).

Propagation

Coconut palm is propagated by seeds collected through series of selections on gardens, mother palms, seeds and seed lings *etc.* The mother palms are selected based on certain standard traits that are known to contribute to the superior performance of progenies. The mother palm selection is followed by seed selection, seedling production and seedling selection. Irrespective of the nature of variety or hybrid, these basic selections are followed with additional selection criteria for the choice variety.

Selection of Seed Gardens

Generally, in every coconut growing region, certain centers are well known for good quality seed nuts and seedlings. The seedlings that are produced from those centers, are always in great demand among the farmers. Within the selected seed centers themselves, individual gardens consisting of a high proportion of high yielding palms are found (Nair and Iyer, 1990). When seed nuts are selected from such high yielding seed gardens, there is a greater chance of female flowers being pollinated by pollen from nearby high yielding palms through open pollination. The seedlings produced from seeds of such high yielding parental palms, will usually be superior for high yield as compared to others (Bavappa *et al.,* 1986). Hence, the seed garden selection is important first step in coconut propagation (Chowdappa *et al.,* 2016).

Mother Palm Selection

The mother palm selection procedure was evolved to enable identification of potential high yielders using some of the easily observable traits, which are considered to be associated with yield potential such as presence of straight stout trunk with even growth and closely spaced leaf scars, spherical crown shape with short fronds, short and stout bunch stalks without the tendency to drooping, more number of leaves and large number of inflorescences carried evenly on the crown, higher number of female flowers in each inflorescence, high, consistent fruit set in bunches, high copra out turn per palm and absence of disease and pest incidences. Among all these characters, maximum attention is paid to the regular yield of nuts (Patel, 1938).

Regular and heavy bearers are selected as these are known to be less affected by adverse weather conditions. In India, 80 nuts/palm/year is taken as the standard. Other factors to be considered are, size and shape of nuts and weight of bunches and nuts, as palms bearing medium sized, spherical/round nuts give higher copra and nut yield. Mother palms should be of 25-60 years old to ensure the better performance over the years. Wherever possible, middle-aged trees are preferred as they will be in their prime of life and it is easier to spot good yielders from mediocre/poor yielders and to ensure better performance and continuous seed production from selected palms (Subramanian *et al.,* 2018).

Selection of Seed Nuts

Seed nuts can be collected from selected mother palms, the mature nuts should be harvested when at least one nut in the bunch starts becoming dry. In Talls, it takes 11-12 months to become a mature seed nut whereas in dwarfs, nuts will mature in 10-11 months after fruit set. In general, dwarf palms are to be harvested once in four weeks and tall palms are to be harvested once in six to eight weeks. They produce a resonant and ringing sound when hit with the harvesting knife or finger indicating that the husk is dry whereas, immature nuts will produce a dull sound. Rope harvest of seed nuts is recommended from the trees that are very tall and located in places where the underneath surface is hard to avoid damage due to impact.

Storage of Seed Nuts

Harvested seed nuts are stored in shade, to prevent drying of nut water, till their husks become completely dry. Seed nuts of the tall variety can be stored up to two months after harvest whereas, the seed nuts of dwarfs should be sown within 15 days of harvest (Liyanage, 1950). Nuts without the splashing water sound indicate that the nut water has become dry and hence they should not be used for sowing. For storing, arrange the seed nuts with the stalk-end up over an 8 cm layer of sand in a shed and cover with sand to prevent drying of nut water. The nuts can also be stored in plots, provided the soil is sandy and the ground is sufficiently shaded (John and Narayana, 1942).

Nursery Management

Selection of the site soils with a minimum depth of 1.2 m and good water holding capacity are preferred for coconut cultivation. Shallow soils with underlying hard rock, low lying areas subject to water stagnation and clayey soils with impeded drainage are to be avoided. However, in lands reclaimed by heaping alternate layers of sand and clay, coconut thrives well. Proper supply of moisture either through evenly distributed rainfall or irrigation and proper drainage are essential for coconut (Subramanian *et al.,* 2018).

Nursery can be raised in the interspaces of the coconut plantation. The nursery area is to be provided with shade using 50-75 per cent shade net if the nursery area is in an open space. About 120m^2 area would be required to sow 1000 nuts in flat or raised beds whereas larger area of 200m^2 would be required to maintain 1000 poly bag coconut seedlings. The time of sowing seed nuts in the nursery will vary depending on the location and the rainy season. Sowing of seed nuts coincided with commencement of the rainy season will reduce the frequency of irrigation required for getting good germination. Seed nuts can be sown all round the year under favorable climatic conditions and good irrigation facilities. Seedbeds should be prepared in an area having loose and well-drained soil. Raised beds of 10-20 cm height are made to provide good drainage. Seedbeds are generally of 1m width and convenient length with 75cm space between beds.

The seedbeds should be treated suitably to avoid damage by termites and fungal diseases. The seed nuts are generally sown in the nursery beds with a spacing of 40cm between rows and 30cm between nuts. Generally, four or five rows of seed nuts can be sown in one bed. The nuts may be planted either horizontally with the widest of the segments at the top or vertically with stalk-end up. The nuts are covered with soil, such that the top portion of husk alone is visible as it helps to prevent possible infection of the collar region of the emerging seedlings. Only seed nuts with nut water should be selected for sowing whereas, some seed nuts even though they are without nut water, may germinate and such seed nuts may be soaked in water for 24 to 36 hours prior to sowing.

The seed beds should be irrigated regularly to ensure that the soil is moist hence, after sowing, the seed beds should be irrigated thoroughly to saturation levels and repeated as frequently as necessary. During summer months, the beds are to be irrigated on alternate days. Irrigation can be provided through hose/micro sprinklers. The seed beds can be covered with suitable mulch (coconut leaves, straw or green leaves *etc.*) to conserve moisture and check the weed growth. The nursery should be kept free of weeds to allow good growth of the seedlings.

A record should be maintained indicating the name of variety sown, date of sowing, number of nuts sown, seedbed number and date of seed nut harvest. Seed nuts of tall varieties begin germination within 60-130 days after sowing and seed nuts of dwarf varieties germinate 30-95 days after sowing. Generally, germination is recorded till the fifth month of sowing and a good seed lot will give 80 to 90 per cent germination. Seed nuts that do not germinate within 5 months after sowing as well as those with dead or malformed sprouts can be removed from the nursery.

Generally, chemical fertilizers are not applied to the seedlings in the nursery since the seedlings are nourished by the endosperm which is again indicating the potential of the seedling. Application of chemical fertilizers may influence the true genetic potential of seedlings, making the selection of genetically superior seedlings difficult.

Poly Bag Seedling Production

Poly bag nursery can be adopted for producing more vigorous seedlings with better root system. Compared to the nursery in the field, watering, weeding and rogueing operations for the elimination of unwanted seedlings are easier in poly bag nursery. The seedlings can be raised in black polythene bags (500-gauge thickness) of 60 cm x 45 cm size for bigger nuts and 45 cm x 45 cm for smaller nuts. The bottom of the bags is to be provided with 8-10 holes for draining the excess water. To fill 100 bags, around 2-2.5m^3 of potting mixture will be required. The commonly recommended potting media are top soil mixed with sand in 3: 1 ratio or fertile top soil, sand and well rotten and powdered cattle manure/vermicompost in the ratio of 3: 1: 1. Red earth, well rotten and powdered cattle manure/vermicompost and sand in 1: 1: 1 ratio can also be used (Samsudeen *et al.,* 2019). The nut has considerable reserves of essential plant nutrients in it. However, the roots are capable of absorbing the nutrients one month after their initiation. Since the potting mixture made of fertile topsoil, cattle manure or vermicompost can provide the nutritional requirement of seedlings, it may not be necessary to add any fertilizer. However, if growth of seedlings is not found satisfactory, application=of 20 g ammonium sulphate and 25 g muriate of potash per bag after two months may be adopted. The fertilizer is to be spread around the seedlings and forked into the medium. The polybags must be irrigated on the same day after the application of fertilizers (Krishnakumar *et al.,* 2006). Application of 25 g each of biofertilizers such as *Azospirillum* spp. and *Phosphobacterium, Bacillus* sp., to the polybags results in production of vigorous seedlings. Use of Plant Growth Promoting Rhizobacteria (PGPR) based bioinoculants, *'Kera Probio'*, (talc formulation of Bacillus megaterium) @ 25 g/seedling and *'KerAM'* (Arbuscular Mycorrhizal bioinoculant) @ 50 g/ seedling also helps in producing robust coconut seedlings.

In order to produce poly bag seedlings, the seed nuts are initially sown very closely and allowed to germinate in a pre-nursery bed. The germinated nuts are picked out from nursery once weeks, until 80 per cent of nuts have germinated or up to 5 months from sowing, whichever is earlier lay the seed nuts in a pre-nursery bed to overcome die germination problem (Perera *et al.,* 1996). The germinated nuts are placed in half filled poly bags with the sprout positioned upwards in the center of the bag and sufficient potting mixture is added to fill the bags up to two-third portion and the sides slightly pressed to keep the nut firm.

The advantage of polybag seedlings is that, there is no transplanting shock since the entire ball of earth with the root system can be placed in the pits and the seedlings establish early and more vigorously. But the disadvantages include difficulty for transportation and higher cost of seedling production (Reddy *et al.,* 2001).

Selection and Handling of Seedlings

Selection of seedlings is an important step for ensuring better performance of plantations. Quality seedlings should be selected through rigorous selection based on seedling traits. One year old good seedlings will have a minimum of six leaves with a short petiole and large number of roots and should have a collar girth of 10cm along with seedlings which show early splitting of leaves may be preferred for planting. The colour of the petiole and vigour of the seedlings can be used as a selection criterion for dwarfs and hybrids. The dwarfs should exhibit the petiole colour of the mother palm while, the petiole colour of hybrid seedlings may range from green/brown/intermediate shades of the parents. Generally, 65 per cent of good quality seedlings can be expected from a well managed nursery. It is advisable to raise a few seedlings of the parental varieties wherever possible, side by side with progeny seedlings to compare during selection of true to type dwarf and hybrid seedlings. Seedlings of dwarf varieties can be easily identified by their early germination, short height, short and sturdy leaves with short and narrow leaflets. Different dwarf varieties are easily recognized by their colour of petiole and split leaves. Seedlings of tall varieties usually grow tall with long leaves and long and broad leaflets. Since early germination is one of the criteria for the selection of seedlings, the storing and sowing of seed nuts should be in lots according to the harvest and nuts from different harvests should not be bulked (Jack and Sands, 1929 and Liyanage, 1955).

Seedlings should be removed from the nursery by lifting with a spade, should never be lifted from the soil by pulling the leaves or petiole. The seedlings should be planted as early as possible after removal from the nursery. Seedlings can be compactly packed and transported. For very long distance transportation, special care should be taken to pack the seedlings in moss/coir pith/other moisture retaining material. Poly bag seedlings can be transported as such and planted directly in the field, after cut and removal of the base of the poly bag to facilitate growth of roots.

Arecanut

Arecanut, scientifically known as *Areca catechu* L., is a significant plantation crop cultivated in several Indian states, including Kerala, Karnataka, Assam, Meghalaya, Tamil Nadu, and West Bengal (Bavappa *et al.,* 1982). It plays a crucial role in the economic and socio-religious aspects of the country. Over the past four decades, there has been a consistent upward trajectory in both the cultivation area and production of arecanut in India. It is a customary practice to chew arecanut, either in its natural state or after going through a processing stage (Priyaa *et al.,* 2023).

Mother Palm Selection

Mother palms should be more than 12 years old and should be a regular bearer, early bearing (36-40 months after planting) with partially drooping or drooping crown should be selected. Palms should have a greater number of leaves (> 10) and shorter internodes, high fruit set (>55 per cent set) around 350-400 fresh nuts/ palm/year, with consistent yield of about 3 kg or more dry kernel (chali)/palm/

year and high recovery of chali from fresh fruit (>25%). Palms should be free from diseases and pests incidence, should be selected based on the requirements for processing chali or tender nuts (Elain and Jaganathan, 2011).

Selection of Seed Nut

Fully ripe heavy nut weighing more than 35 gram will give better germination (96%) than lighter nuts and give a greater number of quality seedings. The mature nuts should be harvested when at least few nuts in the oldest bunch starts falling. Normally it takes 11 -12 months to become a mature seed nut after pollination. Rope harvest of seed nuts is recommended from the trees which are very tall and in places where the ground is hard. The nuts, which float vertically with calyx end pointing upwards in water will produce more vigorous seedlings. Harvested seed nuts can be stored only for about 3 to 6 days since the nuts are recalcitrant, *i.e.*, viability will be lost soon (Raja *et al.*, 2008). Nagavekar *et al.* (1999) reported that arecanut seeds stored in plastic bags (88.13%), moist sphagnum moss (85.00%) and moist sawdust (86.25%) gave significantly higher germination percentages.

Primary Nursery

Sowing the nuts immediately after harvest in soil/sand and watering once in two days will result in early and good germination. Selected seed nuts are sown with their stalk end pointing upwards, 5 cm apart in sand beds of 1.5 m width and convenient length. Thick mulching is to be done with straw/areca leaves, beds are to be watered daily either by using hose or micro sprinklers. Germination of nuts usually commences by 45 days and completed by 95 days.

Secondary Nursery

For raising the seedlings in secondary nursery, beds of about 150 cm width and 15 cm height are suggested. A spacing of 30-45 cm is considered optimum for planting three months old sprouts in secondary nursery and repotting in poly bags is also preferred. The secondary nursery should be given a basal dose of decomposed farmyard manure or vermicompost @ 5 tonnes per ha and the nursery should be partially shaded to get good seedlings.

Polybag Nursery

Seed nuts may be sown directly in poly bags of 6"x9" size and 250 gauge thickness, with holes for drainage. Potting mixture should be of Soil: FYM: Sand in the ratio 7: 3: 2. Well decomposed farm yard manure or vermicompost and sieved sand should be used for potting mixture preparation. Solarization of soil by covering with black polythene sheet and sun drying of potting mixture for one week may be practiced to avoid any soil borne diseases. Daily watering during rainless period is needed to ensure desired growth (Morton *et al.*, 1988). Bohra *et al.* (2008) indicated that use of rooting mixtures such as sand + soil + FYM (1: 1: 1) or sand + soil + vermicompost (1: 1: 1) or sand + vermicompost (2: 1) mixture was superior in comparison to others to improve seedling growth and the final seedling vigour.

Selection of Seedlings

Seedlings should be of 1-1.5 years of age having six or more leaves (early leaf splitting), 90 cm height and 26 cm collar girth, having more than 5 nodes after two years are the best seedlings to get better yield. The seedling should have well established root system with 5-8 main fibrous roots intact and active while transplanting.

The seedlings have to be uprooted with a ball of earth adhering to roots if, they are raised in nursery bed. Poly bag seedlings are preferred for long distance transport. Care should be taken to prevent damage/breakage of seedlings at the collar region during transportation. Seedlings should be free from leaf blight, leaf spot and collar rot (Bhagavan and Nair, 1989).

Formula to calculate seedling index: (Leaf number) X 40 – (Seedling height)

Seedlings with higher selection index values are chosen. Seedlings with poor collar girth and fewer nodes are removed one or two years after planting.

Nursery Requirement

An area of 4,000 square meters is required to maintain 50,000 seedlings. Areca sprouts and seedlings are very delicate and do not withstand exposure to direct sunlight. Hence, proper shade should be provided to the nursery. The shade may be either of coconut or arecanut leaves spread over a pandal or by covering with 50-75 per cent shade net (green or black) or by planting some fast growing green manures or banana around nursery. The nursery should be watered regularly during summer and proper drainage should be provided during rainy season. Periodical weeding and mulching is required. Nursery can be raised in the interspaces of coconut plantation and also in widely spaced arecanut gardens. Sprinkler/micro-jet/hose irrigation systems are well suited to arecanut nursery beds/polybag seedlings.

Palmyra Palm

Palmyra Palm (*Borassus flabellifer* L.) belongs to family Arecaceae and native of tropical Africa but cultivated throughout India. Traditionally the different part of the plant such as root, leaves, seed and fruit is used for various purposes. Integrated development of Palmyra products and creating awareness regarding the crop are needed to maximize the economic value and area of the crop (Sankaralingam *et al.,* 1999). Palmyra palm like the coconut, is an important multipurpose tree of great utility as each and every part of it and hence the palm is equated to the "Kalpa Vriksha" in the mythology. It is referred as tree of life with nearly 800 uses including food, beverage, fibre, fodder, medicinal and timber (Sridevi *et al.,* 2020).

Propagation

Mother Palm Selection

Middle aged about 30 to 40 years with Stature of canopy, dwarf and stout palms are selected. Trees with compact leaves are preferred to long slender stemmed trees, should be high nut yielder and high padaneer yielder. It must be dwarf in stature, early and regular bearer and must be free from pest and diseases.

Selection of Seed Nuts

Seeds are selected from mature fruits bunches which are yellow colour at stylar region and are 80 to 90 per cent ripe. Heap the selected fruit bunches for 5 to 6 days for automatic stripping from bunches then select plumy and healthy seed nut, allow fruits to ferment for easy removal of mesocarp. While removing mesocarp, the fibre adhering to seed nut should be retained which help in absorption of water and leading to better germination and weightless and bored seeds should be rejected.

Sex of Nuts

Seeds of single nut give rise to female trees, double nuts give rise to one female and one male and Trinuts will give two male and one female palms. To maintain male and female ratios it is better to collect 10 to 15 per cent of double nuts. George *et al.* (2007) revealed that the male-specific DNA fragment of size 600 bp produced by the primer OPA-06, is tightly linked to the male-sex locus and is useful for sex determination in Palmyra.

Nursery Management

Fully mature fruits are stored in shade for three weeks then sown directly sown in field or sown in nursery to raise the seedlings, seeds can be sown in mound by heaping sand to form a bed of 1 m width x 30 cm height and convenient length. Masonry beds are built with bricks of 2 m broad 60 cm height and convenient length are prepared. Seeds are sown at 10 cm spacing and covered by layer of 5 cm thick sand. Germination will commence after three weeks of sowing and first leaf will appear after 150 to 160 days. In palmyra palm first sheath of the cotyledon elongates and after it has gone into soil the radicle comes out and produces the root. It forms tuber and eventually the plumule bursts the plumule sheath (*i.e.*, coleoptile) and forms shoot (Kumar *et al.*, 2020).

The seeds of Palmyra should be either planted in the permanent place or in deep container, because emerging shoot will go deep into the soil as tuber, up to 1 m depth in the beginning and later it comes upwards and produces leaves when tubers start rooting the plumule coming out initially utilizes the food material stored in the seed nut. This germinated part goes down into the earth as a tuber and when the tuber matures rooting starts, which absorbs the soil nutrients this give rise to leaf emergence. In the earlier stages of germination only the underground portion of the stem increases in thickness (Palmyra seedlings have tuber above the root). If there is restriction the growth of the seedling will be affected. For direct sowing 3 - 4 whole fruits are planted in pits half filled with sand or sand and soil mixture and for nursery transplantation about 1 year old seedlings are transplanted polybags, after roots are formed the seedlings can be successfully transplanted in the main field.

REFERENCES

Abubakar Ahmed Kasim, Susilawati Ishak and Mohd, 2023. Maximizing Oil Palm Yield: Innovative Replanting Strategies for Sustainable Productivity. *Journal of Environmental and Earth Sciences*. **5**: 61-75.

Agrifarming, 2024. *Coconut cultivation project report, farming profits*. https: //www. agrifarming.in/coconut-cultivation-project-report-farming-profits.

Bavappa KVA, Nair MK and Premkumar T, 1982. *The arecanut palm* (Monograph series). CPCRI. Kasaragod.

Bavappa KVA, Pillai RV and Rethinam P, 1986. *Guidelines for establishing coconut seed gardens in India.* (Technical Bulletin: 17). ICAR - Central Plantation Crops Research Institute, P-9.

Bhagavan S and Nair BP, 1989. Vigour index as an additional parameter in identifying elite palms in arecanut. *Journal of Plantation Crops*, **16**(Suppl.): 389-394.

Bohra A, Ray S, Ravi B, Maheswarappa HP, Subramanian P and Krishnakumar V, 2008. Evaluation of potting mixtures on arecanut seedlings. *Journal of Plantation Crops*, **36(2)**: 1-3.

Chowdappa P, Niral V, Jerard BA and Samsudeen K, (Eds.). 2017. *Coconut*. Daya Publishing House, A Division of Astral International Pvt. Ltd.

Chowdappa P, Samsudeen K, Thamban C and Rajesh MK, (Eds.). 2016. *Planting material production in coconut*. Today and Tomorrows Printers and Publishers.

Elain AS and Jaganathan D, 2011. *Calendar of coconut* (Technical Bulletin No. 67). ICAR–CPCRI.

George J, Karun A, Manimekalai R, Rajesh MK and Remya P, 2007. Identification of RAPD markers linked to sex determination in Palmyra (*Borassus flabellifer* L.). *Current Science*, **93(8)**: 1075-1077.

Ghazali NM, Sinniah UR and Namasivayam P, 2024. Oil palm (*Elaeis guineensis*) seed characteristics and germination potential as influenced by maturity stage and fruitlet position on a bunch. *Journal of Oil Palm Research*, **36(1)**: 94-103.

Heriansyah and Tan CC, 2005. Nursery practices for production of superior oil palm planting materials. *Planter*, **81(948)**: 159-171.

Jack HW and Sands WN, 1929. Observations of the dwarf coconut palm in Malaya. *Malay Agricultural Journal*, **17**: 140-170.

John CM and Narayana GV, 1942. A simple method of preserving seed coconuts. *Madras Agricultural Journal*, **30**: 148-149.

Johnson DV, 1998. Non-Wood Forest Products 10: Tropical Palms. Food and Agriculture Organization of the United States (FAO).

Krishnakumar VK and Reddy DVS, 2006. Production of polybag seedling in coconut. *Indian Coconut Journal*, **36(12)**: 17-19.

Kumar N, 2020. Introduction to spices, plantation crops, medicinal and aromatic plants (3rd ed., pp. 178-189).

Kurian A and Peter KV, 2007. *Commercial crops technology*. New India Publishing Agency.

Liyanage DV, 1950. Sex life of the coconut palm. *Ceylon Coconut Quarterly,* **11(2)**: 33-35.

Liyanage DV, 1955. Planting material for coconuts. Paper presented at the Coconut Conference, Bandirippuwa Estate, Lunuwila Coconut Research Institute, Ceylon.

Morton JF, 1988. Notes on distribution, propagation, and products of Borassus palms (Arecaceae). *Economic Botany,* 42: 420-441. https: //doi.org/10.1007/ BF02860166

Nagwekar DD, Haldankar PM, Rajput JC and Gunjate RT, 1997. Longevity of areca seed nuts as influenced by storage media and storage period. Indian Cocoa, *Arecanut and Spices J.,* **21**: 63-65.

Naik B, Tarai RK, Sahoo AK and Sethy BK and Samal S, 2022. *A textbook of plantation crops.* Nipa Genx Electronic Resources and Solns Pvt Limited. https: //books. google.co.in/books?id=JZF0EAAAQBAJ.

Nair M.K and Iyer R, 1990. *Elite coconut seed garden and commercial production of hybrids* (Technical Bulletin: 22). ICAR - Central Plantation Crops Research Institute, P-12.

Patel JS, 1938. *The coconut: A monograph.* Government Press.

Perera L, Peries RRA and Jayatileke R, 1996. Improvement of seedling quality in poly bags through manipulating of potting media. *Cocos,* **11**: 69-78.

Priyaa R, Hassan N, Devi M, Pangayar Selvi R and Selvanayaki S, 2023. Challenges in adopting value addition technologies in arecanut. *Asian Journal of Agricultural Extension, Economics and Sociology,* 41, 819-823. https: //doi.org/10.9734/ ajaees/2023/v41i92108.

Raja K, Palanisamy V and Selvaraju P, 2008. Studies on the performance of arecanut (*Areca catechu* L.) seedlings to the nursery treatments. *Madras Agricultural Journal,* **92(1-3)**: 106-113.

Rankin IR and Fairhurst TH, 1998. *Nurseries, Field Handbook, Oil Palm Series* (Vol. 1). PPI-PPIC.

Reddy DV, Naresh Kumar S and Prabhu SR, 2001. Evaluation of alternative media to potting mixtures for raising coconut seedlings in polybag. *Journal of Plantation Crops,* **29**: 62-65.

Samsudeen K, Niral V, Ranjini TN and Sudha R, 2019. Nursery practices and seedling selection in coconut. *Technical Bulletin No.* **62(4)**: 18-21.

Sankaralingam A, Hemalatha G and Ali AM, 1999. *A treatise on Palmyra*. All India Coordinated Research Project, Agricultural College and Research Institute, Tamil Nadu Agricultural University and Central Plantation Crops Research Institute, Indian Council of Agricultural Research.

Sridevi TR, Arunachalam R, Chandrakumar M, Parthasarathi G and Nisha R, 2020. Potential review on Palmyra (*Borassus flabellifer* L.). *Advances in Research, 21*(9), 29-40. https: //doi.org/10.9734/air/2020/v21i930229

Subramanian A, Maheswarappa HP, Jerard BA, Sumitha S, Sivakumar V, Rajamanickam K, and Venkatesan K, 2018. *Documentation of elite coconut mother palms for augmenting quality planting material production in coconut growing regions* (Technical Bulletin). AICRP on Palms Centre, Coconut Research Station, Tamilnadu Agricultural University; ICAR - Central Plantation Crops Research Institute. P.40.

Subramanian P, Thamban C, Vinayaka Hegde, Hebbar KB, Ravi Bhat, Krishnakumar V, Niral V and Josephraj Kumar A, 2018. *Coconut* (Technical Bulletin No. 133). ICAR–CPCRI, Kasaragod.

Tabi KM, Ebongue GN, Ntsomboh GN and Youmbi E, 2017. Effect of dry heat treatment along with some dormancy breaking chemicals on oil palm seed germination. *South African Journal of Botany*, **112**: 489-493.

Thampan PK, 1981. *Handbook of coconut palm*. Oxford and IBH Publishing Co.

Chapter 16

Orchard Management in Edible Palms

Maheswarappa H.P. and Mahendra Belagumpi

Directorate of Research, University of Horticultural Sciences, Bagalkot–587104, Karnataka

Palms have played a central role in tropical agriculture and human livelihoods for thousands of years. Edible palms, especially the coconut, arecanut, oil palm and palmyra palm, each bring unique characteristics that have made them indispensable to both local communities and global industries. The coconut palm (*Cocos nucifera*), known for its exceptional versatility, is used almost entirely – from the edible meat, water, and milk to the oil extracted from the kernel and even the fibrous husk and hard shell. Coconut oil is widely prized in cooking and cosmetics for its health benefits, while coconut water has become popular globally as a natural hydration drink. Coconut trees thrive in sandy, coastal areas and saline conditions, enabling widespread cultivation across tropical and subtropical regions. Arecanut (*Areca catechu*), on the other hand, is a key cultural and economic product in South and Southeast Asia, where it is traditionally chewed with betel leaves. Known as betel nut, arecanut contains mild stimulants and tannins, valued not only for its social and ceremonial uses but also for its medicinal applications in traditional practices. It plays a significant role in local economies, providing an essential source of income for farmers in regions where it is cultivated.

Coconut

The coconut palm (*Cocos nucifera* L.), is a monocotyledonous plant belong to the family Arecaceae and the sub family Cocoideae (Nampoothiri *et al.,* 2018). The palm is considered as one of the important plantation and subsistence crops and is the most extensively grown and used nut, for thousands of years, coconut products have held a respected and valuable place in Indian folk medicine. It is believed to be antiblenorrhagic, antibronchitis, febrifugal, and antigingivitic (DebMandal and Mandal, 2011) The Coconut palm is large, growing up to about 30 m height, with pinnate leaves that are 4-6 m long, pinnae 60-110 cm long; old leaves break away cleanly leaving the trunk smooth with leaf scars.

The coconut palm is grown throughout the tropical world, for decoration as well as for its many culinary and non-culinary uses. It is widely cultivated in humid tropical regions occurring in coastal areas between the latitudes 22°N and 22°S of the equator and at altitudes up to 1200 m from mean sea level. Although the coconut palm is widely grown for its nut bearing fruits, which is a source of vegetable oil, sweet water, milky kernel *etc.*, almost every part of the palm is useful and hence, it is often referred to as "*Kalpavriksha*" or "*Tree of Heaven*" or "*Tree of life*" (Ahuja *et al.,* 2014). It is one of the most versatile palms with each and every part finding a use in human life one way or the other. The palm is well known to provide all essential needs of human life *viz.* income, livelihood opportunity, food supplement, energy requirement, environmental stability and raw material for a number of enterprises. In many cultures, it is considered a symbol of auspiciousness, blessing and prosperity. The coconut palm has significant influence on the rural economy and environmental protection particularly in the coastal regions and fragile island ecosystems where it provides livelihood security to millions of people. It is grown in more than 93 countries around the world.

Coconut is an essential component of several rituals and worships in Hindu tradition throughout in India and one can easily say there is no religious function is complete without coconut. Offering coconut during worship to a Hindu god or goddess is a traditionally followed practice in India most particularly in southern India, Sri Lanka and many other countries. There are several occasions where coconut is used and or believed to bring good fortune and blessings of God.

The main coconut growing areas are located in Asia, Oceania, West Indies, Central and South America and West and East Africa. In India, the coconut palm is grown in an area of around 1.935 million hectares. Kerala, Tamil Nadu and Karnataka account for about 88 per cent of this area, while Andhra Pradesh, Orissa, West Bengal, Maharashtra, Goa, Diu and Daman, Assam, Pondicherry, Tripura, Andaman and Nicobar Islands and Lakshadweep account for the remaining 12 per cent of the coconut area in the country (Augustine Jerard *et al.,* 2022).

Climate

The coconut palm is grown under varying climatic. It is essentially a tropical plant, growing mostly between 20° N and 20° S latitudes.

Temperature and Humidity

The ideal mean temperature for coconut growth and yield is 27±5°C and relative humidity more than 60 per cent. Temperature below 10°C hindering reproductive growth of coconut and failing of emergence of inflorescence during this period. Even those emerged inflorescences aborted and fail to produce nuts because of low temperature during the months of December to February is one of the constraints for achieving higher productivity.

Altitude

The coconut palm grows well up to an elevation of 600 m above mean sea level. However, near the equator, productive coconut plantations can be established up to an elevation of about 1000 m above mean sea level. The palms tolerate a wide range in intensity and distribution of rainfall.

Rainfall

A well distributed rainfall of about 2000 mm per year is the best for proper growth and high yield. In areas of inadequate rainfall and uneven distribution, irrigation is needed. However, waterlogging during monsoon period is the common problem in the coconut cultivated region and necessitates proper drainage measures for reaping the full productivity potential.

Sunlight

The palm requires plenty of sunlight and it does not grow well under shade or in too cloudy regions. About 2000 hrs of sunshine in a year is considered necessary for the healthy growth of the palm.

Soil

The natural habitat of coconut is the coastal belt of the tropics where sandy and red sandy loam soils are predominant. The cultivation of coconut has slowly extended to inland areas, even to the hill tops, having varied soil conditions. It grows well in almost all types of soils including sandy, laterite, swampy, alluvial, black and saline soils, provided they have proper drainage system, permitting unrestricted root development, aeration and absence of rock or a hard substratum within 2 m of the surface. The predominant soil type is alluvial where coconut is cultivated. This soil is more fertile and helps in attaining higher productivity. It tolerates salinity and a wide pH ranging from 5.0–8.0 (Sathyaseelan *et al.*, 2022).

Cultivation

Species and Commercial Varieties

Cocos nucifera is the only species under the genus *Cocos* and is a diploid with 32 chromosomes (2n=32) (Sudha *et al.*, 2023). Analysis of genetic diversity and population structure in worldwide coconut germplasm (Cocos nucifera L.) using microsatellite markers. *Scientia Horticulturae, 309,* 111681. The different populations of coconut present a wide range of variability for number of traits and several workers have attempted a systematic classification of the various forms of coconut. The widely accepted classification assigns the present-day cultivars into two groups *viz.*, Talls and Dwarfs based on a few important characters like palm stature, growth characteristics of the palm, nature of flowering and fruit characteristics (Chan *et al.*, 2006; Jerard *et al.*, 2018).

Tall Group

Palms of this type are referred to as var. *typica*. They are widely planted both for household and commercial use and grow to a height of 20-30 m. They are slow maturing and flower 6-10 years after planting. They are long-lived with an economic life of about 60-70 years, although much older palms are known to exist and yield well. They are normally cross-pollinating and therefore, considered to be heterozygous. Tall varieties are the common type that occurs throughout the world. They produce copra of good quantity and quality, and have fairly high oil content as compared to dwarf cultivars. The different cultivars of the Talls are generally named after the place where they are largely cultivated. The tall cultivars are most commonly cultivated for commercial production in all coconut growing regions of the world. The commonly grown tall populations in India include West Coast Tall, East Coast Tall, Tiptur Tall, Andaman tall and Lakshadweep Tall.

Dwarf Group

Palms of this group are referred to as var. *nana*. These are reported to be natural mutants from tall types with short statured thin stem, 8-10 m when 20 years old. They begin bearing about the third year at less than 1 meter high. They have a short productive life of 30-40 years. They are normally self-pollinating and therefore, considered to be more homozygous than talls. They start bearing from 3-4 years after planting. The nuts are smaller and the copra soft, leathery and low in oil content. The dwarf cultivars are generally grown for tender nuts and also used as parent in hybrid production. The common Dwarfs available in India are Chowghat Orange Dwarf, Chowghat Green Dwarf, Kenthalli Orange Dwarf and Ganga bondam green dwarf.

Table 16.1: Basic Characteristics of Tall and Dwarf Coconut

Trait	*Tall*	*Dwarf*
Stem circumference	Sturdy with bulbous base	Thin cylindrical stem
Initiation of flowering	Late (6-10 years)	Early (3-4 years)
Mode of pollination	Highly cross pollinated	Predominantly self-pollinated
Colour of fruits and petioles	Generally, mixtures of greens and browns	Either pure green, yellow, red (orange) or brown
Height increment per year	Greater than 50 cm	Lesser than 50 cm
Nut size	Very big to very small	Very small to medium

Trait	*Tall*	*Dwarf*
Phenotypic variation		
Within cultivar	High	Low
Between cultivars	High	High
Root distribution	Generally more dense and plentiful	Less dense and few
Leaf and bunch attachment	Very strong	Fragile
Copra	Thick, Firm having more oil content	Thin, leathery with lesser oil content
Leaves	Longer, strong with broader leaflets	Shorter, weak with narrower leaflets
Susceptibility	Generally more hardy to biotic and abiotic stress	Generally susceptible to biotic and abiotic stress

The tall and dwarf types have been utilized for development of hybrid varieties, combining the early flowering trait of dwarfs with the hardiness and high yielding character of tall parents and also exploitation of hybrid vigour (Kurian and Peter, 2007).

The tall and dwarf types have been utilized for development of hybrid varieties, combining the early flowering trait of dwarfs with the hardiness and high yielding character of tall parents and also exploitation of hybrid vigour.

In addition, certain variants have also been observed. One is the male coconut tree, which produces only male flowers and another, is spike let less coconut palm or spicata where the male and female flowers borne directly on the main peduncle of spike. Variants such as Sweet coconut (called *Mohacho narel*) with solid sweet endosperm, Soft endosperm coconut with buttery textured endosperm (*Thairu Thengai*), Horned coconut with multiple horn like structures over fruits, Mini Micro coconuts with very small husked fruits are also reported.

Planting

Soils with a minimum depth of 1.2 m and good water holding capacity are preferred for coconut cultivation. Shallow soils with underlying hard rock, low lying areas subject to water stagnation and clayey soils with impeded drainage are to be avoided. However, in lands reclaimed by heaping alternate layers of sand and clay, coconut thrives well. Proper supply of moisture either through evenly distributed rainfall or irrigation and proper drainage are essential for coconut.

Coconut seedlings are selected at one-year age for planting in the main field. The seedlings are planted in 1 x 1x 1 m pits at a spacing of 7.5 x 7.5 m accommodating 175 palms per hectare. The two third of pit is filled with soil, sand and FYM and the seedlings are planted at a depth of about 60 cm from ground level. The pits are mulched and watered regularly. The young seedlings in the main field need to be protected from harsh sun light by providing shade during summer months. Water logging is to be avoided. After three years, when the palms grow to develop the stem, basins are to be formed around the stem and regular management practices are followed (Kumar *et al.,* 2020).

Figure 16.1: Varieties and Hybrids Released by ICAR–CPCRI (Images: ICAR–CPCI, Kasaragod).

Figure 16.2: Varieties and Hybrids Released by ICAR–CPCRI (Images: ICAR–CPCI, Kasaragod).

Generally underplanting is done in plantations where the palms have become senile, unproductive and uneconomic to the farmer. Old palms are removed in stages over a period of 3 to 4 years. First the area to be underplanted is peg marked. To start with, very poor yielders (less than 20 nuts per palm per year) and those trees which are very close to the peg marked point for underplanting are to be removed. Other trees are to be removed at the rate of one third each year during 2nd, 3rd and 4th year after underplanting. If the existing garden is irregularly spaced, old palms within 1 m radial distance from the newly planted seedlings areto be removed in the first year of underplanting, 2 m distance in the second year, 3 m distance in the third year and the rest in the fourth year.

In well drained soils, seedlings can be transplanted with the onset of south-west monsoon during June or with the onset of north-east monsoon during October-November. In low lying areas subject to inundation during monsoon periods, it is preferable to plant the seedlings after the cessation of the monsoon.

Figure 16.3: Coconut Planting (Image: ICAR–CPCRI, Kasaragod).

Tree Architecture and Pruning

Coconut palms are monocotyledons with single growing bud with solid single stem and hence pruning or training are not practiced. However, if the palms are to be allowed to grow straight with good crown, the proper spacing *i.e.* 7.5m x7.5 m is to be maintained. Pruning of leaves are not advocated as the bunches are borne in the axils of the leaves and damage to the leaves will have adverse effect on the subtending bunch in growth and development of fruits. Damage to the lower leaves may cause of buckling of bunches and premature fruit drop.

Nutrition and Water Management

Adequate care should be taken during the early years of growth of young palms for realizing high yield. The field planted seedlings should be shaded and irrigated adequately during the summer months. Irrigation with 45 litres of water once in 4 days has been found satisfactory in all soil types. If it is drip irrigation, daily 10 litres of water need to be provided. Provision of proper drainage is important in areas prone to water logging. The pits should be cleared of weeds periodically. Soil washed down and covering the collar region of the seedlings during the rainy days should also be removed. The pits should be widened every year before the application of manure. The pits should be gradually filled up as the seedlings grow. By fourth year, the basin should be fully prepared to a radius of 1.8 m from the trunk. The palms should be frequently examined for any insect or fungal attack and necessary remedial measures should be taken up promptly.

Regular manuring right from the first year of planting is essential for good vegetative growth, early flowering and bearing and high yield of coconut palms.

It is always advisable to test soil in the coconut garden (once in 3 years) based on the results of which, type and dosage of chemical fertilizers can be decided.

Nutrient deficiency can be diagnosed based on visual symptoms and common deficiency symptoms. However, it is not advisable fertiliser application based on visual symptoms as it will be often too late to correct the problem. It is always advisable to analyse the soil and leaf once in three years and based on the results, fertilizer application should be done.

Quantification of nutrient exhaust usually provides useful guidelines for optimizing the quantity of the individual nutrients and their proportion in which they have to be applied to sustain the soil fertility and coconut productivity. The nutrient exhaust, studies showed that the proportionate requirement of NPK of the palm in terms of N, P_2O_5, K_2O is 02: 1: 3. The quantitative order of requirement of major nutrients for adult bearing palm is K>N>Ca>Mg>P. Potassium dominates the nutrient requirement of the palm while P requirement is found to be the least (Pillai and Davis 1963, Ramadasan and Lal 1966).

The first application of chemical fertilizer should be done three months after planting and the quantity of fertilizer to be applied is approximately one tenth of the recommended dose of fertilizer for adult palms. During the second year, one third of the dosage recommended for adult palms may be applied in two split doses in May June and September-October. This dosage may be doubled during the third year. From the fourth year onwards, fertilizers may be applied at the rate recommended for adult palms.

Table 16.2: General Fertilizer Recommendation for Coconut (g/palm)

Age of Coconut Palm	*May–June*			*September–October*		
	N	*P_2O_5*	*K_2O*	*N*	*P_2O_5*	*K_2O*
First year	Planting in May-June			50	40	135
Second year	50	40	135	110	80	270
Third year	110	80	270	220	160	540
Fourth year onwards	170	120	400	330	200	800

Application of 500 g N, 320 g P_2O_5 and 1200 g K_2O per palm per year is generally recommended for adult plantations. To supply the above quantity of nutrients for an adult palm, it is necessary to apply about 1 kg urea, 1.5 kg rock phosphate (in acidic soil) or 2 kg super phosphate (in other soils) and 2 kg of muriate of potash (MOP). It can be also applied through 700 g Di ammonium phosphate (DAP), 815 g of Urea and 2 kg of MOP. After the receipt of summer showers, in May-June one third of the recommended dose of fertilizers may be spread around the palms within the radius of 1.8 m and forked in. Circular basins of 1.8 m radius and 20 cm depth may be dug during August-September and green leaf or compost or farmyard manure may be spread at the rate of 50 kg per palm basin. The remaining two third of the recommended dose of fertilizers may be spread over the green leaf or compost and covered. Wherever irrigation facilities are available, it is advisable to go for a greater number of split doses, preferably four split doses (March, June, September and December)

Figure 16.4: Nutrients Application in Coconut Basins.

Drip Fertigation

If there is drip irrigation facility, then the water-soluble fertilizers like urea, DAP, phosphoric acid (commercial grade) and muriate of potash can be applied along with drip irrigation in 6 equal splits. Through fertigation, it is recommended to provide 91 g urea, 33 ml phosphoric acid and 170 g muriate of potash per palm per application. When DAP is used it is recommended to provide 70 g urea, 60 g DAP and 170 g muriate of potash for a single dose per palm. When the recommended dose of phosphatic fertilizers are applied continuously, the available phosphorus in the soil goes up. When it is more than 20 ppm, application of phosphatic fertilizers can be skipped for a few years until the level goes below 20 ppm. If the level is 10 to 20 ppm, half of the recommended dose of phosphorus can be applied. A study on the effect of fertigation on the productivity of coconut was conducted at Horticulture Research Station, Arsikere (Karnataka). The mean nut and copra yield per palm over five years was significantly higher with the application of 100 per cent NPK through drip irrigation which was on par with 75 per cent NPK compared and application of 25 and 50 per cent NPK through drip irrigation, net returns and benefit-cost ratio in 75 per cent NPK through drip irrigation were on par with the application of 100 per cent NPK through drip irrigation, but significantly higher than 100 per cent NPK through soil application. Hence, indicated the possibility of saving 25 per cent of the recommended fertilizers by adopting fertigation which ensures higher productivity in coconut (Basavaraju *et al.*, 2014).

Biomass Recycling and INM

Around 6 to 8 tonnes of coconut wastes in the form of leaves, spathe, bunch waste, husk of nuts are available from one hectare of well managed coconut garden. Along with coconut wastes, wastes from other annuals and perennials grown as inter/mixed crops in the coconut gardens can also be recycled. When animal components are included in the coconut mixed farming system, more biomass with varying nutrient status are available.

Figure 16.5: Composting of Coconut Biomass (Image: ICAR–CPCRI, Kasaragod).

The natural decomposition of organic byproducts resulting from coconut cultivation and the nutrient release is very slow due to the presence of lignin and polyphenols in it. CPCRI has identified a local strain of earthworm related to African Night Crawler (*Elldrilus* sp.) which is quite efficient in converting coconut leaves into granular vermicompost (Prabhu *et al.*, 1998). Hence, biomass recycling through vermicomposting is an efficient method of coconut waste recycling. Coconut leaves vermicompost has C: N ratio of 9.95, 1.8 per cent N, 0.21 per cent P and 0.16 per cent K and organic carbon content of 17.84. As much as 4000 kg of good quality vermicompost can be produced from the wastes generated from 1 ha of healthy coconut garden every year by this earthworm that can meet a considerable percentage of nutrient need of the coconut palm. The clear brown coloured liquid collected after the passage of water through a column of actively vermicomposting substrate with earthworms is the vermi wash. Coconut leaf vermi wash is alkaline and contains N, P, K, Ca, Mg, Zn, Mn in appreciable quantities. It also has sugars, amino acids, and phenols along with the plant growth promoting hormones *viz.*, indole acetic acid, gibberellic acid and humic acid. Fresh vermi wash harbours a large number of fluorescent pseudomonad bacteria that help in plant growth promotion and protection (Gopal *et al.*, 2007; Selvamani *et al.*, 2017).

Application of Soil Amendments

In soils with acidic nature, in addition to the recommended level of fertilizers, 1 kg of dolomite or 1 kg of lime may be applied per palm per year. Dolomite/ lime may be broadcasted and incorporated in the basin one month prior to the application of chemical fertilizer. For coconut palms showing yellowing of leaves due to magnesium deficiency, 0.5 kg of magnesium sulphate can be applied in the basin along with other fertilizers during September October. Deficiency of the micronutrients, especially boron, is also observed in coconut palms in certain localities. About 75 g of borax is to be applied at bi monthly intervals till the symptom disappears.

Basin Management with Legume Cover Crops

Agro-technique has been developed to generate significant quantities of organic manure and nitrogen in coconut gardens, utilizing the leguminous cover crops. It involves cultivation of leguminous plants having symbiotic association with efficient Rhizobium strains in coconut basins and interspaces during the monsoon period and incorporation of biomass generated to the palms at the maximum vegetative growth stage of legumes.

Figure 16.7: Basin Management with Cover Crops (Image: ICAR–CPCRI, Kasaragod).

Field experiments on basin management with legumes in adult coconut plantations have revealed the effectiveness of this technique to substitute nitrogen fertilizer for coconut up to 30 per cent. The effectiveness of the legume treatment as a component in the management programme for root (wilt) disease of coconut has also been well demonstrated. *Pueraria phaseoloides, Mimosa invisa, Calopogonium mucunoides,* cowpea (*Vigna unguiculata*), sunhemp (*Crotolaria juncea*), horse gram (*Macrotyloma uniflorum*), daincha (*Sesbania aculata*) and *Sesbania spinosa*are the species of legumes superior in biomass and nitrogen contribution in coconut basins. They contribute about 15-25 kg of biomass and 100-200 g of nitrogen in coconut basins during a growth period of 60-120 days in monsoon season. Pelleting of inoculated seeds with neutral or inert materials enhances nodulation by introduced *Rhizobia* in acidic soils.

Growing *Glyricidia* as Green Manure Crop

Generation of large quantities of nitrogen rich biomass is also possible through the cultivation of fast growing perennial leguminous green leaf manure tree crop, Glyricidia in the coconut plantations. This can be easily grown along the borders of coconut plantation and can generate adequate amount of nitrogen rich green leaves.

It can also be raised in littoral sandy soils where no other green manure can be established. The tree is propagated either through vegetative cuttings or seeds. One-meter long stem cuttings or 3 to 4-month-old seedlings raised in poly bags/ raised beds can be used for planting. It is preferable that the planting season should coincide with the monsoon (South-West/North-East monsoon) for better establishment. Spacing of 1 m x 1 m can be adopted. Two rows of glyricidia can be planted along the boundary of coconut garden in a zig zag manner. Stem cuttings or seedlings can be in an upright position in pits of 30 cm x 30 cm x 30 cm. For better establishment, a basal dose of 50 g of rock phosphate per pit may be applied. Height of the plants when grow-up should always be maintained at 1 m by pruning. Pest and disease are not a major problem for glyricidia and hence, no plant protection measures are required. Pruning can be started one year after planting and should be done at least thrice a year. Studies conducted at ICAR–CPCRI have indicated that the best growth and biomass of leaves could be obtained with planting of three rows of Glyricidia at 1 m x 1 m spacing between two rows of coconut and pruning of leaves during February, June and October. This could produce around 8 t of biomass annually. Application of Glyricidia prunings could supply around 90 per cent, 25 per cent and 15 per cent of the requirement of N, P_2O_5 and K_2O, respectively (Subramanian *et al.,* 2000).

Bioinoculants

Bioinoculants are carrier-based preparations containing beneficial microorganisms in a viable state intended for seed or soil application and designed to improve soil fertility and help plant growth by increasing the number and biological activity of desired microorganisms in the root environment Bioinoculant formulations containing *Azospirillum* spp. and Phosphate solubilising bacteria having 10_{+} colony forming units of bacteria prepared in carriers such as talc or vermicompost are to be applied @100 g per palm. '*Kera Probio*', a talc formulation of *Bacillus megaterium*, effective for raising robust coconut seedlings has been developed at ICAR–CPCRI. Similarly, an Arbuscular Mycorrhizal Fungal (AMF) bioinoculant, '*Ker*AM', has been developed at ICAR–CPCRI, which is a soil-based AMF bioinoculant for coconut seedlings. The bioinoculant contains *Claroideoglomus etunicatum*, one of the dominant AM species isolated from coconut agro-ecosystem with high potential to increase the growth parametersof coconut seedlings. The bioinoculant should be mixed with one kg vermicompost and applied to soil and incorporated. Care should be taken to use only biofertilizer containing adequate number of living microorganism and before the expiry period mentioned in the packet. It has been observed that in many instances desired results are not obtained due to the use of preparation not containing the required number of metabolically active microorganism. Optimum soil moisture is essential after biofertilizer application to ensure the survival of the introduced microbial inoculum in the soil. Hence, biofertilizer application should coincide with the onset of monsoon, especially when the palms are maintained under rainfed condition. However, under irrigated conditions, it can be applied at any time, since maintaining optimum moisture is not a problem. Bhalerao *et al.* (2021) concluded that use of organic manures, such as vermicompost, green manuring and biofertilizers along with chemical fertilizer is essential to sustain the productivity over a period of time.

Figure 16.8: Soil Amendments Application in Coconut Basin (Image: ICAR–CPCRI, Kasaragod).

Irrigation Methods

Irrigation methods commonly adopted in coconut gardens are flooding, basin irrigation, sprinkler or perfo-sprays and drip irrigation. In certain areas of Tamil Nadu, East and West Godavari Districts of Andhra Pradesh where adequate supply of water is available, coconut gardens are flood irrigated. There is considerable wastage of irrigation water under flood irrigation. In addition, flood irrigation leads to poor aeration especially in clay soil leads to water stagnation and spread of disease like Ganoderma. Hence flood irrigation for coconut is not advisable and should be avoided. In basin irrigation, water is applied in the basins of 1.8 to 2.0 m radius which is the active root zone of coconut. Irrigation channels are provided in between two rows and each basin is connected with the channel. In this method there will be some loss of water due to deep percolation, seepage and evaporation. However, this loss is reduced when basins are irrigated through hose pipes. This is being advocated to reduce water loss in transit. Still, there is loss due to deep percolation and surface evaporation. ICAR- CPCRI r e c o m e n d s application of 200 l of water once in four days. For the WCT palms in red sandy loam soils on the west coast, perfo irrigation with 20 mm water, when cumulative pan evaporation is 20 mm is found to be the best irrigation schedule.

Figure 16.9: Perfo Spray Irrigation in Inter or Mixed Cropping Systems (Image: ICAR-CPCRI, Kasaragod).

Sprinkler irrigation or perfo sprays are most suited for inter or mixed cropping systems where the entire surface requires wetting. The quantity of water applied should be at least 75 per cent of open pan evaporation (Eo). Drip irrigation is ideally suited for widely spaced crops like coconut as it saves water, energy and labour and the WUE is high. Drip irrigation is a micro irrigation system in which the water is applied to the root zone at the rate at which the palm can take up. It is ideal considering the advantage of water saving. Four pits with a size of 30 x 30 x 30 cm have to be dug one meter away from the bole of the palm at equidistance and the pits filled with raw coir pith. The water has to be delivered to the pit through conduit tube placed in slanting position. Based on a study conducted at ICAR- CPCRI, it was concluded that yield of coconut with drip irrigation daily @ 66 per cent of the Eo from December to May was adequate (32 liters/palm/day when the evaporation rate is 4 mm day) and comparable to basin irrigation @ 200 litres per palm once in four days. Thus, there is 34 per cent saving of water in drip irrigation. This is applicable to varieties and hybrids and also in different soil types. The number of dripping points should be six for sandy soils and four for other soil types. The rate of water application should be 2-4 litres per hour per emitter (Maheswarappa *et al.*, 2019).

Soil and Water Conservation

Proper soil and moisture conservation practices are essential for ensuring sustainable production especially when coconut is grown under rainfed condition with undulating terrain and sloppy conditions.

Mulching

In order to conserve soil moisture in the coconut plantations, mulching with various types of organic materials can be practiced. The best time for mulching is before the end of the monsoon and before the top soil dries up. For mulching, cut coconut leaves into two or three pieces. To cover 1.8 m radius of coconut basin, 10 to 15 fallen coconut leaves are required and can be spread in two to three layers. Mulching with composted coir pith to 10 cm thickness (approximately 50 kg/palm) around coconut basin is also an ideal method to conserve moisture. Coir pith can hold moisture five times its weight.

Figure 16.10: Muching with Coconut Husk.

Due to its fibrous and loose nature, incorporation of coir pith considerably improves the physical properties and water holding capacity of soil. The applied material may last for about 1 to 2 years. Coconut husks are also used as surface mulch around the base of the palm. It can hold moisture to the tune 3 to 5 times of its weight. Approximately 250 to 300 husks will be required for mulching one coconut basin. Mulching is usually done up to a radius of 2 m leaving approximately 30 cm near the palm. Two layers of husk may be buried in the coconut basin with the concave side facing upwards. These layers facilitate absorption of moisture. Above this, another layer of coconut husk is placed with the convex side facing upwards to arrest evaporation. Effect of this mulch lasts for about 5-7 years. Ranasinghe *et al.* (2010) concluded that when the manure circle was mulched with dried coconut fronds, and the rest of the land was also mulched with a five cm thick layer of coir dust, the palms showed an improved gas exchange ability than palms growing in bare land or with cover crops during the dry period.

Husk Burial

Burial of husk in trenches in between the rows of palms is also effective for moisture conservation in coconut gardens. Husk burial is to be done at the beginning of the monsoon, in linear trenches of 1.2 m width and 0.6 m depth between rows of palms with concave side of husks facing upwards and each layer is to be covered with soil. Catch pit filled with coconut husk Catch pits can be constructed at slopes to conserve soil and water. Though there are no standard dimensions for catch pits, catch pits of 1.5 m length x 0.5 m width x 0.5 m depth can be constructed. A bund is to be made at the downside using the excavated soil and pineapple suckers may be planted on it. This pit is also to be filled with coconut husk (Kumar and Rajagopal, 2007).

Figure 16.11: Husk Burial in Coconut Garden (Image: ICAR-CPCRI, Kasaragod).

Contour Trench Filled with Coconut Husk

This measure is to be taken up where the land slope is high. Trenches of 50 cm width x 50 cm depth and convenient length are to be made in between two rows of coconut palms. These trenches are to be filled with coconut husk. Coconut husks need to be filled in layers with the bottom layers facing up and top layer facing down. A bund of 20 cm height and suitable width (>50 cm) is made at the downstream using the excavated soil. Two layers of pineapple plants are to be planted on the bund with a spacing of 20 cm x 20 cm. Pineapple plants would stabilize the bund and provide additional income to the farmer. The runoff water from the upper side would be collected in the trenches. Soil particles would also get deposited in the trench along with the runoff water. Coconut husk retains the moisture and makes it available for plants during summer months (Subramanian *et al.,* 2018).

Ripening, Harvesting and Handling

Usually 11-12-month-old nuts are harvested. Coconuts are harvested at varying intervals in a year. The frequency varies depending upon the yield of palms. Usually, the nuts are harvested 6 to 10 times in a year. In well maintained and high yielding gardens, bunches are produced regularly and harvesting is done once in a month. Nuts which are 11 months old give fibre of good quality and can be harvested in the tracts where husk is utilized for manufacture of coir fibre. Skilled personnel are traditionally employed for climbing palms for harvesting nuts. Nowadays, lack of availability of skilled climbers for harvesting operations is a serious problem experienced by coconut farmers. A simple palm climbing device invented by a farmer from Kannur district of Kerala is gaining popularity (Subramanian *et al.,* 2018).

Figure 16.12: Mature Nuts of COD.

The harvested matured nuts can be stored up to 3 to 5 months depending on the variety and storage conditions. Storing the fruits with the stalk end at top will prolong the storage life as it delays the germination. For long distance transport, husking is done partially to avoid nut cracking. For cup copra purpose, the matured fruits are husked, split and dried. For ball copra or edible copra, the husked fruits are stored in dry warm enclosures for over nine months and then split to take the ball copra.

The tender coconuts which are meant for drinking purpose are harvested at 5 to 7 months maturity stage wherein the five-month-old fruits will have only water and the subsequently with varying thickness of tender kernel. For long distance transport, the tender coconuts are sent packed as such or trimmed and packed, refrigerated. Unrefrigerated tender coconuts may last up to 4 to 5 days depending upon the locality and storage conditions, whereas refrigerated ones can be stored up to about three months.

Arecanut

Arecanut (*Areca catechu* L.) is an extensively cultivated tropical monocot palm belonging to the tribe Arecae and sub-tribe Arecinae in the family Arecaceae (Palmae). The palm is an unbranched erect, medium-sized, monoecious tree growing in hot and humid tropical regions of the world and its centre of origin is considered to be South East Asia (Bavappa and Nair, 1982). The Areca palm, which is highly cross-pollinated, is an allotetraploid, with chromosome number 2n = 32. The genus *Areca* includes 76 species, *A. catechu* being the only cultivated species (Bharath *et al.,* 2015). However, it has been observed that there is a wide range of variation existing in plant morphology, size, shape and colour of fruits between the different areca growing regions and between different palms/accessions of the same region (Ananda, 1998).

The major arecanut growing countries are India, Bangladesh, Indonesia, Myanmar, China (Taiwan) and Thailand, whereas countries like Sri Lanka, Bhutan, Nepal, Malaysia and Maldives also cultivate in smaller area. The present production of arecanut in the world is 12.14 lakh tonnes from an area of 9.58 lakh hectares (FAO Statistics 2016). In India, arecanut is cultivated in an area of 4.94 lakh hectares with a production of 8.09 lakh tonnes with a productivity of 1636 kg/ha (Directorate of Economics and Statistics, 2018). In the world, our country ranks first both in area and production and is predominantly cultivated in states like Karnataka, Kerala, Assam, Meghalaya, West Bengal and Mizoram. Other states where arecanut is grown are Tamil Nadu, Tripura, Andaman and Nicobar group of Islands, Maharashtra, Goa, Andhra Pradesh, Nagaland and Pondicherry. Arecanut is providing livelihood security for millions of people in rural and urban areas (Balasimha and Rajgopal, 2004).

Importance

Arecanut plays a prominent role in the religious, social, cultural and economic life of people in India. Arecanut, along with several other ingredients, is being chewed by many in several countries as it is believed to have numerous medicinal values. It has an important place in the ancient Indian system of medicine such as Ayurveda, Unani and Homeopathy and in clinical practices in certain other countries such as Philippines, China and other South and South-East Asian countries (Aman, 1969). The World Health Organization has listed out as many as 25 beneficial effects of *A. catechu.* It was also reported that all the alkaloids present in arecanut possess drug-like properties. Most of the folklore medicinal properties of arecanut are now validated and authenticated with proper scientific data. It has antioxidant, anti-inflammatory and analgesic, anti-diabetic, hypolipidemic, antibacterial, anti-fungal, anti-malerial, anti-viral, anti-HIV and AIDS, anti-aging by reducing wrinkles formation in skin, learning and memory improvement, wound healing. anti-ulcer, anti-migraine, antihypertensive, anti-depressant, anti-allergic, anthelmintic, aphrodisiac, hepatoprotective, cytoprotective and anti-cancer properties. Arecanut leaf sheath is also used as alternate fodder for livestock (Bhat *et al.*, 2016; Bhat *et al.*, 2017a; Bhat *et al.*, 2017b)

Ecological Requirement

Climate

Climate plays an important role in the growth and yield of arecanut and generally more than 50 per cent of variation in yield is due to climatic differences. Though, arecanut is grown under different agro-climatic conditions, it is very sensitive to extreme climatic conditions (Bhat and Abdul Khader, 1982). The most important climatic factors that influence the growth and development are altitude, relative humidity and rainfall. Temperature also influences the crop growth and yield to some extent. Regression analysis of weather variables of 12 years indicated that arecanut yield is influenced by relative humidity, evaporation and rainfall (Vijaya Kumar *et al.*, 1991).

Altitude

The cultivation of arecanut is mostly confined to 28° North and South of the equator and can be cultivated up to an altitude of 1000 m MSL in deep and well-drained soils with low water table. Due to its susceptibility to low temperature, the palms do not come up at an altitude of more than 1000 m MSL and hence in general it is mainly grown in low altitudes. In the high altitude areas like Wynad (Kerala) and Coorg (Karnataka), the endosperm of the fruit does not develop sufficient hardness. It is also reported that high altitudes affect the germination of seeds and quality of Chali (dry kernel) (Pillai and Murthy, 1973).

Temperature

Areca palms grow well within the temperature range of 14°C-36°C and are adversely affected by temperatures below 10°C and above 40°C. However, the crop is being grown in temperatures ranging from 5°C (as in places like Mohitnagar, West Bengal) and at 40°C (Vittal in Karnataka and Kannara in Kerala).

Rainfall

Arecanut requires abundant and well distributed rainfall. However it is grown in areas with wide variations in rainfall such as Malnad of Karnataka where the annual rainfall may go up to or even more than 4500mm as well as in low rainfall areas like Maidan parts of Karnataka or parts of Coimbatore district in Tamil Nadu where the annual rainfall is about 750mm.

Relative Humidity

Very high or low relative humidity is not conducive for growth and development of arecanut and it directly influences the water relations of palm and indirectly affects leaf growth, photosynthesis, pollen dispersal, occurrence of diseases and finally economic yield. High humid conditions provide congenial conditions for the rapid spread of diseases like fruit rot, bud rot *etc.* It has considerable influence on evapotranspiration and hence on the water requirement of arecanut.

Soils

Arecanut is predominantly grown in gravelly laterite soil of red clay type and it can also be grown in fertile clay loam soils with special care of microsite improvement. Sticky clay, sandy, brackish and calcareous soils are not suitable for areca cultivation. Laterite, red loam and alluvial soils are suitable (Aiyer, 1966). Laterite soil belt in arecanut growing region is characterised by high rainfall, undulating topography, presence of kaolinite clay mineral, leaching of K^+ and Ca^{2+}, P fixation and high zinc fixing capacity. It is not advisable to grow arecanut on clayey soils and paddy converted lands due to water stagnation, poor soil aeration and lack of fine root production. In Malaysia and Fiji, arecanut is cultivated in the hot, moist, rich alluvial areas of the coastal belt. Arecanut needs deep soil preferably not less than 2 meters for well-developed root system.

Cultivation

Site Selection and Layout

Arecanut thrives well in humid areas when protected against hot sun and heavy wind whereas it cannot withstand both drought and water stagnation. Thus the site selected should have sufficient drainage to drain away excess water and irrigation facility for irrigation during dry weather. The palms also cannot withstand high temperature and exposure to direct sun causes scorching effect on stems and the palms become weak and susceptible to wind fall.

Planting methods like square, triangle and quincunx may be followed. Aligning the rows in north-south direction and planting with angling 35° towards west lowers the incidence of sun-scorching.

Spacing

Systematic studies were conducted at different areca growing regions to find out suitable spacing for arecanut. Under Vittal conditions, palms spaced at 2.7 m x 2.7 m produced maximum yield (the number and weight of fruits) per unit area. The root distribution studies in relation to individual palm yield and unit area yield indicated that 2.7 m x 2.7 m spacing is optimum for arecanut (Bhat and Leela, 1969). It was found that under wider spacing the resource utilization was not full and at closer spacing there is heavy concentration of roots in the lower layers of soil resulting in reduction in yield. Thus a spacing of 2.7 m x 2.7 m is recommended in majority of areca growing regions. However, 3.3 m x 3.3 m spacing is advisable for accommodating high value inter/mixed crops.

Figure 16.13: Arecanut Plantation
(Image: Plmflora | PepperHub)

Planting Season

Planting of seedlings is done either in May-June or September-October, depending on the intensity of rains. In river banks, where there is likely danger of inundation, it is advisable to plant seedlings at the end of the south-west monsoon in the month of September. In other places, where the south-west monsoon is not severe, planting may be done in May-June.

Planting Method

Depth of planting is mainly decided by the soil type and the height of water table. In laterite soil with good drainage, the seedlings planted at 90 cm depth were more vigorous and flowered early (Bhat and Leela, 1968) and gave increased yield than those planted at 30 cm or 60 cm. Pit size of 60 x 60 x 60 cm is recommended when the water table is high. Pits should be filled with top soil, farm yard manure and sand up to half of the pit. Seedling should be planted at the centre of the pit with a ball of earth and covered with soil up to the collar region. Planting depth is important to avoid growth of aerial roots above the ground. Banana can be raised as a shade crop in the interspaces during the initial years.

Drainage

For better growth and development of the plants proper drainage is essential since areca cannot withstand water stagnation. The number of drainage channel depends upon the soil type. In light soils the number of channels may be less and in heavy soils the channels should be dug in each row for proper draining of the excess water. The channels should be at least 15–30 cm deeper than the depth at which the seedlings are planted. The channels have to be cleaned every year for easy flow of water. The planted pits also should be provided with outlets to drain away the water.

Figure 16.14: Arecanut Seedling Planting (Image: ICAR–CPCRI, RS, Vittal).

Nutrient Management

In laterite soils application of 100 g N, 40 g P_2O_5 and 140 g K_2O/palm recommended as a general dose of fertilizer every year. In addition, application of 12 kg each of green leaf and compost per palm is also recommended. In heavy soils general recommendation of nutrients is 50 g N, 40 g P_2O_5 and 140 g K_2O along with green manures. Application of only organics or chemical fertilizer may lead to imbalance in nutrition of arecanut. Thus, nutrient management strategies need to be planned for arecanut taking into account of the soil fertility status. It is advisable to consider nutrient deficiency/toxicity before the development of visual symptoms with the help of plant and soil analysis.

Method and Time of Nutrient Application

Fertilizers and organic manures should be applied when the soil has sufficient moisture but not during heavy rainfall and dry periods. In unirrigated crop, fertilizers can be applied just before monsoon coinciding the months of May-June and after monsoon (September-October). When the crop is irrigated the pre-monsoon application can be advanced to February-March. As far as possible the fertilizers should be applied during nut formation period during December-May. The organic manures are applied during September-October in basins around the base of each palm. In the first year of planting 1/3rd of the recommended dose of fertilizer should be given. In the second year, 2/3rd of the recommended dose and from third year onwards, full dose of fertilizer should be given.

In soils with acidic pH of below 6.0, the fertilizer may be applied in the form of urea, rock phosphate and muriate of potash. In soils with pH above 7.0, single super phosphate or DAP can be used as source of phosphorus. It is important to note that organic manure application alone cannot meet K demand of arecanut as organic manures contain very less K except gliricidia, coir pith, cocoa husk compost and arecanut husk compost. For high yielding arecanut palms with kernel or *chali* yield more than 2.5 kg/palm/year, double dose of fertilizers especially N and K can be applied. Lime application is needed only if soil pH is below 5.

Basin opening is important for application of manures and fertilizers wherever soil hardens after monsoon to provide soil aeration. Soil aeration is most important for production of fine roots, which are required for uptake of nutrients and water. Majority of the feeding roots of arecanut are concentrated within 30 cm depth and 60 cm distance from the trunk. Only fine roots of less than 1 mm thickness will take up water and nutrients. The basins of the palms may be covered with thick layer of soil or forked for incorporation of manures and fertilizers with the soil.

Figure 16.15: Fertilizer Application in Arecanut (Image: ICAR–CPCRI, RS, Vittal).

Drip-Fertigation

With adoption of ferti-drip irrigation, the advantages are reduced labour charges on fertilizer application, weeding and irrigation and diesel charges due to less operational hours. This technology can be profitably adopted in arecanut, which can save fertilizer up to 25 -50 per cent. The initial results have shown that for pre-bearing arecanut palms 50 per cent of recommended nutrient is sufficient when it is given through drip irrigation thus saving considerable amount of fertilizer cost. Annual maintenance cost could be reduced considerably through saving in labour and fertilizer input over normal practice of basin application of fertilizers and irrigation. The highest root dry weight and root/shoot ratio were observed in ferti-drip method (Sujatha and Abdul Haris, 2000). The fertilizers can be injected into the system by different methods. Important techniques are using the main pump, using a separate small pump or by using a fertilizer tank when area of irrigation is large, whereas, when area is small, vacuum injection (venturi) technique can be followed.

Liquid fertilizers are readily soluble in water and can be used for fertigation. However, easily available fertilizers like urea, diammonium phosphate (DAP) and potassium chloride (MOP) are water soluble and cost effective sources of NPK. Fertilizers should be applied during post-monsoon season from December to May. The fertilizers can be applied preferably once in 10 or 20 days, but can be applied once in 30 days also. Application of fertilizers should be avoided during monsoon season. Total quantity of fertilizer to be given is to be split into 9 or 18 parts and each part to be applied once in 20 or 10 days respectively.

For pre-bearing palms 50 per cent of recommended fertilizer is sufficient when supplied through drip irrigation. For bearing palms 75 per cent of the recommended fertilizer dose is sufficient (Bhat *et al.,* 1989). Recommended nutrient dose to be given through fertigation is 75: 30: 105 g N, P_2O_5, K_2O per palm per year. Quantity of fertilizer required to supply recommended dose per palm per year is 136 g Urea, 65 g DAP and 175 g MOP.

Figure 16.16: Fertigation System in Arecanut.

Green Manure Crop

Trials conducted at Vittal, Hirehalli and Mohitnagar to screen the suitable green manure crop for arecanut garden showed that *Pueraria javanica* and *Mimosa invisa* are good green manures and cover crops in arecanut gardens form the point of view of their green manure yield and nutrient addition capacity (Mohapatra *et al.,* 1970). *Sesbania* was found to be a good crop which can withstand water logging and drought. It can be grown in valleys of Assam, Karnataka and Kerala receiving high rainfall. *Calopogaonium* and *Pueraria* may be used as cover crops also in hilly slopes to prevent the soil erosion. *Tephrosia* is used mainly in some parts of Tamil Nadu for its hardiness and high nitrogen content. It comes up well in loamy soils also. A study was also conducted in Maidan part to screen suitable green manure crop for arecanut garden (Sannamarappa, 1976). The results indicated that *Mimosa invisa* and *Centrosema pubescens* produced significantly more green matter, compared with *Pueraria javanica* [*P. phaseoloides*], *Calopogonium mucunoides*, *Crotalaria anagyroides* and *Sesbania speciosa*. However, *M. invisa* and *C. mucunoides* gave the best improvement in soil organic C status.

Figure 16.17: Green Manuring in Arecanut planation (Image: Dreamstime).

Organic Matter Recycling

On an average, 5.0 to 8.5 tonnes of leaf wastes are available from one ha of areca garden per year. Direct application of these wastes in the garden will take long time for decomposition and will not meet the immediate nutrient requirement of the crop. Hence, these materials can be composted using earthworms effectively and used as organic manure in areca gardens (Hussain *et al.,* 2008).

To prepare vermicompost, areca wastes are chopped into small pieces of 10 cm and heaped which can be sprinkled with water daily and kept for two weeks. Then the chopped material is arranged in beds of one metre width and convenient length. Cement tanks or trenches can be used for this purpose. A layer of 10-15 cm waste material is alternated with 2 cm layer of cowdung over which earthworms are released @ 1000 numbers per square metre. The wastes are converted into fine granular, odourless vermicompost within 60 days. About 80 per cent recovery of vermicompost from these wastes is expected. During this period, the earthworm population is doubled. Major nutrient (NPK) and micronutrient (Cu, Zn, Fe and

Mn) contents were slightly higher in all the vermicompost samples than in normal compost (Chowdappa, *et al.,* 1999). Vermicompost had lower C: N ratio and pH than normal compost. Microbial population was considerably higher in vermicompost than in normal compost. About 4 kg of vermicompost per palm per year meets the N and P nutrient demand of arecanut. Potassium needs to be added through other sources like MOP, arecanut husk or *gliricidia*. Two species of earthworms *Eudrilus eugeniae* and *Eisenia foetida* can be used. The recycling of wastes added the organic matter to the soil thereby increasing the organic carbon and available nutrient contents to the extent of 30 per cent (Bhat Ravi *et al.,* 2002). The yield of the crops was not negatively affected by reducing the chemical fertilizer with the addition of composted wastes.

Nutritional Disorders

Common nutritional disorders in arecanut are crown choking, crown bending, oblique nodes and nut spitting. These problems are seen in paddy converted lands and areas with excess soil fertility and water stagnation leading to lack of fine root development. Crown choking can be identified at initial stages with appearance of dark green colour of leaves and reduction in size of old leaves. Zinc deficiency is mainly responsible for development of disorders (Sujatha and Bhat, 2013). Soil application of zinc sulphate @ 10 g per palm is ideal if disorder symptoms are in the initial stages. In case of severe reduction in leaf size and crown choking, spraying of 0.5 per cent zinc sulphate (5 g per litre of water) mainly on fresh foliage can be done so that new leaves will emerge normally. Nut splitting is due to less potassium and boron deficiency which can be managed with identifying the cause and regular application of required nutrients.

Water Management

Arecanut cannot withstand drought for a long time and invariably needs irrigation during long water stress periods. Being a perennial crop, once affected by water stress, it may require two-three years to regain the normal vigour and yield. Irrigation equivalent to pan evaporation is optimum for arecanut during water stress period.

In places with high sub-soil moisture and in areas where the rainfall is well distributed, throughout the year no irrigation is practiced. A three-fold increase in yield of areca was reported with irrigation from Palode (South Kerala). When irrigation was scheduled based on IW/CPE ratio, it was found that basin irrigation of 30 mm of water when the CPE is 30 mm (IW/CPE ratio of 1) is optimum (Yadukumar *et al.,* 1982). This works out to an irrigation frequency of once in 7-8 days during November-December, once in 6 days during January-February and once in 4-5 days during March-April and May. The quantity of water to be applied per irrigation was about 200 l.

To overcome scarcity of water in summer and to stop ground water exploitation, irrigation methods like sprinkler and drip are ideal. Irrigation efficiency is high with drip irrigation (90-95%) compared to 50-60 per cent in conventional irrigation and 70 per cent in sprinkler irrigation. Sprinkler and drip irrigation save about 20 and 44 per cent of water, respectively compared to conventional methods of irrigation like flooding and splashing. Water required to be given through sprinkler irrigation is about 40-60 l per day per palm with an irrigation interval of 3-4 days (Mahesha, 1987).

In drip irrigation, about 16-30 litres of water per day per palm is required based on evaporation rate. Micro tubes/drippers (2-3nos.) should be placed in the basin opposite to each other or in a triangle. Lateral lines of drip system should be folded and tied to stem of the palm during rainy season to avoid clogging of microtubes or drippers. Drip irrigation significantly increased the height, girth, number of nodes, leaf area, number of spadices, percentage of spadices to leaf fall and percentage of nut set compared to palms irrigated by conventional method (Abdul Khader, 1988). Yield increase by 44 per cent by drip irrigation has been observed over conventional method of irrigation. It was concluded that areca palm needs 20 litres of water per day when drip irrigation method is followed.

Harvesting

The areca palm usually initiates fruit-bearing between 4 to 6 years of growth, achieving full productivity within 10 to 12 years and maintaining this level for up to 40 to 45 years. The timing of the harvest is determined by the intended purpose of the palm's yield. Skilled climbers, equipped with extended bamboo or aluminum poles featuring attached knives, carry out the harvesting of semi-tall palm trees.

Different Maturity Stages for Harvest

Age of the Nut	*Stage of Maturity*
5-6 months	Half matured
6-7 months	Mature but green
8 months	Half ripe
9 months	Ripe

Yield

Under effective management practices using high-yielding varieties, the areca palm can yield 3000 to 4500 kg per hectare. In contrast, under standard or average management conditions, the yield drops to approximately 1500 kg per hectare. The national average productivity for areca palms stands at 1268 kg per hectare, while at the state level, it's slightly lower at 1077 kg per hectare

Palmyra Palm

Palmyra Palm (*Borassus flabellifer* L.) belongs to family Arecaceae and native of tropical Africa but cultivated throughout India. Traditionally the different part of the plant such as root, leaves, seed and fruit is used for various purposes. Integrated development of Palmyra products and creating awareness regarding the crop are needed to maximize the economic value and area of the crop. Palmyra palm like the coconut, is an important multipurpose tree of great utility as each and every part of it and hence the palm is equated to the "Kalpa Vriksha" in the mythology. It is referred as tree of life with nearly 800 uses including food, beverage, fibre, fodder, medicinal and timber.

According to Tamil culture, Palmyra palm trees played an important role in water management in the ancient times. The ancestors had a planned vision for the future water needs of the generations that followed, but most importantly a well-balanced ecosystem was maintained. Unlike, other trees, Palmyra palm's roots go vertically into the ground, which in turn bring the ground water to various levels along the path. In irrigated land, it played a significant role in the circulation of water in the aquifers, river beds and rivers. Rivers remained perennial those times and these trees had its own significance. That is why these trees are considered as 'celestial trees'. You could find all the traditional ponds or lakes they had dug with Palmyra palms around like a fence (Aman *et al.,* 2018).

Figure 16.18: Fully Grown Palmyra Palm.

Economic and Nutritional Importance

Its leaves can be used for the making of mats, baskets, fans, hats, umbrellas, and also as writing material and as thatching for kachha houses in the rural areas. Its trunk which is black in colour, hard and strong hence is used for the construction of buildings and bridges. Stalks of the Palmyra palm is used in making of fences. Sap can be collected from tapping the inflorescences and after that it can be used for drinking or can also be processed into sugar. Endosperm of the fruit is soft and gelatinous which is freshly eaten. Nutritionally Palmyra palm parts consist of carbohydrates, vitamins, proteins which are necessary for diet of human health.

The hard outer wood of the trunk is used as pillars, furniture and supporting tool for kutcha houses. The trunk is also used as pipes to supply water in agricultural land and streams. The wood is used to make walking sticks and windows grills. Dried and holed trunk is used for make boats in coastal region. In a nutshell, the trunk is used appropriately depending on the size, texture and condition of the trunk and in no condition does it go completely waste. The tree wood will be stronger in old age trees.

Comparison of the Nutritional Status of Palmyra Palm

Fruit	*Moisture (%)*	*Fibre (g/100 g)*	*Calcium (g/100 g)*	*Iron (mg/100 g)*	*Vitamin C (mg/100 g)*
Mango	86.1	0.9	0.01	0.3	13
Papaya	89.6	0.8	0.01	0.4	46
Palmyra	77.2	13.5	0.09	-	27

Palmyra fruits are considered as drupes and are highly variable in many morphological characteristics. They are most commonly utilized as a source of therapeutic agents. Palmyra fruits are coconut like structure having three seeds, oval in shape with 12-15 cm wide and capped at the base with overlapping sepals (Morton, 1988). The palmyra palm fruit has been used in both forms (tender and matured). At an early stage the tender endosperm part is edible, while after ripening, the yellow colored fibrous mesocarp squeezed to collect the pulp. The ripe fruit pulp contains beta-carotene and has anti-inflammatory effects. Traditionally the matured pulp is used for making pitha and fermented drink preparations (toddy) (Nadkarni, 1954; Vaidyaratnam, 1994). Traditional cuisines have been employed the fruit pulp of Borassus flabellifer L. and the sap that was collected from the flower section as a sweetener for diabetes patients. Proteins, lipids, carotenoid, vitamin B complex, ascorbic acid and other minerals are found in the palmyra apart from sugars. Besides all palmyra fruits are also rich in immunosuppressive effects (Kapoor, 2000).

Role of Palmyra in Indian Economy

Sugar yielding palms are known for their economic potentialities. Palm belt in the world, comprising Palmyra, the date, the sago and the coconut extends from 44° South latitude to 45° North latitude and widely encompasses the tropical countries like Congo, Sri Lanka, India, Bangladesh, Indonesia and Malaysia. The Genus *Borassus* is one of the most widely distributed of the Palmae with a range extending in a broad belt from Western Africa to Eastern Indonesia. Palmyra, *Borassus flabellifer* L. is grown extensively in drought prone areas of India. It has been estimated that 8.50 crores of Palmyra palm exit in India. It provides immense potentialities for food and industry. Immature endosperm, mesocarp, pulp, tuberous seedlings, sweet sap from the inflorescence, toddy, palm sugar, brush fiber, wood, hypocotyl of young seedlings are the economic parts. The major Palmyra growing state in India is Tamil Nadu where the palm is cultivated in 22,600 hectares. The state tree of Tamil Nadu, Palmyra is a drought tolerant tree, which is cultivated in marginal and wastelands and as a bund crop in cultivated lands. The palms are distributed in Tirunelveli, Thoothukudi, Kanyakumari, Ramanathapuram, Madurai, Salem, Vellore, Cuddalore and Chengleput. districts of Tamil Nadu. The two districts *viz.*, Thoothukudi and Tirunelveli constitute major area under Palmyra. The Palmyra industry is very important in terms of productivity and income generation. The jaggery obtained from neera is superior to cane jaggery. On an average a tree gives about 150 litres of neera. The total generation of wealth would amount to about Rs.852 crores per annum (1985). The palm fiber has great export potential. It is exported to 33 countries with Japan, UK, West Germany, Belgium and Poland accounting to over 80 per cent of export.

Ecological Requirement

Soil and Climate

Palm grows well in fallow lands and all soil types such as sandy soil, red soil, black soil and alluvial soil. It was found that acid soil is not suitable for the Palmyra palm in Sri Lanka but in India it can grow in all types of soil. Deep sandy and loamy soil is preferable.

Palms mainly distributed to tropical and subtropical regions. Minimum sunlight is required for its growth and development. An annual rainfall of 75 mm is required for its growth. Palms can be grown at an altitude of sea level up to 800 m above mean sea level.

Cultivation Practices

Propagation

Crop is propagated through seeds. Ripe fruits are collected from the mother palms which are healthy, piled up for four weeks under the shade then used as seed material.

Seed Germination

Unlike other palms, the Palmyra palm is typically grows downwards before first leaf emergence, which takes 170-180 days after sowing. In Palmyra palm first sheath of the cotyledon elongates and after it has gone into soil the radical comes out and produces the root. Seed nuts collected from a identified mother palm (30 to 40 years age) are stored in shade for 3 weeks. These seeds may be directly sown *in situ* or sown in nursery to raise the seedlings. The seeds of Palmyra should be either planted in the permanent place or in deep container. Seeds start germinating in 3 weeks and emerging shoot will go deep into the soil as tuber in the beginning and later it comes upwards and produces leaves when tubers start rooting. The plumule coming out initially utilizes the food material stored in the seed nut. This germinated part goes down into the earth as a tuber and when the tuber matures rooting starts, which absorbs the soil nutrients. After this leaves come out one after the other.

During germination, the single cotyledon enlarges and emerged as a germ tube breaking the endocarp. The germ tube elongates as a pale yellow sheath enclosing the embryo and carries it down. This hypocotyle is referred as the apocolon, a storage organ of first rudimentary leaf, which gets initiated at the centre of the apocolon, which, although referred as a "tuber" is neither a modified stem nor a root. It is modified hypocotyle enclosing the coleoptile and coleorhizae. During early stages of apocolon formation, the spongy haustorium that is formed in the centre of the seed, stores the hydrolysed sugars of the endosperm.

The results from Pandirimamidi centre also confirmed that, 10-15 days time taken for emergences of plumule from stone under pot irrigation in raised beds. Further it was observed that time taken for maximum tuber development varied between 99-104 days and mean length of apocolon and tuber is 46.7 and 6.5 cm, respectively. Time taken for leaf emergence is about six months.

Nursery Transplanting

Seeds can be sown in mound formed by keeping sand to form a bed of 1 m broad, 60 cm height or nursery bed built with bricks of 2 m broad, 60 cm height. Seeds are sown in 10 cm space and covered with sand. About one year old seedling is lifted from the nursery and containerized in polythene bags. After rooting, transplant can be done in the main field.

Spacing: 4m x 4m (625palms/ha).

Manuring

Generally farmers adopt sheep penning to the palms. Application of 10 kg FYM per pit before planting and dosage may be increased biannually till reaching 60 kg FYM/tree/year.

Composting of Palmyra leaf base pith with the following technology resulted in production of suitable organic manure in 100 to 105 days period with sufficient nutrient content. Palmyra pith (100 kg) + Poultry manure (10 kg) + Urea (1kg) + Pleurotus fungus 2 bottles (1kg) + Ligno Cellulolytic Fungi (1kg) + *Trichoderma viride* (1 kg) (turning for 15 days interval up to 45 days. Application of *Trichoderma viride* on 46th day and again turning at 15 days interval) (Anonymous, 2019).

Figure 16.19: Established Palmyra Palm Plantation.

Inter Cropping

Cow pea, green gram, red gram, Bengal gram or even Ber, Amla, Pomegranate, Karonda, Guava (semi arid zone fruit crops) appears to be good.

Figure 16.20: Palmyra Palms on Bunds.

Pruning

Effect of Defoliation on Inflorescence Sap and Tender Endosperm (Nungu) Yields in Palmyra

To know the effect of defoliation on inflorescence sap and tender endosperm (Nungu) yields in Palmyra a study was conducted at Horticultural Research Station, Pandirimamidi with four different levels of defoliation *i.e.*, 30, 50, 70 and 90 percent defoliation from the years 2010 to 2012. The control plants were left without defoliation. The results revealed that total yield of inflorescence sap was the highest in 30 per cent defoliated palms (194.8 l) followed by Control (158.8 l) in male palms. In female palms also 30 per cent defoliated palms recorded the highest inflorescence sap (210.5 l) followed by Control (193.6 l). Lowest inflorescence sap yields were recorded with 90 per cent defoliated palms in both male (119.5 l) and female (105.4 l) palms. The yield of tender fruit endosperm (Nungu) per bunch was maximum in control palms (2.16 kg/bunch) which were followed by 30 per cent defoliated palms (1.95 kg). Minimum tender fruit endosperm yield per bunch was recorded with 90 per cent defoliated palms (0.98 kg). The number of fruit bunches produced plant^{-1} was maximum (8.61) in control plants followed by 30 per cent defoliated palms (7.96). Minimum number of bunches were recorded with 90 per cent defoliated palms (Murthy *et al.*, 2015).

Figure 16.21: Per cent Defoliated Palm.

Growth and Yield

Palmyra is slow in growth habit, where in first frond appears in about 5 months. First fan shaped tree leaves appears only in the second year. When it attains the height of 12-18m, comes to flowering (13-15 years) for padaneer (sweet sap) purpose. Average of 100-200 litres are obtained for a period of 4 months from February to May. Padaneer and fruit yield are highly variable in individual palms.

Tapping

Extraction of sap (Neera/Padaneer) from inflorescence is called tapping. According to sex of the palm and age of the inflorescence, different kinds of tapping are available.

1. Aripanai

In male palm sheath covering the young inflorescence (2 week old) is removed and allowed to dry for 3 days. The end is cut every time and pot is tied to the inflorescence. This is called Aripanai and practiced for one to one and half month

2. Vallupanai

One month old inflorescence is selected and each male spike bearing sessile flower is pretreated by pressing and stroking of inflorescence. 3-6 such spikes brought together wrapped with Palmyra leaves and fitted to a pot.

3. Thattapalai

It is practiced in female plant. The tapper soften the tissue by hitting the inflorescence, main axis with the iron rod and fork is used to press the regions from which the fruit develop

4. Kaivetty

Is employed when inflorescence is about 2-3 month old. The inflorescence at this stage is matured and bear fruits and the fruits are sliced as the tapping progress. Normally female palm is tapped for a for longer period (April-Dec.) compared to male Dec-Feb.

Neera Tapping

The chief product of the Palmyra palm is the sweet sap obtained by tapping the tip of the inflorescence, as is done with the other sugar palms and the coconut palm. Toddy can be obtained from the young inflorescence of either male or female trees. The Palmyra neera is sweet when quite fresh, but bitter when fermentation sets in. The spathes of the Palmyra trees appear in January and the toddy-tapper commences his operations straightaway. He climbs these trees with the help of a loop made of strips of Palmyra stalk and binds the spathes tightly with strips of Palmyra stalk to prevent them from further expansion and descends, after having thoroughly bruised the flowers within to facilitate the exit of juice. For several mornings this operation of crushing is repeated, and each time a slice is taken off the end of the spathes to facilitate the flow of the sap. In about a week the sap begins to flow freely. When this occurs, an earthen pot is inserted at the tip end of the spathe and this pot is tied to the spathe to collect the juice that flows from this spathe.

Every morning and evening these vessels are emptied, and for a period of six months the Palmyra will continue to pour its sap at the rate of four to seven litres a day. The juice if allowed to ferment is slightly intoxicating and unpalatable. Neera is collected every morning just at sunrise. Neera is brought down from the top of the tree in either earthen post or vessels, and then poured into stainless steel containers and bigger vessels, after being filtered through a fine mesh cloth or wire-mesh. All neera collected as above is brought to a central place where it is processed. It contains all the constituents of a cool and healthy drink with food and mineral value. It builds the body; keeps the human system cool and improves digestion. It can be consumed in fairly large quantity without any harm to the system. Fresh neera, as it trickles from the tree, is easily susceptible to fermentation at ordinary temperature, unless it is quickly treated with some preservative.

Figure 16.22: Palmyra Palm Neera Tapping.

Harvesting

At the age of 25-30 years, tapping can be done. Sap is harvested just before the inflorescences open. The leaves are removed at the time of tapping so that tapper can easily reach the influences. After removing the flower buds in male palms, inflorescences (usually about 12) are tied together. Usually three to four days after cutting the tops of the stalks of the inflorescences, bucket is hung which is supported by a basket. Individual inflorescences are handled in female palm where flowers are broken off and bent towards the bucket to increase the sap flow. This was carried out early in the morning. To prevent fermentation, some slaked lime is added in the sap but this also affects the flavour. Yield of the sap depend on the skill and technique used by the tapper (Sankaralingam *et al.,* 1999).

Yield

(a) 150 litres of padaneer/tree/year

(b) 24 kg jaggery/tree/year

(c) Jaggery recovery/litre of padaneer- 180-250 g of jaggery.

REFERENCES

Abdul Khader KB, 1988. Effect of drip irrigation on yield and yield attributes of arecanut. In *Proceedings of the National Seminar on Drip Irrigation*. Water Technology Centre, TNAU, Coimbatore.

Ahuja Subhash, Ahuja S and Ahuja U 2014. Coconut - History, uses, and folklore. Asian Agri-History. **18**: 221-248.

Aiyer AKYN, 1966. *Field crops of India* (6th ed.). Bangalore Printing and Publishing Co. Ltd.

Aman A, Rajan R and Sinha S, 2018. The Palmyra palm (*Borassus flabellifer* L.): Overview of biology, uses, and cultivation. *Biomolecule Reports,* 4: 1-5.

Aman, 1969. *Medicinal secrets of your food - Arecanut* (pp. 700-702). Secretary, Indo-American Hospital, Mysore, India.

Ananda KS, 1998. Exploitation of dwarf - A natural mutant in arecanut breeding. In N. Muralidharan and R. Raj Kumar (Eds.), *Proceedings of PLACROSYM XIII* (p. 9). Coimbatore, Tamilnadu, India.

Ananda KS, 2007. *Arecanut descriptors Part-I.* Central Plantation Crops Research Institute.

Ananda KS, 2010. Collection, evaluation and utilization of genetic resources in *Areca catechu* L. In *Abstracts of National Conference on Horticultural Biodiversity for Livelihood, Economic Development, and Health Care* (*Swadesh Prem Jagriti Sangosthi-2010*) (pp. 11-12). UHS, Bangalore.

Anonymous, 2019. Annual report for 2018-19. AICRP on palms, Central Plantation Crops Research Institute, Kasaragod, Kerala, India.

Atapattu Anjana J, Udumann and Shashi S, 2024. Leveraging Agroforestry Principles for Nature-Based Climate-Smart Solutions for Coconut Cultivation. 10.1007/978-3-030-98067-2_166-1.

Augustine Jerard B, Maheswarappa HP and Dhillon WS, 2022. *Coconut.* In W. S. Dhillon (Ed.), *Plantation Crops of India* (p. 438). Narendra Publishing House. ISBN: 978-93-91063-83-2.

Balasimha D and Rajagopal V, 2004. *Arecanut.* Central Plantation Crops Research Institute.

Basavaraju TB, Bhagya HP, Prashanth M, Arulraj S and Maheswarappa HP, 2014. Effect of fertigation on the productivity of coconut. *Journal of Plantation Crops* **42(2)**: 198–204.

Bavappa KVA, 1977. Mangala – A superior arecanut variety. *Arecanut and Spices Bulletin,* **8**: 55-56.

Bavappa, K. V. A. (1990). Agronomic desirability of coconut based farming systems. In S. de Silva (Ed.), *Proceedings of Coconut Based Farming Systems*. Proceedings of the XXVII COCOTECH Meeting, 25–29 June 1990, Manila, Philippines (pp. 57–68).

Bavappa KV A and Jacob VJ, 1982. High intensity multi-species cropping: A new approach to small scale farming in the tropics. *World Crops*, **34(2)**: 47–50.

Bavappa KVA, Abdul Khader KB, Biddappa CC, Khan HH, Kasturi Bai KV, Ramadasan A, Sundararaju P, Bopaiah BM, Thomas GV, Misra LP, Balasimha D, Bhat NT and Shama Bhat K, 1986. Coconut and arecanut based high density multi-species cropping systems. *Journal of Plantation Crops*, **14(2)**: 74–87.

Bavappa, KVA and Nair MK, 1982. Cytogenetics and breeding. In K. V. A. Bavappa, M. K. Nair, and T. Prem Kumar (Eds.), *The Arecanut Palm* (pp. 51-96). Central Plantation Crops Research Institute.

Beccari O, 1919. The palms of the Philippine Islands. *Philippine Journal of Science*, **14**: 295-362.

Bhalerao PP, Maheswarappa HP and Sumitha S, 2021. Effect of integrated nutrient management on coconut (Cocos nucifera) based cropping systems in south Gujarat. *Current Horticulture*, **9(2)**: 52-55.

Bharath BG, Ananda KS, Rijith J, Nagaraja NR, Chandran KP, Karun A and Rajesh MK, 2015. Studies on genetic relationships and diversity in arecanut (*Areca catechu* L.) germplasm utilizing RAPD markers. *Journal of Plantation Crops*, **43(2)**: 117-125.

Bhat Ravi, Reddy VM and Jose CT, 2001. Effect of intercrops on arecanut yield. *Journal of Plantation Crops*, **29(1)**: 66-67.

Bhat Ravi, Reddy VM and Khader KBA, 1999. Areca-based high-density multispecies cropping system in coastal Karnataka. *Journal of Plantation Crops*, **27(1)**: 22-26.

Bhat Ravi, Sujatha S and Balasimha D, 2007. Impact of drip fertigation on productivity of arecanut (*Areca catechu* L.). *Agricultural Water Management*, *90*(1-2), 101-111. https: //doi.org/10.1016/j.agwat.2007.02.016

Bhat Ravi, Sujatha S, Khan HH, Sivakumar K and Siju Antony, 2002. Nutrient recycling in areca-based cropping system. In K. Shreedharan, P. K. Vinodkumar, Jayarama, and M. Chulaki Basavaraj (Eds.), *Proceedings of PLACROSYM XV* (pp. 11-14). Mysore.

Bhat KS and Leela M, 1968. Cultural requirement of arecanut. *Indian Farming*, **18**: 8-9.

Bhat KS, and Leela M, 1969. The effect of density of planting on the distribution of arecanut roots. *Tropical Agriculture*, **46**: 55-61.

Bhat KS, Ashwin D and Mythri S, 2016. Analgesic, anti-inflammatory and wound healing properties of arecanut, *Areca catechu* L. *International Journal of Ayurveda and Pharma Research*, **4(12)**: 10-15.

Bhat KS, Ashwin D and Mythri S, 2017. Antidiabetic potential of arecanut, *Areca catechu* L., and certain arecanut formulations available for treating diabetes. *Indian Journal of Arecanut, Spices and Medicinal Plants,* **19(1)**: 23-31.

Bhat KS, Mythri S and Ashwin D, 2017. Antilipidemic activity of arecanut, *Areca catechu* L.: A valuable herbal medicine. *International Journal of Herbal Medicine,* **5(1)**: 35-38.

Bhat NT and Mohapatra AR, 1989. Effect of supplying nutrients through organic manures, inorganic fertilizers, and their combination on arecanut crop. *Journal of Plantation Crops,* **16**(Supplement): 443-447.

Blume, 1836. *Rumphiassive Commentations Botanicae de Plantis Indiae Orientalis.* Leyden.

Chan E and Elevitch CR, 2006. *Cocos nucifera (coconut).* Species profiles for Pacific Island agroforestry, **2(1)**: 1–27.

Chowdappa P, Biddappa CC and Sujatha S, 1999. Efficient recycling of organic wastes in arecanut (*Areca catechu*) and cocoa (*Theobroma cacao*) plantation through vermicomposting. *Indian Journal of Agricultural Sciences,* **69(8)**: 563-566.

DebMandal M and Mandal S, 2011. Coconut (*Cocos nucifera* L.: Arecaceae): In health promotion and disease prevention. Asian Pacific journal of tropical medicine. 4. 241-7. 10.1016/S1995-7645(11)60078-3.

Directorate of Economics and Statistics, 2018. *Agricultural situation in India.* Ministry of Agriculture and Farmers Welfare, New Delhi.

Dransfield J, Baker WJ and Lange CBA, 2008. *Genera Palmyra: Evolution and classification of palms.* Royal Botanical Gardens, Kew.

FAO Statistics, 2016. *http: //www.fao.org/faostat/en/#data*

Fazio S, 2011. *Handbook on Biofuels and Family Agriculture in Developing Countries.*

Gawankar MS, Haldankar PM, Malshe KV, Arulraj S and Maheswarappa HP, 2018. Fertigation schedule for the cultivation of *Tenera* oil palm in Konkan Coastal Region of Maharashtra.

Goh K, Wong CK and Ng PHC, 2016. Oil palm. In *B. 978-0-12-394807-6, Handbook of Oil Palm* (pp. 176-201). doi: 10.1016/B978-0-12-394807-6.00176-3

Gopal Murali, Alka Gupta and George V Thomas, 2007. Coconut leaf vermicompost and vermiwash. Extension folder no. 178, CPCRI, Kasaragod.

Hussain M, Ray AK, Maheswarappa HP, Krishnakumar V, Ravi Bhat, Subramanian P and George V Thomas, 2008. Recycling of organic biomass from arecanut-based high-density multi-species cropping system models under Assam condition. *Journal of Plantation Crops,* **36(1)**: 53-57.

Hussain M, Ray AK, Maheswarappa HP, Subramanian P, Ravi Bhat, Krishnakumar V and Acharya GC, 2011. Sustainable productivity and economics of arecanut-based high-density multi-species cropping system under Brahmaputra valley region of Assam. *Journal of Plantation Crops*, **39(1)**: 68-72.

Jerard BA, Damodaran V, Jaisankar I, Velmurugan A and Swarnam TP, 2018. Coconut biodiversity: Nature's gift to the tropical islands. In C. Sivaperuman, A. Velmurugan, A. K. Singh and I. Jaisankar (Eds.), *Biodiversity and climate change adaptation in tropical islands* (pp. 145–186). Academic Press. https://doi.org/10.1016/B978-0-12-813064-3.00007-2

Kapoor LD, 2000. *Handbook of Ayurvedic medicinal plants: Herbal reference library press*. Florida, USA, p. 82.

Kovoor A, 1983. The Palmyra Palm: potential and perspectives. FAO Plant Production and Protection Paper, FAO, Rome.

Kumar N, 2020. *Introduction to spices, plantation crops, medicinal and aromatic plants* (3rd ed., pp. 178–189).

Kumar SN and Rajagopal V, 2007. Drought management in coconut garden through soil moisture conservation.

Kurian A and Peter KV, 2007. *Commercial crops technology*. New India Publishing Agency.

Mahalakshmi M, Kumar N, Jayakumar P and Soorianathasundaram K, 2001. Fertigation studies in banana under normal system of planting. *South Indian Hort.*, **49**: 80-85.

Mahesha A, 1987. Agro-climatological studies of southern Dakshina Kannada district in relation to plantation crops (M. Tech. Thesis). KREC, Surathkal.

Maheswarappa HP, Dhanapal R, Subramanian P and Palaniswami C, 2013. Evaluation of coconut based high density multi-species cropping system under organic and integrated nutrient management. *Journal of Plantation Crops*, **41(2)**: 130–135.

Maheswarappa HP, Hegde MR, Dhanapal R, Sairam CV and Vidhan Singh T, 2001. Impact of integrated mixed farming system in coconut (Cocos nucifera) garden on coconut yield and economic analysis. *Indian Journal of Agronomy*, **46(1)**: 56–63.

Maheswarappa HP, Sairam CV, Dhanapal R, Vidhan Singh T and Hegde MR, 2000. Economic analysis of coconut based mixed farming systems. *CORD*, **16(2)**: 37–49.

Maheswarappa HP, Krishnakumar and Varatharajan, 2019. An overview on water management in coconut (*Cocos nucifera*). *Indian Journal of Agronomy*. **64**: 431-439. 10.59797/ija.v64i4.5298.

Manorama K, Behera SK, Suresh K, Prasad MV, Mathur RK and Harinarayana P, 2021. Mulching and technological interventions avoid land degradation in an intensive oil palm (*Elaeis guineensis* Jacq.) production system. *Land Degradation and Development*, **32(13)**: 3785-3797.

Marcelino JP and Diaz EV, 2016. Frond pruning enhanced the growth and yield of eight-year-old oil palm (*Elaeis guineensis* Jacq.). *Annals of Tropical Research,* **38(2)**: 96-105.

Mohanaraj SN and Donough CR, 2016. Harvesting practices for maximum yield in oil palm: Results from a re-assessment at IJM Plantations, Sabah. *Oil Palm Bulletin,* **72(72)**.

Mohapatra AR, Bhat NT and Aboobaker VO, 1970. Some aspects of green manuring for arecanut. *Arecanut and Spices Bulletin,* **2**: 6-10.

Morton JF, 1988. Notes on distribution, propagation, and products of *Borassus* palms (Arecaceae). *Economic Botany,* **42(3)**: 420-441.

Murthy GN, Ramana KTV, Raju MS and Vengaiah PC, 2015. Effect of different levels of defoliation on inflorescence sap yield and tender fruit endosperm yield in Palmyra (*Borassus flabellifer* L.). *Journal of Research ANGRAU,* **43(3 and 4)**: 58-62.

Nadkarni KM, 1954. *Indian Materia Medica*. Popular Book Department, Bombay, India, **3(4)**: 2571-2575.

Nair PKR, 1979. *Intensive multiple cropping with coconuts in India*. Verlag Paul Parey.

Nair PKR and Balakrishnan TK, 1976. Pattern of light interception by canopies in a coconut + cocoa crop combination. *Indian Journal of Agricultural Sciences,* **46(10)**: 453–462.

Nampoothiri KUK, Krishnakumar V, Thampan P and Nair A, 2018. The Coconut Palm (*Cocos nucifera* L.) - Research and Development Perspectives. 10.1007/978-981-13-2754-4.

Nelliat EV, 1973. Multiple cropping or multi-storeyed cropping in plantation crops. *Journal of Plantation Crops,* **1**(suppl.): 204.

Nelliat EV, Bavappa KVA and Nair PKR, 1974. Multi-storeyed cropping: A new dimension in multiple cropping for coconut plantations. *World Crops,* **26(6)**: 262–266.

Palaniswami C, Thomas GV, Dhanapal R, Subramannian P, Maheswarappa HP and Upadhyay AK, 2007. Integrated nutrient management in coconut based cropping system. *Technical Bulletin No. 49*. CPCRI.

Pillai NG and Davis TA 1963. Exhaust of macronutrients by coconut palm-preliminary study. *Indian Coconut Journal,* **16**: 81–7.

Pillai RSN and Murthy KN, 1973. Altitude and areca cultivation. *Arecanut and Spices Bulletin,* **4**: 12-13.

Pirker J, Mosnier A, Kraxner F, Havlík P and Obersteiner M, (2016). What are the limits to oil palm expansion?. *Global Environmental Change,* **40**: 73-81.

Prabhu SR, Subramanian P, Biddappa CC and Bopaiah BM, 1998. Prospects of improving coconut productivity through vermiculture technology. *Indian coconut J*, pp. 79-94.

Ramachandrudu K and Manorama K, 2013. Oil palm cultivation. In M. V. Prasad, B. Srinu, and J. Jameema (Eds.), *Compendium of Lectures on Oil Palm Production Technology* (pp. 55-60). Directorate of Oil Palm Research, Pedavegi, Andhra Pradesh.

Ramadasan A and Lal SB, 1966. Exhaust of nutrients from coconut garden – Factors affecting production. *Coconut Bulletin*, **20**: 173–5.

Ranasinghe CS, Premasiri RDN and Silva LRS, 2010. Effect of mulches and cover crops on water status and gas exchange of coconut (*Cocos nucifera* L.) palms in gravelly soils. In *Cocos* (Vol. 15).

Rao KK, 2009. Effect of different methods of irrigation and nutrient requirement on yield of oil palm. *International Journal of Oil Palm*, **6(1)**: 31-34.

Ray AK, Acharya GC, Maheswarappa HP and Krishnakumar V, 2011. Arecanut-based cropping systems in North East Region of India. In G. V. Thomas, V. Krishnakumar, H. P. Maheshwarappa, Ravi Bhat, and D. Balasimha (Eds.), *Arecanut-based cropping/farming systems* (pp. 45-54). CPCRI, Kasaragod.

Reddi SG, Maheswarappa HP, Chandravathi B and Patil DR, 2016. Effect of fertigation on yield and economics of oil palm (*Elaeis guineensis* Jacq.). *Karnataka Journal of Agricultural Sciences*, **29**: 200-202.

Reddi S, Patil GDR, Chandravathi B and Maheswarappa HP, 2015. Studies on vegetables as intercrops in juvenile oil palm plantation and its economics. *Karnataka Journal of Agricultural Sciences*, **28(4)**: 494-496.

Reddy DVS, Subramanian P and Gopalasundaram P, 2002. Coconut based high density multispecies cropping system under different levels of fertilizer in red sandy loam soils. In P. Rethinam, K. K. Khan, V. M. Reddy, P. K. Mandal and K. Suresh (Eds.), *Plantation crops research and development in the new millennium* (pp. 106–111). Coconut Development Board.

Reddy, V. M., and Prasad, M. V. (2011). Banana—A potential intercrop in juvenile oil palm plantation. *International Journal of Oil Palm*, **8(1 and 2)**: 55-56.

Reddy VM, Suresh K, Saraswathi MS, Bijimol M and Nagamani Ch, 2004. Intercropping in irrigated oil palm in India. Paper presented at the *PLACROSYM XVI—Reorientation of Plantation Crops Research in the WTO Regime*, December 14-17, 2004.

Rethinam P, 1990. Cropping systems involving plantation crops. In A. K. Bandhopadhyay, S. Michel Raj, B. Gangwar and J. C. Dagar (Eds.), *Plantation based multiple cropping system*. CARI.

Rethinam P, 2014. Increasing vegetable oil production through oil palm cultivation in India—Status and strategies. *Technical Bulletin of Society for Promotion of Oil Palm Research and Development* (*SOPOPRAD*), 1-18.

Sankaralingam A, Hemalatha G and Ali AM, 1999. *A treatise on Palmyra*. All India Coordinated Research Project, Agricultural College and Research Institute, Tamil Nadu Agricultural University and Central Plantation Crops Research Institute, Indian Council of Agricultural Research.

Sannamarappa M, Kumar SNS and Nagaraj B, 1976. Effect of tillage operations in arecanut gardens of Maidan parts of Karnataka on productivity of palms. *Journal of Plantation Crops,* **4**: 58-59.

Sathyaseelan N and Ravi B, 2022. Coconut planting and aftercare. *Indian Coconut Journal*, 20–25.

Selvamani V, Maheswarappa HP and Chowdappa P, 2017. Soil health management in coconut. In J.-I. P. Maheswarappa and P. Chowdappa (Eds.), *Soil health management in plantation crops* (pp. 27–65). Today and Tomorrow's Printers and Publishers.

Solamalai A, Bhaskar M, Sadasakthi A and Subburammu K, 2015. Fertigation in high value crops—A review. *Agricultural Reviews,* **26(1)**: 1-13.

Subramanian P, Biddappa CC, Maheswarappa HP, Dhanapal R and Palaniswami C, 2000. Growing of Gliricidia as green manure crop in coconut garden under littoral sandy soil. *Journal of Plantation Crops,* **28(3)**: 212–217.

Subramanian P, Dhanapal R, Thomas GV, Maheswarappa HP and Thamban C, 2014. *Coconut based integrated farming system* (Technical Bulletin No. 78). CPCRI.

Subramanian P, Thamban C, Hegde V, Hebbar KB, Bhat R, Krishnakumar V, Niral V and Kumar J, 2018. *Coconut* (Technical Bulletin No. 133). ICAR–CPCRI.

Sudha R, Rajesh MK, Niral V, Samsudeen K, Shil S, Ramchander S and Diwakar Y, 2023. Analysis of genetic diversity and population structure in worldwide coconut germplasm (*Cocos nucifera* L.) using microsatellite markers. *Scientia Horticulturae,* **309**: 111681.

Sujatha S and Abdul Haris A, 2000. Root distribution pattern as influenced by different irrigation methods in young arecanut (*Areca catechu* L.) palms. *Journal of Plantation Crops,* **28(2)**: 117-122.

Sujatha S and Bhat Ravi, 2010. Response of vanilla (*Vanilla planifolia* A.) intercropped in arecanut to irrigation and nutrition in humid tropics of India. *Agricultural Water Management,* **97**: 988-994. https: //doi.org/10.1016/j.agwat.2010.01.031

Sujatha S and Bhat Ravi, 2013. Impact of vermicompost and nitrogen-phosphorus-potassium application on biomass partitioning, nutrient uptake, and productivity of arecanut (*Areca catechu* L.). *Journal of Plant Nutrition,* **36(6)**: 975-988.

Sujatha S and Ravi Bhat, 2015. Resource use and benefits of mixed farming approach in arecanut ecosystem in India. *Agricultural Systems, 141*, 126-137. https: // doi.org/10.1016/j.agsy.2015.10.005

Suresh K and Rethinam P, 2001. Determination of canopy size in oil palm plantations—A survey. *International Journal of Oil Palm*, **2(2)**: 55.

Suresh K, Reddy VM, Sarma KN, Bhanusri A and Sivasankara Kumar KM, 2003. Quantification of root biomass in oil palm grown under basin irrigation. *International Journal of Oil Palm*, **2(3 and 4)**: 39-41.

Tabi KM, Ebongue GN, Ntsomboh GN and Youmbi E, 2017. Effect of dry heat treatment along with some dormancy breaking chemicals on oil palm seed germination. *South African Journal of Botany*, **112**: 489-493.

Vaidyaratnam PS, 1994. *Indian Medicinal Plants: A Compendium of 500 species.* Reprint 2002, Varier Saryavaidya Sala, Orient Longman, India; **4**: 293-296.

Vengaiah PC, Murthy GN, Prasad KR and Kumari KU, 2012. Post-harvest technology of Palmyra (*Borassus flabellifer*): Present practices and scope. In *International Conference on Food Processing* by Omics Group, India.

Vijaya Kumar BG, Veerappadevaru G, Balasimha D, Abdul Khader KB and Ranganna G, 1991. *Journal of Plantation Crops,* **19**: 33-36.

Yadukumar N, Abdul Khader KB and Bhat KS, 1982. Scheduling irrigation for arecanut with PAN evaporation. In *Proceedings of SIJAR, CPCRI, Vittal* (pp. 33-37).

Chapter 17

Nursery and Plantation Management in Oil Palm

K. Ramachandrudu, K. Manorama, A. Subbanna, M. Amrutha Lakshmi and R.P. Premalatha

ICAR–Indian Institute of Oil Palm Research, Pedavegi, Eluru district, Andhra Pradesh

Oil palm is a remarkable plantation cum oil seed crop which has revolutionized the vegetable oil industry globally. The productivity of oil palm is the highest among vegetable oil yielding crop among vegetable oil seeds crops. It is a perennial crop with productive life of 30 years and generic life for about 200 years. Introduction and commercialization of oil palm has lead to profound effect on global food security and economic development of tropical regions like Malaysia, Indonesia, Thailand *etc.* in the world. It is very hardy crop with wider adaptation under various agro-climatic conditions. Oil palm was introduced to India in 1848 as an ornamental palm in Botanical Garden, Kolkata. Organized cultivation of the crop was started in Kerala during 1960s and at present, it is grown commercially in 16 states in an area of 5.00 lakh ha in India. Indonesia, Malaysia, Nigeria, Thailand and India are the leading producers of oil palm in the world while Andhra Pradesh, Telangana, Karnataka, Tamil Nadu and Mizoram are the major growing states in India. Major global exporters of palm oil are Indonesia and Malaysia and the major countries importing palm oil are India, China, Netherlands and Pakistan.

Key Facts about Oil Palm

- ☆ Highest vegetable oil yielding crop per unit area: 4-6t/ha/year.
- ☆ Only oil producing crop which produces two types of oil such as crude palm oil (4-6t/ha) and palm kernel oil (0.4-0.6t/ha)
- ☆ Capable of meeting oil demand in future
- ☆ Long productive life span: 27 years
- ☆ Short gestation period among palms: just 3 years
- ☆ Assured/established marketing system
- ☆ Good support from Government of India through subsidies
- ☆ Strong technical support from layout to harvesting of the produce
- ☆ Intercropping options during both juvenile and matured phases
- ☆ Year round income to farmers
- ☆ Less labour intensive crop
- ☆ Less water consumption compared to rice, sugarcane and banana
- ☆ Generation of huge on farm biomass: 8t/ha/year
- ☆ Highly amenable to organic farming
- ☆ Establishment of processing industries and employment generation.
- ☆ Wide adaption to various agro-climatic conditions and
- ☆ Good response to crop management.

There are three different fruit types in oil palm *viz., Dura, Pisifera* and *Tenera* which look similar morphologically but clearly different in fruit characters.

Dura: Dura palms have fruits with thick shell (2-8 mm in thickness), low to medium mesocarp content (33-55%). These are used as female parent in developing *Tenera* hybrid.

Pisifera: *Pisifera* palms are shell-less and exhibit varying degree of fertility. Female sterile ones are used as male (pollen) parent in development of *Tenera* hybrids.

Tenera: *Tenera* is a commercial hybrid variety which is a cross between *Dura* (female parent) and *Pisifera* (male parent). *Tenera* fruits have thin shell (0.5-4.0mm) with medium to high mesocarp content (60-95%). There is a prominent fibre ring in its mesocarp.

Though the crop can withstand 3 month dry period or water stagnation, continued moisture stress drastically affects the palm growth and yield. Similarly under irrigated conditions, the crop requires around 150 to 300 liters per day per palm depending upon the season and soil type. Areas with very hot days and cool nights give reduced yields due to floral abortion.

In addition to soil moisture, atmospheric humidity strongly influences photosynthetic capacity of oil palm. Oil palm being a tropical crop requires relative humidity of 60-80 per cent throughout the year. Another climatic factor that is critical to oil palm growth is the solar radiation. It requires about 1800-2000 sunlight hours per annum for producing good yield. To receive required amount of radiation, oil palm leaves should be exposed to at least five hours of bright sunlight every day. Solar radiation of < 350 Cal $cm^{-2}day^{-1}$affects the growth and yield of oil palm. In India sunlight is not at all a limiting factor as the cloud cover is restricted to a few days during rainy season and receives more than 5 hours of light daily except on cloudy days.

Soils

Though oil palm can be grown on a variety of soils, moist, lateritic, deep loamy or alluvial soils with rich organic matter and good water permeability are found ideal for its cultivation. The preferred pH is ranged from 4.5-7.0 but tolerates up to 8.2. The electrical conductivity of soils should not exceed 2 dS/cm as it influences the uptake of nutrients (Paramananthan, 2013). It is predominantly cultivated on tropical soils that belong to soil orders ultisols, oxisols and inceptisols. These soils are highly acidic in nature and have low buffering capacity. Care must be taken to avoid coastal sandy soils with poor water holding capacity, saline and alkaline soils. Being hardy in nature, it tolerates periodic flood water or high water table and water stress to a considerable extent. A soil depth of about 100 cm is very much required for its growth and yield. Oil palm grows well up to an elevation of 500m above mean sea level. If the slope of the land is more than 20°, terracing is must for its cultivation.

Selection of Site

Site suitability plays a crucial role in overall performance of oil palm and any wrong decision while selecting the site bears huge penalty as it affects the crop growth and yield throughout its life cycle. While selecting a suitable site for oil palm cultivation, it is essential to consider weather, soil type, soil depth, water availability and land slope. Recognizing importance of site selection, ICAR–IIOPR has developed Decision Support System (DSS) software "Oil Palm Site Suitability Evaluator" (OPSSE) to evaluate the site suitability by providing information pertaining to the site. This considers all above parameters and calculates the suitability class. It provides evaluation report with site suitability grade and expected yield range. Further, OPSSE is expected to aid in proper decision making at individual farm level while selecting a suitable site for growing oil palm both under irrigated and rain fed conditions. Suitability is classified as most suitable, moderately suitable, marginally suitable and not suitable. Depending on comparative advantage with other crops, decision can be taken whether to go for oil palm or not to aid in identifying suitable locations for oil palm cultivation. This DSS is available for use on personal computers and it is providing valuable guidance through a computer-assisted expert system (Manorama *et al.,* 2024).

Hybrid Seed Production and Seed Processing

Oil palm is mainly propagated by seedlings and to some extent clonal plantlets through tissue culture. Oil palm hybrid seed production is a difficult task which involves high technical skills and it is completely different from commercial cultivation. It requires careful selection of mother palms based on a set of standards, crossing desirable parental material under strict quality control measures and finally monitoring the seed germination.

Nursery Management

Oil palm is propagated by seedlings which are raised from seed sprouts for 12 months under specific conditions in nursery. Direct planting of seed sprouts in field is not recommended for the fear of pest damage and lack of genetically uniform stand subsequently. That is why nursery is inevitable in oil palm as it lays the foundation for successful plantations and there are 117 nurseries in India. At the beginning, oil palm seedlings were raised on nursery beds in African countries and then poly bag culture came into practice during 1960's in Malaysia and at present it is a common method globally. Of late, pro tray technology is under practice to some extent in Malaysia, Thailand and Indonesia and now it is widely followed in India during the primary nursery stage. Whatever may be the nursery method, production of quality planting material in oil palm is the basic requirement for successful establishment and productive plantations.

Improper nursery management may lead to unproductive or under productive palms in farmers' plantations. There is a need for intensive care during initial period to have maximum growth as leaf area is directly related with yield. So, any shortcoming in planting material and short cuts in nursery will have long term consequences on yield. In India, oil palm processors are mainly involved in raising seedlings at nurseries and distribution to farmers. Thus, production of genetically uniform seedlings mainly depends on planting of good quality sprouts, proper nursery management as per the set standards and perfect culling of abnormal seedlings. Information generated from series of experiments conducted at ICAR–Indian Institute of Oil Palm Research on management aspects of oil palm nursery is provided under various sub headings in this chapter for successful nursery programme with optimum inputs under Indian conditions.

Nursery Establishment

Criteria for Site Selection

Oil palm seedlings are commonly raised in nursery for 12 months before actual field planting. The selected site must be appropriate for efficient, effective and economical design of the nursery in order to produce quality seedlings throughout the nursery period.

- ✰ Area should preferably be plain or with slope less than 3 per cent without depressions/humps.

- Area must not be prone to flooding/water logging which in turn affects seedling growth due to water stress and nutrient imbalance.
- There must be a permanent water source (bore well/perennial river) for ensuring uninterrupted and adequate water supply. Site must be located on the bank of rivers or water streams in North Eastern States.
- Area should be free from shade from trees as oil palm needs continuous sunlight particularly at secondary nursery site.
- Nursery site need not have any agricultural value but should be well connected by roads for easy access and transport of seedlings.
- Availability of adequate labour force near the vicinity of nursery.
- Availability of good quality water.
- Protection from cattle and wild animals as per the location.
- Assessment of weather data of the location.

Land Requirement

Area/size of nursery depends on the spacing at secondary nursery, duration of nursery, ease of operations and minimum facilities. An optimum spacing recommended for healthy and vigorous growth of seedlings for 12 month duration is 75cm x 75cm x 75cm. So, area can be calculated keeping in view of having 20,512 seedlings/ha at main/secondary nursery, if spaced at 0.75m in equilateral triangular method. Closer spacing *i.e.*, <75cm causes lankiness/etiolation in seedlings at the end of the nursery period. There must be a sufficient area for working area, roads, paths, office cum store room, rest place for supporting/casual nursery workers, pump set room, farm pond/sump, shade net pandal/house, security cabin, gate, fence *etc.* There must be 7 ha land for raising one lakh seedlings for 12 months period under comfortable maintenance by incorporating all the above components. Out of 7ha, minimum of 0.25 ha (2500m^2) area is required for primary nursery under shade net house, about 5ha for secondary nursery and the rest (1.75-1.78 ha) for other purposes (Ramachandrudu *et al.,* 2023).

Site Preparation, Layout and Infrastructure Development

The selected site must facilitate healthy growth of seedlings and easy movement. Felling of trees and clearing of bushes should be carried out in advance. Depressions or humps need to be leveled to avoid the water stagnation which is detrimental to the growth of seedlings. Clearing, designing, lining, fencing *etc.* are important activities during the site preparation. Though oil palm nurseries are big in size and spread in many acres but there is no standard blue print for designing of oil palm nursery.

Farm roads are to be made for easy movement of vehicles or transportation of seedlings by trucks/tractors. Low cost main road and cross roads are to be laid out according to the land shape and purpose. Semi permanent and low cost structures like office room for nursery staff, store room for keeping fertilizers, pesticides and nursery tools *etc.* rest area for supporting staff/daily workers and security cabin for watch and ward have to be constructed. It is better to divide whole area into one acre/hectare blocks for ease of management. There is a need for construction of shade net house or pandal with roof top micro sprinklers or over head micro sprinklers prior to planting of seed sprouts. Similarly, watering system *i.e.*, drip system or over head sprinkler should be installed at secondary nursery area prior to shifting of primary seedlings to secondary stage of nursery.

It is better to have fence around the nursery to avoid the tress passing of animals and it can be either live or barbed wire fence. However, fence is compulsory if the site is nearby forest to avoid entry of wild animals. There must be a farm pond/water harvesting pond in rain fed areas like North Eastern States for watering seedlings during rain free months. Watch and ward is must for safeguarding the nursery round the clock and watch tower is good option for monitoring big nurseries.

Type of Nursery

Oil palm is propagated by seedlings which are raised from seed sprouts in poly bags in double stage nursery system in two distinct stages *i.e.*, primary/pre nursery in small poly bags/pro trays with 3-4 months duration under shade net whereas secondary/main nursery in big poly bags for 9-8 months in open condition. This is the most commonly followed system in all oil palm growing countries.

Poly Bags

Black coloured, recycled, box type (provide level base) and perforated (3-4mm holes at the lower half with a gap of 7.5cm) polyethylene bags are used for raising oil palm seedlings. Two types of poly bags *i.e.*, small size bags (15cm x 23cm/6" x 9" with 250-300 gauge) at the primary nursery stage and big size bags (38cm x 45cm/15" x 18" with 600-800 gauge) during the secondary nursery stage are used.

Pro/Pot Trays

Recycled and black coloured plastic pro or pot trays with 12 or 24 cavities (cavity size: 10cm depth x 6cm top x 4.5cm bottom) with drainage hole at the bottom can be used for raising oil palm seedlings during the primary stage of nursery.

Advantages of Pro Trays over Poly Bags

- More number of seedlings (2.5times higher than poly bags) per unit area can be raised as it needs lesser space.
- Less medium requirement and it is 10times lesser than conventional method. For example, poly bag accommodates about 1.5kg medium while pro tray needs only 150g medium.
- Reduction of water requirement: mostly micro sprinklers (like mist) are used for watering in seedlings in pro trays. The frequency of watering is once a day and twice a day during summer months with 15 min duration.

Retention of moisture is more in organic manure due to more surface area of particles.

- Less labour consumption and all the nursery operations become easier, faster and more convenient. Even supervision and monitoring also become easy. For example, a single worker can plant about 250 primary seedlings in conventional method whereas 750 primary seedlings in pro tray method.
- Less effort and energy are required to fill pro trays as quantity of medium required is 10 times lesser than poly bags. Similarly, less effort is needed for lining and arrangement of trays in the nursery.
- Soilless media can be used which in turn avoid soil borne diseases and insect pests and weed growth.
- Ease of loading/unloading and transport of seedlings to the main nursery.
- Ease of application of liquid manures and fertilizers.
- Absence of transplanting shock after shifting of seedlings to the secondary nursery.
- Though cost of pro trays is higher than poly bags, reusability of trays can compensate the cost and in fact they are more economical over a period of time. Good quality trays can be reused for 4 or 5 times if they are handled carefully.
- Culling of abnormal seedlings is quite easy and more convenient.
- Overall, pro tray technology is quite cost effective when compared with poly bags.

Disadvantages of Pro Trays over Poly Bags

- The main limitations of pro tray method are less cavity size and closer spacing between/among seedlings as compared with poly bags.
- Delay in shifting of primary seedlings beyond 3 or 4 months leads to lankiness due to very close spacing among the seedlings in trays.

Oil Palm Seed Sprouts

The main source of planting material in oil palm is either indigenously produced or directly imported from other countries. Only 20 per cent requirement is met through indigenous source and the rest from exotic sources. Mostly oil palm seed sprouts are imported from Costa Rica, Malaysia, Thailand, Indonesia and Papua New Guinea. Mortality (2-5%) and abnormality (15-25%) of seedlings during the nursery stage and mortality (3-5%) in farmers' plantation during the first year are important factors to be considered while placing order for seed sprouts. There is no standard number of seed sprouts as it mainly depends on the source/variety and nursery management practices. On an average, 180-200 sprouts are needed for planting in one hectare land.

Collection and Preparation of Growing Medium/Mixture

In India, village tank silt, top soil and forest soil are commonly used for raising oil palm seedlings. Proper solarisation of soil is required for killing harmful microorganisms/tiny insects/eggs or grubs *etc.* About 4.5m^3 soil mixture is required for filling 1000 small poly bags. About 10 kg DAP (10g/bag) and 10 kg neem cake powder (10g/bag) have to be incorporated with the medium at the time of filling for promoting more root growth and avoiding root grub incidence, respectively. Similarly, DAP@100g and neem cake powder @100g/big bag should be added to the growing medium before filling. Growing medium *i.e.*, soil must contain organic manure for vigorous and healthy growth of seedlings. Location specific manures like cattle manure, vermicompost, pig manure, goat/sheep manure and poultry manure can be used as growing medium in 2: 1 (v/v) ratio for oil palm. Fully decomposed manure assures the production of good quality and healthy seedlings.

Responsibility of raising oil palm seedlings at nurseries is mainly vested with oil palm processors in India. Oil palm industry is generating massive amount of waste whose disposal is a challenging task to the processors. Requirement of soil can be reduced significantly by using coco peat and palm oil mill wastes which are ligno-cellulosic in nature and good source of plant nutrients like N, P, K, Ca and Mg. Among mill wastes, there is a great scope for using of empty fruit bunch fibre compost, decomposed decanter cake, decomposed/treated POME sludge and boiler ash for raising seedlings in oil palm nurseries as the availability and collection of good quality soil is becoming problem and difficult. There is no need to apply recommended dose of chemical fertilizers when growing mixture is prepared with soil and decanter cake/palm oil sludge in 4: 1 ratio (Ramachandrudu and Suneetha, 2020) and POME sludge in 2: 1 ratio and EFB compost in 2: 1 ratio.

Filling and Arrangement/Lining of Poly Bags

Fill the bags with thoroughly mixed medium without any gap up to top of the bag, water and keep the bags for a week for proper settlement of soil. If any gap in bags, top up with medium. There should be at least an inch (2.54cm) gap in primary bag above the surface of the medium and similarly 5cm or 2" gap in secondary bag. Gap over the medium in bags is required for carrying out various nursery operations conveniently. Average weight of filled primary bag and secondary bag is about 1.5-2.0kg and 20-25kg, respectively.

Care should be taken to provide enough space for vigorous and healthy growth of seedlings during the nursery period in oil palm. For convenience and ease of management in the primary nursery, bags should be arranged in beds with 1.2m width and 0.45-0.60m wide paths among beds and length of the bed as per availability of space inside the shade net house/pandal. Ten bags can be placed across width of the bed and bags of particular cross code should be kept together. An area of 2500m^2 area required for accommodating one lakh primary bags/2.5lakhs pro trays conveniently in shade net pandal/house.

Lining has to be carried out at secondary nursery site to space out seedlings to have an equal access for light and space. There must be a sufficient space between bags and rows to avoid competition for light and have easy access for nursery operations like fertilizer application, weeding, mulching, plant protection, maintenance of drip system *etc.* Bags have to be arranged in equilateral triangular pattern with 0.75m x 0.75m x 0.75m spacing which accommodates about 20,512 bags per hectare. Equilateral triangular method is suggested for accommodating more number (15%) of seedlings/unit area as compared with the square method. Seedling rows must be in straight line to each other and parallel to watering line. Bags should be placed on leveled surface in upright position to avoid the falling of water over the ground.

Selection and Planting of Seed Sprouts

Oil palm seed sprouts with clearly differentiated plumule and radicle are to be selected for ensuring proper seedling growth in the nursery. Optimum stage of sprouts for planting is when the radicle is about 10-15mm long. Physically damaged seed sprouts *i.e.*, seeds without radical or plumule and brown germ affected ones should not be planted. One should be very careful while planting undifferentiated seed sprouts. If sprouts without proper differentiation of plumule and radicle are planted, there is a chance for planting them upside down which in turn results in twisted seedlings. Similarly, over grown sprouts *i.e.*, >30mm radical length are subjected to physical damage to growing portion while trying to push their lengthened radical into the medium at the time of planting. Damaged radicle either by physical force or brown germ disease may affect the root system which may show collante symptoms. There is a dire need to discard genetically abnormal, very weak and diseased/insect pest infested sprouts. Sprouts have to be handled carefully and should not be allowed either to get desiccated or damaged. At the time of planting, they can be carried to nursery site in a small plastic trays covered with moist gunny/cotton cloth. Planting should be taken up immediately after arrival of consignment without any delay. Sprouts must be treated with carbandazim@0.2 per cent (2g/l) solution prior to planting to avoid or kill fungal infections if any.

Water the medium filled bags 2 or 3 days prior to planting of sprouts to keep the medium in moist condition at the time of planting. A hole of 2-3 cm deep should be made in the middle of the bag and then sprout can be placed in a hole gently by keeping plumule pointing upward and radicle downward. Then cover sprouts with medium up to 10mm depth and firm the medium gently to fill the air gaps around the sprout. Good contact with medium is important but avoid pushing it too hard/ excessive firming after planting to avoid physical damage. Tip of the plumule need not be visible and planting in this manner is not recommended as it increases the chance of damage to the growing point and may affect the development of a stable root system due to movement of medium. Deep planting of seed sprouts beyond 3.8cm depth result in higher proportion of juvenile seedlings. And also, it delays emergence of shoot, bifurcation and pinnation of leaves and finally decreases leaf production. Similarly, superficial/shallow planting may lead to twisted seedlings.

Normally shoot emerges from medium in a week time after planting. Monsoon is the right time for initiating the nursery activities though it can be taken up round the year.

There are three types of seed sprouts in oil palm *i.e.*, singles (one sprout/nut), doublets (2 sprouts/nut) and triplets (3 sprouts/nut) and among these singles are preferred planting material. Only the vigorous sprout has to be retained by removing rest of the sprouts in doublets and triplets as the growth of seedlings from weak sprouts has been found significantly lesser when compared with singles in both and primary and secondary nursery stages (Ramachandrudu, 2024).

Shading

Shade for oil palm seedlings is necessary only in primary stage during summer season particularly in places like Andhra Pradesh, Orissa, Gujarath, Karnataka, Tamil Nadu *etc.* wherein temperature goes beyond 40° C in summer months. Shade net pandal or house should be erected by using wooden/bamboo poles/concrete pillars/iron poles/angles or GI frame. Top of the structure must be covered with 75 per cent or 50 per cent green colour UV stabilized HDPE shade net (Ramachandrudu *et al,* 2023) for providing shade to seedlings in primary stage. Sides of the structure should be covered with 35 per cent green shade net to have free aeration which provides comfortable environment for workers and reduces the severity of fungal infections in seedlings due to accumulation of air in the shade net house.

Shifting of Seedlings from Primary Stage (Small bags) to Secondary (Big bags) Stage

Seedlings grown for 3 or 4 months in primary nursery should be transferred to secondary nursery without further delay. Continuation of seedlings beyond 4 months in primary stage leads to entry of roots from bags into land below the bag and all the roots penetrated to the land cut off while lifting seedlings and it leads to shocking of seedlings in secondary stage. A small trowel or cylindrical core can be used for making holes in secondary bags. Primary bags should be cut and removed without making injury to roots. Then place the seedlings in hole made in the centre of secondary bags and consolidate soil around it. Collar of the primary seedling should be leveled with soil surface in the secondary bag. Medium in secondary bags should be in moist condition at the time of planting of primary seedlings. Watering in secondary bags has to be done 2 days in advance of planting and at the same time watering in primary seedlings has to be stopped 2 days before shifting so that medium remains in moist condition in secondary bags and in contact with root system in primary seedlings, respectively.

Deep planting of seedlings beyond 7.6cm from soil surface results in higher proportion of juvenile seedlings. And also, it delays emergence of shoot, bifurcation and pinnation of leaves and finally reduces the leaf production. Similarly, shallow planting leads to exposure of roots and then tilting/lodging/toppling of seedlings. In both the cases, resulted seedlings become not fit for field planting. So, proper handling and planting while shifting can avoid transplanting shock to a great extent. Transplanting shock is not commonly noticed during the rainy season due to presence of enough humidity in the air. Transplanting shock as well as sun scorching can be reduced significantly by keeping primary seedlings intact in open condition for a week.

Mulching

Mulching in secondary nursery stage with oil palm kernel shells, fruit mesocarp fibre, empty fruit bunch fibre, rice husk, saw dust, coir pith, groundnut shell husk, dry grass *etc.* can be done for conserving moisture, regulate soil temperature, preventing weed growth and soil compaction in bags. Thickness of mulch above soil surface should be at least one inch. Better growth and vigour was observed under mulching with oil palm empty fruit bunch fibre@250g/seedling as compared with palm kernel shells and fruit mesocarp fibre (Ramachandrudu, 2024). Covering of nursery site with poly film/mulching sheet or empty fruit bunch fibre (2-3" thickness) in oil palm nursery can be followed to avoid weed growth in inter space of the nursery.

Nutrient Management

Growth and productivity of oil palm in farmers' plantations is mainly dependent on the quality of seedlings which are raised for a period of 12 months under specific nutrient management. There is a direct relation between nutrient management and growth and vigour of oil palm seedlings and mostly primary nutrients like nitrogen, phosphorous and potassium are required during the nursery stage. If any nutrient deficiency like boron observed, relevant fertilizer like borax can be applied to correct as per the requirement. Nutrients availability and uptake depends on pH of the medium and watering. Being heavy feeder and lengthy nursery period, oil palm needs continuous supply of nutrients till they reach farmers' plantations. Growing medium mostly with soil used for oil palm nursery cannot meet the nutrient demand of seedlings. So, there is an immense need for external application of chemical fertilizers to meet the requirement.

Fertilizer quantity and frequency of application depend on the nature of growing medium used and growth of seedlings in oil palm. During the first few weeks of nursery, oil palm seedling is entirely dependent on the supply of stored food material in endosperm. Leaf area of the seedling in primary nursery is significantly correlated with kernel weight, indicating importance of kernel reserves for early growth, but after 6 months in the main nursery the correlation is no longer significant. It is observed in most of the cases that primary nursery seedlings have shown significant response to chemical fertilizers. Chemical fertilizers *i.e.*, DAP and NPK complex (17: 17: 17) fertilizers can be applied during the primary and secondary stages of nursery, respectively.

Fertilizer has to be applied after the complete appearance of first leaf which takes about 3-4 weeks after planting of seed sprouts. Di ammonium phosphate

(DAP) fertilizer has to be applied manually at fortnight interval from 2nd month onwards in primary nursery. Oil palm seedlings need only nitrogen and phosphorous during the primary stage of nursery. The following is the recommended fertilizer schedule during the primary nursery stage (Table 17.1):

Table 17.1: Fertilizer Schedule during the Primary Nursery Stage

Age of Seedling (Month)	*Fertilizer*	*Quantity (g/seedling)*
2-1 fortnight	DAP	1.5
2-2 fortnight	DAP	1.5
3-1 fortnight	DAP	2.5
3-2 fortnight	DAP	2.5
4-1 fortnight	DAP	4.0
4-2 fortnight	DAP	4.0

Di ammonium phosphate and NPK complex fertilizers (15: 15: 15/17: 17: 17/19: 19: 19) can be applied manually at monthly interval during the secondary nursery stage (Table 17.2). Applied fertilizer should be evenly spread on the surface of the medium at least 2cm away from the base of the seedling and incorporated into medium by scrapping the surface of medium and watered immediately. Application of Mg and Boron should be need based. Magnesium deficiency has not been observed so far in oil palm nursery seedlings in India. However, boron deficiency is becoming regular during nursery stage and can be corrected by applying 0.25, 0.5g and 1.0g/seedling with bimonthly gap whenever symptoms are noticed.

Table 17.2: Fertilizer Schedule in the Secondary Nursery Stage

Age of Seedling (months)	*Fertilizer*	*Quantity (g/seedling)*
5	DAP	12
6	NPK complex	16
7	NPK complex	20
8	NPK complex	25
9	NPK complex	30
10	NPK complex	35
11	NPK complex	40
12	NPK complex	45

Foliar Application of Fertilizers

Application of water soluble or dissolvable fertilizers in liquid form over the foliage can be done only in primary nursery as it is not feasible in secondary nursery due to wastage of fertilizer. Utmost care is very much needed in foliar application method. Concentration of fertilizer solution should not be more than 1 per cent for macro nutrients and 0.5 per cent for micro nutrients and excess concentration can lead to burning or death of seedlings. Application must be either in the morning or late afternoon and the whole seedling should be drenched with fertilizer solution. It has to be taken up at weekly or 10days intervals and watering of seedlings has to be stopped for one or two days after foliar spray.

Fertigation is the best option as it is quite cost effective, saves 50 per cent of recommended fertilizer dose and increases fertilizer use efficiency during the secondary stage of oil palm nursery (Ramachandrudu *et al.,* 2009). Chemical fertilizers *i.e.*, DAP and NPK complex fertilizers can be applied@50 per cent of recommended dose of fertilizers/seedling through drip system of watering at monthly interval. **Liquid organics** like fermented palm oil mill effluent, cattle urine and liquid manure@10 per cent concentration can be used at monthly interval for getting good growth and vigour of oil palm seedlings.

Biofertilizers are the formulation of living microorganisms which are capable of fixing atmospheric nitrogen, convert insoluble phosphorus to available form, mobilize other nutritive elements from non usable to usable form through biological processes. Integrated application of biofertilizers (*Azotobacter chroococcum+Bacillus megaterium+Glomus aggregatum+Frateuria aurantia*) and 25 per cent recommended dose of fertilizers (RDF) improved the seedling growth and vigour in oil palm seedlings in nursery. Biofertilizers (solid/liquid) can be applied to oil palm seedlings@ 10g/1ml, 25g/2.5ml and 5g/0.5ml at the time of seed sprout planting, shifting of primary seedlings to secondary stage and after 8 months (booster dose) of the nursery, respectively.

Watering

Water is the most important factor for successful growing of seedlings at plant nursery. Regular watering to seedlings must be ensured to have optimum growth of seedlings in oil palm. Amount of water required depends on stage of nursery, age of seedlings, nature of growth medium and weather of the location. Water requirement of oil palm seedlings at different stages has been estimated @4mm/day in 0-2 months, 5mm/day from 2-4 months, 7mm/day during 4-6 months and 10mm/day at 6-8 month stage. The quantity of water required in liters/hectare can be obtained by multiplying the above depths of water with 10,000. Watering in primary nursery can be done through either roof top micro sprinklers or over head micro sprinklers in shade net house/pandal. It is advised not to do watering through hose pipe as it causes the damage to the seedlings by exposing roots and toppling of seedlings due to more pressure. When hose or sprinklers are used, it should be ensured that water droplet size is fine enough to avoid soil erosion and exposure of the root system. During the primary stage, water@50ml/day/seedling during the first and second months and 75ml in 3rd and 4th month can be given for getting good growth of seedlings.

Watering in secondary nursery stage can be done by either over head sprinkler or drip system depending on nursery size and economic feasibility. The major advantage of sprinkler system is less establishment cost and labour consumption. In addition, it builds up humidity in air in nursery area which in turn reduces atmospheric drought to some extent. But sprinkler system wastes/consumes lot of water which encourages luxuriant weed growth in inter space and leads to regular weed control. Less consumption of water, zero wastage of water, weed free inter space, provision for fertigation, more uniform growth of seedlings are the main advantages of drip system as compared to the sprinkler system. There must be one online or inline dripper/emitter with a discharge capacity of 4lph (litres per hour) in each bag. Higher cost of establishment and maintenance for uninterrupted water supply are somewhat the disadvantages of the drip system. Water@1litre/ day or 2l on alternate days per seedling can be given to obtain optimum seedling growth during the secondary stage of nursery.

Watering should be daily except in rainy season or on rainy days in order to maintain sufficient moisture in growing medium. Watering has to be done daily preferably in the morning. It should be done both in morning and evening during summer months. Care must be taken to avoid water stagnation and washing of soil from the top of the nursery bags which otherwise exposes the growing roots/ seed nuts. Both excess and deficit watering affect the growth of seedlings. Excess watering is quite common during the primary stage in oil palm nurseries and it is the main reason for mortality of sprouts and seedlings due to fungal infection. Visible symptoms of over watering are slight to severe yellowing of leaves and stunted growth of seedlings. Similarly, deficit watering leads to stunted growth and death of sprouts and seedlings. Wilting of leaves is an early sign of under watering and further it hinders opening of leaves resulting in many spear leaves and ultimately leaf production is affected.

Weeding

It is known that weeds compete with seedlings for water and nutrients and suppress the growth of oil palm seedlings as weeds are more vigorous and grow at faster rate. Weeding has to be done manually in primary nursery both in poly bags and inter space between beds. It is better to avoid herbicide usage in inter space between beds because faulty application can lead to damage or death of seedlings in primary nursery. Rather plastic mulch can be used to control weeds between the beds in primary nursery. Good quality plastic mulch sheet can last for 1.5-2 years if handled carefully.

Weeds have to be removed manually in bags and chemically in inter space and around the secondary nursery to prevent harbouring of pests which may act as a source for secondary infection. Glufosinate ammonium or gramoxone @2.5ml/l of water can be sprayed at 3 month interval to check weed growth. While spraying herbicide in inter space, nozzle of the sprayer should be fitted with spray guard to avoid the drift of spray liquid over oil palm seedlings. Do not go for watering after the application of herbicide for 2 days particularly in nurseries wherein sprinkler system is followed. Mulching in combination with drip method in secondary nursery can check weed growth to a great extent. Weed growth in inter space of nursery can be avoided by mulching with plastic mulch or oil palm empty fruit bunch fibre when the nurseries are located nearby oil palm processing mills.

Mortality and Culling of Abnormal Sprouts and Seedlings

Death of seed sprouts before- and after emergence and seedlings during the primary as well as secondary stages of nursery is common in oil palm and the mortality on an average is ranged from 2-5 per cent. Sometimes, mortality goes up to 10-15 percent under poor management conditions.

Oil palm is a highly cross pollinated crop with heterozygous genetic constitution. Genetic segregation takes place during the hybridization process between *Dura* x *Pisifera*. No two individual palms are completely alike both in phenotype and genotype in heterogenous *Tenera* population. Removal of abnormal/ off types with inferior genetic makeup and undesirable/unwanted seedlings during the nursery stage is called culling. It is the most important and crucial operation because negligence or improper culling leads to significant yield loss in farmer's plantations. Only trained staff has to be engaged for identification of abnormal seedlings in nursery.

Genetic and environmental factors are mainly responsible for incidence of abnormalities in oil palm nursery. Sometimes, faulty planting, handling and cultural operations produce abnormal seedlings. So, enough care has to be taken before resorting to culling operation. Recommended frequency of culling in oil palm nursery is three times and the type of abnormalities commonly observed in oil palm nurseries in India are mentioned below:

1. The first round of culling has to be taken up at the time of seed sprouts planting:
   - ☆ Abnormal sprouts
   - ☆ Brown germ affected sprouts
   - ☆ Sprouts with damaged radicle and plumule
   - ☆ Sprouts devoid of either plumule or radicle.
2. The second round of culling, most crucial, should be done at the time of shifting of seedlings from the primary to secondary stage of nursery. The following are abnormalities normally noticed during the primary nursery stage. Seedlings with:
   - ☆ **Narrow or grassy leaves**: Leaf becomes very narrow, erect and finally appears like grass leaves and growth of the seedling becomes slow.
   - ☆ **Rolled leaves**: Leaf lamina becomes rolled which gives spike like appearance and seldom opens correctly.
   - ☆ **Twisted leaves**: Leaves become twisted, role around the seed nut. Upside down planting is mainly responsible for seedlings with twisted leaves.

- ☆ **Crinkled leaves**: Corrugations on leaf lamina are developed. Boron deficiency also produces the same symptom. There is always scope for confusion and it can be tested through application of borax. It is boron deficiency if the seedlings are recovered or else it is a genetic problem. Also, heavy dose and frequent application of pesticides cause crinkling in leaves.
- ☆ **Collantes**: Leaf lamina becomes laterally compressed and forms a constriction in middle of the leaf. Heavy dose and frequent application of pesticides lead to collantes.
- ☆ **Chimera**- Leaves with chlorotic/pale yellow or milky white colour strips on leaf lamina and sometimes it is confused with white stripe.
- ☆ **Albinism**-All the leaves turn to pale yellow or milky white colour.
- ☆ **Runts**-Stunted or very weak in growth as compared with normal seedlings.

3. The third and final round of culling has to be done after 12 months of planting or at the time of dispatch of seedlings to farmers. The following are the abnormal seedlings commonly observed in the secondary stage of nursery:
   - ☆ **Runts**: Seedlings with stunted or weak growth and smaller than normal seedlings.
   - ☆ **Juvenile seedlings**: Older leaves remain undivided and remain as primary leaves even after 12 months of age. Normally root system is week in juvenile seedlings.
   - ☆ **Upright or sterile seedlings**: Leaf bases are widely spaced from one another and leaves are erect and most vigorous in growth. Sometimes, shallow planted seedlings tend to show erect growth habit.
   - ☆ **Flat top seedlings**: Distinct flat top appearance with newly developed leaves rarely projecting beyond old leaves.
   - ☆ **Seedlings with short internodes**: Formation of pinnae closer to one another from the rachis and appears compact.
   - ☆ **Seedlings with wide internodes**: Gap between two leaflets of pinnae is wider.
   - ☆ **Narrow pinnate**: Seedlings having leaves with narrow or rolled leaflets which appear like needles.
   - ☆ **Limp form seedlings**: Leaf has limp and hangs down instead of being erect.
   - ☆ **Chimera**: Appearance of chlorotic or milky white stripes on leaf lamina.
   - ☆ Seedlings with **crinkled/puckered** leaves.
   - ☆ Seedlings with **short** and **broad pinnae**.

In addition, the following type of seedlings must be removed from the nursery:

- ☆ Severely diseased/infested with insect pest/nutrient deficient seedlings.
- ☆ Even healthy and normal seedlings with very weak or slender stem.

On an average, percentage of abnormal seedlings during the primary nursery is varied from 10-15 while it is ranged from 5-10 during the secondary nursery and the total per cent of abnormality in oil palm nursery is about 15-25 under Indian conditions. Good nursery management certainly reduces the per cent of abnormalities. Culled seedlings must be destroyed immediately and should not be allowed to maintain separately.

Preparation and Handling of Seedlings and after Care

At least, a month prior to distribution to farmers, seedlings must be tilted clock- and anticlockwise to cut penetrated roots into the ground below the bags. Prior and after rotation of seedlings, enough water has to be given to the seedlings to avoid sudden shock. This practice is highly helpful as it reduces transplanting shock to seedlings in farmers' fields. Dried leaves at the bottom of the seedlings have to be removed at the time of distribution. It is also suggested to give water to seedlings before dispatching for field planting. Seedlings should be lifted from the bottom and side of the bag with gentle hold on leaves. Proper handling at the time of loading onto vehicles and proper arrangement in vehicles should be done to facilitate easy transport to long distances without any physical damage to seedlings. If any delay in transplanting, seedlings must be placed in partial shade and watered regularly till planting. Or else, all the leaves will dry within a week and affect the survival of seedlings after planting in the field.

General Quality Standards

- ☆ Primary seedlings (4 months old) should have 20-25cm height, 4-6 leaves and 4-6cm collar girth.
- ☆ Secondary seedlings (12 months old) should have 100-120cm height, 12-14 leaves and 20-25cm stem girth.
- ☆ Seedlings must be free from severe incidence of diseases, insect pests and nutrient deficiencies.
- ☆ Seedlings should be normal and vigorous in growth.
- ☆ Normal seedlings with very week or slender stem can also be avoided.
- ☆ After lifting and transporting, seedlings should not show symptoms like leaf drying or yellowing.

The purpose of nursery management is production of good quality seedlings and supplying the same to farmers but not for profit making business. One should not aim for area coverage with discards/seconds but with ordering extra sprouts and raising extra seedlings at nursery.

Micropropagation (Tissue Culture)

Vegetative propagation in oil palm is possible only through *in vitro* regeneration or tissue culture. As compared to conventional breeding programmes and seed production, tissue culture of oil palm offers several advantages. It allows rapid multiplication of uniform planting materials with desired characters. This enables improvement of planting materials using existing individuals which have all or most of the desired qualities such as good oil yield and quality, slow vertical growth and pest resistance. Additionally, it also opens new avenues for producing novel planting materials *via* genetic engineering because tissue culture is the means for regeneration of tissues transformed with genes for traits of interest. Tissue culture is employed as a means for producing good *Tenera* palms for commercial planting and multiplying desired parents *i.e. Dura* and *Pisifera* for seed production. It is also practiced to expedite the exploitation of progenies from inter specific (*E. oleifera* x *E. guineensis*) crosses. Tissue culture in oil palm is practiced in many oil palm growing countries wherein commercial labs producing millions of clonal plantlets. Oil production can be boosted by 25-30 per cent by using tissue culture technology.

The *in vitro* regeneration through somatic embryogenesis of oil palm involves explants collection, callogenesis, somatic embryogenesis, organogenesis and hardening of plantlets and this process takes about 4 years to produce clonal material. Explants used in oil palm tissue culture programme are immature and mature zygotic embryos, young plantlets, immature male and female inflorescence, immature leaves and seedling leaves. Uniform growth, true to type, increase in yields by 15-20 per cent, mass multiplication and disease free and desired propagules are the main advantages of tissue culture in oil palm. However, limited availability of explants, low callusing rate, low embryogenesis rate and mantled abnormality are the main constraints.

Land Preparation and Planting

Land preparation varies with the type of soil chosen for planting oil palm. If possible it is better to prepare the land at least 3 months in advance of planting. In sloppy lands, circular platform with a diameter of 3-4m and 7-8° slope back to the hillside is required. Terracing has to be undertaken in lands where the slope is >20°. Drainage channels have to be provided in advance where drainage is a problem to avoid stagnation of water in upper layers of soil. Oil palm can be planted either in equilateral or hexagonal method with 9m x 9m x 9m x spacing or square method with 9m x 9m spacing. Plant density per hectare is 143 and 123 palms in hexagonal and square method of plantings, respectively. Layout can be done with the help of triangles made up of iron/plastic wire. Mutual shading of fronds/leaves under high density of population reduces palm yield by inducing more male inflorescences and affecting photosynthesis. Rows should be oriented in North-South direction for efficient utilization of solar radiation and avoiding shading.

Pits of 60cm_3 should be dug prior to planting and allowed to season at least a month in advance. Pit dimensions can be extended up to one meter in clay type soils like deep black soils or black soils. Though rainy season is the best for planting of seedlings but it can be done round the year under irrigated conditions. At the time of planting, 30kg of any organic manure/15kg vermicompost, 400 g of SSP or 250g DAP under irrigated conditions or 250g of rock phosphate under high rainfall conditions and 50g of phorate granules/pit are to be incorporated with soil at the time of planting. Twelve month old seedlings are used for planting in the main field. Gap filling has to be done within a year with aged seedlings and special care should be given for gap filled seedlings so that they can catch up the growth with rest of palms quickly. Polythene bag is cut open and the seedling with entire ball of earth is placed in the pit such that top surface of the ball of soil is in level with surrounding soil level as planting below or above soil level is not congenial for later development of the seedling. Then the remaining pit space is filled with soil and compacted. Gap filling/replanting has to be done with healthy and robust seedlings (18 – 24 months) during rainy season. There is a need for special care for replanted palms so that they can catch up with the other palms.

Lay out for triangular system of planting: Mark the base line A in N-S direction at one end of a block and put the stakes at 9m spacing on this line. From this base line, another line B1 in E-W direction is made on left hand corner and stakes are placed at 7.8m on this line. Similarly, another line B2 is made after 54m (6x9) from line A. Now a chain is used on which marks are made at every 4.5m with two different colours so that the distance between similar coloured marks is 9 m. Thus the marks on even numbered positions have one colour and odd numbered positions have another colour. The first (odd) mark is placed on the 2nd stake of B1 and last mark on second stake of B2. Now mark palm positions by placing stakes on another colour (even). Remove the chain and place it on 3rd stake of B1 and B2 and mark the odd numbered positions by fixing stakes and continue till the block is completed.

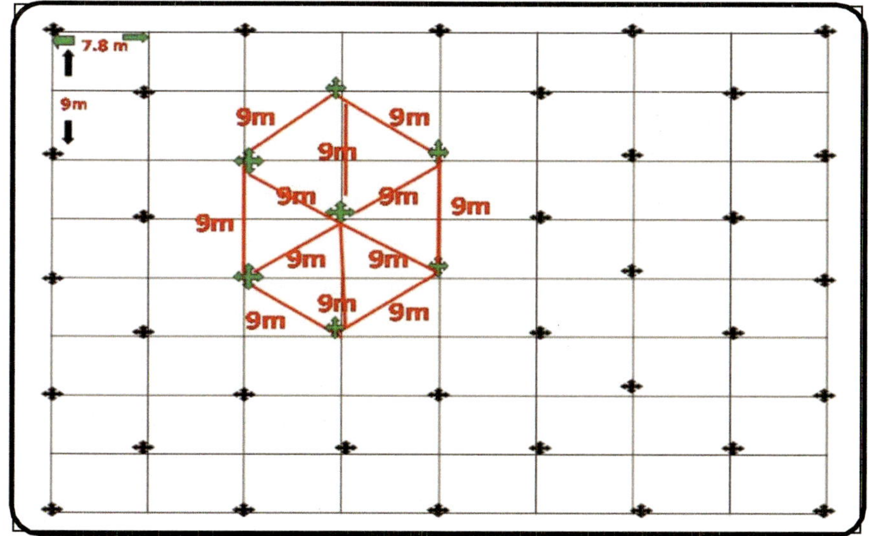

Plant Density in Oil Palm

Crop growth and yield in oil palm mainly depend on the plant density. Optimum plant density is the population wherein interplant competition for light, water and nutrients is almost absent. Mainly, agro-climate and growth habit of genotype decide the optimum plant density in oil palm. Further, management practices and duration of productive life also play a crucial role in plant density. Spacing adopted varied from 7.5 to 10m with usually equilateral triangular arrangement is employed in oil palm, giving between 115-205palms/ha (Hartley, 1988). The recommended plant density in coastal, inland and peat soils of Malaysia is normally 136, 148 and 160palms/ha, respectively. The choice of planting density is thus a compromise between maximizing early yields which demand high density and avoiding excessive inter palm competition at maturity, needing a low density.

Oil palm canopy is almost hemispherical in shape and it does not grow irregularly to fill the available space as dicotyledonous trees do. Equilateral triangular method is followed commercially to accommodate more number of palms/unit of land in oil palm growing countries of the world. The spacing recommended in India is either 9m x 9m x 9m for equilateral triangular method or 9m x 9m for square method of planting which is common for all the oil palm growing states with diversified agro-climatic conditions. Planting density or spacing in oil palm depends upon farmer's choice. If farmer wants inter/companion cropping, he can opt for less plant density (124palms/ha) or else he can go for more plant density (143palms/ha).

Though plant density can significantly impact overall yield and plantation management, but exact number can vary depending on specific conditions and cultivation practices. The following are some general guidelines (Corley and Tinker, 2016):

- ✰ Standard density planting: 136-160 palms/ha
- ✰ High density planting: 160-200 palms/ha
- ✰ Low density planting: 100-120 palms/ha

The plant density of 160 for Peninsula coastal alluvium, 170palms/ha for riverine alluvium was reported as optimum. On organic soils, the optimal plant density found to be around 180-190palms/ha. Palm height increment appeared unaffected by density until 7 to 8 year after planting. Thereafter, increasing density increased height increment (Donough and Kwan, 1991).

Plant density is inversely proportional to the total leaf area which means when leaf area is less, optimal plant density of oil palm can be denser (Corley, 1973). At optimal plant density, leaf area index is 5-6 and the light interception is about 85 per cent when LAI is 6. The optimal plant density changes with age.

Closer planting/high density improves light interception and increases total dry matter production by the crop. However, closer planting decreases dry matter production/palm due to increasing inter palms competition. Up to a point, the reduction in yield will be more than compensated for by increasing number of palms/ha, so that yield/ha rises, even though yield/palm falls. Beyond the optimum planting density, however, yield starts declining due to inter palm competition, less leaf production and self shading on lower leaves (Corley and Tinker 2016).

Breure *et al.* (1990) reported in Papua New Guinea that the maximum yield level with highest density (186 palms/ha) during the yield stabilization period 5-6years. However, the yield level started downfall and recorded the lowest at highest density. The highest bunch yield was observed at 110palms/ha. When palms become 10 year old, leaf area stops increasing and canopy size stabilizes.

Optimal planting density depends mainly on leaf area and light interception (Smith *et al.,* 1996). A density of about 150 palms/ha is reasonable for most environments in Malaysia and Indonesia. A higher density may be needed in parts of West Africa and a lower density may be optimal on volcanic soils of PNG. Fertile soils can have high density as compared to poor soils. It varies between environments and type of planting materials (Corley and Donough, 1992).

Corley and Tinker (2016) opined that curves of cumulative yield plotted against density were quite flat topped and they estimated that a deviation of 10 palms/ha on either side of the optimum would not reduce cumulative yield by more than 1 per cent. Yield was better correlated with distance between palms if the distance between palms within the row and the distance between rows was constant (Corley and Tinker, 2016).

Basin Management

Basins should be formed around the palm after planting of seedlings. In the first year, basin of one meter radius has to be formed and it must be widened to 2 m radius in the second year and 3m radius from the third year onwards. Basin area represents active feeder root zone, therefore, it must be kept weed free to avoid competition for nutrients and water. During the juvenile period, green manuring crops like sun hemp, dhaincha, sesbania, pillipesara can be grown in palm basins and incorporated in the basin itself when they reach flowering stage. Green manuring in the juvenile phase and oil palm leaves and male inflorescences during bearing phase in the form of mulch helps in conservation of moisture, control of weeds, regulation of soil temperature, improvement of physico-chemical and biological properties of soil, addition of organic matter and nutrients in basins. Basins around palms help in demarcation of active root zone, carrying out various cultural operations both in juvenile and matured phase of oil palm. Further, basins facilitate proper utilization of water and nutrients by avoiding wastage. Care must be taken while forming a basin to avoid depression around bole which leads to water stagnation around the bole and finally it will lead to death of palm due to stem wet rot because of continuous water stagnation during rainy season.

Cover Cropping

Growing of leguminous creepers in inter space of oil palm plantations is mainly recommended for rain-fed areas with hilly/mountainous terrains. The main advantages of cover cropping are protection of soil from erosion, smothering of weeds, enriching soil with nitrogen and conservation of moisture. When they are incorporated into the soil, they improve the soil fertility by adding considerable amount of organic matter and plant nutrients. The commonly recommended cover crops are *Pueraria phaseoloides, Calapogonium mucanoides, Centrosema pubescens, Mimosa invisa* and *Mucana bracteata*. Cover crops can be sown immediately after planting of oil palm seedlings in plantation leaving palm basins. Cover cropping is an option for farmers or growers who are not interested in inter cropping.

Ablation

Removal of male and female inflorescences produced during the juvenile phase is called ablation. Flowering in oil palm starts after 12-18 months of planting in the field. Ablation should be done immediately after appearance of inflorescences by pulling out by hand wearing gloves or ablation tool developed by ICAR–IIOPR. tI has to be stopped by middle of the third year or else it will delay the fruiting stage. Chisel with narrow blade can be used for removal of matured inflorescences due to formation of fibre in stalk of inflorescence. Ablation improves the vegetative growth of palm *i.e.*, stem girth, root system and vigour by retaining nutrients which are supposed to be used by developing inflorescences or fruit bunches. It also improves drought resistance capacity of young palms by improving shoot and root growth especially in low production areas wherein dry conditions exist.

Mulching

Mulching is an application of plant biomass on the surface of palm basin to mainly conserve soil moisture and suppress weed growth. In addition, it improves physico-chemical and biological characters of soil, regulates soil temperature and adds a good amount of organic matter and nutrients particularly organic carbon, nitrogen, phosphorous, potassium and magnesium after decomposition. Biomass of oil palm plantations such as leaves (7.2t/ha) and male inflorescences (0.4t/ha), oil palm mill waste like empty fruit bunch fibre can be used for mulching in palm basins (Ramachandrudu, 2024). Empty fruit bunches/fibre and pruned leaves can be used for mulching basins @ 250 and 150 kg/palm/year, respectively. Repeated practice of mulching with oil palm leaves in oil palm improved the soil organic carbon significantly as compared with non mulched palms (Manorama *et.al.*, 2021). Mulching with empty fruit bunches improved soil fertility and oil palm yield significantly (Denang *et al.*, 2020) as compared to non mulched palms. It is the most essential practice which can cut down the frequency of irrigation and water requirement under irrigated areas and significantly extends the soil moisture availability in palm basins under rain fed areas.

Nutrient Management

Oil palm, being a heavy feeder, requires a balanced and adequate supply of nutrients for its optimum growth and yield. Nutrient management should ensure adequate and continuous supply of both macro and micronutrients for getting consistent higher bunch yields. It is estimated that 93.5kg Nitrogen, 11.0 kg Phosphorus, 92.7kg Potassium, 19.3kg Magnesium and 20.3kg Calcium are required for getting 25 tonnes fruit bunches/ha/year. Improving nutrient use efficiency (NUE) is vital for meeting growing food demands and reducing environmental impacts. Utilizing the right source, dose, timing, and method of placement of fertilizers enhances NUE. Straight fertilizers allow precise nutrient application while mixtures are suitable for areas with labor constraints. In acidic soils, rock phosphate is preferred for phosphorus, while polymer-coated fertilizers reduce nutrient losses in light-textured soils with high rainfall. Selection of fertilizers depends on factors such as market availability, soil condition and grower preference.

Fertilizer Dose, Frequency and Method of Application

Fertilizer dose, frequency and method of application are to be site specific and should be based on soil moisture retention characteristics and climatic conditions especially the prevailing rainfall pattern of a locality. A fertilizer dose of 1200g N, 600g P_2O_5 and 1200g K_2O per palm per year from third year on wards is recommended. One third of this dose may be applied during the first year of planting and two-third of the dose during the second year of planting. Fertilizers are to be applied in two equal splits during May-June and September-October under rain fed conditions while in 3 or 4 equal splits under irrigated conditions. Application of biomass or compost is advantageous especially when the soil is poor in organic matter content. Magnesium sulphate @500g per palm is recommended where magnesium deficiency is observed. Borax @ 100-150g per palm is recommended if boron deficiency is noticed. For the newly planted crop, the first dose of fertilizers has to be given three months after planting. Neem cake @5kg can be applied along with second dose of fertilizers. Nutrient losses are to be minimized by adopting suitable agronomic measures. The fertilizer schedule recommended for both rain fed and irrigated conditions are furnished in Table 17.3 and 17.4. For high yielding plantations under irrigated conditions, an additional dose of 25 per cent of recommended dose of fertilizers can be applied to maintain the productivity. However, efforts must be made to apply the fertilizers as per the soil and leaf nutrient status especially under irrigated conditions.

Table 17.3: Recommended Nutrient Schedule for Oil Palm

Age of Palm	*Nutrients (g/palm/yr)*				
	N	*P_2O_5*	*K_2O*	*Mg SO_4*	*Borax*
I Year	400	200	400	125	25
II Year	800	400	800	250	50
III Year onwards	1200	600	1200	500	100

Table 17.4: Nutrient Requirement in Terms of Fertilizers

Age of Palm	*Nutrients in Terms of Fertilizers (g/palm/year)*						
	N as Urea AS	*P_2O_5 as DAP*	*SSP*	*RP*	*K_2O as MOP*	*$MgSO_4$*	*B as Borax*
I Year	870 (or) 2000	–	1250 (or)	1000	667	125	25
			(or)				25
	700 (or) 1610	435	–	–	667	125	
II Year	1740 (or) 4000	–	2500 (or)	2000	1333	250	50
			(or)				
	1400 (or) 3215	870	–	–	1333	250	50
III Year onwards	2610 (or) 6000	–	3750 (or)	3000	2000	500	100
			(or)				
	2100 (or) 4825	1305	–	–	2000	500	100

Fertilizer can be applied through broadcasting in a basin leaving 60cm gap from palm trunk and then mixed with soil by forking. Irrespective of irrigation method, fertilizers should be applied in entire basin during rainy season whereas round the year in case of basin system of irrigation. Spot application is suggested under drip and micro jet system of irrigation. Weeding is must in palm basins prior to application of fertilizers to avoid the competition for nutrients from weeds and irrigation has to be given after fertilizer application.

Fertigation with 50 per cent RDF (600: 300: 600g NPK/ha) at monthly interval can be followed for obtaining optimum growth and fruit bunch yield. Water dissolvable fertilizers like urea, DAP and MOP @ 5-3-5kg, respectively, along with 1.2 kg $MgSO_4$ and 240 g of Borax can be given per month per acre. Prior to and after application of fertilizers through drip system, system has to be run for about 30 minutes to maintain equal pressure as well as flushing of left over fertilizer reminiscences. Each fertilizer has to be dissolved in water separately (in a drum) and pumped through the system. Fertigation offers numerous benefits over soil application methods in oil palm cultivation, including precision, enhanced fertilizer use efficiency, flexibility, 50 per cent reduction in fertilizer dose and environmental sustainability, ultimately contributing to improved crop performance and profitability (Rs.13000/ha) for growers.

Based on the availability, organic manures such as farm yard manure/cattle manure (100 kg) or poultry manure (100 kg) or pig manure (50 kg), goat/sheep manure (50 kg) or vermicompost (25 kg) per palm/year can be applied with onset monsoon every year. Neem cake @10kg/palm/year can also be applied along with organic manures.

Soil and leaf analysis can reveal nutrient imbalances, such as excessive levels of certain nutrients or antagonistic interactions between nutrients. Adjusting fertilizer formulations and ratios based on nutrient balance recommendations can help prevent toxicity or nutrient antagonism and promote optimal nutrient uptake. Therefore, assessing nutrient status via soil and leaf analysis provides valuable insights for determining optimal fertilizer doses. Soil analysis helps evaluate nutrient-supplying capacity, influenced by factors like parent material, climate, rainfall, and human activities. Soils are classified into three fertility classes based on their nutrient status (Table 17.5), guiding fertilizer application rates.

Table 17.5: Fertilizer Recommendation Based on Soil Nutrient Status

Parameter/ Nutrient Element	*Soil fertility status*		
	Low	*Medium*	*High*
Soil organic carbon (%) (SOC)	<0.50	0.50 – 0.75	>0.75
Available N (kg ha^{-1})	<280	280 - 560	>560
Available P_2O_5 (kg ha^{-1})	<20	20 - 50	>50
Available K_2O (kg ha^{-1})	<150	150 - 300	>300
Exchangeable Ca (meq100g^{-1})	<1.5	1.5	>1.5
Exchangeable Mg (meq100g^{-1})	<1.05	1.05	>1.05
Available B (mg kg^{-1})	<0.5 deficient		
	Fertilizer Recommendation		
	Nutrient deficiency symptoms may occur. Fertilizer response is likely. Increase fertilizer dose by 25 per cent	Hidden hunger is likely. May respond to fertilizer. Maintain fertilizer dose	No response to fertilizer. Reduce fertilizer dose by 25 per cent

Leaf nutrient concentrations in oil palm are influenced by various factors, including leaf age, fruiting cycle and soil properties. Developing fertilizer recommendations based on both soil and leaf analyses results are therefore advisable. Optimum nutrient ranges recorded in different locations in India can guide the interpretation of leaf analysis values and formulation of fertilizer recommendations (Table 17.6). This integrated approach ensures efficient nutrient management tailored to specific soil and environmental conditions, promoting sustainable oil palm cultivation in the country.

Table 17.6: Optimum Leaf Nutrient Ranges for Oil Palm Plantations in different States of India

State	*N (%)*	*P (%)*	*K (%)*	*Mg (%)*	*B (mg kg^{-1})*
Andhra Pradesh	1.57-2.63	0.08-0.16	0.48-0.88	0.25-0.71	22.6-60.2
Goa	1.64-2.79	0.31-0.52	0.37-0.75	0.35-0.63	3.10-13.9
Gujarat	2.63-2.85	0.16-0.18	0.56-0.88	0.34-0.84	9.10-32.5
Karnataka	2.34-2.97	0.08-0.40	0.78-0.91	0.25-0.98	5.71-31.0
Mizoram	1.91-2.95	0.46-0.65	0.63-1.00	0.48-0.88	9.41-31.0
Tamil Nadu	1.61-2.11	0.11-0.12	0.33-0.89	0.23-0.73	30.6-54.8

Behera *et al.*, 2022.

Android-based mobile application developed by ICAR–IIOPR incorporates nutrient norms for seven Indian states *viz.*, Andhra Pradesh, Karnataka, Gujarat, Tamil Nadu, Mizoram, Telangana and Goa to generate nutrient advisories based on soil and leaf analysis data (Manorama *et al.*, 2021). This application provides recommendations for both soil application and fertigation, facilitating precise nutrient management tailored to local conditions. Fertigation technique, standardized at ICAR–IIOPR, reduced nutrient requirements by 50 per cent through monthly applications of readily available nutrient solutions, with recommended doses of 600: 300: 600 g of N, P_2O_5, K_2O per palm per year in 12 equal splits (Rao *et al.*,)

Integrated Nutrient Management

A balanced approach combining mineral and organic manures is essential for maintaining soil health. While mineral fertilizers are commonly used to enhance soil fertility, long-term reliance on nitrogenous (N) fertilizers like urea and ammonium sulfate can lead to soil acidification and decreased yields (Phicot *et al.*, 1981). Intensive use of mineral fertilizers may also disrupt soil cation balance as observed in oil palm plantations in Benin, Nigeria (Ogendengbe, 1991). Conversely, organic manures can mitigate the adverse effects of mineral fertilizers on soil health. Compost, for instance, fosters beneficial microorganisms, enhances nutrient conversion, improves soil structure, and aids in pollutant degradation, thereby reducing reliance on inorganic fertilizers and lowering costs. A field experiment was also conducted to evaluate the compost in mature palms involving different combinations of inorganic and organic fertilizers. It is revealed that a combination of one-third nutrient requirement as inorganic fertilizers and two-third requirement as palm waste compost has been found better when compared with only organics. This integrated approach did not affect yield performance of palms (Sunitha and Varghese, 2002) and this combination was also found as cost effective (Sunitha and Varghese, 2004).

Composting oil palm mill waste such as empty fruit bunches (EFB), decanter cake, mesocarp fiber, palm oil mill effluent (POME) and boiler ash, is a recent innovation in Southeast Asia's palm oil industry. Conventional open and closed systems are used for composting, with the addition of mature compost and nitrogen-rich products like urea or mature compost accelerating decomposition. Application of compost has significantly reduced mineral fertilizer requirements in oil palm plantations, resulting in cost savings and improved fertilizer efficiency (Tohiruddin and Foster, 2013).

Soil and Leaf Sampling in Oil Palm Plantation

Nutrient analysis in soil reflects the nutrient status while the leaf analysis indicates uptake of nutrients by palms. Density of soil and leaf sampling in the plantation is one percent. Soil sampling up to 0-30cm is sufficient for general fertilizer recommendations keeping a distance of 1.5m from the trunk of oil palm in the basin. In oil palm, 17th leaf is an indicator leaf for nutrient analysis and sampling has to be done in the forenoon. There must be 2 months gap after fertilizers application for collecting soil and leaf samples. Leaf analysis complements soil analysis and helps in predicting the nutritional disorders before the appearance of visual symptoms in plant tissue. However, the rate of nutrient requirement and crop removal varies widely depending on soil type, palm age, planting material, climatic conditions etc. As it is a high nutrient demanding crop, the net mining of nutrients from soil is enormous and judicious use of fertilizers promises optimum economic and sustainable yields. Calibrated soil and leaf tests help in achieving balanced and precise fertilizer recommendations. It is advised to generate recommendations based on soil and leaf nutrient analysis.

Water Management

Traditionally, oil palm is a rain fed crop in Malaysia, Indonesia, Nigeria and Ghana where the annual rainfall is high (1200-1800 mm) and spread uniformly round the year. But this type of situation prevails only in Andaman and Nicobar Islands in India. It is grown under both rain fed as well as irrigated conditions in India. Relatively, yield level under rain fed conditions is very low as the rainy season is confined to a few months. Irrigation is compulsory under Indian conditions as it is a fast growing crop with high productivity and biomass production. Deficit or surplus of water can create stress to oil palm and adversely affect its growth and yield. Irrigation without any deficit is considered optimum for oil palm, which means that irrigation should be given at such rates and frequencies, so that water is readily available for the plants with minimal losses. To realize its full yield potential, irrigation should be given to ensure adequate moisture in soil throughout the year in tandem with other agronomic practices. Quality of irrigation water has a significant influence on soil properties as well as growth and yield of palm. Water with pH between 6.0 and 7.0 and electrical conductivity <1.0dS/m is generally considered to be desirable for irrigation to oil palm.

When sufficient water is not available for evapotranspiration, oil palm like any other plant regulates transpiration losses through stomata closure. Once the stomata are closed, the photosynthetic activity will be affected which in turn affects both growth and yield. Prolonged drought results in inward leaf curling *i.e.* rolling of leaflets. In extreme cases of drought, necrosis of older leaves, young leaves and ultimately roots. In farmers plantations, commonly observed symptoms in palms are folding of primitive leaves, splitting of spear leaf, non-opening of spear leaves, rolling of leaflets of mature leaves, decreased leaf production due to decreased photosynthesis, yellowing and bronzing of leaf tips, wilting and drying of leaflets. Water deficit suppresses female inflorescence formation and increases maleness and abortion of female inflorescences which in turn reduce bunch yield. So, there is an inverse relation between water deficit and bunch yield *i.e.*, higher the water deficit, lesser is the bunch yield.

Moisture Sensitive Stages

Effect of a drought period on yield components occur with long time-lags (2-3 years) because drought- sensitive processes such as floral sex determination or early inflorescence abortion take place months or years before the maturity of a given bunch. There are two drought sensitive phases in oil palm. The first one is sexual differentiation (approx. 30 months before harvest) and abortion (approx. 10 months before harvest) and the second drought-sensitive phase may coincide with the photoperiod sensitive phase, making their distinction difficult. When moisture stress occurs at above stages, there will be more production of male flowers resulting in decreased yields. Irrigation at these stages can effectively mitigate the adverse effect of water deficit on bunch production. Water stress diminishes FFB yield by inducing inflorescence abortion and reducing the sex ratio, ultimately leading to fewer bunches.

Micro irrigation (drip/micro jet) and check basin methods are normally followed in oil palm plantations in the country. Water use efficiency is low in check basin method due to high water losses in conveyance, evaporation and more leaching of nutrients particularly in light soils whereas it is reverse under micro irrigation methods. There must be 4 drippers of 8 LPH (liters per hour) and 2 micro jets of 35 LPH in each basin to provide the required quantity of water to palms. Oil palm requires high atmospheric humidity for normal growth and yield. A low atmospheric humidity in winter and summer seasons results in high vapour pressure deficit (VPD) above the crop canopy. High atmospheric VPD even under well irrigated conditions prompt the closure of stomata and thereby reduce the photosynthetic activity which in turn leads to leaf breaking with the onset of monsoon.

Oil palm water requirement depends on age of plantation, soil type and weather conditions particularly potential evapotranspiration (PET) of the location. It requires on an average 100 to 250 litres water/palm/day from the third year onwards. The amount should be increased to 300l/day/palm during harsh summer months. As compared with rice (150lakh l/ha/yr), banana (120lakh l/ha/yr) and sugarcane (133lakh l/ha/yr), oil palm requires less water (67lakh l/ha/yr) for its optimum yield. Irrigation is not required during rainy season and it must be need based as per the moisture status and type of soil. When mulching is practiced in palm basins, regular irrigation/amount of water is not needed as it conserves moisture considerably. Water requirement of mature oil palm in typical coastal soil (sandy loam) of West Godavari district in different seasons is furnished in the following Table. Water requirement during the first year and second year should be 1/3rd and 2/3rd of the recommended dose for mature palms.

Season	*Water Requirement in litres*
Summer (March-May)	215-265
Rainy (June-September)	100-150 need based
Winter (October-February)	160-170
Sever summer	300-320

Water stress at different developmental stages affects FFB yields disparately, with continuous flowering palms susceptible to disruptions throughout fruit development stages. Seasonal yield variations persist even under irrigated conditions (Manorama, 2024).

Stage	*Water Deficit (mm/year)*	*Symptoms*	*Yield Loss (%)*
1	< 200	Not a serious problem	0-10
2	200-300	Non opening of immature and younger leaves, defective old leaves	10 - 20
3	300-400	Increased unopening of younger leaves and defective leaves, drying of older leaves	20 -30
4	400-500	Unopened immature leaves and dried leaflets	30 -40
5	> 500	Young leaves may not open, leaf bud cracks and breaks	> 40

Water Requirement of Oil Palm Based on Crop Factor

Water requirement of a crop is the quantity of water required by the crop in a given period of time for its optimum growth under field conditions. It is a function of rainfall, soil water reserves and evapo- transpiration. Water requirement varies from place to place depending on climatic conditions like sunshine hours, temperature, relative humidity, wind velocity *etc.* This is the best available method to estimate crop water requirement from direct measurement of evapotranspiration. In this method, pan evaporation or Penman's estimate of evaporation is multiplied by an appropriate "crop factor".

Water use of crops is very closely related to evaporation. In fact, crop water use is composed of evaporation of water from the soil surface and transpiration of water through the leaves. Combined together, these two factors are named as evapotranspiration. While evaporation is easily measured, transpiration is not. Therefore, it is much simpler to relate the crop evapotranspiration to daily evaporation via a crop factor. A crop factor is related to the percent of ground covered by the crop canopy and therefore will vary depending on the crop stage. For an adult oil palm, 0.7 is considered as "crop factor".

The following simple method of calculation has been devised based on the evaporation rates prevailing in the area especially during summer months:

Evaporation from open pan: 6.70 mm Crop factor : 0.7

Potential evapotranspiration (PE) = Pan evaporation x Crop factor = 6.70 x 0.7 (Crop factor being assumed as 0.7) = 4.69 mm/day

= 46,900 l/day/ha (1 mm of rainfall is equal to 1 l/sq. m)

Since 143 palms are accommodated in one hectare area, the quantity of water per palm per day works out to be 328 l.

Water holding capacity at not less than 70 per cent of the field capacity is acceptable and will not affect the FFB yield of oil palm significantly.

Therefore, the minimum quantity of water to be applied will be: 4.69 mm x 70 per cent = 3.283 mm/day or 32,830 l/ha/day or 220 l/palm/day

By providing irrigation regularly at the above quantities coupled with proper nutrient management, FFB yield @20 tonnes/ha/year can be ensured.

Irrigation trials conducted in different countries indicated that yield responses to irrigation increased with increased level of water deficits. The increase in yield is due to increase in bunch number with little or no effect on bunch weight. Increase in bunch number is due to change in sex ratio and reduced abortion. Effect of irrigation could be seen after several months (as many as 28 months) of irrigation treatment.

At ICAR–IIOPR, research has focused on estimating water requirements at different growth stages of oil palm and comparing irrigation methods for optimal water productivity. Micro irrigation with drip systems has been identified as effective, with a crop factor of 0.7 deemed optimal. Additionally, a web-based application has been developed and made available on the Google Play store for estimating water requirements in specific locations, initially covering Andhra Pradesh, Tamil Nadu, and Karnataka.

Weed Management

Basin area of palm represents active root growth zone which should be kept weed free to avoid competition for nutrients and water or effective utilization of applied inputs. Control of weeds in young plantations is more critical. Weeding in oil palm can be done manually, mechanically and chemically. Uprooting of bushes/

tall statured weeds before flowering has to be taken up regularly to control them permanently during rainy season. Glufosinate ammonium@2.5ml/litre of water or Gramoxone @2.5ml/litre of water can be sprayed periodically with 3months gap to control weeds in plantations. Herbicides such as 2, 4-D, 2, 4, 5-T, halogenated aliphatic acids, dalapon and TCA should not be used as they are found to produce physiological abnormalities in oil palm. Herbicide spraying has to be taken up when weeds are in early growth stage for effective control of weeds and also it may be ensured that spray should not fall on palms during the juvenile phase. And also, petrol operated brush cutter, tractor mounted grass cutter and rotavator can be engaged for cutting/removal of weeds in oil palm plantations. Intercropping, mulching with oil palm leaves/empty fruit bunch fibre and sod culture can reduce weed flora considerably.

Canopy Management

Oil palm tree structure consists of fronds/leaves, inflorescences, fruit bunches, trunk, bole and roots. Oil palm is single stemmed and at the end of trunk is crown portion or canopy consisting of only fronds. So, the canopy of oil palm is restricted to the crown of leaves surrounding apical meristem. Young leaves are erect and matured leaves are horizontal and senescent leaves are somewhat bending and finally canopy looks hemispherical in shape.

When palms are 5-6 year old, they produce about 35-40 leaves/year and then declines to 24-26 leaves after 10 years. On average, mature oil palm tree produces 2leaves/month. Leaf pruning in oil palm is not a recommended practice but it is done for the ease of fruit bunch harvesting. It has to be done as per the requirement, because severe pruning adversely affects palm growth, causes abortion of female inflorescences and finally reduces FFB yield. Pruning should be done by giving clean cut to the petiole as close the stem as possible with the help of a sharp chisel during early phase and aluminium pole with sickle in mature phase. It is to be done during lean crop season preferably after rains. In addition, crown cleaning by removing male inflorescences and bunch sheaths has to be done at the time of leaf pruning to prevent disease infection in the crown portion of palms.

Oil palm possesses a solitary growing point that generates a fresh leaf primordium every fortnight, with a complete leaf unfolding over a span of two years. The inflorescence primordium, emerging from leaf axils, may develop into either male or female inflorescence, with the development process spanning 2.5 to 3 years. At the initial stage of fruiting starting from 2nd to 6th year after planting, sex ratio (number of female inflorescences to total number inflorescences) is very high and the sex ratio starts declining from the 7th year onwards. There is a need to cut or prune one or two leaves prior to harvesting of fruit bunch and harvesting in oil palm is carried out round the year and the frequency is varied from 25 to 40 times per annum. This is a common practice carried out in oil palm plantations. On an average 15 leaves are pruned from a single palm in a year. Canopy management in oil palm is just confined to removal of excess leaves devoid of fruit bunches in their axils, male inflorescences and fruit bunch sheath.

Frond pruning in oil palm significantly affects fruit bunch production specifically number and weight of fruit bunches. However, no significant effect was noticed on its agronomic characteristics. Retention of 32-40 fronds per palm produced more and heavier bunches compared to palms with 24 fronds and palms without frond pruning. Palms without frond pruning and those with excessive pruning below 32 fronds produced lesser and lighter bunches. Hence, there must be 33 leaves or 4 whorls of leaves in the crown of oil palm trees round the year as leaves are source for photosynthates (Suresh *et al.,* and Junito and Elvie, 2016).

Intercropping

Oil palm can be planted either in an equilateral triangular/hexagonal method with 9m x 9m x 9m spacing or square method with 9mx9m spacing. Such a wide spacing is required for oil palm canopy development at peak vegetative growth phase which comes after 10 years of planting. In a pure stand of oil palm, basic resources like soil, light and aerial environment are not being fully utilized. Based on the growth, the life span of oil palm is divided into three distinct phases *i.e.,* first phase called juvenile/pre bearing spreads over 3 years, second phase 4-10 years and third phase starts from 11th year. Estimates have shown that roots of oil palm explore about 5 and 15 percent of available soil volume indicating that 95 and 85 per cent of the soil will be available for the intercrops during the first and second years, respectively. Suresh and Rethinam (2001) reported that the average canopy size was 13.19, 47.86 and 63.58m^2/palm during the first, second and third year of juvenile phase.

The main objective of inter cropping is utilization of space left between two rows of the main crop and produces more output per unit area. Often, inter cropping is chosen as an insurance against the risk of main crop failure or to ensure a stability of crop yields in variable environments or price fluctuations of main crop. Inter cropping is not a priority in major oil palm producing countries like Indonesia and Malaysia as land is not a limiting factor in those countries. However, size of land holding and economic condition of Indian farmers are changing the mindset of farmers to go for inter cropping in oil palm plantations. Further, encouragement of diversified cropping in place of mono cropping under *Atmanirbhar Bharath* for doubling farmers income by the Govt. of India has been promoting cropping systems approach in the country. Inter cropping with location specific annuals, biennials and some perennials in juvenile phase and crops like cocoa, Java long pepper, banana, pineapple *etc.* during mature phase is common in oil palm plantations in India. Like coconut and areca nut plantations, there is a scope for utilizing both horizontal and vertical space by growing suitable crops in mature/grownup oil palm plantations.

Principles of Intercropping in Oil Palm

- ☆ Crops grown in association should have complementary effects rather than competitive interaction.

- Component crops should have similar agronomic practices.
- Component crops should have different root systems so that they do not compete for nutrients and water.
- Standard plant population of the main crop has to be maintained.
- Palms like coconut, areca nut *etc.* cannot be grown as companion/ intercrop in oil palm plantations as all the cultivated palms belong to the same family Arecaceae.
- Component crops of similar pest infestation should not be chosen.
- Planting method and crop management should be simple, less time taking and less cumbersome, economical and profitable so that it may have wider adoptability.

Advantages of Intercropping in Oil Palm

Juvenile Phase

- Provides income to farmers during the juvenile/non bearing phase of oil palm.
- Better utilization of available natural resources like space and light.

Mature Phase

- Better utilization of resources *viz.*, space, light, water, nutrients, labour and time in the system.
- It offers good opportunity to farmers to adopt nature's principle of diversity in their plantations.
- Increased productivity or total biomass production/unit area/unit time due to effective utilization of space and light which otherwise would have been used for the growth of weeds in the system.
- Improves soil fertility, productivity and health by changing the physico-chemical and biological properties of soil significantly.
- Reduces soil and water erosion in the cropping system.
- Control of weeds in the system due to presence of more shade.
- Effective utilization of nutrients from different layers of soil.
- Offers better employment to farm workers and farmers' family.
- Provides crop yields in installments and security during the period of fluctuating market prices of oil palm.
- Provides additional income to the farmers.
- Overall, net profit/unit area is more due to higher productivity of the system.
- Oil palm-cocoa combination can lead to self sustainable system due to generation of huge biomass.

Disadvantages of Intercropping in Oil Palm

Inter cropping in juvenile phase can be practiced without any problem but it creates the following problems during the mature phase:

- There is a risk of pest incidence with certain crop combinations and quick spread of pests due to more density of cropping. Ex. Incidence of bag worm, rugose spiralling white fly *etc.* in oil palm–cocoa combination. There is always a possibility of one component of the system acting as a trap crop which may damage another component subsequently.
- Oil palm leaves at the time of FFB harvesting and leaf pruning fall over the intercrops and can damage intercrops physically which ranges from 10-25 per cent.
- Mechanization of FFB harvesting cannot be possible.

Suitable Intercrops for Oil Palm

Juvenile phase

Except rice, all the location specific sole/pure crops *viz.*, annuals, biennials and perennials-papaya, pine apple, guava, fodder crops can be grown as inter crops in juvenile stage of oil palm. Systematic intercropping can not affect the growth and development of oil palm. Rice also can be taken up on undulated areas or hill slopes (North Eastern States) under rain-fed conditions. But it is strictly prohibited in plains under irrigated conditions wherein water stagnation is must for rice. Studies in Nigeria indicated that oil palm/cassava system is the most attractive system followed by oil palm/maize + pineapple mixture, oil palm + coco-yam, oil palm + banana, oil palm + yam. In Zaire, planting of coffee along with oil palm and removal of coffee after 7 years is recommended.

Mature Phase

Common method of planting in oil palm is equilateral triangular method which accommodates 143 palms per hectare of land. The microclimate of oil palm is totally different from its sister crop coconut. Coconut is quite porous/quite sunny and many crops can be grown whereas oil palm is quite shady which gives scope for some crops only. Sunlight infiltration in grown up oil palm plantations is normally ranged from 22-25 per cent. Temperature is about 3-5°C lesser and relative humidity 10-15 per cent higher when compared with open conditions. Except light, the microclimate of oil palm is suitable for many inter crops but too much shade or less light highly restricts the crop choice significantly. Keeping in view the microclimate of oil palm, series of experiments have been conducted at ICAR–IIOPR, Pedavegi, Andhra Pradesh and its Regional Centre, Palode, Kerala to identify successful inter crops. The list of selected inter crops belong to various categories for mature oil palm plantations is furnished in the Table 17.7.

Table 17.7: Recommended Intercrops for Mature Oil Palm Plantations

Sl.No.	*Category*	*Recommended Intercrops*
1	Plantation crop	Cocoa and coffee
2	Fruit crops	Banana and pineapple.
3	Vegetables	Colocasia and elephant foot yam.
4	Cut flowers	Red ginger lily, heliconia, torch ginger, beehive ginger, spiral ginger, pineapple ginger, shampoo ginger, red button ginger, tulip ginger, shell ginger, Siam rose and Malay rose.
5	Perennial spices	Java long pepper, bush pepper, cinnamon, nutmeg and black pepper (border rows only).
6	Annual spices	Turmeric and ginger.
7	Medicinal crops	Mango ginger, black turmeric, white turmeric, wild turmeric, ginger, black ginger, East Indian arrow root, galanga, aromatic ginger, snap ginger, bitter ginger and brahmi (perennial).
8	Ornamental plants	Asparagus, fish tail fern, philodendron, ti plant, dracaena, aglaonema and dieffenbachia.
9	Aromatic crops	Citronella, lemon grass, palmarosa, coleus *etc.*

It is clear from experimental results that only shade loving and shade tolerant crops can be grown in oil palm since the availability of light in plantations is less than 30 per cent. Among the inter crops, oil palm-cocoa has been identified as the most compatible and synergistic combination. The major advantage with oil palm-cocoa combination is same economic life span *i.e.*, 30years each and hence both crops can be uprooted at the same time.

Cropping Systems

Farmers can cultivate above listed inter crops either individually or in combination in various cropping pattern (double/multiple/multilayer/multitier cropping systems). Successful cropping models standardized at ICAR–IIOPR, Pedavegi are oil palm-cocoa, oil palm-banana, oil palm-Java long pepper +bush pepper, oil palm-red ginger+heliconia, oil palm-Java long pepper + cut foliage plants, oil palm-Java long pepper+ medicinal plants, oil palm-Java long pepper+ pine apple, oil palm+ornamental ginger plants (premium cut flowers), oil palm-fodder grasses (Guinea and Napier) and oil palm-aromatic plants for Andhra Pradesh and Telangana (Ramachandrudu *et al.,* 2018). Oil palm-cocoa/cinnamon and black pepper on glyricidia+black pepper trailed on oil palm+anthurium/kacholam are recommended oil palm based cropping systems under Kerala conditions (Varghese and Sunitha, 2005) and the same can be replicated under high rainfall areas. All the recommended crops are tropical in nature and can come up well in other oil palm growing states of the country. Hence, only the shade loving and shade tolerant crops may be chosen as inter crops in grown up oil palm plantations. Studies clearly indicate the need for systems approach for enhanced productivity as well as profitability in oil palm plantations. Finally, the success of inter cropping in oil palm plantations mainly depends on the selection of suitable crops and efficient management.

Interaction between Oil Palm and Intercrops/Effect of Intercropping on Oil Palm

When crops are grown together, they compete with each other for nutrients, water and space when they are limiting. It is to be ensured that there is no overlapping of active root zones of main crop and inter crops. In general, the nutrient and water requirement of crop combination is substantially lesser than the sum of individual requirement. So, it is recommended that the nutrient requirement of all the crops should be met separately in order to avoid competition.

Studies during juvenile phase revealed that there was no adverse effect on growth of oil palm and almost all inter crops have slightly improved growth of oil palm though differences were found non-significant except for plant height. Banana and guinea grass enhanced the plant height significantly as compared to oil palm without inter crop. Similarly, initial yield of (bunch number and bunch weight/palm) of oil palm was also not adversely affected by inter crops. The uptake of nitrogen by guinea grass was the maximum followed by maize and chillies whereas uptake of potassium was more in banana followed by chillies and guinea grass. All intercrops added substantial quantities of (0.47-17.19t/ha) of organic matter to the soil and thus adding about 7-121 kg N, 0.6-42.4 kg P and 8-194.5 kg K/ha/year (Reddy *et al.,* 2004). Intercropping with banana, pine apple and elephant foot yam in juvenile phase promoted the growth and early yield of oil palm (Gawankar *et al.,* 2018).

The crop combination in the multi-tier system (NRCOP Regional Centre, Palode) *viz.*: oil palm+cocoa or cinnamon or glyricidia with black pepper+pepper trailed on oil palm+anthurium or kacholam under Kerala conditions enhanced the output of the system. Leaf nutrient values of N, P, K, Ca and Mg were found within optimum range irrespective of the treatments. This showed that the system does not have adverse effect on the nutritional health and performance of oil palm. The study also revealed that system helped in reducing the run off of water and soil (Varghese and Sunitha, 2005).

Experimental results at Indian Institute of Oil Palm Research, Pedavegi, indicated that interaction between oil palm and inter crops was complementary when compared to mono cropping. Oil palm FFB yield was higher (3-5t/ha) in intercropped plots as compared with mono cropped one (Table 3). Higher root density of oil palm has been observed in inter cropped (13.29 kg/m^3) area as compared with oil palm alone (10.78 kg/m^3). Roots of oil palm have a tendency of accumulation in areas wherever water and nutrients available. Availability of more moisture must be responsible for higher root density in inter cropped area. Microflora population *i.e.,* bacteria, fungi and actinomycetes was found significantly higher in intercropped area as compared to oil palm alone. Among the inter crops, there was a tremendous improvement in microbial population in cocoa plot. Organic carbon (OC), P and K levels were recorded higher in palm basin as compared with inter space in all the treatments. Among the treatments, significant improvement in OC, P, K and Ca in soil was noticed in inter space of oil palm+cocoa. Higher nutrient levels and microbial population in cocoa plot must be due to addition of more biomass from cocoa. Thus, experiments conducted both in abroad and India revealed that inter cropping in mature phase of oil palm did not affect the yield of oil palm. Further, crops like cocoa and banana add lot of organic matter to the system. Banana adds about 2.5t biomass/ha which in turn adds 11.9kg nitrogen, 3.74kg phosphorous and 96.56kg potassium to the system. In terms of fertilizers, it is equal to 25.87 kg urea, 23.5kg SSP and 160kg MOP (Ramachandrudu *et al.,* 2018).

Cocoa has been found biologically most compatible and physiologically adaptive crop in oil palm plantations. There is a symbiotic association between oil palm and cocoa. Oil palm provides shade to cocoa which in turn provides nutrients to oil palm. Both the crops are known to add lot of biomass to the system. Leaf biomass added by oil palm is 7.2 tonnes/ha/year which adds 67.75 kg nitrogen (147 kg urea) 5.25 kg phosphorous (32.8 kg SSP) and 83.52 kg potassium (139.20 kg MOP). Similarly cocoa adds 1.5 - 2.0t biomass (pruned material) per hectare annually and which supplies on an average 45kg N, 5.5kg P and 27kg K/ha/year which is equal to 98kg urea, 34kg SSP and 45kg MOP, respectively. Also, it provides 250-300kg of pod husk which is rich in potassium (2.5-5.3%) content and adds about 10-15kg potassium. Cocoa sheds leaves which carpet soil, help in conservation of moisture, control of weeds, regulation of soil temperature, water and soil erosion. Further, it improves microclimate of cropping system which in turn reduces temperature and evapo-transpiration, increases microflora and fauna (millipedes and earth worm activity) in the system. Oil palm (fibrous root) and cocoa (tap root) do not compete with each other due to their different root systems. Nutrient values of N, P, K, Ca and Mg in oil palm leaves were found within optimum range. This shows that this cropping system does not have adverse effect on the nutritional health and performance of oil palm rather improves oil palm yield significantly (Ramachandrudu *et al.,* 2024).

Steps for Successful Intercropping in Oil Palm

Many farmers have been cultivating inter crops indiscriminately in oil palm plantations. Cutting/tying of leaves for the sake of more area, ploughing of land close to the palms, lack of basins around the palms, sowing of inter crops in oil palm basin itself and negligence of oil palm during the juvenile phase and more density of intercropping during mature phase are the major mistakes done by the farmers. As a result, oil palm growth and yield are severely hampered and in many cases the yielding stage is delayed by one year. The following are the suggestions for the successful inter cropping in oil palm plantations:

Juvenile Phase

- ✰ There must be a basin around the palm as per the recommendation when inter cropping is practiced.

- ✰ Ploughing of land close to the stem should be avoided as it destroys the root system of oil palm and affects the growth of palms as it reduces absorption of water and nutrients and favours entry of harmful microbial organisms and sometimes leads to death of palms due to severe damage to the root system.
- ✰ Tying of leaves should not be done as it affects the growth of oil palm by reducing photosynthetic area.
- ✰ Wherever inter cropping is practiced in oil palm, it is better to have separate irrigation facility for the main crop and inter crops. This arrangement can meet the irrigation requirement of the system and make the cropping systems successful without any problem.
- ✰ Both the crops should be managed as per the recommended package of practices.

Mature Phase

- ✰ First of all, inter cropping has to be practiced only under assured water supply conditions.
- ✰ Care should be taken to avoid planting of inter crops in oil palm basins to avoid the competition for nutrients and water. A minimum of 3m radius has to be kept free around each palm to avoid any competition with the main crop.
- ✰ Leave intra/columnar space vacant for ease of cultural operations and free movement.
- ✰ Inter crop density must be optimum in order to have free ventilation and filtered light required for maintaining health of the system.
- ✰ Every fifth inter space of oil palm can be left vacant which is needed for running of tractor for loading of fruit bunches or else bunch collection becomes laborious and tedious.
- ✰ There must be a separate irrigation systems for oil palm and inter crops as requirement of water is more in oil palm when compared with inter crops. For example, excess irrigation in cocoa is leading to excess vegetative growth with sparse fruiting.
- ✰ It is noticed in some plantations that leaves of oil palm are pruned to provide light to inter crops. This practice definitely reduces the oil palm yield as it reduces source of food material to fruit bunches. So, unnecessary leaf pruning in oil palm should be stopped.
- ✰ Physical damage by pruned leaves at the time of harvesting of fruit bunches and leaf pruning is unavoidable and damage is varied from 10-25 per cent. Damage is more on tall statured crops like cocoa and banana whereas it is lesser in short statured crops like turmeric, ginger and asparagus. The fallen leaves have to be removed immediately just to avoid more impact on inter crops.

- Combined yield of the system must be higher in monetary terms than that of the main crop when grown alone.
- As a rule, the fertility and productivity of the land has to be taken care of under intensive cropping systems. Only economic produce of the crops has to be taken from the system. As far as possible, crop residues should be recycled to soil by way of mulching in the system.
- Both the crops have to be maintained as per the recommended package of practices.

Integrated/Mixed Farming in Oil Palm

Introduction of livestock component in cropping system is known as integrated or mixed farming system. Forage production in oil palm plantation is the base for integration of cattle. Integrated farming involves the creation of biodiversity in particular location and the repeated use of waste matter of one enterprise for the production of another enterprise and vice versa in the same agro-ecosystem. Recycling of waste products minimizes investment and increases productivity. This system conserves and enriches energy, soil, water, plant nutrients and microorganisms as well as storing energy in the form of biomass. It is a holistic and versatile approach suitable for any kind of social, economic, physical and biological environment.

Cattle, sheep and goat have been studied and found suitable for integration with oil palm in East Asian Countries. It offers forage grown in inter space, crop residues and oil palm mill by-products for livestock. Mixed farming system enables synergistic interactions, which have a greater total contribution than the sum of their individual effects, in which both ecological and economic sustainability can be addressed in a mutually reinforcing manner (Edwards *et al*, 1988). Lot of work has been done in major oil palm growing countries but its application in oil palm estates or farmers' plantations is very limited. Farmers consciously diversify the use of resources to produce a mix of activities that are economically rewarding. Within this broad variety of agricultural activities, opportunities are created that enable switching of practices within and between crops and animals. Diversification rather than specialization is the primary consideration in integrated farming systems in oil palm.

Benefits of Integrated Farming in Oil Palm

Integration of oil palm with livestock is symbiotic in nature and offers the following benefits to plantation and farmers:

- Diversification and efficient use of resources like land, water, light *etc.*
- Efficient use of biological and chemical energy in the system and less dependence on external inputs.

- ✰ Improvement in soil fertility and health through enhancement of micro flora and fauna in the system which in turn lead to enhanced productivity of forage and oil palm.
- ✰ Livestock control weeds through grazing which is eco-friendly as compared to herbicide usage and quite cheaper than manual control of weeds.
- ✰ Establishment of sustainable systems which use recycling of on farm resources, involves no pollution and consistent with environmental protection.
- ✰ Better use of farm labour for higher productivity and increased income.
- ✰ Increased economic output/productivity, reduction of risk and development of stable farm households.

Cattle rearing (grazing and staying) is an integral part of oil palm plantations in Andhra Pradesh and Telangana. Cattle are sheltered either in sheds constructed in plantations or inter space of the plantations. Cattle are supplemented with fodder brought from outside since the forage/weed flora grown in oil palm plantation is not sufficient due to dense shade.

Plant Protection

Insect Pest Management in Oil Palm

In India, around 60 species of insects are reported to feed on oil palm (Kalidas *et al.*, 2014). Among them major defoliators are bag/case worm, leaf web worm, slug caterpillar *etc.* and major sucking pests are invasive whiteflies (Rugose spiraling white fly and Bondar nesting whitefly) and mealy bugs *etc.* Above all, Rhinoceros beetle should have a special mention by being a key pest and omnipresent. In instances, specific insect pest occurrences seen which are confined to nursery stages (for example shoot borer) and as well as adult stages of oil palm. Insect pest occurrence and their control measures are discussed under this section.

Pests of Juvenile Stages of Oil Oalm

Cockchafer beetle and leaf eating caterpillars (bagworm-*Metisa plana*, leaf webworm - *Acria meyeriki*) are the major defoliators causing loss of plant vigour during juvenile stage of oil palm. Their incidence is chiefly observed during the monsoon periods particularly where light soils exist. Bagworm or case worm occurrence can be seen during nursery stages causing heavy defoliation. Early instars of bagworm nibble the leaf chlorophyll causing brown patches and later stages cause irregular holes. They are seen protected around by a case which is hanging from underside of leaf. Leaf web worm, slug caterpillar, grass hopper, aphids, rhinoceros beetle and mealy bugs are other occasional pests noticed during nursery stage of oil palm. The leaf feeding caterpillars can be controlled by spray application of Lambda Cyhalothrin 5EC at 1ml/L whereas the cockchafer beetles and grasshoppers can be managed by spray application of Deltamethrin 2.8EC at 1ml/L. A second spray is also recommended after 15-20 days of first spray if live pest population is still present.

Rhinoceros Beetle

In recent past, incidence of Rhinoceros beetle (*Oryctes rhinoceros*) during juvenile stages of oil palm is at increasing trend causing either death of individual palms or delaying the palm growth by damaging the central shoot. Though the pest incidence is observed round the year, but peak infestation is observed during the monsoon period. Infestation is severe where plantation sanitation is poor. Presence of dead logs, farm yard manure pits, dung heaps and any type of decaying organic matter *etc.* act as breeding sites for Rhinoceros beetles which support the pest buildup ultimately aggravating the damage. Adult beetles are the damaging stage of the pest. They bore into the base of the spear to chew the tender tissues causing the spindle to break at the base and leading to death of central shoot. The central shoot can be easily pulled out from the palm and the tip of the pulled-out shoot looks like chewed fiber. In instances, a secondary shoot may emerge leaving the central shoot. A permanently marked hole at the base of palm frond and 'V' shaped cuttings of leaflets on both sides of the frond are the chief symptoms of damage. In early infection, presence of chewed up fiber near the base of central shoot and beetle inside the bored hole can be observed. The physical damage caused by the beetle to the central shoot is a predisposing factor for the incidence of bud rot and red palm weevil.

For managing Rhinoceros beetle, an integrated pest management has to be adopted. Since the pest is an active flier, management should be adopted at community level for effective management of the pest. It is much easier to control juvenile stages of beetle in breeding sites rather than adults. Breaking up palms into smallest possible chunks helps them to rot down quickly thereby reduces breeding habitats. Similarly, compost pits need to treat with either biocontrol agents like *Metarrhizium anisopliae* or dust formulation of Chlorpyriphos. Either of them is thoroughly mixed with the compost. Predators like reduviid bugs, *Platymeris laevicollis* and parasitoids like *Scolus* sp. are also found to naturally control the grub population of rhinocerous beetle. If the presence of these natural enemies is observed application of insecticide should be avoided. The adult beetles from fresh and active burrows in crowns of young palmcan be extracted by using a iron hook placed at the end of metal rod. After extraction, the bore hole should be filled with a mixture of mancozeb (fungicide) and fine sand at the ratio of 3g: 1kg. *Metarhizium anisopliae* and *Orcytes rhinoceros* virus (baculovirus) are the major microbial biocontrol agents against rhinocerous beetle. Baculovirus solution dipped adult beetles are released in to the natural environment. Both these biocontrol agents cause disease in adult, grub and pupa stages and kill the pest within 15-20 days. Damaged or waste fish nets can also be used for trapping the beetles. A lump of fish nets is placed around the innermost whorl of the palm. Thus, placed nets entangle the incoming population of beetle at the plant itself, which ultimately die due lack of food. Pheromone traps are used for monitoring the pest population. When the daily catch is increasing progressively application of lambda cyhalothrin @ 2ml/L is done. If the catch is still increasing trend, the insecticide spray should be repeated after 15 days of first spray. After insecticide spray, palms with recurrent infestation in the plantation are identified and a perforated plastic sachet containing Naphthalene balls is hanged at the central shoot. Similarly, placing a mixture of neem seed powder and sand in 1: 2ratio @150g/palm at the base of 3 inner most leaves in the crown limits the damage by the beetles.

Pests of Adult Stage Oil Palm

Invasive Whiteflies

In India, recent invasion by exotic whiteflies (Rugose spiraling whitefly-RSW and Bondar nesting whitefly-BNW) created havoc among the oil palm growers due to their wide spread occurrence and damage potential. RSW is an invasive whitefly of neo tropical region and it is highly polyphagous pest having a wide range of host range of more than 200 plant species. Rugose white fly adults are about three times larger than the commonly found whiteflies (approx. 2.5 mm in length) and are lethargic by nature. Males have long pincer-like structures at the end of their abdomen. Females lay eggs on underside of leaves which are elliptical and yellowish in colour, 0.3mm long, translucent with a short stalk and are laid singly. The eggs are covered with irregularly spiralling deposits of white flocculent wax which is seen in a semi-circular spiraling fashion. The first instar nymphs after emergence from the egg settle as a colony and suck the plant sap using needle-like mouth parts. The nymphal stages are immobile, oval become more convex with age. They are light to golden yellow in color, and will produce a dense, cottony wax as well as long, thin waxy filaments which get denser over time and protect the nymph from external environment.

Points to be Considered while Planning Pest Management

Identification of the pest is the most important point to be considered while planning any pest management program. Without correct identity of the pest, the management options adopted may go wrong.

While planning any insecticidal spray, presence of active live stages should be ensured.

Existing and predicted environmental conditions especially the rainfall and temperatures should be taken in to consideration.

Oil palm plantation is of perennial nature. So any application of insecticides disrupts the existing ecosystem with respect to pest and natural enemy and especially the pollinator activity. So insecticide application should be minimized to possible extent.

Insecticide application should be done during evening hours to safe guard the pollinators.

Application of bio control agents should be avoided during hot summer winds.

Split doses of fertilizer application especially nitrogenous fertilizers should be ensured to avoid buildup of sucking pests.

Migration of the pest from locally grown annual crops should be considered.

Oil palm ecosystem sustains variety of natural enemies. Conservation of natural enemies by way of providing plants with nectar and pollen sources for sustaining their growth during off periods of host insects should be considered.

In general, intercropping of oil palm restricts pest movement and distribution in a given plantation. So, farmers should go for intercropping in oil palm plantations.

Neem based formulations or neem seed kernal extracts are better alternatives to chemical pesticides when the population of pest is low. They can be used to avoid the buildup of pest.

Disease Management in Oil Palm

Pests have emerged as significant challenges in oil palm production, leading to diminished quality and yield. Diseases affect the crop at different stages *viz.* seed, nursery and plantation are discussed in this section. Diseases like seed rot and brown germ pose threats during the seed stage, while leaf spots, leaf rots, bud rot, and spear rot are significant constraints in the nursery. In oil palm plantations, basal stem, rot, bud rot, orange spotting, stem wet rot, bunch rot, and spear rot emerge as major threats. Proper monitoring and accurate identification of diseases early is mandatory for successful management strategies. Though some of the diseases are of minor importance, may become severe, if timely diagnosis and corrective measures are not taken up.

Seeds and Seed Sprouts

Brown germ disease in oil palm sprouts begins with light brown discoloration on the radicle, leading to rotting those spreads to the micropyle and halts embryo development, ultimately making the seeds non viable. This is caused by various fungi, such as *Aspergillus niger*, *Aspergillus terreus*, *Botryodiplodiatheobromae*, *Penicillium* species, *Rhizopus nigricans*, *Trichoderma* species, among others, which penetrate the germ pore and infect the embryo and kernel before emergence. Additionally, *Schizophyllum commune* causes seed infections that initially appear as small white mycelial patches on the seed's surface, eventually covering the shell and penetrating the germ plug, blocking respiration and causing seed rot. Fungal infections often occur during fruit bunch harvesting or seed germination due to mesocarp damage or humid, unhygienic conditions. Effective management of these seed-borne diseases relies on seed treatments. Preventive methods include prophylactic treatments with 0.1 per cent Captan, 0.2 per cent Thiram, or 0.1 per cent Propiconazole 25 EC or Difenconazole25 EC for pre- and post-emergence periods. Wet seed dressing with broad spectrum systemic fungicides for 20 minutes helps prevent brown germ disease. Additionally, dipping seeds in bioagent suspensions, such as *Trichoderma* species, *Pseudomonas fluorescens*, or *Bacillus* species for 30 minutes not only controls pathogens but also enhances seed germination. These treatments are vital in curbing fungal infections and ensuring healthy seed development for improved crop yields.

Nursery

Early Leaf Disease/Anthracnose

Leaf diseases caused by fungi *viz. Botryodiplodia theobromae* and *Colectotrichum gleosporoides* (*Glomerella cingulata*) occur in the primary nursery stage. In case of *Botryodiplodia,* small light brown spots develop at the digital portions of the leaves. Lesions enlarge, turn dark brown and are surrounded by a yellow halo resulting in tip blight. Leaves become pale, extremely brittle and get killed. *Pycnidia* are also seen on the spots as small raised black dots. In case of *Glomerella,* yellow to brown lesions develop between veins, elongate and subsequently turn black surrounded by a yellow halo. Fungal fruiting bodies acervuli are formed in the necrotic area.

Leaf Spots

Leaf spot is a prevalent and potentially severe disease affecting oil palm seedlings in the nursery. If left unmanaged, it can pose a significant threat to seedling health and development. These infections typically manifest during both the primary and secondary nursery stages and are triggered by a range of fungal pathogenssuch as*Cercospora, Curvularia, Drecshlera, Neopestalotiopsis, Nigrospora, Oxydothis, Pestalotiopsis, Phoma, Helminthosporium, Phyllosticta etc.*

One of the most prevalent is *Curvularia* leaf spot, caused by *Curvularia lunata, Curvularia oryzae,* and *C. geniculatus.* This disease is characterized by small, circular, translucent yellow spots that appear on both leaf surfaces, which later turn dark brown with a yellowish-orange halo and a sunken center. A distinctive feature is the narrow rim of raised, oily tissue around the lesions. As the disease progresses, it can lead to necrotic lesions that damage leaflets and, in severe cases, destroy the entire frond. Seedlings between two and twelve months of age are particularly susceptible. Another significant leaf spot is caused by *Pestalotiopsis* species, notably *Pestalotiopsis palmarum, P. microspore,* and *P. versicolor*. The lesions typically appear as large, irregular orange spots with dark brown centers, often coalescing to form necrotic patches. This disease is exacerbated by magnesium deficiency and is most common in older fronds. It affects both nursery seedlings and palms in the plantation, leading to drying and brittleness of the leaves.

Cercospora leaf spot, caused by *Cercospora elaedis,* produces similar symptoms, with orange to brown sunken lesions surrounded by a yellow halo. These spots tend to be larger and more frequent on older fronds, particularly along leaf margins, petioles, and the rachis, eventually leading to the death and desiccation of affected tissues. In addition to these, several other fungal diseases contribute to leaf spot conditions. *Helminthosporium carbonum* causes eye spot disease, while *Cochliobolus heterostrophus* leads to leaf rot. Anew, unidentified species of *Helminthosporium* has been found responsible for leaf speckle. Eye spot infection begins as chlorotic yellow patches on the leaf, which eventually turn brown with a circular or oval-shaped halo. Although not lethal, this disease delays transplantation by weakening nursery seedlings. Leaf rot often occurs in tandem with eye spot and causes the necrotic spots to coalesce, resulting in leaf decay. The affected portions of the leaf turn dark brown and become brittle, easily detaching from the plant. Leaf speckle, while less economically significant, manifests in a similar manner but without the necrosis or chlorosis typical of the other infections.

Leaf Rot

Leaf rot in oil palm is attributed to *Rhizoctonia solani*. The infection typically starts at the base of the spear, presenting as an irregular pale green patch bordered by a brown zone that eventually darkens before drying out. Severely affected tissues become frayed and display a shot hole appearance. A hallmark symptom of this condition is the extensive desiccation and fragmentation of dead tissues. Moreover, the risk of plant mortality increases when leaf spot is present alongside anthracnose disease. Leaf rot can also be considered an advanced form of anthracnose caused by *Colletotrichum gloeosporioides* which can lead to the complete deterioration of seedlings.

Management of Leaf Spots and Rots

Effective management of nursery leaf spots and leaf rots involves several key strategies. First, regular monitoring of plant health is crucial to swiftly identify diseases and insect damage, as pathogens often exploit compromised tissues. Sanitation plays a vital role in prevention; therefore, it is essential to hygienically remove diseased leaves and eliminate potential inoculum sources like weeds and mature palms. All nursery tools should be disinfected and using concrete floors instead of soil can help prevent algal growth. Additionally, avoiding overcrowding and excessive shade enhances air circulation and reduces humidity, which is conducive to disease spread. Water management is equally important; keeping leaves dry by avoiding overhead irrigation and scheduling watering during early morning hours limits leaf wetness, thus reducing fungal growth. Fertilizers should be applied judiciously to maintain healthy seedlings while avoiding excessive nitrogen use. Finally, implementing plant protection measures is vital.

Application of fungicides before and after transplanting, as well as during periods of elevated humidity, is critical for preventing disease onset in oil palm seedlings. Due to the heightened sensitivity of these seedlings to copper toxicity, use of copper-based fungicides is strongly discouraged. Instead, effective management can be achieved by utilizing contact fungicides such as Mancozeb at a rate of 2.75 g/liter or Chlorothalonil 78.12 per cent WP at 3 g/liter. These can be alternated with systemic fungicides including Propiconazole 25 EC, Hexaconazole 5 per cent SC, Tebuconazole 25.9 per cent EC, Triflumizole 42.14 per cent SC and Azoxystrobin 23 per cent SC at a concentration of 1 g/liter. Additionally, combination products such as Carbendazim + Mancozeb at 2.5 g/L, Propiconazole 13.9 per cent + Difenoconazole 13.9er cent, Fluopyram 17.7 per cent w/w + Tebuconazole 11.4 per cent, Tebuconazole 50 per cent + Trifloxystrobin 25 per cent w/w WG (75 WG), and CabrioTop (Pyraclostrobin 5 per cent + Metiram 55 per cent WG) at 1 g/ liter can also be employed to enhance efficacy against various fungal pathogens.

Nursery Bud Rot

Young leaf rots from the base upward and dries rapidly. Rotting originates in the bud without previous necrosis of spear, destroys the non emerged leaves and apical meristem. It can be controlled by cleaning and removal of affected tissues using a sharp knife and drenching the crown with 0.1 per cent Carbendazim 50 per cent WP/Chlorothalonil 78.12 per cent WP.

Diseases in Plantation

Bud Rot

Incidence of bud rot is more during monsoon season, particularly June to September. Rhinoceros beetle infestation is a predisposing factor in mature palms. If affected palms are identified in early stage and treated, disease control is assured. Spear leaves initially turn yellow and subsequently brown. Affected spear bends at base and hangs down as rotting starts at the basal portion close to meristem and spear comes out with a gentle pull. Rotten tissues emit an offensive smell. Rotting leads to total destruction of meristem and big crater forms in the centre of crown and causes death of palm. Outer leaves remain green for long period even after death of palm.

Though association of *Botryodiplodiatheobromae,Drechslera*sp, *Cladosporium* sp*Erwinia lathyri, Phytopthorasp*have been reported as pathogenic but the exact cause of the disease is not proven. Prophylactic crown drenching with Carbendazim 50 per cent WP 1g or Chlorothalonil 78.12 per cent WP 3g per 300ml+2g Streptomycin Sulphate 9 per cent and Tetracycline Hydrochloride 1 per cent. once with onset of monsoon and another during post-monsoon season, complete removal of rotten tissues, cleaning of the crown and pasting with 1 per cent Carbendazim 50 per cent WP solution in affected palms.

Basal Stem Rot

Basal stem rot (BSR) is the major disease (*Ganoderma* sp) causing significant losses to the oil palm industry. *Ganoderma* sp have a wide host range such as coconut, toddy palm, areca nut, rubber and several forest trees. Disease incidence is high in coastal plantations where the disease has already affected old coconut plantation and the inoculums is available in debris. Necrosis in lower leaves extend progressively to younger leaves and palm looks like suffering from malnutrition. Appearance of light brown lesions/rotting at the stem base and dry rot in inner portion of stem are some of the symptoms. Roots become friable and disintegrate easily, turn dry and powdery. Formation of brackets can be noticed at the base of palms at advanced stage.

Management

Effective management of Ganoderma should align with the disease's incidence and severity.

1. *Management of dead/collapsed palms and infected debris*

Sanitation: Remove all dead palms, stumps, boles, and brackets, including major root masses, and chip them to <10 cm thickness. Disinfect the area with a 4 per cent formalin solution (40 ml/litre) to eradicate mycelial propagules. Cover the treated area with plastic for 7-10 days, and incinerate the debris.

2. *Management of palms with Ganoderma brackets*

- ✰ **Surgery**: Remove Ganoderma brackets and a small portion of nearby tissue, applying a mixture of copper sulfate and Mancozeb (10 per cent, 100 g in 1 litre per palm).
- ✰ **Isolation by trenches**: Dig trenches 0.6-1 m deep, 0.3-0.5 m wide, and 1.5 m from the trunk.
- ✰ **Chemical treatments**:

  Basin drenching with Hexaconazole 5 EC or Propiconazole 25 EC at 0.2 per cent (80 ml in 40 L) monthly, applied in a ring around the palm.

  Trunk injection of fungicides at 5 per cent (10 ml in 200 ml of water) via a small drilled hole at a 45-degree angle.

3. *Management of abnormal foliar symptoms:*

- ✰ Early detection is crucial. Implement quarterly basin drenching and apply Trichoderma-enriched manure (5-15 kg/palm) or fortified neem cake (5 kg/palm). Combine chemical and biological treatments by initially applying fungicides, followed by Trichoderma applications after a one-month residual period.

4. *Replantation*

- ✰ Excavate soil, old stumps, and roots in a pit (1.5 m x 1.5 m x 1.5 m) and disinfect with a 4 per cent formalin drench (80 ml in 2 liters per pit). Prior to planting, deep ploughing once or twice during summer months expose the soil to temperature which kills soil borne pathogens inoculam like spores and mycelium. Replant 45-60 days later, ensuring seedlings are distanced from previous rows.

Good Agronomic Practices

- ✰ Use 2 kg of Trichoderma-fortified neem cake and maintain soil moisture with mulch. Prevent cross-contamination by ensuring water does not flow from diseased to healthy palm basins and rely solely on micro irrigation for watering. Organic matter in soil should be increased by addition of manures or mulching with oil palm leaves *etc.* Good drainage system has to be provided in poor drained soils.

Stem Wet Rot

The causal organism has not been identified so far. Lower whorls of leaves turn to chocolate brown colour and then it will lead to necrosis and progress to younger whorls of leaves at advanced stage. Sudden and simultaneous death of unexpanded spear leaves and appearance of yellow discoloration in remaining fronds rapidly wither and die. Complete destruction of internal tissues of stem except a very narrow band around stem periphery. Affected internal tissues show fibrous mass which is brown, yellowish or grey in colour become putrified and emit very strong odour. With extensive rotting in centre of the affected stem, usually large cavity will form and finally palm dies.

Management

Early detection of affected palms based on foliar symptoms and test probing of rotting by auguring at bole region. Trunk surgery can be practised by removing rotten tissues should be removed at the base or just above the ground level. After careful scraping, the entire area should be given a prophylactic coat with Hexaconazole 5 per cent EC + Monocrotophos 36 per cent SL followed by hot coal tar. Fill the cavity formed in the stem with sand and cement mixture. Root feeding/ stem injection of Hexaconazole 5 per cent EC 5ml + Streptomycin sulphate 2 gin 100 ml of water after the surgery and prophylactic treatment with a fungicide to the cavity ensures 100 percent recovery.

Bunch Rot

Bunch rot is an important disease in all oil palm growing countries. Initial colonization of fruits is by *Marasmius* followed by secondary invasion by other microorganisms. The most frequently occurring inoculums source is dead fruit bunch. Where diseased bunches are left on palm, spreading of mycelium from one part to the other parts of the palm can often seen.

The disease becomes severe during the monsoon season. During the early stages of infection, strands of mycelium can be seen spreading over the bunch surface affecting only the tissues between individual fruits. Mycelium development is more at back of the bunch against the subtending frond, where conditions are very moist. Such early stage of infection can be seen on unripe bunches of palm when harvested. In the later stages of development, the mycelium grows over the fruit surface and penetrates the mesocarp to produce wet rot. The rotting results in sharp increase in the fattyacid content of the fruit. If affected fruits are left on the palm, the rot ultimately rises leaving only the fibrous tissues of the mesocarp.

Crown cleaning by removing the dead inflorescences, bunch stalks, aborted bunches *etc.* before the onset of monsoon help in reducing the inoculums build-up and harbouring of pathogen. To check the spread of the disease and to eradicate the inoculum of the fungus, crowns of the infected palms should be thoroughly cleaned and sprayed with Carbendazim 50 WP (0.1%) or Hexaconazole 5 EC or Propiconazole 35 EC(0.1%)or Chlorothalonil (0.3%) solution.

Upper Stem Rot

It is caused by *Phellinus noxius* and infected debris in the plantation and air borne spores are the source of this disease. Frond butts of affected palms fall off early. Gum exudation and bleeding occur during early stages. Dry rot appears in irregular pattern underneath bleeding patches. Palm breaks suddenly at the point of rotting. Removal of rotten tissues followed by swabbing of Hexaconazole 5 EC @ 0.1 per cent , application of hot coal tar to the exposed cavity and filling the cavity with sand, cement and insecticide mixture can control the problem.

Nutrient Disorders

The following are commonly observed nutrient deficiencies or disorders in oil palm plantations grown in India.

Potassium Deficiency

This is very common in oil palm plantations grown throughout the country. Soils with low K status, peat soil, soil with poor moisture retention capacity and heavy application of calcium ammonium nitrate, magnesium sulphate, rock phosphate fertilizers and inadequate application of potassic fertilizers cause potassium deficiency. Firstly, pale green spots appear on the leaflets and then turn bright orange colour with necrosis at the centre at advanced stage. Confluent orange spotting, diffused mid crown yellowing, orange blotch and premature desiccation are the symptoms occur in older leaves (Rankine and Fairhurst, 1999). When spotting becomes extensive, tissue becomes grayish brown, brittle and fragment. Application of 3kg muriate of potash (MOP) in three equal splits can correct the deficiency.

Boron Deficiency

It is wide spread in oil palm plantations in the country (Behera *et al.,* 2016). It normally appears under heavy application of N and K fertilizers, high rainfall conditions, in low and high pH soils, sandy and peat soils. Crinkling/puckering in leaf, hook leaf, round leaf tip, blind leaf (a pair of pinnae) does not terminate the apex of leaf), fish bone leaf, leaflet shattering (fracture of midrib and pinnae droop sharply), little leaf(Shortening of leaf and leaflets giving a characteristic flat top appearance), stunted growth, turning/twisting of leaves are the major symptoms. Balanced application of N and K fertilizers and 150g of borax/palm/ year can correct the deficiency.

N/K Imbalance or White Stripe

Excess application of N and B deficiency leads to visibility of white/chlorotic stripes in leaflets of young leaves. Stripes are formed on both sides of leaflet midrib. In severity, the crown becomes upright and obconical in shape. It appears when leaf N: K ratio is more than 2.5: 1. Application of 2.5-4.5kg MOP/palm/year with temporary cessation of N fertilizer rectifies the problem.

Magnesium Deficiency

Of late, Mg deficiency has been noticed in many of the plantations in India. Heavy application of potassic and ammonical fertilizers and inadequate application of $MgSO_4$ leads to Mg deficiency. It is common in lateritic, acidic and peat soils. Olive green and ochre coloured areas appear in older leaves and merge gradually with healthy green leaves. This chlorosis is more pronounced in the central portion of the leaflet as the portion at the distal end experience partial shading. Yellow colour spreads down towards midrib until whole pinnae affected and becomes deep orange in severe stage. Shading effect (absence of chlorosis on shaded portion of leaves) is the characterstic feature of Mg deficiency. Application of 2 kg Mg SO_4 is recommended to control the problem.

Oil Palm is a crop of the future with rich source of nutrition, health, value addition, prosperity, waste utilization, environment friendly, cogeneration, import substitution and sustainability. There is an enormous scope for oil palm development in India due to availability of vast stretches of land under diversified agro-climatic conditions and with untapped underground water potential. Contribution of ICAR–IIOPR to oil palm research and development in India strengthens and aids in India achieving self sufficiency in vegetable oil in the country.

REFERENCES

Babu BK, Mathur RK, Kumar PN, Ramajayam D, Ravichandran G and Venu MVB, 2017. Development, identiûcation and validation of CAPS marker for SHELL trait which governs dura, pisifera and tenera fruit forms in oil palm (*Elaeis guineensis* Jacq.). PLoS One **12(2)**: e0171933. https: //doi.org/10.1371/ journal.pone. 0171933.

Babu BK, Mathur RK, Ravichandran G, Anitha P and Venu MVB, 2019. Genome-wide association study (GWAS) for stem height increment in oil palm (*Elaeis guineensis* Jacq) germplasm using SNP markers. *Tree Genet. Genom.* **15**: 40. https: //doi.org/10.1007/s11295-019-1349-2.

Babu BK, Mathur RK, Ravichandran G, Anitha P and Venu MVB, 2020. Genome wide association study (GWAS) and identification of candidate genes for yield and oil yield related traits in oil palm (*Eleaeis guineensis*) using SNPs by genotyping-based sequencing. *Genomics*, https: //doi.org/10.1016/j.ygeno.2019.06.018.

Baudouin L, Baril C, Clement Demange A, Leroy T and Paulin D,199. Recurrent selection of tropical tree crops. *Euphytica*, **96**: 101-114.

Behera SK, Rao BN, Suresh K, Ramachandrudu K and Manorama K, 2016b. Soil fertility, leaf nutrient concentration and yield limiting nutrients in oil palm (*Elaeis guineensis*) plantations of Surat district of Gujarat. *Ind. J. Agril. Sci.* **86(3)**: 409-413.

Behera SK, Suresh K, Rao BN, Manoja K and Manorama K, 2016c. Soil Nutrient status and leaf nutrient norms in Oil Palm (*Elaeis guineensis* Jacq.) plantations grown in West coastal area of India. *Communications Soil Sci. Plant analysis,* **47(2)**: 255-262.

Behera SK, Suresh K, Rao BN, Ramachandrudu K, Manorama K and Harinarayana P, 2017. Soil fertility and yield limiting nutrients in oil palm plantations of north-eastern state Mizoram of India. *J. Plant Nutri.* **40(8)**: 1165-1171.

Behera SK, Shukla AK, Suresh K, Mathur RK 2022. Nutritional imbalances and nutrient management in oil palm. In: Subhra Saikat Roy, Poonam Kashyap,Tarun Adak (ed) Natural Resource Management in Horticultural Crops, Today and Tomorrow's Printers and Publishers, NewDelhi-110002, pp 161-185.

Breure CJ, Menendez T and Powell MS, 1990. The effect of planting density on the yield components of oil palm. *Exptl. Agri.* **26(1)**: 117-124.

Chan KW, Watson I and Lim KC, 1981. Use of oil palm waste material for increased production. *Planter,* **57**: 14-37.

Chen CP and Baker MJ, 1993. Pastures as the secondary component in tree-pasture systems. grasslands for our world, published by SIR Publishing, Wellington, pp.756-762.

Corley RHV, 1973. Effects of plant density on growth and yield of oil palm. *Exp. Agric.* **9**: 169-180.

Corley RHV, Lee CH, Law IH and Wong CY, 1996. Abnormal flower development in Oil Palm clones. *Planter.* **62**: 233-240.

Corley RHV and Tinker PB, 2016. *The Oil Palm.* 4th Edn, A Blackwell Publishing Company, UK.

Culman M, Claudio MdeF, Bayonac C and Cruza JDC, 2019. Using agrometerological data to assist irrigation management in oil palm crops: A decision support metyhod and results from crop model simulation. *Agril. Water Manage.* **213**: 1047-1062

Dahlan I, Islam M and Rahman MA, 2000. Nutrient intake and digestibility of fresh, ensiled and pelleted oil palm fronds by goats. Asian Australian J. Animal Sci. **13**: 1407-1413.

Denang CKHCK, Daljit S, Arifin A, Keeren S and Rajoo S, 2020. Effects of empty fruit bunch application on oil palm yield. Soil properties and cost benefit analysis. *J. Envi. Sci, Pollut. Res.* **6(3)**: 435-437.

Devendra C, 2008. Integration and integrated systems: enhancing potential impacts with ruminants in oil palm plantations. *Planter*, **84**: 355-367.

Donough CR and Kwan B, 1991. Oil palm planting density: results from trials in Sabah and the possible options. *Planter*, **67**: 483-506.

Edwards CA, 1988. Mixed farming systems: A review of the issues. *Agrl. Syst.* **27(1)**: 1-28.

Egonyu JP, Baguma J, Martínez LC, Priwiratama H, Subramanian S, Tanga CM, Anankware JP, Roos N and Niassy S, 2022. Global advances on insect pest management research in oil palm. *Sustainability*, **14(23)**: 162-188.

Ferand M, 1960. Le noli (Elaeis melanococca ou Corozo oleifera en Columbia). *Oleagineux*, **15**: 823-827.

Fawzi HA, Zainudin MN and Abdul Wahab AA, 1998. Cattle integration in oil palm-establishment and financial implications. *Planter*, **74**: 319-334.

Gascon JP and de Berchoux C, 1964. Caracteristiques de la production d'Elaeis guineensis (Jacq.) de diverse origins et leurs croisements. Application a la selection du palmier ahuile. *Oleagineux*, **19**: 75-84.

Gawankar MS, Haldankar PM, Salvi BR, Haldavanekar PC, Malshe KV and Maheswarappa HP, 2018. Inter cropping in young oil palm plantation under Konkan Region of Maharastra. *Int. J. Curr. Microbiol. App. Sci.* **7(12)**: 2752-2761.

Gerritsma W and Wessel M, 1997. Oil palm: domestication achieved? *Neth. J. Agric. Sci.* **45**: 463-475.

Haji Baba AS, Azillah A, Mukherjee TK and Abdullah RB, 1998. Growth and reproduction performance of small ruminants under integrated livestock-oil palm production systems. *Asian Australian J. Animal Sci.* **11**: 573-579.

Hartley CWS, 1988. *The Oil Palm*, 3rd Edn, Longman, London.

Heller L, 2005. Palm oil reasonable replacement for trans fats, say experts. *Food Navigator.*

Jalaludin A, 1998. Quantification of weed flora in oil palm plantations. *J. Oil Palm Res.*, **10(1)**: 23-30.

Jourdon C, Ferriere NM and Perbal G, 2000. Root system architecture and gravitropism in oil palm. *Annals Bot.*, **85(6)**: 861-868.

Junito P Marcelino and Elbie V Diaz, 2016. Frond pruning enhancing growth and yield of eight year old oil palm. *Annals Tropi. Res.* **38(2)**: 96-105.

Kalidas P and Subbanna ARNS, 2022. Pests and their management in oil palm. *Trends in Horticultural Entomology.* pp. 1457-1475.

Kalidas P, Rao SC and Rao KP, 2014. Oil palm cultivation in India: past, present and future scenario. *J. Oilseeds Res.* **31(1)**: 1-12.

Kawamoto H, Mohamed WZ, Shukur NIM, Ali MSM, Ismail Y and Oshio S, 2001. Palatability, digestibility and voluntary intake of processed oil palm fronds in cattle. *Japanese Agril. Res.* **25**: 195-200.

Keerthana J, Lakshmi MA, Aditya S, Ajesh BR and Manyam P, 2023. Detection and management of basal stem rot of oil palm: Classical to modern approaches. In: *Detection, Diagnosis Manage. Soil-borne Phytopathogens* Springer Nature Singapore. pp. 225–267.

Kochu Babu M, 2008. Utilization and recycling of palm oil mill effluent. *Annual Report*, Published by NRC for Oil Palm, pp. 62-63.

Kovachich WG, 1954. *Cercospora elaeidis* leaf spot of the oil palm. *Transactions British Mycol. Soc.* **37(3)**: 209-211.

Lakshmi MA, Rajashekara H, Vanitha K, Ajesh BR, Mohan H and Reddy MG, 2022. Major diseases of cashews and oil palms. In: *Diseases of Fruit and Plantation Crops and Their Sustainable Management.* Nova Science Publications. pp. 530.

Lakshmi MA, Suresh K, Ajesh BR and Keerthana S, 2023. Panoramic overview of advancements in the arena of detection and management of basal stem rot of oil palm caused by *Ganoderma* spp. *The Andhra Agric. J.*, **70(1)**: 001-017.

Lee M, Xia JH, Zou Z, Jian Y and Rahmadsyah YA, 2015. A consensus linkage map of oil palm and a major QTL for stem height. *Sci. Rep.* **5**: 8232. https: //doi.org/10.1038/srep08232.

Lisma Safitri, Sri Suryanti, Valensi Kautsar, Agung Kurniawan and Fajar Santiabudi, 2018. Study of oil palm root architecture with variation of crop stage and soil type vulnerable to drought. *IOP Conf. Ser.: Earth Environ. Sci.* 141012031.

Lynch J, 1995. Root architecture and plant productivity, *Plant Phy.* **109**: 7-13.

Manorama K, Behra SK, Suresh K, Prasad MV, Mathur RK and Harinarayana P, 2021. Mulching and technological interventions avoid land degradation in an intensive production system. *Land Degrad. Dev.* **32**: 3785-3797.

Manorama K, Behera SK and Suresh K, 2021. Establishing optimal nutrient norms in leaf and soil for oil palm in India. *Ind Crops Products,* **174**: 114223.

Manorama K, 2024. Water management in oil palm. Training Manual: Compendium of Lectures on Oil Palm Production Technology. Published by Director, ICAR–IIOPR, Pedavegi. pp: 108-118.

Manorama K, Obi Reddy GP, Suresh K, Ray SS, Behera SK, Nirmal Kumar and Mathur RK, 2024.Characterization and Mapping of the Potential Area of Oil Palm Using Multi-Criteria Decision Analysis in a Geographic Information Systems Environment. *Agriculture,* **14(7)**, 986; https: //doi.org/10.3390/agriculture14070986.

Mary Rani KL, 2024. Indian vegetable oil and oil palm scenario. Training Manual: Compendium of Lectures on Oil Palm Production Technology. Published by Director, ICAR–IIOPR, Pedavegi. pp: 1-11.

Mathur RK, Suresh K, Murugesan P, Parimala K, Nair S and Ravi Kumar M, 2003. Genetic variability in oil palm hybrids for morpho-physiological characteristics. In: Abstract of Symposium on Sustainability of Plantations Crops through Integrated Approaches for Crop Production and Product Diversification during December 10-13, 2002 at Mysore. pp. 34.

Mathur RK, Manorama K, Kalyanababu B, Ravichandran G, Ramachandrudu K and Prasad MV, 2021. Innovations to Make India Self-Reliant in Edible Oil Production: The Story of Oil Palm. In: *Innovations in Agriculture for A Self-Reliant India*. Edited by Ghosh P K *et al.,* NIPA Publications, New Delhi. pp: 83–102.

Mayes S, James C, Price Z, Groves LM, Jack PL and Corley RHV, 1996. Integration of DNA markers into oil palm breeding programmes. PIPOC- competitiveness for the 21st century, pp. 49-54.

Mohd Sukri I and M Abdullah Fazal, 1998. Utilization of oil palm by-products as livestock feed. National Seminar on Livestock and Crop Integration in Oil Palm, Johore.

Ogendengbe HD, 1991. The nutrient status of Okomu oil palm plantation. MSc thesis, University of Ibadan, Nigeria.

Olivin J, 1968. Etude pour la localization d'un bloc industriel de palmiers ahuile. II. Les criteria de jugement. *Oleagineux* **23**: 499-504.

Olivin J, 1986. Study for setting of a commercial oil palm plantation. *Oleagineux,* **41**: 113-118.

Paramanathan S 2000. Soil requirements of oil palm for high yields. In: *Managing oil palm for high yields: agronomic principles* (Ed. Goh K.J.), *Malaysian Soc. Soil Sci. Param Agric.Surveys,* Kaula Lumpur, pp. 18-38.

Paramananthan S, 2013. Managing marginal soils for sustainable growth of oilpalms in the tropics. *J.Oil Palm Envi. Health,* **4**: 1–16.

Phicot JMP, Sedego MP and Arrivers HF, 1981. Evaluation de la fertilite dun sol ferrugineux tropical sous l influence de femures milnerales et oganiques. *L' Agron Trop.* **36(2)**: 122–133.

Pootakham W, Jomchai N, Areerate P, Shearman JR, Sonthirod C, Sangsrakru D, Tragoonrung S and Tangphatsornruang S, 2015. Genome-wide SNP discovery and identiûcation of QTL associated with agronomic traits in oil palm, using genotyping-by sequencing (GBS). *Genomics* **105**: 288–295.

Prathap, V. 2024. Botany of oil palm. Training Manual: Compendium of Lectures on Oil Palm Production Technology. Published by Director, ICAR–IIOPR, Pedavegi. pp. 12-20.

Rajanaidu N, Tan YP, Ong EC and Lee CH, 1986.The performance of inter origin commercial DxP planting material. In: *Proc. Int. Workshop Oil palm germplasm and utilization,* PORIM, Kual Lumpur, pp. 155-161.

Ramachandrudu K, Suresh K, Sekhar G and Suneetha V, 2009. Effect of fertigation on growth and vigour of oil palm seedlings. *Int. J. Oil Palm,* **6(2)**: 25-28.

Ramachandrudu K, Suneetha V and Mahesh Babu M, 2019. Millipede-A potential bio-agent for recycling of biomass in oil palm gardens. Abstract published in Book of Abstracts during international conference on "Crop Residue Management" organized by Dept. of Agriculture, Govt. of Bihar and Bihar Agricultural University in Patna during 14-15 October 2019. pp. 5-7.

Ramachandrudu K and Suneetha V, 2020. Evaluation of fertilizing capacity of palm oil sludge on growth and biomass production of oil palm seedlings. *Indian J. Horti.* **77(3)**: 548-552.

Ramachandrudu Kummari, Narasimha Rao Bezawada and Manorama Kamireddy, 2023. Nutritive screening and forage yield of Guniea and hybrid Napier grass varieties under dense shade of mature oil palm plantation. *J. Oil Palm Res.* DOI: https: //doi.org/10.21894/jopr.2023.0051.

Ramachandrudu K, Prasad MV and Mathur RK, 2023. Studies on economic aspects of oil palm nurseries in Andhra Pradesh. *Ind. J. Agril. Marketing,* **37(1)**: 39-50.

Ramachandrudu K, Suresh K and Manorama K, 2023. Morpho-physiological changes in oil palm tenera hybrid (*Elaeis guineensis* Jacq.) seedlings raised under different shade levels. *J. Horti. Sci.* **18(2)**: 394.401.

Ramachandrudu K, Suneetha V, Suresh K, Rao BN and Manorama K, 2018. Systems approach in oil palm for higher productivity and profitability. *Technical Bulletin* ISBN No. : 81-87561-53-X, Indian Institute of Oil Palm Research, Pedavegi. pp. 1-52.

Ramachandrudu K, 2024a. Production of quality seedlings in oil palm. Training Manual: Compendium of Lectures on Oil Palm Production Technology. Published by Director, ICAR–IIOPR, Pedavegi. pp: 58-69.

Ramachandrudu K, 2024b. Systems approaches in oil palm. Training Manual: Compendium of Lectures on Oil Palm Production Technology. Published by Director, ICAR–IIOPR, Pedavegi. pp: 119-128.

Rankine I and Fairhust TH, 1999. Management of phosphorus, potassium and magnesium in mature oil palm. *Better Crops Int.* **13(1)**: 10-15.

Ravichandran G, Somasundaram G, Anitha P and Bhagya HP, 2019. Recent advances in oil palm seed production, seed processing and sprout production. In: Recent advances in oil palm production and special emphasis on emergence of new pest and its management. (Eds) K. Manorama and K. Suresh. ISBN: 81-87561-58-0. Model Training Course manual. pp: 10-16.

Reddy VM, Suresh K, Saraswathi MS, Bijimol G and Nagamani Ch, 2004. Inter cropping in irrigated oil palm in India. *J. Planta. Crops,* **32**: 218-220.

Rethinam P and Murugesan P, 2000. *Oil Palm Nursery Manual.* National Research Centre for oil palm, Pedavegi, Andhra Pradesh.

Singh R, Abdullah MO and Low E, 2013. Oil palm genome sequence reveals divergence of interfertile species in old and new worlds. *Nature*. **500**: 335–339.

Smith BG, Donough CR and Corley RHV, 1996. Relationship between oil palm clone phenotype and optimal planting density. In: *Proc. 1996 PORIM Int. Palm Oil Congr. Competitiveness for the 21st Century*, pp. 76-86.

Sparnaaij LD, Menendez T and Blaak G, 1963. Breeding and inheritance in the oil palm (Elaeis guineensis Jacq.). Part I. The design of a breeding programme. J.W.Afr. *Inst. Oil Palm Res.* **4**: 111-125.

Sunitha S and Thomas Varghese P, 1999. Composting of oil palm wastes for efficient recycling of nutrients in palm plantations. *Planter*, **75(885)**: 677-681.

Sunitha S and Varghese PT, 2004. Yield of oil palm as influenced by integrated nutrient management. *J. Planta. Crops*, **32**: 263-265.

Sunpapao A, Kittimorakul J and Pornsuriya C, 2014. Disease Note: Identification of *Curvularia oryzae* as cause of leaf spot disease on oil palm seedlings in nurseries of Thailand. *Phytoparasitica*, **42**: 529-533.

Suresh K and Rethinam P, 2001. Determination of canopy size in oil palm plantations-A survey. *Int. J. Oil Palm,* **2(2)**: 55-56.

Suresh K, Reddy VM, Sharma KN, Bhanusri A and Shivashankar Kumar KM, 2003. Quantification of root biomass in oil palm grown under basin irrigation. *Int. J. Oil Palm*, **2** (3-4): 39-41.

Suresh, K. 2024. Compendium of lectures on *Oil Palm Production Technology.* ICAR–Indian Institute of Oil Palm Research, Pedavegi, Andhra Pradesh.

Tejuddin I and Chong D, 1995. Ruminants in plantation systems in Malaysia. Integration of ruminants into plantation systems in Southeast Asia, Lake Toba, North Sumatra, Indonesia.

Tohiruddin L and Foster HL, 2013. Superior effect of compost derived from palm oil mill by-products as a replacement for inorganic fertilizers applied to oil palm. *J. Oil Palm Res.* **25(1)**: 123–137.

Varghese PT and Sunitha S, 2005. Natural resource management-soil and water conservation techniques and land use systems. *Annual Report*, National Research Centre for Oil Palm, pp. 44-45.

Veeramachaneni Suneetha and Ramachandrudu K, 2018. Phyto-morphological and growth variations in oil palm seedlings under influence of microbial fertilizers and agrochemicals. *Int. J. Current Microbil. Applied Sci.* **7(7)**: 2356-64.

Wong CC, 1998. Soil compaction under cattle grazing in oil palm. *J. Tropi. Agril. Food Sci*, **26**: 203-210.

Yusof Basiron, Jalani BS and Chan KW, 2001. Advances in oil palm research Vol. I and II, Malaysian Palm Oil Board, Malaysia.

Yusof Basiron, 2007. Palm oil production through sustainable plantations. *European J. Lipid Sci. Tech.* **109(4)**: 289-295.

Zeven AC, 1967. The semi-wild oil palm and its industry in Africa. *Agric. Res. Rep.* 689, Wageningen Univ., Wageningen.

Chapter 18

Nursery Production and Canopy Management in Cashew

J.D. Adiga, G.N. Manjesh, H.P. Bhagya and G.L. Veena

ICAR–Directorate of Cashew Research, Puttur–574202, Karnataka

Cashew is native of tropical America; from Mexico it was introduced to India by the adventurous Portuguese. Cashew came, conquered, and took deep root in the entire coastal region of India. The crop found the Indian soil more homely than its homeland. The first introduction of cashew into India was made in Goa from where it spread to other parts of the country. It was during 1960s that cashew gained commercial importance with the beginning of export of cashew kernels to European countries. Developmental programmes related to cashew developmental activities started in the beginning of 1992 onwards. At present, cashew represents 12th position among the agricultural commodities exported from India in terms of value which accounts for 0.24 per cent of total export earnings and 1.4 per cent of total agricultural export earnings.

Cashew is a highly cross-pollinated crop and plants propagated using seeds do not produce true to type and they exhibit variation in their progenies. The clonal progenies are true-to-type of the mother plant and give relatively more uniform yield and come to bearing earlier than the seedling origin plants. The earlier efforts made on different vegetative propagation techniques (layering, grafting, and budding) at ICAR–Directorate of Cashew Research, different cashew research stations in the country as well as All India Coordinated Research Project on Cashew (AICRP-C) Centres have concluded the superiority of the "Softwood grafting" technique over other methods of propagation and hence, this is recommended for commercial reproduction of cashew varieties for large scale spread. Improved cashew planting material are the most important input in cashew production because they determine the potential yields and the ultimate productivity per unit area. The cashew nursery management and production of quality grafts plays an important role. In nursey cashew grafts are raised with special care until they are ready for transplanting into the field.

Grafting or graftage is the art of producing plants vegetatively by joining tissues of plants so as to continue their growth together. The upper part of the combined plant is called the scion (bud, bud stick) while the lower part is called the rootstock having a root system. Various steps involved in the commercial-scale production of cashew varieties by "Softwood grafting".

Raising of Rootstock Seedlings

To raise the healthy rootstock, good quality seeds should be collected, soaked and sown.

Selection of Seed Nuts

- Collect the freshly harvested seed nuts (February-April) preferably VTH-174, V-4, MDK-2.
- The seed nuts should be sun-dried for 2-3 days.
- Select medium size nuts (6-7g)
- Every year required to collect the seed nut freshly for raising rootstock
- The seeds gradually lose viability after 8-10 months of storage.

Preparation of Potting Mixture and Filling of Polythene Bags

- The potting mixture should be prepared in the proportion of 1: 1: 1 (red soil; coarse river sand: compost) in heavy rainfall areas and 1: 1 (red soil: Compost) in low rainfall areas. Rock Phosphate @ 5g/kg of the potting mixture may also be added for getting good root growth of seedlings.
- Polythene bags of 25 cm x 15 cm size and 300 gauge thickness should be used for filling potting mixture.
- In heavy rainfall areas, 30-40 drainage holes of 0.5 cm diameter may be punched uniformly on the polythene bags. In low rainfall areas, the number of drainage holes can be few.
- Potting mixture should be filled up to the brim level (Figure 18.3).
- The filled bags should be arranged in beds of 1000 numbers each (100x10).

Figure 18.1: Preparation of Polybags Filled with Media and Arranged in Beds.

Maintenance of Seedlings in the Nursery

The seedlings should be watered daily depending upon the weather conditions. During monsoon season, if water stagnation is observed, the polythene bags may be pressed from sides, so that the excess water drains out of the bags. This reduces the incidence of collar rot/damping-off of seedlings. During the rainy season, the seedlings should be sprayed with Bordeaux mixture (1%) at 10 days intervals to control collar rot/damping-off of seedlings. The seedlings may be sprayed with systemic insecticides such as acetamiprid (0.5 g per lit.) or λ- cyhalothrin (0.003%) or thiamethoxam (0.2g per lit.) to control tea mosquitoes and other pests at regular intervals. The side shoots arising from the axils of leaves should be removed frequently in order to get vigorous seedlings with a single main stem.

Softwood Grafting Technique

Selection of Scion Sticks and Precuring

The important steps to be followed while selection of scion sticks are

- Avoid selection of thin and short scion sticks
- Scion shoots free from pest damage should be selected
- Avoid collection of scions from shade areas/canopy
- Select non-flowered lateral shoots of the current season's growth (3-5 months old)
- The leaves are to be removed using a secateur (precuring or defoliation) 10 days in advance of grafting (Figure 18.2).
- The selected shoots should be 10-12 cm long, straight, uniform round with brown coloration, and of pencil thickness.
- The precured scions should be separated from the mother tree well before they sprout.

Figure 18.2

Collection of Scions

The scion sticks should be separated from the mother tree preferably during morning hours. The precured scion sticks should be collected from the mother trees before the terminal buds sprout, the length of the scion should be 10-12cm. Further, the detached scion sticks may be dipped in cold water and then placed in a polythene bag of 100- gauge thickness and brought to the grafting shed and utilized for grafting. For distance transport of scion requires the use of sphagnum mass for packing the scions and icebox for transport is desired during hot days.

Preparing the Rootstock Seedling for Grafting

The grafting operation should be carried out under the shade either in a grafting shed or makeshift shelters. About two-month-old seedlings (of 25-30 cm height) are selected as rootstocks. The girth of the seedling should be about 1cm at the place of grafting. On selected rootstock seedling, two pairs of bottom leaves are retained and the other leaves are removed with a sharp grafting knife. Then the terminal shoot is decapitated at a height of 15 cm (where the softwood portion is available for grafting). A cleft of 6-7 cm deep is made on the decapitated stem. From the inner sides of the cleft a little portion of the wood is removed at the tip. Preparing the scion stick for grafting. If the scion stick is longer, then it should be reduced to a length of 10-12 cm by cutting off the excess portion at the bottom. Very short scion sticks should be discarded. The scion stick should be carefully mended into a wedge shape of 6-7 cm long. However, care must be taken to retain the bark on either side of the wedge. The cut surface should be smooth and it should not be soiled at the time of preparing the wedge.

Grafting Technique

The wedge of the scion is inserted carefully into the cleft of the rootstock. If the scion is thicker in girth, then care must be taken to match the cambial layers of both rootstock and scion at least on one side with the help of a finger. Then the graft joint is secured firmly with a polythene strip (2.0 cm wide, 30 cm long, and 100 gauge thickness). Then a long and narrow white polythene cap (20 cm x 4 cm size and 200 gauge thickness) is inserted on the grafted plant. After 3 weeks, the polythene caps are removed gently and the grafts are shifted to an open area in the nursery. At the time of removing the caps, about 70-80 percent of the grafts show sprouting.

Season of Grafting

The grafts in cashew can be produced almost throughout the year with a mean graft success of 65-70 percent. Grafting may be suspended from December to January as there will be poor graft success in winter.

Management of Grafts in the Nursery

The important management practices to be followed in cashew nursery are listed below

- The Cashew grafts produced in the monsoon season (June- November) are to be maintained at the nursery till the next planting season (July-August). Consequently, the following points should be considered:
- In the nursery, the grafts should be arranged in beds after spreading black polythene sheets on the ground to prevent the grafts from striking roots into the ground. Otherwise, frequent shifting of grafts from one place to another place is required which is labor-intensive.
- The grafts should be watered daily depending upon the prevailing weather conditions. During the heavy rainy season, if water stagnation is observed in the polybags, then the excess water must be removed by pressing the sides of the bags.
- During the summer months, the grafts should be provided with partial shade by erecting pandal and covering with agro shade nets (35-50 per cent shading). As soon as the monsoon starts, the shade should be removed. Complete shading should be avoided as it results in lanky and weak growth of grafts. Laminated shade nets (35 per cent shading) can also be used.
- During the rainy season, the grafts may be sprayed with Bordeaux mixture (1%) at 10-15 days intervals. Spraying should be done when the weather is clear.
- The grafts may be sprayed with systemic insecticides such as monocrotophos (1.5 ml/liter) or acetamiprid (0.5 g per lit.) or λ-cyhalothrin (0.003%) or thiamethoxam (0.2g per lit.) to control insect pests such as tea mosquitoes, leaf minor, leaf-eating caterpillars, *etc.*
- Frequent removal of side shoots arising from the rootstock portion of the grafts should be carried out. The flower panicles if any from the grafts in the nursery should be removed frequently.
- Practice of removing polythene strip should be followed from the graft joint after 4-5 months of grafting. If not, there will be girdling at the graft

joint. However, the polythene strip should be removed before selling the grafts.

- ☆ After three-four months of grafting, the bottom leaves on the rootstock portion of the graft should be removed.

Estimated Cost of Graft Production

- ☆ On average 65-70 percent of saleable grafts can be realized on a commercial scale. The estimated cost of production depends upon the cost of inputs (cost of materials and labor), for producing about 50,000 saleable grafts, and the estimated amount required at Puttur,.

Seasonal Influence on Graft Success

- ☆ Higher success in the case of softwood grafting is achieved in the seasons in which fluctuation in the day and night temperatures are minimum and high humidity is prevalent throughout the day. This type of weather condition is prevalent during the monsoon season (June-November) on the West Coast. During other seasons, the graft success is slightly reduced due to the non-availability of proper scion sticks and unfavorable weather conditions. However, grafts can be produced almost throughout the year with a mean success of 65-70 per cent.

Factors to be Considered for the Establishment of Cashew Nurseries

- ☆ Cashew is normally a season-oriented crop. Therefore, the nursery operations require a proper planning and time schedule for timely delivery of grafts to growers. The important factors to be considered for the establishment of cashew nurseries are
- ☆ Site selection
- ☆ Wind and shade condition
- ☆ Topography
- ☆ Water sources
- ☆ Soil type
- ☆ Transportation costs
- ☆ Protection
- ☆ Accessibility to public

Canopy Management in Cashew

Canopy management is the handling of tree canopies to optimize the production of quality fruits. The canopy management comprises of its components like training and pruning, use of growth regulators, selection of rootstocks/variety, *etc.* Canopy in a fruit yielding tree refers to its physical composition comprising of stem, branches, shoots and leaves. Canopy density is determined by the number and size of the leaves. Canopy architecture is determined by the number, length and orientation of the stem, branches and shoots. Canopy management refers an interpretation of physiology of light penetration and interception which are critical components of overall tree productivity. The main controlling factors are amount of incoming radiation and percentage of radiation intercepted by tree canopies. The productivity of fruit crops depends on several factors, management of canopy architecture being the most important one (Goswami *et al.,* 2014).

Principles of Canopy Management

Fruit trees produce fruit regardless of human intervention. However, for human consumption, it is important to manage fruit tree canopies to optimize the balance between vegetative growth and fruit production, and also to keep fruit picking manageable. The basic concept in canopy management of a perennial tree is to make the best use of land and the climatic factors for an increased productivity in three dimensional approaches. Canopy management includes a range of techniques to after the position and the amount of leaves, shoots and fruits in space which determines, to a large extent the plant geometry structure including spatial distribution of leaf area and leaf orientation.

Canopy management refers an interrelation of the physiology underlying the relationship between vegetative growth and production.

1. Orchard architecture largely depends upon orchard production system which is a combination of variety, rootstock, tree spacing, training and pruning
2. These factors strongly interact to develop a specific production system and determined yield, fruit quality and longevity of the trees.
3. In perennial fruit crops, cultural practices like nutrition, irrigation, planting density, rootstocks training system, pruning and growth retardants can be used as potential means to alter the shoot vigor, size and shape of the canopy and the microclimate at the canopy and thereby increase yield and quality (Vandana *et al.,* 2017).

Cashew is a vigorous evergreen perennial woody plant having long juvenility and high heterozygosity. Canopy development in cashew is a seasonal and continuous process. However, the varieties which are available nowadays are of semi vigorous to vigorous type. Canopy management has a direct influence on plant vigour which ultimately influences the cashew nuts yield. The manipulation of the canopy by training and pruning, plant growth inhibitors and dwarfing rootstocks plays an important role in management practice to regulate vegetative growth, flowering and yield in fruit trees [Srilatha *et al.,* 2015].

The response to pruning depends on age, growth habits, tree vigour, varieties, location and cultivation practices of cashew. Heavy pruning promotes excessive vegetative growth and often reduces the yield due to the dense canopy with reduced flowering [Balamohan *et al.,* 2016]. The application of plant growth retardants such as paclobutrazol (PBZ) has been found useful in the manipulation of vegetative growth, vigour and yield of cashews [Meena *et al.,* 2018; Babli *et al.,* 2019]. The optimized application of PBZ to obtain maximum benefit with minimum undesirable impact on food and environmental safety aspects [Kishore *et al.,* 2015].

In cashew canopy development, two types branching exists, in which one is intensive and the other is extensive (Saroj *et al.,* 2014). In high yielding trees more than 60 *per cent* intensive branches are seen whereas low yielders possess less than 20 *per cent* intensive branches. The 'intensive shoot' grows to a length of about 25- 30 cm and ends in a panicle, while in the 'extensive type', the shoot grows to 20-30 cm length and rests. Concurrently in the intensive type that tends to give bushy appearance to the tree, 3 to 8 lateral shoots come up below 10-15 cm of the apex and few of these laterals may also bear panicles. On the other hand in the 'extensive type', a bud sprouting 5-8 cm below the apex gives rise to further growth which continues for two or three years without giving flowers and results in spreading tree habit.

The canopy management, its components like tree training and pruning, affects the quantity of sunlight intercepted by trees, as tree shape determines the presentation of leaf area to incoming radiation. An ideal training strategy focuses around the arrangement of plant parts, especially, to develop a better plant architecture that optimizes the utilization of sunlight and promotes productivity. Light is critical to growth and development of trees and their fruits and productivity. The green leaves harvest the sunlight to produce carbohydrates and sugars which are transported to the sites where they are required – inflorescence, buds, flowers and fruits. Better light penetration into the tree canopy improves tree growth, productivity, yield and fruit quality. The density and orientation of planting also impact light penetration in an orchard. Generally, in close planting, quicker shading becomes a problem. An east-west row orientation results in more shading as compared to the western and southern orientation of trees. Strong bearing branches tend to produce larger fruits. The problem of a fruit grower is initially to build up a strong and balanced framework of the trees, then equips them with appropriate fruiting. Obviously, pruning in the early years has to be of a training type to provide strong and stocky framework with well-spaced limbs or any other desired shape.

Some of the basic principles in canopy management are:

- ✰ Maximum utilization of light.
- ✰ Avoidance of built-up microclimate which is congenial for pest and diseases infestation.
- ✰ Convenience in carrying out the inter-cultural operations.
- ✰ Maximizing productivity with quality fruit production.
- ✰ Economy in obtaining the required canopy architecture (Gorakh Singh, 2010).

In cashew trees, the canopy needs to be fully exposed to the sunlight for harvesting good yields. Canopy management in cashew could be achieved through training and pruning of cashew orchards, selection of rootstocks and by using growth regulators.

Training of Young Cashew Plantation

In case of the new plantations with the grafts the plants should be trained in the early years *i.e.*, 2-3 years so as to provide better plant architecture which facilitates the easy inter-cultural operations. Training indirectly assists in ease of other operations such as weeding, manuring and pest and disease management (Satpathy, 1988). The lateral shoots arising from rootstocks need to be removed periodically till 2-3 years. This will assure the proper growth of the scion portion of the grafts. The grafted plants should be shaped by removing the branches and water-shoots growing from the main stem up to a height of 0.75m to 1.0m from the collar region during first 3-4 years. Besides, weak and interloping branches should also need to be removed. After the age of 4-5 years, in tall type of cashew plants the main trunk may be de-topped at a height of 4-5 m from the ground region. This will ensure a round globular canopy which helps to harvest maximum sun light for photosynthesis. Severe pruning of the young grafts may be avoided at it may extend the juvenility and the pre-bearing period of the plants (Nayak *et al.,* 1996).

In general, two types of training systems are being practiced in cashew, a) Modified leader system and, b) Open center system.

Modified Leader System

In this system, cashew grafts are allowed to grow as single stem up to a height of 75 to 100 cm by removing side sprouts. Then lateral branches are allowed to grow at desirable direction by de-topping. De-topping height varies from 2.5 to 4 m depending on spacing. Under normal spacing (8m x 8m), de-topping at 4 m from ground level is recommended. Whereas, for high density planting (5m x 5m), de-topping at 2.5 m from ground level is recommended. Removal of cris-cross branches and trimming of branches has to be resorted to get dome shape canopy and the same should be maintained in later years by imposing mild pruning. This kind of canopy helps in reducing week shoots and water shoots development. Modified training system is suitable for both normal and high-density planting system.

Open Center System

Cashew grafts are allowed to grow straight up-to 50-60 cm from ground level. The terminal growing point is pinched off to form lateral branches. The branches are regulated to grow in four directions at equal distance. Because of fast vegetative growth, the canopy spreads rapidly. To avoid this, canopy centre needs to be opened once in a while to support more light interception to the interior plant parts. This encourages flowering at inner and outer surface of canopy and thus increases the yield (Nayak *et al.,* 2019).

Pruning in the Established Plantations

The trees which are not received any training and pruning in the initial years grow haphazardly and resulting in canopies without desirable shape and size. Besides, the development of deadwood, inter- mingling of branches with neighboring trees, crisscross branches, development of water shoots *etc.* will bring down the productivity of the tree (Nayak *et al.,* 1996).

Deadwood/dry branches: The dead wood/dry branches develop mainly because of the effect of shade on lower branches caused by overlapping of the upper branches. Deadwood will be an additional burden to the plants. Furthermore, the dead and decaying woods may invite the entry of pathogenic organisms or saprophytic growth which may spread to heal their parts of the plant in due course of time.

Criss-cross branches: The lower branches remain crawling on the ground for want of space and sunlight, where the plants are not trained or pruned in the initial years. Similarly, the branches at higher level also grow haphazardly in search of sunlight resulting in irregular canopy architecture.

Intermingling of branches: The problem of entangling of branches starts after 10-12 years in regularly spaces (8x8 m) plantations. The exterior branches get entangled with neighboring trees as a result only a portion of canopy (crown portion) remains exposed to sunlight. Such a development inside the plantation is a hindrance to the regular intercultural operations and general maintenance of the orchard.

Water shoots/sprouts: Water shoot are vegetative shoots which are extraordinarily vigorous growing from dormant buds at higher points on main stem in upright direction. They grow at the expense of parent branches from which they arise. They are erect in growth and much thicker in size than the normal branches and bear much longer and coarser leaves. These branches outgrow the rest of the neighboring drooping branches. If water shoots are not removed in time, they soon cover the Centre of the canopy and obstruct sunlight.

The old trees with deadwood, crisscross branches, watershoots and inter mingling branches should be pruned at least once in 2-3 years. Pruning can be taken up in dormant season *i.e.* at least 2-3 months earlier to productive flushing. All the types of unwanted growth mentioned before are to be pruned off. However, the plant should have a better look and structure after pruning. This can be achieved using one's discretion and experience in pruning and orchard management.

Leader Shoot Pruning

Cashew trees enter a brief resting period after the harvest of the crop (May - June) and it continues up to next productive flushing season (September - November). The flushes or flower bearing twigs are known as lateral shoots. These shoots usually form the terminal portion of a leader shoot will give a single shoot (lateral) from its terminal bud. If the terminal bud is disturbed by means of pruning the dormant lateral buds will sprout resulting in more number of lateral shoots per unit area. This will result in increased number of productive inflorescences.

Pruning the leader shoots can be taken up at least 2-3 months (July to August) before flushing. In a tree about 50-60 per cent of the leader shoots may be headed back to one-third of their original length. A pair of leaves may be retained while pruning wherever possible. While pruning, the leader shoot should be of a pencil thickness and should not have turned to ash color before taking up pruning.

Precautions to be taken while Pruning

As a natural response the cuts resulting from pruning will heal faster if cuts are smooth and non-jagged. While attending pruning the following points are to be considered

a) While removing the deadwood the cut must be made back to living tissue as good callus formation and healing cut end is only possible form properly made cut end only.

b) It is essential to make the cut close to the branch. The cut should be nearly even along the stem or trunk leaving a minimum stub and clear wound for faster healing.

c) In the case of pruning off the diseased part care should be taken to remove all the infested parts.

d) Wherever larger branches are being removed, care should be taken to avoid breaking way of bark or wood portion of the plant.

e) When a cut is made a considerable amount of hardwood will be exposed and it should be protected from pests and pathogens. All the larger cuts may be treated with 10 per cent Boredeaux paste while the leader shoot pruned canopy may be sprayed with 1 per cent Boredeaux mixture.

f) It is essential to relate the appearance of the plant while pruning. Plant should have a balanced and natural appearance after pruning.

Yield Increase in Pruned Trees

The past season leader shoots can produce only one lateral from its terminal. Pruning enhances the production of lateral shoots, thus the yield can be increased. Pruning intensity and time varies for different specific agro-climatic regions. Pruning of dead wood and crisscross branches can increase yield by 30-40 per cent (Khan *et al.*, 1987). Leader-shoot pruning doubled the yield in cashew (Mohan and Room Singh, 1988). Results of pruning on 28-year old trees revealed that trees with three branches pruned recorded the highest number of panicles/sq.m (39), highest number of flowers/panicle (588.70) and fruit-set to an extent of 14.42 per cent, while unpruned trees recorded only 7.75 per cent increase in yield (Panda, 1990). Under Jhargram conditions, pruning of leader-shoots during July enhanced the number of productive laterals, increased the number of bisexual flowers per panicle, fruits per panicle and yield per tree (Chattopadhyay and Ghose, 1994).

Pruning treatment increased the number of laterals/leader but did not affect duration of flowering and harvest (Mohan and Rao, 1995).

Effect of the pruning in different shoots in two varieties namely, BPP-4 and BPP-6 was conducted at the Cashew Research Station, Bapatla, Guntur district (AP). The shoots were decapitated back to 5 cm in mid-July, mid-August and mid-September months of the leader shoots, lateral shoots and leader as well as lateral shoots pruned separately and different growth parameters on individual trees were studies. The response to the pruning, the variety 'BPP-4' performed better as compared to BPP-6. The production of flowering shoots and nut yield as influenced by the cultivar, level of pruning and time of pruning that a moderate incremental growth with large number of flowering shoots could be obtained by pruning the leader shoot in mid-August under local agro-climatic condition. The study further indicated that the vigorous cultivar 'BPP-4' and off-season production cultivar 'BPP-6' performed well during a rainy year compared to the dry year which was associated with prolonged dry spell and delayed rains in August-September months. Another important observation from the study indicated that the off-season cultivar of cashew needs essentially the pruning of the leader shoot in mid-August so as to avoid the off-season flowering and to increase productivity in the normal season. Pruning of leader shoots in mid-July was found to be beneficial during both the years of study to produce higher tree yield of nuts (Prasanna Kumar *et al.,* 2015).

Effect of Rootstocks in Cashew Canopy Management

Rootstocks play a very important role in propagation of plants, it may modify form or stature and adopt a variety to a soil in an incompatible climatic condition and also build up the resistance to biotic and abiotic stresses meanwhile increase the production and productivity. Rootstocks plays a very important role in improving production, canopy architecture, flowering and fruiting quality and tolerance to stress. Mainly in canopy management or tree structure rootstocks plays a major role.

The root system of the young dwarf cashew is one very well-developed main root that branches many times and can grow to 10 m or more in deep sandy soils. Lateral roots develop in the upper soil layers between 15 and 32 cm deep. The length of the superficial roots may reach twice the diameter of the crown in dry-land conditions (Barros, 1995). When irrigated the lateral roots are concentrated around the wet area of soil. The characteristics of the tap and lateral roots are of importance in relation to the fertilization of cashew (Crisóstomo *et al.,* 2007). Great variation exists in the depth of the main root and distribution in depth and length of the lateral roots due to the effects of topography, soil texture, stoniness and the presence of a hardened soil layer on the development of the cashew root system (Falade, 1984).

Using dwarfing rootstocks offers the possibility to manipulate tree vigour, better anchorage, nutrient uptake, tolerance to biotic and abiotic stress, as well as yield and productivity without increasing input costs [Webster *et al.,* 2004].

Rootstock selection is a critical tool for the management of vegetative and reproductive growth of scion in perennial fruit crops, which are propagated by grafting or budding. Numerous studies have shown that they offer the advantage of rootstocks in the cultivation of tropical and temperate fruit crops on aboveground tree growth and yield [Balamohan *et al.,* 2016; Webster *et al.,* 2004; Nibolkar *et al.,* 2016]. Very limited studies have been investigated on cashew to select suitable rootstocks to modify scion vigour and increase productivity. The preliminary results reported by Adiga and co-workers (2014), provided the background information for the performance of vigorous cashew cultivars as influenced by dwarf rootstocks (2008-12). NRC-492 could be used as a rootstock to induce semi dwarfism with a higher nut yield. Although cashew is a scion dominant species, the effect of rootstock is reflected in terms of stionic combination in particular, to control the plant vigour of the plant.

Different rootstocks differentially influence the morphology of grafted cultivars, including tree height, trunk cross-sectional area (TCSA), internodal length and yield. In one of the studies by Janani and co-workers (2020) reported that VRI3/ Taliparamba-1 had low growth vigour based on low means of tree height, plant volume, TCSA and canopy spread. The stionic combinations of VRI-3/NRC-492 recorded the highest cumulative nut yield of 16.77 kg/tree (five seasons of cropping). This showed the possibility of manipulating cashew nut productivity through rootstock. Based on the observations on growth and yield of various stionic combinations, it was revealed that NRC-492 could be used as a rootstock to induce semi dwarfism with a higher nut yield. Although cashew is a scion dominant species, the effect of rootstock is reflected in terms of stionic combination in particular, to control the plant vigour of the plant.

Use of Growth Retardants in Canopy Management of Cashew

The canopy management by pruning in later stages of growth often affects orchard life and performance of trees. High density planting system (HDP) has been attempted in cashew to obtain early benefits in terms of yield during initial years of planting. Under HDP, maintenance of tree and canopy growth becomes important due to closer spacing and shading of canopy of trees. In cashew, due to non-availability of dwarf clones and dwarfing rootstocks, use of growth retardants, paclobutrazol (PBZ), assumes importance. Hence, the study aimed to evaluate the morpho-physiological responses of cashew to PBZ treatments under field trials. PBZ treatments resulted in reduced vegetative growth and enhanced reproductive growth with most striking responses at PBZ 3 g a.i./tree treatment. PBZ treatments altered cashew tree physiology by modifying tree size, canopy growth, internodal length, branching pattern and overall ground coverage of the tree. Higher total leaf chlorophyll content, better photo assimilation and enhanced leaf photosynthesis contributed in inducing early flowering and development of more flowering panicles with perfect flowers. Enhanced fruit set and increased number of nuts/m2 canopy contributed in yield increment. Regression analysis showed leaf pigments, nut number and number of inflorescences as the most contributing traits for yield enhancement under PBZ. These findings may highlight the exploitation of morpho-physiological traits for better canopy growth and yield maximization by PBZ in cashew under the HDP.

PBZ treatments are effective in arresting vegetative growth and promoting reproductive growth of cashew. The PBZ treatments altered cashew tree physiology through reduction in vegetative growth, enhancement of flowering, production of more fruits and more fruit set due to efficient distribution of photosynthates, enhanced total leaf chlorophyll contents and increased leaf photosynthesis. These ultimately resulted in enhanced nut yield. Therefore, the findings may provide useful insights on finding solutions to tackle low productivity of cashew by proper regulation of endogenous growth hormones that can relate to enhanced nut yield. In addition, these findings may also throw light on induction of the desired physiological effects in cashew trees that can help in modifications of canopy growth and tree vigour. These in turn can be exploited well under the HDP system to harness early benefits with enhanced yield in cashew.

Effect of different Planting Density on Growth and Yield of Cashew

In India, the established processing capacity of raw nuts is around 15-20 lakh tonnes, where the domestic contribution is around 7-8 lakh tonnes. In the recent years, there is an increase in the domestic demand for cashew. Thus, India has been importing nearly half of the raw cashew nuts processed in the country mainly from the African countries at the cost of ' 8839 crores annually (Anonymous, 2017). Of late, the import possibility from many of these African countries is dwindling, as these countries have setup their own processing facilities and also the competition for import of nuts from these countries by the major cashew processing and exporting countries like Vietnam is increasing. The major problem for deficit of raw nuts for processing by Indian cashew industries is the low productivity (800 kg/ha). It is mainly due to large area of old senile orchards, low plant population per unit area, poor canopy management and non-adoption of improved package of practices. In recent times, demand for cashew in both domestic and international market is growing every year. In India, cashew consumption has increased by about 5.5 times in the last decade and is expected, grow further in the future. It has been estimated that the domestic demand for raw cashew nut is about 50 million MT or more by 2050 (Saroj and Nayak, 2016).

Hence, to meet this huge demand for cashew there is urgent need for increasing the productivity per unit area. This can be achieved easily by the adoption of ultra and high-density planting systems. In recent times, there is a shift in farmers' perception from production to productivity and profitability which can be achieved through accommodating a greater number of plants per unit area. Studies on high density planting systems in fruit crops such as guava, mango and cashew have been shown to be more economical compared to the traditional planting system Yadukumar *et al.,* 2001, Bal and Dhaliwal 2003, Sousa *et al.,* 2012, Gaikwad *et al.* (2017). Efforts have been made to standardize the high-density planting in cashew

(Rejani *et al.*, 2013), and mango (Gunjate *et al.*, 2009) and some pruning techniques for improving nut yield in cashew (Mohan and Singh 1988, Kumar *et al.*, 2015, Murali *et al.*, 2015). The results of these studies have revealed that closer planting will help in increasing the productivity. However, the responses of the varieties to the pruning varied. Therefore, it is very much essential to identify varieties suitable for ultra and high-density planting which respond to pruning. study revealed that interaction effect of varieties by spacing was observed for most of the growth and yield related characters except plant height and nut traits. Among the three varieties, H-130 performs better under ultra-high-density conditions compared to the other two popular cultivars with respect to nut yield per ha suggesting H-130 is highly suitable for ultra-density planting among the tested varieties. However, in the long run, development of dwarf varieties should facilitate highly efficient ultra-density planting in cashew.

Conclusion

Canopy management is an 'art' of fruit growing - it is much more than cutting off a few branches. In fact, removing wood/branch from a tree is one of the last things you want to do. To optimize your fruit production and productivity, thoughtful canopy management is one of the most important subjects to sustain the yield and quality of fruits. To establish an ideal plantation the young grafts are to be trained from the first year of planting itself which helps in facilitating easy and effective intercultural operations like base cleaning, trench making, fertilizer application, irrigation, pesticide spray against TMB, swabbing against stem and root borer, harvesting and picking nuts. In old and unthrifty plantations, the development of deadwood, water shoots crisscross branches, intermingling branches with the neighbouring trees and the branches crawling on the ground should be pruned to enhance nut yield. The leader shoot pruning should also be attended at least once in 2-3 years along with the removal of the above-mentioned unwanted growth which will be of help in boosting the nut yield. Meanwhile dwarfing rootstocks, planting density, use of growth retardants and varieties selection also play an important role in cashew canopy management aspect.

REFERENCES

Adiga JD, Kalaivanan D, Meena RK, Mohana GS and Lakshmipathi P, 2014. Performance of vigorous cashew cultivars as influenced by dwarf rootstocks. *Vegetos*. **27(2)**: 242-248

Anonymous, 2017. Directorate of cashew nut and cocoa development board. http://www.dccd.gov.in. Anonymous. 2017. Indian Horticulture Database. http:// www. nhb.gov.in.

Babli Mog, Janani P, Nayak MG, Adiga JD and Meena R, 2019. Manipulation of vegetative growth and improvement of yield potential of cashew (*Anacardium occidentale* L.) by Paclobutrazol. *Sci. Hort.*, **257**: 108748.

Bal JS and Dhaliwal GS, 2003. High density planting studies in guava. Haryana *J. Hort. Sci.,* **32**: 19-22.

Balamohan TN, Mekala P, Rajadurai S, Priyadarshini G, Prakash K, Soorianathasundara K and Kumar N, 2016. Canopy management in mango. TNAU Offset and Printing Press, Coimbatore

Barros LM, 1995. Botânica, origem e distribuição geográfica. In: Araújo JPP, Silva VV (eds) Cajucultura: Modernas Técnicas de Produção. EMBRAPA/CNPAT, Fortaleza, pp 55–72.

Bhat MG, Nagaraja KV and Rupa TR, 2010. Cashew research in India. *J. Hortl. Sci.* **5(1)**: 1-16.

Chattopadhyay N and Ghose SN, 1994. Studies on the effect of time and extent of pruning in increasing the yield of cashew. *J. Plant. Crops,* **22**: 111-114.

Crisóstomo LA, Pimentel CRM, Miranda FR and Oliveira VH, 2007. Cashew – Dwarf Variety. In Fertilizing for High Yield and Quality: Tropical Fruits of Brazil, A.E. Johnston, ed. (Switzerland: International Potash Institute), p. 50–69.

Falade JA, 1984. Variability in soils and cashew tree size. *J. Plant. Crops* **12**: 30–37.

Gaikwad SP, Chalak SU and Kamble AB, 2017. Effect of spacing on growth, yield and quality of mango. *Journal of Krishi Vigyan* **5(2)**: 50–53.

Gorakh Singh, 2010. Practical Manual on Canopy Management in Fruit Crops. Department of Agriculture and Cooperation, Ministry of Agriculture, Govt. of India, New Delhi-110001

Goswami AK, Nagaraja A, Jai Prakash and Madhubala Thakre, 2014. https: // www.biotecharticles.com/Agriculture-Article/Canopy-Management-of-Fruit-Crops-3199.html.

Gunjate RT, Kumbhar AR, Thimaiah IM and Amin SM, 2009. Growth and fruiting of some mango cultivars under high density plantation in arid conditions of Gujarat. *Acta Hort.,* **820**: 463–468

Janani P, Adiga JD and Kalaivanan D, 2020. Effect of certain rootstocks on vegetative reproductive growth and yield of cashew cultivars. *CJAST,* **39(16)**: 57982.

Khan MM, Kumar DP and Hiremath IG, 1987. Pruning studies in cashew. *The Cashew,* **1**: 1-12.

Kishore K, Singh Hari Shankar, Kurian RM, 2015. Paclobutrazol use in perennial fruit crops and its residual effects: A review. *Indian J. Agri. Sci.,* **85**: 863-872.

Kumar PB, Hari Babu K and Srihari D, 2015. Effect of time and level of pruning on flowering, fruit and yield of cashew (Anacardium occidentale L.). *Acta Hort.,* **1080**: 245–262.

Meena RK, Saroj PL, Adiga JD, Nayak MG, Meena HR, 2018. Effect of Paclobutrazol on Flowering, Fruiting and Yield of Cashew in West Coast Region of Karnataka. *Int. J. Curr. Micro. and Appl. Sci.,* **7(10)**: 380-391.

Mohan E and Singh R, 1988. Effect of time and severity of pruning in cashew. *Proc. of VIII Symposium on Plantation Crops*, pp. 7-10.

Mohan E and Rao MM, 1995. Effect of growth regulators and pruning on the growth and yield of cashew. *Environ, Eco.,* **13**: 675-679

Mohan E and Room Singh, 1988. Effect of time and severity of pruning in cashew. Proc. of VIII Symposium on Plantation Crops. 28 – 30 December 1988, Kochi, Kerala, India. pp. 7–10.

Murali K, Prasanna Kumar P and Aneesha Rani MS, 2015. Effect of tertiary shoot pruning and foliar spray of nutrients on flowering and yield of cashew under high density planting system. *Bioscan* **10**(1): 411–415.

Nayak MG, 1996. Training and pruning practices for cashew. The Cashew, **10**: 5-9.

Nayak MG, Preethi P, Muralidhara BM and Shamsudheen M, 2019. Training, Pruning and Aftercare in *Cashew. Bulletin*, ICAR–Directorate of Cashew Research, Puttur, Dakshina Kannada, Karnataka.

Nimbolkar P, Awachare Chandrakant, Reddy Y, Chander Subhash, Hussain Firoz, 2016. Role of rootstocks in fruit production – A Review. *J. Agricultural Engineering and Food Technology*, **3(3)**: 183- 188.

Panda JM, 1990. Effect of pruning severity on yield of cashew nut. *Cashew Bull.,* **27**: 15-17.

Prasanna Kumar B, Hari Babu K and Srihari D, 2015. Effect of time and level of pruning on flowering, fruit and yield of cashew (*Anacardium occidentale* L.). *Acta Hort.* **1080**: 245-262. DOI: 10.17660/Acta. Hortic.2015.1080.32

Rejani R, Adiga JD and Yadukumar N, 2013. Performance of different varieties of cashew under high density planting system. *J. Plant. Crops.* **41(1)**: 28–33.

Saroj PL and Nayak MG, 2016. High density orcharding in cashew. *Cashew Week* **16**: 13–14.

Saroj PL, Nayak MG and Meena RK, 2014. Physiology of flowering, fruit and nut development in cashew. In Souvenir, National Seminar-cum-Workshop on Physiology of Flowering in Perennial Fruit Crops, H. Ravishankar, V.K. Singh, A.K. Misra, and M. Mishra, eds., p. 105–114.

Satpathy B, 1988. Thinning, training and pruning of cashew to increase production and productivity of cashewnut. *The Cashew,* **2**: 6-8.

Sousa CAFD, Cavalcanti MILG, Vasconcelos LFL, Sousa HUD, Ribeiro VQ and Silva JALD, 2012. 'Tommy Atkins' mango trees subjected to high density planting in subhumid tropical climate in north eastern Brazil. *Pesquisa Agropecuária Brasileira* **47(1)**: 36–43.

Srilatha V, Reddy YTN, Upreti KK, Jagannath S, 2015. Pruning and paclobutrazol induced vigour, flowering and hormonal changes in mango (Mangifera indica L.). *The Bioscan* **10(1)**: 161- 166.

Vandana, Shivendu Pratap Singh Solanki and Kunzang Lamo, 2017. Canopy management: way to develop fruit tree Architecture. *Biomolecule Reports- An International eNewsletter*. BR/10/17/39. ISSN: 2456-8759

Webster AD, 2004. Vigour mechanisms in dwarfing rootstocks for temperate fruit trees. *Acta Hort.*, **658**: 29- 41.

Yadukumar N, Rao EVVB and Mohan E, 2001. High density planting of cashew. *Tropical Agriculture,* **78**: 19–28.

Chapter 19

Multiple Cropping System in Plantation Crops

Subramanian P.[1], Anitha Karun[2], Bhavishya[3], Nihad[4], Surekha[1], Sujatha, S.[5], Ravi Bhat[1] and Abin C. Ajith[1]

[1]*ICAR–Central Plantation Crops Research Instiute, Kasaragod*
[2]*Former Director, ICAR–Central Plantation Crops Research Instiute, Kasaragod*
[3]*ICAR–CPCRI, RS Vittal*
[4]*ICAR–CPCRI, RS Kayamkulam*
[5]*ICAR–Indian Institute of Horticultural Research, Bangalore, Karnataka*

Introduction

Plantation crops are commercially important and contribute to the economies of many countries through export earnings, employment generation, and poverty reduction, particularly in rural areas (Anandaraj 2015). Smallholder plantation crops play a substantial role in the economy and livelihood of millions of rural people. Important smallholder plantation crops cultivated in India are coconut (*Cocos nucifera* L.), and arecanut (*Areca catechu* L.).

Coconut is a multipurpose palm predominantly grown in small holdings in the tropical region. It is agriculturally and economically important, interwoven with the lives and cultures of people in Southeast Asia, the Pacific region, Africa, and parts of Latin America. The Asia and Pacific region accounts for a significant share of coconut cultivation, covering 10.67 million hectares (87%) and producing 10.11 million metric tons of copra equivalent (86%) (ICC, 2020). In India alone, coconut cultivation spans approximately 2.1 million hectares, producing around 21 billion nuts annually, significantly contributing to rural incomes.

Economic produce of arecanut is its fruit known as betel nut or supari, used for masticatory purposes. Globally, arecanut is cultivated on 1.57 million hectares with a production of 2.44 million tons (2021–2022 statistics, www.faostat.com). Major arecanut-growing countries, including India, Bangladesh, Indonesia, and Myanmar, contribute 94.5 per cent of the total area and 90 per cent of global production. India leads in arecanut production, with 1.56 million tons from 0.79 million hectares during 2021–2022, accounting for 50.6 per cent of the total area and 64.2 per cent of global production (www.faostat.com).

The volatility of price of coconut and arecanut is a serious concern for the smallholder farmers. A strategy to mitigate this risk is the adoption of diversified cropping systems as income from an array of crops would contribute towards stabilizing the whole farm-income.

The smallholder plantation crop sector is characterized by fragmented small and marginal holdings and resource-poor farm environments, often with less marketable and marketed surplus. According to the Agriculture Census data, the average size of operational holdings has decreased from 2.28 hectares in 1970-71 to 1.84 hectares in 1980-81, to 1.41 hectares in 1995-96, and to 1.08 hectares in 2015-16. Considering the declining trends in the size of agricultural holdings and the prospective increase in population over time, the fragmentation of holdings is likely to continue, further decreasing the average size of operational holdings in the country. Therefore, implementing a multistoried cropping system in plantations of coconut and arecanut is essential for sustainable agricultural development and economic stability.

In the future, the availability of land for cultivation will be a major impediment due to the rapid increase in urbanization and the degradation of fertile land from soil erosion and the overuse of chemical inputs. Effective utilization of available land with a proper diversified cropping system for producing field and horticultural crops is the only approach to achieve nutritional security for the growing population and sustainable income for the farmer community. Introducing a diversified cropping system along with advanced agro techniques including organic farming would realize sustainable productivity and maximize income per unit area while maintaining soil fertility through the recycling of natural resources.

Why multiplecropping is more amenable in coconut and arecanut?

1. **Effective Use of Vertical Space**: Palms have unbranched single stems with tall and sparse canopies, allowing sufficient sunlight to penetrate to the lower layers. This creates an opportunity to grow shorter crops underneath without significant competition for light.
2. **Complementary Root Systems**: Palms have fibrous root systems with a horizontal spread tendency and limited vertical growth, typically to a depth of 1 to 1.5 meters. This allows for the accommodation of

dicotyledonous trees with tap root systems that grow deeper. As a result, there is reduced competition for nutrients and water, enabling a more harmonious coexistence of multiple plant species.

3. **Microclimate Creation**: The canopy of palms provides a shaded environment that reduces temperature extremes and increases humidity. This microclimate benefits many understory crops that require cooler and more stable conditions
4. **Soil Health and Fertility**: Multiple cropping systems produce a significant amount of recyclable biomass, which, when recycled, enhances soil health by improving soil structure, increasing organic matter, and promoting nutrient cycling. The diversity of plants contributes to a more resilient and sustainable agricultural ecosystem.
5. **Economic Diversification**: A major advantage of multiple cropping systems is higher income and acting as insurance against market fluctuations. Growing multiple crops in the same area diversifies income sources for farmers, reducing the risk associated with market fluctuations or crop failures. Farmers benefit from the continuous harvest of different crops throughout the year
6. **Pest and Disease Management:** Multiple cropping can help in pest and disease management by disrupting the habitat and lifecycle of pests and pathogens. The diversity of plants creates a less favorable environment for the spread of specific pests and diseases.
7. **Resource Efficiency**: Palms are widely spaced, and monocropping leads to underutilization of resources. Multiple cropping systems make efficient use of available resources like water, nutrients, and sunlight. The different crops have varying water and nutrient needs, which can be managed more efficiently in a multistoried system.
8. **Sustainability**: Multiple cropping promotes sustainable agricultural practices by mimicking natural forest ecosystems. This approach enhances biodiversity, reduces soil erosion, and improves resilience to environmental stress
9. **Crops grown under shade:** Many crops such as cocoa, pepper, ginger, turmeric, and cardamom thrive under partial shade, making the shaded conditions provided by palms ideal for their cultivation.
10. **Improved Land Use Efficiency:** By growing multiple crops on the same piece of land, farmers can maximize the productivity and profitability of their land, making the most out of limited resources.

Types of Cropping System Practiced in Coconut

Intercropping

Intercropping refers to the practice of growing two or more crops, typically annuals or biennials, together on the same piece of land, particularly in the spaces between two rows of coconut/arecanut.

Experiments conducted under ICAR CPCRI and AICRP, as well as in agricultural universities in India and abroad, have resulted in the selection of suitable intercrops for the interspace of coconut gardens, as listed in Table 19.1.

Table 19.1a: Suitability and Performance of Intercrops under Coconut: Tuber Crops, Rhizomatous Spices, Vegetables, and Fruits

Intercrop	*Yield/ha*	*Experiment-Location*	*Nutrient Management*	*Reference*
Elephant foot yam	13.9 t	Navsari, Gujarat, Dark greyish-black soil	50 per cent RDF + 50 per cent Organic	Bhalerao *et al.*, 2021
Tannia	65605 leaves			
Tapioca	5.08 t	Kasaragod, Kerala, red loam soil	As per the POP of coconut and intercrop	Vargheese *et al.*, 1979
Sweet potato	7.1 t	Kasaragod, Kerala		Gopalasundaram and Nelliat *et al.*, 1979
Chinese potato	5.6 t			
Sweet potato	6.49t/ha	Mondouri, West bengal	As per the POP of coconut and intercrop	Subba *et al.*, 2021
Sweet potato	9.75t/ha	Shakigopal, Bhubaneswar, Orrissa	As per the POP of coconut and intercrop	Nedunchezhiyan *et al.*, 2007
Tapioca	10.86t/ha	Kasaragod, Kerala	As per the POP of coconut and intercrop	Suja *et al.*, 2020
Turmeric	13.14 t	Navsari, Gujarat	Fully organic	Bhalerao *et al.*, 2021
Ginger	7.58 t	Kasaragod, red loam soil	As per the POP of coconut and intercrop	Varghese *et al.*, 1976
Bush pepper	195.9 g/ plant	Peru vannamuzhi, Kozhikode	As per the POP of coconut and intercrop	Thankamani and Ashokan 2010
Snake gourd	10579 kg	Kasaragod, sandy loam soil	As per the POP of coconut and intercrop	Hegde *et al.* (1993)
Amaranthus	6610 kg			
Brinjal	6120 kg			
Pumpkin	10.12 t	Kasaragod, coastal sandy soil		Subramanian *et al.*, 2006

Intercrop	*Yield/ha*	*Experiment-Location*	*Nutrient Management*	*Reference*
Drumstick	1146 kg	Arsikere, Karnataka	Fully organic	Siddappa *et al.*, 2023
Okra	9.85 t/ha	Arasikere, Hassan District, Karnataka	As per the POP of coconut and intercrop	Kumar *et al.*, 2016
Baby corn	4.55 t/ha			
Gherkin	7.4 t/ha			
Cucumber	6.51 t/ha			
Tomato	7.29 t/ha			
Bitter gourd	7.1 t/ha	Bhatye, Ratnagiri, Maharastra	As per the POP of coconut and intercrop	Malshe *et al.*, 2023
Ridge (*Luffa acutangula* cv. Pusa Nasdar)	10.7 t/ha	Port Blair, Andaman and Nicobar Islands, India,	INM	Nai and Nair 2006
Sponge gourd (*Luffa cylindrica* [*L. aegyptiaca*] cv. Local)	11.6 t/ha			
Greater yam (*D. Alata*)	9.28 t	Kasaragod, Kerala	As per the POP of coconut and intercrop	Suja *et al.*, 2022
Lesser yam (*D. Esculenta*)	7.46 t			
White yam (*D. Rotundata*)	5.64 t			
Pineapple	9750 kg	Mondouri, West Bengal	As per the POP of coconut and intercrop	Ghosh and Bandopadhyay 2011
Pineapple	15.96 t	Philippines		Mario E Baliad, 1994
Pineapple	14.67 t	Kasargod, Coastal sandy soil		Subramanian *et al.*, 2009
Banana	24.65 t	Veppankulam, TamilNadu	50 per cent RDF + 50 per cent Organic	Selva Rani *et al.*, 2024
Banana	8.27 t	Navsari, Gujarat	50 per cent RDF + 50 per cent Organic	Bhalerao *et al.*, 2021
Banana	9.7 t	Arsikere, Karnataka	Fully organic	Siddappa *et al.*, 2023

Figure 19.1a: Coconut + Elephant Foot Yam Intercropping System.

Figure 19.1b: Coconut + Turmeric Intercropping System.

Figure 19.1c: Coconut + Ginger Intercropping System.

Figure 19.1d: Coconut + Colocasia Intercropping System.

Figure 19.1e: Coconut + Cassava Intercropping System.

Figure 19.1f: Coconut + Pumpkin Intercropping System.

Figure 19.1g: Coconut + Snake gourd Intercropping System.

Figure 19.1h: Coconut + Brinjal Intercropping System.

Figure 19.1i: Coconut + Pineapple Intercropping System.

Figure 19.1j: Coconut + Banana Intercropping System.

Table 19.1b: Suitability and Performance of Intercrops under Coconut: Cereals, Millets, Pulses Fibre Crops and Oilseeds

Intercrop	*Yield/ha*	*Experiment-Location*	*Nutrient Management*	*Reference*
Rice	850 kg	Kasaragod and Pilicode, Kerala	As per the POP of coconut and intercrop	Gopalasundaram and Nelliat (1979)
Pearl millet	1102 kg			Gopalasundaram and Nelliat (1979)
Finger millet	752 kg			
Varagu	2250 kg			
Cowpea	483.5 kg	Kasaragod, Kerala, red loam soil		
Green gram	275 kg			
Black gram	275 kg			
Redgram	293.5 kg	Pilicode, Kerala		
Ground nut	965 kg			
Soybean	980 kg	Kasaragod, red loam soil		Hegde and Yusuf (1993)
Maize	2.2 tons cob/ha	Santo Island.	As per the POP of coconut and intercrop	Serge Braconnier 1998
Finger millet	2072 kg/ha	Balaramapuram	As per the POP of coconut and intercrop	Sneha *et al.*, 2023
Blackgram var. DBGV 5	1, 183.33 kg	Vellayani, Kerala	As per the POP of coconut and intercrop	Pooja *et al.*, 2021
Ramie	209.6 kg/ha	Goa	As per the POP of coconut and intercrop	Manjunath *et al.*, 2010

Figure 19.2a: Coconut + Flower Intercropping System.

Figure 19.2b: Coconut + Groundnut Intercropping System.

Table 19.1c: Suitability and Performance of Intercrops under Coconut: Fodder Crops, Medicinal Crops and Flowers

Intercrop	*Yield/ha*	*Experiment-Location*	*Nutrient Management*	*Reference*
Pusa Giant	45.14 t	Kayamkulam, Kerala	As per the POP of coconut and intercrop	Ramakrishnan Nayar and Sahasranaman (1979)
Hybrid napierNB-21	53.02 t			
Guinea grass	49.21 t			
Stylosanthes	19.05 t			

Intercrop	*Yield/ha*	*Experiment-Location*	*Nutrient Management*	*Reference*
Hybrid napier CO3	126.9 t/ha/yr	Kayamkulam, Kerala	75 per cent RDN through cowdung slurry + coconut leaf vermicompost + Azospirillum sp	Nihad *et al.,* 2023
Hybrid Bajara Napier (HBN) var.Suguna	132.04 t ha-1	Kayamkulam, Kerala	Fertilizer application based on soil test	Jeena Mathew *et al.,* 2023
Hybrid napier bajra	96.83t/ha	Kasaragod, kerala	As per the POP of coconut and intercrop	Subramanian *et al.,* 2007
Marigold	2.0 t	Kayamkulam, Kerala	vermicompost (16.8 kg N ha^{-1}) +16.8:18:18 kg NPK ha^{-1}	Nihad *et al.,* 2017
Globe amaranth	1.9 t		Grown on residue	
Kalmegh (*Andrographis paniculata*)	3396 kg DH (Dry Herbage)	Arsikere, Hassan District, Karnataka	As per the POP of coconut and intercrop	Basavaraju *et al.,* 2011
Makoi (*Solanum nigrum*)	2926 kg DH			
Coleus (*Coleus forskohlii*)	418 kg dry tuberous			
Garden rue (*Ruta graveolens*)	3596 kg DH			
Lepidium (*Lepidium sativum*)	492 kg seeds			
Tulsi (*Ocimum sanctum*)	4127 kg DH			
Arrow root (*Maranta arundinaceae*)	5341 kg fresh rhizomes			
Kacholam (*Kaemferia galanga*)	1079 kg fresh rhizomes			

Intercrop	*Yield/ha*	*Experiment-Location*	*Nutrient Management*	*Reference*
Cowhage (*Mucuna pruriens*)	2779 kg seeds	Arsikere, Hassan District, Karnataka	As per the POP of coconut and intercrop	Basavaraju *et al.*, 2011
Roselle (*Hibiscus sabdariffa*)	440 kg dry calyces seeds			
Ambrette (*Abelmoschus moschatus*)	368 kg seeds			
Citronella (*Cymbopogon winteranus*)	24937 kg green leaves			
Lemon grass (*Cymbopogon flexuosus*)	45788 kg green leaves			
Vetiver grass (*Vetiveria zizanoides*)	2179 kg dry roots			
Tippili (*Piper longum*)	13.4 g berries/ plant	Walpita Estate, Kotadeniyawa	As per the POP of coconut and intercropt	Gunathilake *et al.*, 2002
Ingurupiyali (*Kaempheria galangal*)	39.2 g rhizome/ plant			
Komarica (*Aloe vera*)	1719 g leaves/plant			
Rathnitol (*Plumbago indica*)	131.9 g roots/plant			
Katuwelbatu (*Solanum xanthocapum*)	24.8 g total plant/plant			
Kapukinissa (*Hibiscus abelmoschum*)	27 g seeds/ plant			
Amukkara (*Withania somnifera*)	48.8 g roots/ plant			
Senehekola (*Cassia angustifolia*)	4.2 g leaves/ plant			

Intercrop	*Yield/ha*	*Experiment-Location*	*Nutrient Management*	*Reference*
Adhathoda (*Adhathoda vasica*)	180 g total plant/plant			
Jasmine	1123.2 kg	Bhatye, Ratnagiri, Maharashtra	As per the POP of coconut and intercrop	V.V. Shinde *et al.* 2017)
Heliconia	96982 no.			
Tuberose	314987 spikes	Kahikuchi, Guwahati, Assam	As per the POP of coconut and intercrop	Nath *et al.*, 2019
Gerbera	743200 spikes			
Bird of paradise	32175 spikes			
Gladiolus	61987 spikes			
Marigold	20810 kg			
Marigold	6053 kg	Aliyarnagar, Tamil Nadu	As per the POP of coconut and intercrop	Rani *et al.*, 2018
Gomphrena	5206 kg			
Chrysanthemum	4502 kg			
Celosia	4067 kg			
Zinnia	3509 ka			

Figure 19.3a: Coconut + Guinea Grass Intercropping System.

Figure 19.3b: Coconut + Marigold Intercropping System.

Mixed Cropping

Growing perennials in the interspaces of coconut. Suitable perennial crops as intercrops in the interspaces of coconut are presented in Table 19.2.

Table 19.2: Suitability and Performance of different Perennial Crops in the Coconut Garden under different Agro-climatic Zones

Intercrop	*Yield/ha*	*Experiment-Location*	*Nutrient Management*	*Reference*
Pepper	41.12kg	Veppankulam, Tamil Nadu	50 per cent RDF + 50 per cent Organic	Selva Rani *et al.*, 2024
Cocoa	300 kg dry beans			
Cocoa	285 kg dry beans	Arsikere, Karnataka	Fully organic	Siddappa *et al.*, 2023
Cocoa	245 kg dry beans	Philippines	100 per cent RDF	Mario E Baliad, 1994
Coffee	13.72 kg	Philippines	As per the POP of coconut and intercrop	
Lime	1205 kg	Arsikere, Karnataka	Fully organic	Siddappa *et al.*, 2023
Lime	800 kg	Mondouri, West Bengal		Ghosh and Bandopadhyay 2011

Intercrop	*Yield/ha*	*Experiment-Location*	*Nutrient Management*	*Reference*
Valencia sweet orange *Citrus sinensis* L. Osbeck	3.54t	Ghana, Africa,	As per the POP of coconut and intercrop	Andoh-Mensah and Ofosu-Budu, 2012
Lemon	700 kg	Mondouri, West Bengal	As per the POP of coconut and intercrop	Ghosh and Bandopadhyay 2011
Guava	1300 kg			
Mulberry	1840 kg leaves (375 kg cocoon yield)	Arsikere, Karnataka		Shanthaamallaiah *et al.* 1982)
Mulberry	48.26 and 11.05 Mg/ ha/y green and dry fodder yield	Vellanikkara, Kerala	As per the POP of coconut and intercrop	Raj Reshma *et al.*, 2019
Noni	6159 kg/ha/ year	Kasaragod, Kerala	Vermicompost + vermin wash + biofertilizer + *In situ* green manuring + husk incorporation + coconut leaf mulch	Maheswarappa *et al.*, 2017
Vanilla	1.87 kg vine^{-1} bean yield	Kasaragod, Kerala	Cow dung slurry -6 tonnes ha^{-1}	Maheswarappa *et al.*, 2016
Nutmeg	Pericarp yield16.07 kg/tree, mace yield 0.320 kg/ tree and nut yield 2.46 kg/tree	TNAU Coimbatore	As per the POP of coconut and intercrop	Ammaan and Paramaguru, 2017

Figure 19.4a: Coconut + Cocoa Intercropping System.

Figure 19.4b: Coconut + Pepper Intercropping System.

Multi Storied Cropping

Multistoried cropping system is an alternative farming approach designed for small and marginal farmers, offering higher economic returns per unit area. As a form of intercropping, this system enables the simultaneous growth of plants of varying heights in the same field. It is primarily practiced in orchards and plantations to optimize solar energy utilization, even in areas with high planting density. By cultivating crops with different heights, root structures, and growth durations, the system efficiently makes use of vertical space. The tallest plants thrive in strong sunlight and high evaporative conditions, while the shorter plants benefit from shade and higher humidity.

As defined by Nair (1979), multistoried cropping involves growing multiple crops with different heights and growth habits on the same land to maximize the use of vertical space and resources. This system integrates various plant species, including trees, shrubs, and ground covers, each occupying a different layer or "storey." The upper layer consists of tall trees, the middle layer includes shorter trees or shrubs, and the lower layer is filled with herbs or ground-covering plants. This method improves the efficiency of light interception, water usage, and nutrient absorption, fostering biodiversity and enhancing overall productivity and sustainability. Multistoried cropping is particularly beneficial in plantation systems like coconut, arecanut, and oil palm, as it helps maintain soil fertility, reduce pests and diseases, and provide additional income through diversified crop production. Some of the successful multistoried cropping systems that have been developed are presented in Table 19.3.

Table 19.3: Coconut Based Multistoried Cropping Systems under different Agro-climatic Zones

Experiment-Location	*Nutrient Management*	*Component Crops and Field*	*Reference*
Philippines, Clay loam soil	Fully organic	Black Pepper 28kg (123 vines) Cocoa 52.1 kg from 329 plants Pineapple 3700 Kg	Margate and magat 1983
Kasargod, India	Recommended POP	Cocoa SH 59 pods per tree Pepper 192 kg/ha Pineapple 777kg/ha	CPCRI 1984
		Cinnamon in SH 162 g/plant Pineapple 1196 kg/ha Pepper 228 kg/ha	CPCRI 1985

High Density Multispecies Cropping System

High-Density Multispecies Cropping System (HDMSCS) involves growing a wide variety of crop species in a single coconut plantation at high planting densities to optimize resource use efficiency and cater to the diverse needs of farmers (Bavappa *et al.,* 1986; Rethinam, 1990). In the study of the HDMSCS model, which included crops such as coconut, black pepper, pineapple, banana, clove, turmeric, ginger, vegetables, sweet corn, and baby corn, various nutrient management strategies were evaluated. Three treatments were tested: two-thirds of the recommended NPK fertilizer combined with biomass recycling (via vermicompost); one-third of the recommended NPK fertilizer with biomass recycling, biofertilizer, green manuring, and vermiwash; and a fully organic approach using biomass recycling, biofertilizer, green manuring, vermiwash, husk burial, and mulching around coconut basins. Coconut yield did not significantly differ among the treatments, suggesting that fully organic management can achieve yields comparable to inorganic fertilizer use, while delivering the highest economic returns (Maheswarappa *et al.,* 2013).

Figure 19.5a: Coconut + Cocoa Intercropping System at Kasaragod Kerala.

Figure 19.5b: Coconut + Cinnamon Intercropping System at Kasaragod Kerala.

Research conducted at ICAR–CPCRI, Kasaragod, India, demonstrated that a coconut-based HDMSCS incorporating coconut, black pepper, banana (cv. Kadali and Robusta), cinnamon, and nutmeg, and maintained under organic management practices (including biomass recycling, biofertilizer application, green manuring, coconut husk burial, and vermiwash), produced yields comparable to those using integrated nutrient management strategies. This finding supports the viability of organic management for such cropping systems. The organic system also exhibited higher net energy, energy use efficiency, energy productivity, and energy profitability, resulting in greater net returns (CPCRI, 2022).

An evaluation of organic nutrient sources in a coconut garden with sandy loam soil intercropped with vanilla showed that the application of cow dung slurry (6 t ha^{-1}) significantly increased the mean fresh yield of vanilla beans. This was closely followed by the application of vermicompost (5 kg per plant), biofertilizers

(Bacillus and Azospirillum), and vermiwash, which produced comparable results (Maheswarappa *et al.,* 2016).

Figure 19.6: Coconut Based High Density Multi Species Cropping System.

Subramanian *et al.* (2016) found that black pepper grown organically in the coconut-based HDMSCS produced a yield of 2.4 kg per vine, similar to that achieved with integrated nutrient management practices. However, soil nutrient analysis revealed significantly lower available potassium (K) content in the organic plots compared to the integrated nutrient management plots, indicating a higher rate of K depletion in the absence of inorganic inputs. This emphasizes the need to replenish K through organic resources when adopting organic nutrient management practice

Table 19.4. Coconut Based High Density Multi Species Cropping Systems

Experiment-Location	*Nutrient Management*	*Component Crops and Field*	*Reference*
Kasaragod	Fully organic	Black pepper 1.7 kg/vine Banana *Njalipoovan* 7.2 kg/plant Banana Grand naine 13.7 kg/plant Pineapple 1kg fruit/plant Clove 1 kg/tree Ginger 862 kg/ha Turmeric 955 kg/ha Baby corn 1854 kg/ha Sweet corn 650 kg/ha Brinjal 865 kg/ha Elephant foot yam 3080 kg/ha	Maheswarappa *et al.,* 2013

Experiment-Location	*Nutrient Management*	*Component Crops and Field*	*Reference*
Kayamkulam	Integrated Nutrient Management	Banana 1619 kg from 310 plants Pineapple 515 kg from 550 plants Black pepper 29 kg from 60 vines Nutmeg mace 7.2 kg from 70 trees	Krishnakumar *et al.*, 2011
Ambajipeta, East Godavari Andhra Pradesh	Recommended POP to both coconut and component crops	Cocoa 216 kg/ha (108 plants) Banana 7.1 t/ha (324 plants) Pineapple 810 kg/ha (854 plants) Elephant foot yam 1362 kg/ha (486 plants) Heliconia 5832 spikes (486 plants)	Padma *et al.*, 2016
Kahikuchi, Assam		Black pepper 136.2 kg/ha Banana 1005.12 kg/ha Assam Lemon 2534 fruits/ha Ginger 657 kg/ha Pineapple 938 kg/ha	Nath *et al.*, 2008

Coconut Based Integrated Mixed Farming

This system involves integrating subsidiary enterprises such as livestock, poultry, and rabbit farming into coconut gardens, using fodder or pasture grown in the interspaces (Bavappa, 1990; sand Gopalasundaram, 1990). By combining crops and livestock, including dairy, poultry, duck rearing, aquaculture, and shade-tolerant fodder cultivation, the system promotes the diversity necessary for organic farming. This approach not only generates additional income but also creates more employment opportunities, offering a buffer against fluctuations in coconut prices (Subramanian *et al.*, 2014). Crop residues and fodder provide valuable animal feed, while manure and litter from livestock supply renewable organic matter and essential plant nutrients, fostering a sustainable and integrated agricultural system. The recycling of crop and livestock waste plays a key role in maintaining soil health.

Incorporating these practices enhances sustainability and maximizes the positive impact of species diversity on soil fertility. Fodder grasses like hybrid Napier and guinea grass, grown under the shade of coconut palms, can yield around 80–120 tons of fodder per hectare annually–sufficient to support 8 to 12 crossbred milch cows. This, in turn, generates enough farmyard manure to meet on-farm organic fertilizer needs. Additionally, installing a biogas plant provides energy for the farmhouse and slurry for fertilizing coconut and other crops.

In mixed farming systems, higher microbial biomass carbon and dehydrogenase activity were observed in the root zones of coconut and hybrid Napier grass compared to monocropping systems with only coconut. The microbial populations in both coconut basins and interspaces were significantly higher, especially under 100 per cent organic farming treatments, with increased bacteria, fungi, and actinomycetes populations (Palaniswami *et al.*, 2008).

Figure 19.7a: Coconut + Fodder in Coconut Based Farming System.

Figure 19.7b: Coconut Based Farming System.

The benefits of organic farming for soil health were clearly demonstrated in the coconut-based mixed farming system, which involved cultivating fodder grass in the interspaces and integrating animal husbandry (dairy, poultry, and pisciculture). Managed under three different treatments (100 per cent organic, 100 per cent inorganic, and a 50-50 combination), the system showed significantly higher populations of soil microbes and function-specific microorganisms under the organic treatment. Increased microbial biomass and enhanced activity of soil enzymes such as phosphatase and dehydrogenase indicated that biological activity in the soil was much more pronounced under organic management compared to inorganic or integrated approaches (Subramanian *et al.,* 2014). A summary of published results are shown in Table 19.5.

Table 19.5: Summary of Reports Published on Coconut Based Integrated Farming Systems

Experiment-Location	*Components*	*Economic Returns (Lakh Rs.)*	*B:C Ratio*	*Reference*
IFS-Thiruvan-antha-puram, Kerala	Coconut on bunds (30 no.) Teak trees along the border (15 no.) Multitier cropping (in interspaces of coconut) papaya (6), clove (1), nutmeg (1), cocoa (1), rose apple (1), mango (1), jack (1), bread fruit (1), spices (ginger + turmeric), tuber crops, fodder crops and azolla (in a shallow pit of size 2 x 1 x 0.2 m^3, lined with silpaulin) Cow unit: Cross bred (1+1) Freshwater fish (Genetically Improved Farm Tilapia) reared in trenches dug between the bunds planted with coconut	2.92	1.37	Sudha *et al.,* 2021
	Coconut + Fodder crops (Hybrid Napier (Co-3) and (*Stylosantheshamata*)) + Cow (5 no. HF)	2.28	3.69	Jagadeesha *et al.,* 2023
Kasaragod, Kerala, red sandy loam	Fodder crops (guinea grass (and hybrid napier, NB 21); Dairy unit: five Holstein Friesian breed cows and one Jersey; Poultry unit: 100 layers and 100 broiler (for each batch); Biogas unit 3 m^3 Rabbits (Ten female and 4 male Russian chinchilla breeds); Aquaculture (625 m^2 surface area)-about 4-6 cm long fingerlings of 4 select- ed species, *viz.* catla (Catlacatla), rohu (Sebeo rohita), mrighal (Cirrhinus mrigha- la) and grass carp (Ctenupharyn godonidellus)	2.86	1.39	Mahesh-warappa *et al.,* 2000

Experiment-Location	*Components*	*Economic Returns (Lakh Rs.)*	*B:C Ratio*	*Reference*
Goa	Coconut; Forage grass; Milch Cows(2 no.)	0.50	1.43	Manjunath and Itnal 2001
Puri, Odisha	Coconut + Vegetables + Cattle 2, Goat 2 + Poultry 6+ Duck5 + fish (Rohu and Silver carp)	0.56	2.00	Bhol *et al.,* 2015
Lunuwila, Srilanka	Coconut + Goat (24 Nos)		2.04	Nuwarapaksha *et al.,* 2023
Kasargod	Coconut (WCT) + Fodder grass (BN CO-3) and guinea grass/ Fodder sorghum, fodder maize, (Stylosanthes - annual and perennial), fodder cowpea + Black Pepper (Panniyur-1) + Banana (Njalipoovan variety) + Dairy (Holstein Friesian and Jersey cross breed cows) + Poultry (Broiler birds 4-6 batches/year) + Japanese quails + 8. Biogas plant (3 m^3) + Azolla unit (4 m^2) + Aquaculture (Catla, Rohu, Mrigal, Grass carp) + Goatery unit	Rs. 2.85	2.36	Subramanian *et al.,* 2014
South Minahasa Regency, Indonesia	Coconut + Dwarf Elephant Grass + Beef cattle			Osak *et al.,* 2018

Arecanut Based Cropping Systems

A well-designed cropping system offers numerous benefits, including providing balanced nutrition, ensuring regular employment, maintaining soil health, mitigating aberrant weather situations, and increasing farm productivity and income from arecanut farming (Bhat *et al.,* 2024).

Arecanut (*Areca catechu* L.) is an important plantation crop grown in India for its kernel obtained from the fruit. Traditionally, arecanut is consumed alone or in combination with betel leaves and slaked lime. It also has economic, socio-cultural, religious, and medicinal importance in India and other South-East Asian countries. Arecanut is a tall growing palm, and there is an excellent opportunity for temporal and spatial distribution of different crop species because of the rooting pattern, low light interception by the palm, and microclimate available in the plantation. Inter-/mixed cropping in arecanut can improve resource use efficiency, which gives ample scope to overcome the soil, weather, and crop constraints in arecanut production (Thomas *et al.,* 2011). Congenial microclimate, enrichment of SOC, increased productivity, and income are the benefits of the cropping system approach (Sujatha *et al.,* 2011b).

Intercropping of Annuals/Biennials/Short-statured Perennials

Crops like banana, ginger, chilli, colocasia, upland paddy, turmeric, elephant foot yam, pineapple, arrowroot, and dioscorea are more adapted for intercropping in arecanut gardens (Shama Bhat, 1974; Sannamarappa and Muralidharan, 1982; Muralidharan, 1990).

Among vegetables intercropped in arecanut in Assam, cabbage, cauliflower, brinjal, and radish were found suitable (Ray *et al.,* 2011). The net return per rupee investment is maximum for brinjal (15.8), followed by cabbage and radish (Ray *et al.,* 2007). In sub-Himalayan West Bengal, vegetable crops like cabbage, spinach, radish, turnip, carrot, tomato, brinjal, chilli, capsicum, dolichos bean, pumpkin, ash gourd, bottle gourd, and snake gourd are ideal for intercropping in arecanut due to higher cash outflows and benefit-cost ratios (Sit *et al.,* 2006 and 2011). Cultivating turmeric varieties like Suguna, Sudarsana, CLS 2A, and Kasturi in the interspace of the arecanut garden was suggested by Sit *et al.* (2004).

Medicinal and aromatic plants like Shatavari (*Asparagus racemosus* Willd), Vetiver (*Chrysopogon zizanioides* L.), Long pepper (*Piper longum* L.), Brahmi (*Bacopa monnieri* L.), *Nilgirianthus ciliatus*, Periwinkle (*Catharanthus roseus* L.), Aloe (*Aloe vera*L.), Lemon grass (*Cymbopogon flexuosus*Steud.), Palmarosa (*Cymbopogon martini* (Roxb.) W.Watson), Basil (*Ocimum basilicum* L.), Davana (*Artemisia pallens*Wall.) and Patchouli (*Pogostemon cablin* Benth.) perform better as intercrops in arecanut (Sujatha *et al., 2011b*). Regarding net profit per rupee investment, all medicinal and aromatic plants are superior, and system productivity can be considerably enhanced with intercropping (Sujatha *et al.,* 2006).

Mixed Cropping with Perennial Crops

At Hirehalli of Karnataka, the yield of arecanut increased by 37 per cent over three years due to intercropping with cinnamon (Sannamarappa and Muralidharan, 1982). They also reported that intercropping of four varieties of coffee *viz.*, arabica S-6, arabica S-1936, San Ramon, and Robusta, also showed similar effects on the yield of arecanut.

Arecanut+Cocoa Mixed Cropping System

Cocoa (*Theobroma cacao* L.) is an ideal intercrop with arecanut. The micro-climate existing in arecanut is congenial for cocoa. The Land Equivalent Ratio of the arecanut–cocoa system can be increased to 1.74, giving a 74 per cent yield advantage over sole arecanut (Balasimha, 2009). It is advantageous to grow cocoa at a spacing of 2.7 m X 5.4 m in arecanut with optimum nutrition and irrigation at 100 per cent ET (Abdul Haris*et al.,* 1999; Balasimha, 2009; Sujatha *et al.,* 2015). The review of cropping system research in arecanut established the importance of including cocoa as a component crop for improved productivity, profitability, and soil fertility (Bhat and Sujatha, 2011a).

Arecanut + Banana Mixed Cropping System

Banana is usually grown as a nurse or shade crop in arecanut plantations, and its advantages are reported earlier (Shama Bhat and Khader, 1970; Bhandary, 1974). The yield of arecanut is not affected by mixed cropping with banana (Bhandary, 1974; Nagaraj, 1974; Waman *et al.,* 2019b). With regard to the suitability of banana cultivars for intercropping in Karnataka and Kerala, Robusta, Mysore Poovan, Red banana, and Karpuravally are ideal (Nayar *et al.,* 1985).

Growing Black Pepper, Betel Vine and Woody Pepper as Mixed Crop with Arecanut

Black pepper is an excellent crop for mixed cropping with arecanut and high economic returns can be expected (Abraham, 1974; Nair, 1982; Nayar, 1982). Experimental data from mixed cropping of arecanut and black pepper for ten years showed no detrimental effect on the yield of arecanut palms due to training black pepper on them. Further, it helped to augment the farmer's income by mixed cropping of black pepper (Nayar, 1982). Black pepper varieties like Panniyur-1 and Karimunda are better for mixed cropping system with arecanut. Dry black pepper yield ranged between 0.75 to 9.25 kg vine^{-1} with an average of 1.00 kg vine^{-1} in the Sub Himalayan Terai Region of West Bengal (Sit *et al.,* 2002). Betel vine can also be grown profitably in arecanut in the Sub Himalayan Terai region of West Bengal (Chenchaiah *et al.,* 2002). Woody pepper (*Piper pendulispicum* C. DC.), an underutilized species in Andaman and Nicobar Islands, can be grown as an intercrop with arecanut (Waman *et al.,* 2019a).

Arecanut+Vanilla Mixed Cropping System

Vanilla can be successfully grown as an intercrop of arecanut plantations in India with proper nutrition and irrigation. The results of a five-year study on water and organic nutrition of vanilla in arecanut plantation indicated that micro-sprinkler irrigation at 1.0 Ep is significantly superior over 0.5 and 0.75 Ep (Sujatha and Bhat, 2010). Organic manure application in the form of vermicompost and FYM, and recommended NPK produce green bean yield at par with recycling of gliricidia prunings. The benefit : cost (B:C) ratio is highest with recommended NPK (2.27) followed by recycling of Gliricidia prunings (2.10), vermicompost (1.87), vermicompost+ arecanut husk mulching (1.80) and FYM (1.64).

Arecanut Based High Density Multispecies Cropping System (HDMSCS)

Successful cultivation of mixed crops like cocoa, black pepper, clove, lemon, and banana in arecanut based High density multispecies cropping system (HDMSCS) is well documented (Bhat, 1978; Bavappa*et al.,* 1986; Abdul Khader *et al.,* 1992; Sannamarappa, 1993; Reddy and Bhat, 1994; Bhat *et al.,* 1999). It increases resource use efficiency, employment opportunities, and income (Bhat *et al.,* 2024) and makes the system more sustainable (Sujatha *et al.*, 2020). Other tangible benefits noticed from the system are improvement in microclimate (Bavappa *et al.*, 1986), reduction in leaching losses of water and nutrients due to runoff (Mohapatra and Bhat, 1990), and improvement in soil fertility due to increased recycling of biomass (Abdul Khader *et al.*, 1992; Bhat and Sujatha, 2007).

Effect of Growing Inter/Mixed Crop on Soil Fertility

Several long-term trials have shown that nutrient cycling, soil fertility, and carbon cycling are influenced by inter–/mixed cropping (Bhat *et al.*, 2024). The impact of arecanut based cropping system on soil quality indicators is well documented (Manikandan *et al.*, 1987; Bhat and Sujatha,2007, 2011a, 2011b; Sujatha*et al.*, 2011a; Sujatha and Bhat, 2015). Increased root proliferation in arecanut due to intercropping would increase organic matter content in the soil. Intercropping of medicinal and aromatic crops increased the soil pH. The SOC content increased in *Aloe vera*, *A. pallens*, *P. longum*, and *B. monnieri*, while it depleted in grasses and rhizomatic medicinal and aromatic plants. Sannamarappa and Muralidharan (1982) also stressed crop rotation when exhaustive tuber and rhizome crops are intercropped with arecanut. This implies the need for crop rotation and the need-based application of organic and inorganic nutrients.

The recyclable biomass available in the arecanut garden contains reasonable quantities of N, P, and K (Biddappa*et al.*, 1996), which can be easily converted to compost using earthworms with a recovery percentage of 74 to 87 (Chowdappa*et al.*, 1999). As reported by Abdul Haris *et al.* 1999 and Balasimha, 2007, cocoa biomass in terms of litterfall and pruned biomass has great potential for recycling. Manikandan *et al.* (1987) noticed reduced soil pH and improved organic carbon status in the arecanut + cocoa system. This might be due to cocoa leaf fall, in situ root decay, and fertilizer application. The available P in the arecanut basins of the arecanut + cocoa mixed cropping system showed a higher concentration than sole arecanut. However, the reduction in Zn under the system indicates the need to include Zn in the fertilizer schedule of the arecanut + cocoa system (Manikandan *et al.*, 1987).

The intensive cropping system involving arecanut, cocoa, black pepper, clove and banana is a successful crop combination that generates a considerable quantity of organic waste. The arecanut based high density multispecies cropping system (ABHDMSCS) produced 14.3 t ha^{-1} recyclable biomass (Bavappa*et al.*, 1986). However, different cropping models produce differing quantities of recyclable biomass, which varies from 8.72 to 10.35 t ha^{-1} $year^{-1}$. The reasonable amount of biomass available from arecanut gardens can be converted to vermicompost with a recovery of 82–87 per cent and nutrient content of 1.71 per cent N, 0.21 per cent P, and 0.43 per cent K (Biddappa*et al.*, 1996; Bhat and Sujatha,2011b; Sujatha *et al.*, 2015). The ABHDMSCS adopted in a clay loam soil at Kahikuchi, Assam, India, showed an increase in SOC, available N, P, and K and a decline in soil pH (Paul *et al.*, 2020). The studies on the nutrient requirement of ABHDMSCS (arecanut +

pepper + cocoa + banana) indicated that 2/3rd of recommended fertilizer dose for each component crop is sufficient for optimum yield levels (Bhat and Sujatha, 2011a). Organic carbon content was significantly higher in arecanut + black pepper rhizospheres (1.87%) than in cocoa (1.62%) and clove (1.66%). The overall results of adopting the arecanut-based cropping system revealed that the system could be self-sustainable over the years with organic matter recycling except for exhaustion of soil test K. The exhaustion of soil K probably indicates the necessity of including K in the fertilizer schedule of the system, as organic matter cannot supplement the K requirement of crops, and all the component crops are heavy feeders of this major nutrient. The nutrient status of the soils at Vittal under mono-crop and HDMSCS indicated improved soil fertility under HDMSCS compared to mono-crop. Arecanut-based HDMSCS improves soil fertility status in terms of soil pH, soil organic carbon, and available P (Bhat and Sujatha, 2011a). The experimental results indicate that the advantages are substantial in multiple cropping due to the prevention of soil erosion and nutrient loss compared to mono-crop systems.

Arecanut Based Integrated Farming System

Arecanut-based integrated farming system (ABIFS) model comprising of areca- nut, fodder grass (cv. C03) as intercrop and sole crop, dairy of four to six milking cows and fishery component (264 m^2) was economically sustainable, especially for the small and marginal farmers, and recycling of wastes could meet the nutrient requirement of the crop, except K (Sujatha and Bhat, 2015). This model was economically sustainable, especially for small and marginal farmers, and could meet the crop's nutrient requirements, except potassium. On average, the organic waste recycling potential of the arecanut + dairy unit was 13.7 t ha$^{-1,}$ and the dairy unit alone contributed to 87 per cent of the manure production. After recycling to the system, the total nutrient supply from ABMS was estimated at 218 kg N, 51.8 kg P, and 33 kg K, which can meet N and P demand of 1.7 and 2.2 ha of arecanut, respectively. Using hard laterite soil for livestock enterprises like dairy, fishery, and fodder cultivation improved resource use efficiency and profits per unit area per unit time (Sujatha and Bhat, 2015)

Future Line of Research Works

Future research in plantation crops should focus on innovative strategies to enhance productivity, sustainability, and resilience, particularly in response to climate change and resource limitations. Key areas for exploration include:

Precision Agriculture and Smart Farming: Advancements in precision agriculture techniques, such as drone monitoring, sensors, and automated irrigation, can optimize resource use in plantation crops. The integration of artificial intelligence (AI) and machine learning (ML) can further enhance crop health monitoring, pest and disease management, and yield forecasting.

Climate-Resilient Cropping Systems: Developing climate-resilient plantation crop varieties, particularly suited for diverse cropping systems, along with effective intercropping strategies, is essential to adapt to changing weather patterns, rising temperatures, and water scarcity. Research should focus on identifying crops and management practices that perform optimally in varied environmental conditions.

Water Use Efficiency: Exploring water-saving technologies suitable for intercropping systems, such as drip irrigation and mulching, and identifying crops with lower water requirements is critical for regions facing water shortages. Improving water use efficiency in both plantation and intercrop systems is key to ensuring long-term sustainability.

Organic and Sustainable Farming Practices: Expanding research on organic farming methods and sustainable practices in plantation crops can reduce dependence on chemical fertilizers and pesticides, promoting healthier ecosystems. Studies should prioritize organic nutrient management, natural pest control, and cover crops to enhance soil health.

Biodiversity and Ecosystem Services: Understanding how multiple cropping systems can enhance biodiversity and ecosystem services in plantation environments is vital. Research should explore the benefits of agroforestry and integrated livestock systems in improving soil fertility, pest control, and supporting wildlife habitats.

Carbon Sequestration and Climate Mitigation: Investigating the role of plantation-based cropping systems in carbon sequestration can help mitigate climate change. Research on the carbon-capturing potential of palm-based systems and their impact on reducing greenhouse gas emissions can support the development of eco-friendly cropping practices.

Integrated Pest and Disease Management (IPDM): Advancing IPDM strategies that combine biological control, disease-resistant crop varieties, and cultural practices is essential for minimizing losses in multiple cropping systems within plantations. Future research should focus on environmentally safe pest control techniques and the use of biocontrol agents.

Genetic Improvement and Biotechnology: Research into genetic improvement through traditional breeding, molecular biology, and biotechnology can enhance traits such as yield, disease resistance, and stress tolerance in plantation crops. Exploring CRISPR and other gene-editing technologies offers new opportunities for crop improvement.

Value Addition and Post-Harvest Technologies: Developing better post-harvest management techniques and value-added processes for plantation crops can reduce losses, improve product quality, and increase farmers' incomes. Research on processing technologies, packaging, and storage solutions is crucial for maximizing the economic potential of plantation crops.

Socioeconomic and Policy Research: Understanding the socioeconomic impacts of multiple cropping systems on smallholder farmers, along with policy frameworks that support sustainable plantation agriculture, is critical. Studies on market access, value chain development, and farmer education can help create an enabling environment for sustainable agricultural growth.

These research priorities are essential to fostering a more resilient, productive, and sustainable future for plantation crops

Conclusion

The adoption of multiple cropping systems in plantation crops offers significant benefits, particularly in terms of maximizing resource use efficiency, enhancing soil health, and boosting economic returns. By integrating diverse crops and livestock enterprises within plantation environments like coconut and arecanut, farmers can effectively utilize both vertical and horizontal space, optimizing the use of light, water, and nutrients. This diversity fosters sustainable agricultural practices, reduces the risks associated with monocropping, and creates additional income streams through crop diversification and the inclusion of livestock, poultry, or aquaculture.

Multiple cropping systems also enhance soil fertility by recycling biomass generated within the system and stimulating microbial activity, resulting in healthier and more resilient soils that support long-term productivity. Furthermore, these systems contribute to environmental conservation by minimizing the need for chemical inputs, promoting biodiversity, and improving ecosystem stability. Overall, the implementation of multiple cropping systems in plantation crops not only increases farm profitability but also plays a vital role in sustainable and eco-friendly agricultural development

REFERENCES

Abdul Haris A, Balasimha D, Sujatha S, Bhat, R and Abdul Khader KB, 1999. The influence of drip irrigation and fertilizer on yield and photosynthetic characteristics of cocoa in mixed cropping system with arecanut. *J. Plantation Crops*, **27**: 131-135.

Abdul Khader KB, Balasimha D and Bhat NT, 1992. Resource use in arecanut based high density multispecies cropping system. *J. Plantation Crops* **20**: 19-24.

Abraham KJ, 1974. Intercropping in arecanut helps to build up farmer's economy. *Arecanut Spices Bull.* **5**: 73-75.

Ammaan M and Paramaguru P, 2017. Evaluation of growth and yield characters of nutmeg (Myristica fragrans Houtt.) ecotypes under coconut ecosystem. *Environment and Ecology,* **35(2A)**: 808-810.

Anandaraj M, 2015. Status of plantation crops sector in India. *J. Plantn. Crops,* **43**: 1-8.

Andoh-Mensah E and Ofosu-Budu GK, 2012. Evaluation of coconut-citrus intercropping systems in the context of lethal yellowing disease of coconut in Ghana. *African Journal of Food, Agriculture, Nutrition and Development,* **12(7)**.

Balasimha D, 2007. Efficacy of pruning in enhancing bean yield of cocoa. *J. Plantation Crops* **35**: 201-02.

Balasimha D, 2009. Effect of spacing and pruning regimes on photosynthetic characteristics and yield of cocoa in mixed cropping with arecanut. *J. Plantation Crops* **37**: 9-14.

Basavaraju TB, Nanjappa HV, Umesha K, Vasundhara M and Arulraj S, 2011. Intercropping of medicinal and aromatic plants in coconut gardens.

Bavappa KVA, Kailasam C, Abdul Khader KB, Biddappa CC, Khan HH, Kasturi Bai KV, Ramadasan A, Sundararaju P, Bopaiah BM, Thomas GV, Misra LP, Balasimha D, Bhat NT and Shama Bhat K, 1986. Coconut and arecanut based high density multispecies cropping systems. *J. Plantation Crops* **14**: 74-87.

Bavappa KVA, 1990. Agronomic desirability of coconut based farming systems. *In: Proceedings of Coconut Based Farming Systems* (ed. Silva de. S.). Proceedings of the XXVII COCOTECH Meeting, 25-29 June, 1990, Manila, Philippines. 57-68.

Bhalerao Pankaj, Maheswarappa HP and Sundaram Sumitha, 2021. Effect of integrated nutrient management on coconut (*Cocos nucifera*) based cropping systems in south Gujarat. Current Horticulture. 9. 10.5958/2455-7560.2021.00025.X.

Bhandary DK, 1974. Study of inter and associated crops in areca gardens of the malnad tract of Karnataka. *Arecanut Spices Bull.* **5**: 67-68.

Bhat KS, 1978. Agronomic research in arecanut - a review. *J. Plantation Crops* **6**: 67-80.

Bhat R and Sujatha S, 2011a. Arecanut based high density multispecies cropping/ farming systems. In: *Arecanut based cropping/farming systems*, (Eds. Thomas, G. V., Krishnakumar, V., Maheshwarappa, H. P., Bhat Ravi and Balasimha, D.), ICAR–Central Plantation Crops Research Institute, Kasargod, India. pp. 27-44.

Bhat R and Sujatha S, 2011b. Organic matter recycling in arecanut based high density multispecies cropping system. In: *Arecanut based cropping/farming systems*, Thomas, G. V., Krishnakumar, V., Maheshwarappa, H. P., Bhat Ravi and Balasimha, D. (Eds.), ICAR–Central Plantation Crops Research Institute, Kasargod, India. pp. 6-26.

Bhat R and Sujatha S, 2007. Soil fertility status as influenced by arecanut based cropping system and nutrient management. *J Plantation Crops*, **35**: 158-65.

Bhat R, Reddy VM and Khader KBA, 1999. Areca based high density multispecies cropping system in coastal Karnataka. *J. Plantation Crops*, **27(1)**: 22-26.

Bhat R, Sujatha S, Bhavishya Priya UK, Gupta A and Uchoi A, 2024. Arecanut (*Areca catechu* L.). In: *Soil Health Management for Plantation Crops*,Thomas, G.V. and Krishnakumar, V. (Eds). Springer, Singapore. https: //doi.org/10.1007/978-981-97-0092-9_4

Bhat KS, 1978. Agronomic research in arecanut – a review. *J. Plantation Crops* **6**: 67-80.

Bhol N, Sarangi SK and Behera UK, 2015. Coconut (*Cocos nucifera*)-based farming system: a viable land use option for small and marginal farmers in coastal Odisha. *The Indian Journal of Agricultural Sciences*, **85(11)**: 1488-1497.

Biddappa CC, Upadhyay AK, Hegde MR and Palaniswami C, 1996. Organic matter recycling in plantation crops. *J. Plantation Crops*, **24**: 71-85.

Chenchaiah KC, Sit AK and Biswas CR, 2002. Performance of annuals and perennials intercrops in areca based cropping system of Terai Region of West Bengal. *J. Plantation Crops* **30**: 41-43.

Chowdappa P, Biddappa CC and Sujatha S, 1999. Efficient recycling of organic wastes in arecanut (*Areca catechu*) and cocoa (*Theobroma cacao*) plantation through vermicomposting. *Indian J. Agric. Sci.* **69**: 563-566.

CPCRI, 2022. Annual report. ICAR–Central Plantation Crops Research Institute, Kasaragod, 104 p

CPCRI, 1984. Annual Report 1982. Central Plantation Crops Research Institute, Kasaragod, India. pp.32.

CPCRI, 1985. Annual Report 1984. Central Plantation Crops Research Institute, Kasaragod, India. pp.28-30.

Ghosh DK and Bandopadhyay A, 2011. Productivity and profitability of coconut based cropping systems with fruits and black pepper in West Bengal. *Journal of Crop and Weed*, **7(2)**: 134-137.

Gopalasundaram P and Nelliat EV, 1979. Intercropping in coconut. 6-23. *In: Multiple Cropping in Coconut and Arecanut gardens* (Nelliat, E. V. and Bhat, K. S. (Eds.).Tech. Bull. No. 3, Central Plantation Crops Research Institute, Kasaragod.

Gunathilake HAJ, Arambewewela L, Ratnayake H and Rajapakse S, 2002. Feasibility of growing medicinal plants in coconut lands of the wet zone of Sri Lanka. *COCOS*, **14**: 87-96.

Hegde MR and Yusuf M, 1993. Performance of soybean varieties as intercrops in coconut garden. *Journal of Plantation Crops* **21**(Suppl.): 29-31.

Hegde MR, Yusuf M and Gopalasundaram P, 1993. Intercropping of vegetables in coconut garden. 407-412. In: *Advances in Coconut Research and Development* (Eds. Nair, M. K., Khan, H. H., Gopalasundaram, P. and Bhaskara Rao, E. V. V.), Indian Society for Plantation Crops, Oxford and IBH Pub. Co. Pvt. Ltd., Kasaragod, India.

ICC, 2020. Coconut statistical yearbook 2020. International Coconut Community, Jakarta

Jagadeesha SK, Chandasheker GS and Swetha, 2023. Coconut-livestock (Cow) based integrated farming system. *The Pharma Innovation Journal*, **12(3)**: 3705-3707.

Jeena Mathew S, Indhuja A, Abdul Haris K, Nihad, Ravi Bhat and Subramanian P, 2023. Leguminous and grass fodder combinations for enhancing soil health and system productivity in coconut gardens of the humid tropics. *Eco. Env. and Cons.* **29**(October Suppl. Issue): S265-S273.

Krishnakumar V, Maheswarappa HP, Jayasekhar S and Shanavas M, 2011. Economics of HDMSCS in coconut Economic evaluation of high density multispecies cropping system in root (wilt) disease affected coconut (*Cocos nucifera*) area in Kerala, pp. 125-130.

Kumar KN, Maheswarappa HP and Basavaraju TB, 2016. Effect of integrated nutrient management practices on growth and yield of vegetable crops grown as intercrops in coconut (*Cocos nucifera*) garden. *The Indian Journal of Agricultural Sciences*, **86(10)**: 1361-1365.

Maheswarappa HP, Dhanapal R, Subramanian P and Palaniswami C, 2013. Evaluation of coconut based high density multi species cropping system under organic and integrated nutrient management. *J Plantn Crops*, **41(2)**: 130–135.

Maheswarappa HP, Krishnakumar V, Gupta A and Geetha Kumari A, 2016. Influence of organic sources of nutrients on vanilla (*Vanilla planifolia*) as an intercrop in coconut garden. *J Plantn Crops*, **44(2)**: 85–89.

Maheswarappa HP, Subramanian P and Dhanapal R, 2000. Root distribution pattern of coconut (*Cocos nucifera* L.) in littoral sandy soil. *J. Plantn. Crops*, **28(2)**: 164-166.

Maheswarappa HP, Subramanian P and Geetha Kumari A, 2017. Influence of organic nutrient management practices for noni (*Morinda citrifolia*) when grown as an intercrop in coconut garden. *Indian Journal of Agricultural Sciences*, **87(9)**: 1155–1157.

Maheshwarappa HP, Krishnakumar V, Gupta A and Geetha Kumari A, 2016. Influence of organic sources of nutrients on vanilla (*Vanilla planifolia*) as an intercrop in coconut garden. *Journal of Plantation Crops*, **44(2)**: 85-89.

Malshe KV, Wankhede SM and Ghavale SL, 2023. Growth and yield performance of bitter gourd (*Momordica charantia*) as intercrop in coconut (*Cocos nucifera*) garden.

Manikandan P, Joshi OP, Khan HH, Mohapatra AR and Biddappa CC, 1987. Nutrient profile in an arecanut-cacao system on a laterite soil. *Tropical Agric.* **64**: 13-16.

Manjunath BL and Itinal CJ, 2001. Farming system options for small and marginal holdings in different topographies of Goa. *Indian Journal of Agronomy*, **48(1)**: 48_1.

Manjunath BL, Ramesh R and Korikanthimath VS, 2010. Feasibility of ramie intercropping in coconut. *J Plantn Crops*, **38(3)**: 207-210.

Margate RZ and Magat SS, 1983. Coconut based multi storey cropping. *Phillippine Journal of Crop Science*, **8(2)**: 81-86.

Mario E Baliad, 1994. *The Philippines Journal of Coconut Studies,* **19(1)**: 1-4.

Mohapatra R and Bhat NT, 1990. Soil research on arecanut in the last decade. *J. Plantation Crops* **17**(suppl.): 54-63.

Muralidharan A, 1990. Intercropping in arecanut. *J. Plantation Crops*, **17**: 25-38.

Nagaraj B, 1974. Intercropping in arecanut gardens of maidan areas of Karnataka. *Arecanut Spices Bull*, **5**: 78-79.

Nair MG, 1982. Intercropping with pepper. *Indian Farming*, **32**: 17-19.

Nair MK and Gopalasundaram P, 1990. Coconut based farm family models in India. In; Proceedings, COCOTECH Meeting on Coconut Based Farming Systems, 27th, APCC, Jakarta, Indonesia, 25-29 Jun. 1990. (Ed. Sumit de Silva): pp. 360-369

Nair Ramachandran, 1979. "Intensive multiple cropping with coconuts in India. Principles, programmes and prospects."

Nair PKR, 1983. Agroforestry with coconuts and other tropical plantation crops. In: Huxley PA (ed) *Plant Research and Agroforestry*, pp. 79-102. ICRAF, Nairobi

Nair AK and Nair SA, 2006. Influence of FYM and nutrients on ridge and sponge gourd yield intercropped with coconut palm in South Andaman. *Internat. J. Agric. Sci.,* **2(1)**: 284-285.

Nath JC, Deka KK, Maheswarappa HP and Sumitha S, 2019. System productivity enhancement in coconut (*Cocos nucifera*) garden by intercropping with flower crops in Assam.

Nath JC, Saud, BK, Chowdhury D, Deka KK and Sarma UJ, 2008. Coconut based high density multispecies cropping system in Assam.

Nayar TVR, 1982. Grow pepper in arecanut gardens. *Indian Fmg.* **32**: 27-31.

Nayar TVR, Muralidharan A and Shama Bhat K, 1985. Evaluation of banana varieties for intercropping in arecanut gardens. In: *Proceedings of Silver Jubilee Symposium on Arecanut Research and Development*, Shama Bhat, K. and Nair, C.P.R. (Eds.), ICAR–Central Plantation Crops Research Institute, Regional Station, Vittal, India. pp. 70-73.

Nedunchezhiyan M, Byju G and Naskar SK, 2007. Sweet potato (*Ipomoea batatas* L.) as an intercrop in a coconut plantation: growth, yield and quality. *Journal of Root Crops*, **33(1)**: 26-29.

Nihad K, Haris AA, Subramanian P, Mathew J, Indhuja S, Neenu S and Shil S, 2023. Efficient nutrient recycling strategy through integrated nutrient management in hybrid napier within a coconut (*Cocos nucifera*)-based system. *The Indian Journal of Agricultural Sciences*, **93(6)**: 647-652.

Nihad K, Krishnakumar V and Abdul Haris A, 2017. Marigold-globe amaranth sequential cropping in coconut plantations of coastal humid tropics. *Journal of Plantation Crops*, **45(2)**: 121-128.

Nuwarapaksha TD, Rajapaksha UN, Ekanayake J, Weerasooriya SA, Atapattu AJ, 2003. Exploring the Economic Viability of Integrating Jamnapari Goat into Underutilized Pastures under Coconut Cultivations in Coconut Research Institute, Sri Lanka. Biol. Life Sci. Forum 27.

Osak REMF, Anis SD and Rumambi A, 2018. Productivity of dwarf elephant grass (*Penisetum purpureum* cv. Mott) and coconut (*Cocos nucifera*) in Coconut-Beef Cattle Integrated Farming System (Coco-Beef IFS) in South Minahasa, Indonesia. *International Journal of Environment, Agriculture and Biotechnology*, **3(5)**: 1874-1878.

Padma E, Ramanandam G, Kumar KR, Rao NC and Maheswarappa HP, 2016. Performance of coconut-based high density multispecies cropping system under East Godavari Dist. of Andhra Pradesh. *Scientist*, **1(3)**: 2.

Palaniswami C, Thomas GV, Subramanian P, Dhanapal R and Selvamani V, 2008. Comparative evaluation of soil health and fertility under organic, inorganic and integrated nutrient management in coconut based mixed farming system. In: Singh HP, Thomas GV (eds) *Organic Horticulture: Principles, Practices and Technologies.* Westville Publishing House, New Delhi, 231 p.

Paul SC, Acharya GC, Chakraborty R, Maheswarappa HP, Hussain M and Ray AK, 2020. Sustainability of soil health and system productivity through arecanut based cropping system in the NE region of India. *J. Plantation Crops*, **48**: 111-119.

Pooja AP, 2021. Physiological response, yield and dry matter partitioning in blackgram varieties under shading stress in coconut garden. *International Journal of Farm Sciences*, **11(4)**: 70-74, 2021; doi: 10.5958/2250-0499.2021.00058.6

Ramakrishnan Nayar TV and Sahasranaman KN, 1979. Mixed Farming in Coconut Gardens Root (Wilt) Affected Tract.

Raj RM, Raj AK, Kunhamu TK and Prakash A., 2019. Fodder yield and nutritive value of mulberry (*Morus indica* L.) under varying plant density and pruning frequency in coconut garden. *Range Management and Agroforestry*, **40(2)**: 255-261.

Rani S, Rajakumar D, Shoba N and Maheswarappa HP, 2018. Productivity and economic advantages of flower crops in coconut based intercropping system. *Indian Journal of Horticulture*, **75(2)**: 279-282.

Ray AK, Acharya GC, Maheswarappa HP and Krishnakumar V. 2011. Arecanut based cropping systems in North-East Region of India. In: *Arecanut based cropping/farming systems*, Thomas GV, Krishnakumar V, Maheshwarappa HP, Bhat R, and Balasimha D, (Eds.), ICAR–Central Plantation Crops Research Institute, Kasargod, India, pp. 45-54.

Ray AK, Borah AS, Maheswarappa HP and Acharya GC, 2007. Economics of intercropping vegetable and flowering crops in pre-bearing arecanut garden under Assam condition. *J. Plantation Crops* **35**: 84-87.

Reddy VM and Bhat R, 1994. Cropping systems in arecanut. *Indian Cocoa, Arecanut and Spices J.* **18**: 13-15.

Rethinam P, 1990. Cropping system involving plantation crops. In: *Plantation Based Multiple Cropping System* (Eds. Bandopadhyay, A. K., Michel, R. S., Gangwar, B. and Dagar, J. C.). CARI, Port Blair.

Sannamarappa M, 1993. Arecanut based cropping system in maidan parts of Karnataka. *J. Plantation Crops* **21**(suppl.): 3-6.

Sannamarappa M and Muralidharan A, 1982. Multiple Cropping. In: *The Arecanut Palm*, Bavappa, K.V.A., Nair, M.K. and Premkumar, T. (Eds.), ICAR–Central Plantation Crops Research Institute, Kasaragod, India, pp. 133-49.

Selva Rani A, Subbulakshmi S, Sudha R, Kavitha K, Nazreen Hassan SH, Muthulakshmi M and Suresh S, 2024. Synergizing Sustainability: Integrated Nutrient Management and Intercropping for Optimal Coconut Cultivation in South India. *Horticulturae*, **10(6)**: 653.

Serge Braconnier, 1998. Maize-coconut intercropping: effects of shade and root competition on maize growth and yield. *Agronomie*, **18(5-6)**: 373-382. ffhal-00885891ff

Shama Bhat K, 1974. Intensified inter/mixed cropping in areca garden – the need of the day. *Arecanut Spices Bull.* **5**: 67-69.

Shama Bhat K and Abdul Khader KB, 1970. Inter and mixed cropping in arecanut gardens. *Indian Fmg.* **20**: 35.

Shanthaamallaiah NR, Venkatesha Gowda TN, Krishna Manohar R, Shivayogishwara B and Balakrishna P, 1982. Note on the intercropping of mulberry in coconut plantation, PLACROSYM-Vm pp. 465-467

Shinde VV, Ghavale SL and Maheswarappa HP, 2017. Intercropping Flower Crops A Best Option In Coconut Based Cropping System pf Maharashtra. *Bioinfolet* **14(2)**: 142-145.

Sneha SR, Raj SK, Jacob D, Pillai PS and Radhakrishnan NV 2023. Weed management in finger millet (*Eleusine coracana* L.) Intercropped in coconut garden. *Indian Journal of Weed Science*, **55(3)**: 349-354.

Siddappa R, Sumitha S, Maheswarappa HP and Basavaraju TB, 2023. Evaluation of integrated nutrient management of coconut (*Cocos nucifera*) based cropping system in the central zone of Karnataka. *The Indian Journal of Agricultural Sciences*, **93(1)**: 110-112.

Sit AK, Chenchaiah KC and Acharya GC 2002. Black pepper- a profitable mixed crop in areca based cropping system under Terai region of West Bengal. *Indian J. Arecanut, Spices and Medicinal Plants*, **4**: 155-157.

Sit AK, Chenchaiah KC, Acharya GC and Ghosh A, 2006. Feasibility of vegetable cultivation in arecanut garden under sub-Himalayan terai region of West Bengal. *J. Plantation Cropsm* **34**: 361-363.

Sit AK, Chenchaiah KC, Acharya GC and Roybarman J, 2004. Comparative study of different turmeric cultivars in areca based cropping system and open field under sub Himalayan terai region of West Bengal. *J. Plantation Crops* **32**: 45-47.

Sit AK, Chenchaiah KC, Acharya GC and Thomas GV, 2011. Arecanut based cropping systems in Sub Himalayan Terai region of West Bengal. In: *Arecanut based cropping/farming systems*, Thomas, G.V., Krishnakumar, V., Maheshwarappa, H.P., Bhat, R. and Balasimha, D. (Eds.), ICAR–Central Plantation Crops Research Institute, Kasargod, India, pp. 55-62.

Subba S, Chattopadhyay N, Bandyopadhyay A and Ghosh DK, 2021. Performance of sweet potato germplasm under coconut based cropping system. *Journal of Crop and Weed*, **17(3)**: 76-82.

Subramanian P, Dhanapal R, Thomas GV, Maheswarappa HP and Thamban C, 2014. Coconut based integrated farming system. Technical bulletin. ICAR–CPCRI, Kasaragod, 32 p.

Subramanian P, Krishnakumar V, Alka Gupta MG, Surekha R and Selvamani V, 2016. Impact of long term organic cultivation practices on soil health and productivity of coconut. In: Abstracts of third international symposium on coconut research and development. ICAR–CPCRI, Kasaragod, p. 54.

Subramanian P, Dhanapal R, Palaniswami C, Maheswarappa HP, Gupta Alka and Thomas George V, 2009. Coastal sandy soil management for higher coconut productivity. Technical Bulletin No. 50. Central Plantation Crops Research Institute, Kasaragod. 20 p.

Subramanian P, Dhanapal R, Palaniswami, C and Joseph Sebastian, 2006. Feasibility studies on growing pumpkin as intercrop in coconut under coastal sandy soil. *Journal of Indian Society of Coastal Agricultural Research*, **24(1)**: 190-192.

Subramanian P, Palaniswami RDC and Sebastian J, 2007. Feasibility studies on growing hybrid Bajra Napier fodder grass as intercrop in coconut under coastal littoral sandy soil. *Journal of Plantation Crops*, **35(1)**: 19-22

Subramanian P, Dhanapal R, George VT, Maheswarappa HP and Thamban C, 2014. Coconut based integrated farming system. *Technical Bulletin*, 78.

Sudha R, John J, Meera AV , Sajeena A, Jacob D and Bindhu JS, 2021. Coconut based integrated farming: A climate-smart model for food security and economic prosperity. *ISPC.*

Suja G, Subramanian P, Surekha R, Jaganathan D, Maria CL and Raj RK, 2022. Organic management for yams (Dioscorea spp.) intercropped in coconut (*Cocos nucifera*) plantation: Evidences from a validation trial in the coastal humid tropics of Kerala, India. *The Indian Journal of Agricultural Sciences*, **92(8)**: 966-969.

Suja G, Surekha R, Subramanian PJaganathan D, Maria CL and Raj RK, 2020. Validation of organic management in cassava intercropped in coconut plantation in the humid coastal tropics of Kerala, India. *Journal of Plantation Crops*, **48(1):** 45-54.

Sujatha S, Bhat R. and Bhavishya, 2020. Arecanut based cropping/farming systems. In: *Arecanut,* P. Chowdappa (Ed.), Today and Tomorrow's Printers and Publishers, India, pp. 145-175.

Sujatha S, Bhat R and Chowdappa P, 2015. Recycling potential of organic wastes of arecanut and cocoa in India: A short review. *Environ. Technol. Rev.* **4**: 97-102.

Sujatha S and Bhat R, 2010. Response of vanilla (*Vanilla planifolia* A.) intercropped in arecanut to irrigation and nutrition in humid tropics of India. *Agric. Water Manage.* **97**: 988-994.

Sujatha S and Bhat R, 2015. Resource use and benefits of mixed farming approach in arecanut ecosystem in India. *Agric. Systems*, **141**: 126-137.

Sujatha S, Bhat R, Balasimha D and Elain Apshara S, 2011a. Arecanut based inter/mixed cropping systems. In: *Arecanut based cropping/farming systems*, Thomas GV, Krishnakumar V, Maheshwarappa HP, Bhat Ravi and Balasimha D (Eds.), ICAR–Central Plantation Crops Research Institute, Kasargod, India, pp. 6-26.

Sujatha S, Bhat R, Balasimha D and Kannan C, 2006. Crop diversification in arecanut through intercropping of medicinal and aromatic plants. *J. Plantation Crops* **34**: 318-322.

Sujatha S, Bhat R, Kannan C and Balasimha D, 2011b. Impact of intercropping of medicinal and aromatic plants with organic farming approach on resource use efficiency in arecanut (*Areca catechu* L.) plantation in India. *Ind. Crops Prod.* **33**: 78-83.

Thankamani CK and Ashokan PK, 2011. Effect of drip irrigation on yield, nutrient uptake and of bush pepper (*Piper nigrum*) intercropped in coconut garden. *Journal of Medicinal and Aromatic Plant Sciences.* **33(3)**: 355-358.

Thomas GV, Krishanakumar V, Maheswarappa HP, Bhat R and Balasimha D, 2011. *Arecanut Based Cropping/Farming Systems.* ICAR–Central Plantation Crops Research Institute, Kasaragod, 138 p.

Varghese PT, Nair PKR, Nelliat EV, Rama Varma and Gopalasundaram P, 1979. Intercropping with tuber crops in coconut garden. pp: 399-415. In: Proceedings, PLACROSYM I (Ed. E. V. Nelliat) RRII, Kottayam.

Varghese TP, Krishnaraj N, Nair PKR, Nelliat EV, Mathew J and Mukherjee C, 1976. Economics of Inter and Mixed Cropping in the Coconut Gardens of Kerala.

Waman AA, Jha SK, Bohra P and Jerard BA, 2019a. Woody pepper: A potential spice crop for intercropping in arecanut gardens. *J. Andaman Sci. Assoc.* **24**: 131-135.

Waman AA, Maheswarappa HP, Jerard BA, Velmurugan A, Senthilkumar S. and Chakraborty G, 2019b. Evaluation of intercrops in arecanut gardens of South Andaman Island. *J. Andaman Sci. Assoc.* **24**: 1-4.

Index

A

B

C

D

E

F

G

H

I

J

K

L

M

N

O

P

Q

R

S

T

U

V

W

Y